SPECTRAL
LINE SHAPES

SPECTRAL LINE SHAPES

Volume 9
13th ICSLS

Firenze, Italy   June 1996

EDITORS
Marco Zoppi
Lorenzo Ulivi
Consiglio Nazionale Delle Ricerche
Firenze, Italy

American Institute of Physics

AIP CONFERENCE
PROCEEDINGS 386

Woodbury, New York

Authorization to photocopy items for internal or personal use, beyond the free
copying permitted under the 1978 U.S. Copyright Law (see statement below), is
granted by the American Institute of Physics for users registered with the Copyright
Clearance Center (CCC) Transactional Reporting Service, provided that the base fee
of $6.00 per copy is paid directly to CCC, 222 Rosewood Drive, Danvers, MA 01923.
For those organizations that have been granted a photocopy license by CCC, a
separate system of payment has been arranged. The fee code for users of the
Transactional Reporting Service is: 1-56396-656-5/ 97 /$6.00.

© 1997 American Institute of Physics

Individual readers of this volume and nonprofit libraries, acting for them, are
permitted to make fair use of the material in it, such as copying an article for use in
teaching or research. Permission is granted to quote from this volume in scientific
work with the customary acknowledgment of the source. To reprint a figure, table, or
other excerpt requires the consent of one of the original authors and notification to
AIP. Republication or systematic or multiple reproduction of any material in this
volume is permitted only under license from AIP. Address inquiries to Office of
Rights and Permissions, 500 Sunnyside Boulevard, Woodbury, NY 11797-2999;
phone: 516-576-2268; fax: 516-576-2499; e-mail: rights@aip.org.

ISBN 1-56396-656-5
ISSN 0094-243X
DOE CONF- 9606258

Printed in the United States of America

Contents

Preface xv
International and Local Committees xvii

HIGH DENSITY PLASMAS

Theoretical and Experimental Advances in Spectroscopy of Strong and/or Dynamic Electric Fields in Plasmas 3
    E. Oks**

A New Multi-Particle Model of Ion Broadening Applicable for Both High and Low Densities 15
    A. Derevianko and E. Oks

Novel Diagnostic for Nonthermal Electric Fields in Plasmas 19
    S. Alexiou, A. Weingarten, Y. Maron, M. Sarfaty, and Ya. E. Krasik

Screening Potential in High Density Plasmas 21
    M. Amari, J. P. Arranz, J. Butaux, and H. Nguyen

Redistribution Function for Resonance Radiation in a Hot Dense Plasma 23
    A. E. Bulyshev, A. V. Demura, V. S. Lisitsa, A. N. Starostin,
    A. E. Suvorov, and I. I. Yakunin

Hydrogen Lines with Central Stark Components as a Sensitive Magnetic Probe 25
    A. V. Demura and E. Oks

The Influence of Charge-Exchange on Ly-Radiation 27
    A. Devdariani, E. Leboucher-Dalimier, P. Angelo, P. Gauthier,
    and P. Sauvan

On the Importance of the Divergence of Ionic Microfields in Plasmas for the Shift and Asymmetry of Hydrogen Spectral Lines 29
    J. Halenka, W. Olchawa, and B. Grabowski

The Frequency Fluctuation Model Applied to the Hydrogen-Like Helium Paschen-α Line 31
    T. Meftah, S. Alexiou, A. Calisti, L. Godbert, R. Stamm, and B. Talin

Redistribution Function of Resonance Radiation for Atoms with Degenerate Structure 33
    V. I. Savchenko, N. J. Fisch, A. A. Panteleev, and A. N. Starostin

Experiments on Hot and Dense Laser-Produced Plasmas 35
    C. A. Back**, N. C. Woolsey, A. Asfaw, S. H. Glenzer, B. A. Hammel,
    C. J. Keane, R. W. Lee, D. Liedahl, J. C. Moreno, J. K. Nash,
    A. L. Osterheld, A. Calisti, R. Stamm, B. Talin, L. Godbert, C. Mossé,
    S. Ferri, and L. Klein

Spectroscopic Diagnostics of 2D Effects in Dense Plasmas. Comparison with 2D Simulations 45
    P. Angelo, H. Derfoul, P. Gauthier, P. Sauvan, A. Poquerusse, T. Ceccotti,
    E. Leboucher-Dalimier, T. Shepard, C. Back, M. Vollbrecht, I. Uschmann,
    and E. Förster

Redistribution of Resonance Radiation in Hot and Dense Plasmas 47
    C. Mossé, S. Ferri, A. Calisti, B. Talin, R. Lee, and L. Klein

X-Ray Spectral Line Shapes for the Excimer-Laser-Produced High Density Plasma Diagnostics 49
    A. Magunov, A. Faenov, I. Skobelev, T. Pikuz, D. Batani, M. Milani,
    A. Conti, A. Masini, M. Costato, A. Pozzi, E. Turcu, R. Allot, N. Lisi,
    M. Koenig, A. Benuzzi, F. Flora, T. Letardi, L. Palladino, and A. Reale

Refractive Fringe and Stark Diagnostics of Laser Induced Plasmas on Solid Targets 51
    S. Siano, G. Pacini, R. Pini, and R. Salimbeni

Time-Resolved X-Ray Spectroscopy of Laser Heated Plasmas 53
    A. Giulietti[*], D. Giulietti, L. A. Gizzi, and O. Willi

High Precision X-Ray Measurements of Polarization Shifts in Dense Aluminum Plasma 57
    O. Renner[*], D. Salzmann, P. Sondhauss, E. Förster, A. Djaoui, and E. Krouský

Quasimolecular Features in Hot Dense Plasma Emission 61
    P. Gauthier[*], P. Sauvan, S. Alexiou, P. Angelo, E. Leboucher-Dalimier,
    A. Calisti, and B. Talin

Quasimolecules and Spectral Line Broadening in Dense Plasmas 65
    J. P. Arranz, J. Butaux, H. Nguyen, and A. Reggadi

Effects of Atomic States Interference on Spectral Line Shapes in MMM Theory 67
    I. N. Kosarev, C. Stehlé, N. Feautrier, A. V. Demura, and V. S. Lisitsa[*]

Line Profile Measurements in a Gas-Liner Pinch Discharge 71
    Th. Wrubel[*], S. Glenzer, S. Büscher, and H.-J. Kunze

LOW AND MODERATELY DENSE PLASMAS

Closing in on Stark Broadening for Weakly Coupled Plasmas 79
    S. Alexiou[**]

Shifted and Asymmetric Profiles of Hydrogen and Hydrogenic Ion Lines 99
    S. Günter[**] and A. Könies

Profiles of High Principal Quantum Number Balmer and Paschen Lines from Alcator C-Mod Tokamak Plasmas 113
    B. L. Welch, H. R. Griem[*], J. L. Weaver, J. U. Brill, J. Terry,
    B. Lipschultz, D. Lumma, G. McCracken, S. Ferri, A. Calisti,
    R. Stamm, B. Talin, and R. W. Lee

Semiclassical Stark Broadening Calculations of Neutral Oxygen Lines 117
    N. Ben Nessib, Z. Ben Lakhdar, S. Sahal-Bréchot

Line Asymmetries and Plasma Statistics 119
    A. V. Demura, D. Gilles, and C. Stehlé

The Shift of Stark Broadened H_γ Line 121
    J. Halenka, W. Olchawa, and T. Wujec

Determination of the Total Number of Absorption Atoms in a Plasma by Parameters of Self-Reversed Spectral Line 123
    G. G. Il'in, O. A. Konovalova, E. V. Sarandaev, and M. Kh. Salakhov

The Polarized Spectrum of Hydrogen in the Presence of Electric
and Magnetic Fields... 125
    R. Casini and E. Landi Degl'Innocenti
A New Source for the Experimental Study of Ions at Low Temperature 127
    F. C. De Lucia**
Rotational Spectroscopy of Unstable Molecules Produced
in a Low Density Plasma... 141
    L. Dore, C. Degli Esposti, and G. Cazzoli
Plasma Broadening and Shifting of Analogous Spectral Lines
Along Isoelectronic Sequences 143
    B. Blagojević, M. V. Popović, N. Konjević, and M. S. Dimitrijević
On the Stark Broadening of B III Spectral Lines........................ 147
    M. S. Dimitrijević and S. Sahal-Bréchot
On the Stark Broadening of Mg II Spectral Lines 149
    S. Sahal-Bréchot and M. S. Dimitrijević
Experimental Stark Widths, Shifts, and Transition Probabilities
of Several ArII Lines... 151
    J. A. Aparicio, M. A. Gigosos, S. Mar, and V. R. Gonzáles
Stark Broadening of Several N II Lines................................ 153
    C. Pérez, I. de la Rosa, M. A. Gigosos, J. A. Aparicio, and S. Mar
A New Deconvolution Method Applied to the Si II (1) Lines Profiles 155
    A. Lesage, M. Depiesse, D. Meiners, J. Richou, and F. Wollschläger
Observation of the Giant Coulomb Broadening in the Gas
Discharge Plasma.. 157
    A. A. Apolonsky, S. A. Babin, S. I. Kablukov, S. V. Khorev,
    E. V. Podivilov, A. I. Chernykh, and D. A. Shapiro*
Classical Descriptions of Rydberg Spectral Line Shapes 161
    L. Bureyeva, V. Lisitsa, H. Van Regemorter, and O. Motapon
Laser Generated Plasma of Li, Zn, and Li-Zn Mixture.................... 163
    S. Gogić and S. Milošević*

ATOMS AND MOLECULES IN STRONG LASER FIELDS

Spectral Analysis of Ultrashort Pulses After Nonlinear Propagation 169
    R. Lange, J.-F. Ripoche, M. Franco, G. Grillon, B. Prade,
    and A. Mysyrowicz**
Optical Collisions in an Atomic Beam Experiment........................ 181
    J. Grosser**, O. Hoffmann, and S. Klose
Interaction of Static and Polarization Channels in Spectra
of Stimulated Emission in Ion-Electron Collisions in a Strong Laser Field..... 195
    V. A. Astapenko and A. B. Kukushkin
HgIn Excimer and Bound-Bound Emission Following Laser
Excitation of Atomic In in an Indium-Mercury Vapor Mixture 197
    P. Bicchi*, C. Marinelli, and R. A. Bernheim

Energy-Pooling Ionization Produced by the Collisions of In Atoms in Presence of a Resonant Laser Field 201
    P. Bicchi, C. Marinelli, E. Mariotti, M. Meucci, and L. Moi

Energy Pooling Collisions in Pure Cadmium Vapors 203
    S. Barsotti, F. Fuso, A. F. Molisch, and M. Allegrini

Multistep Excitation of Autoionizing Yb Atom States and Line Shape of Autoionizing Resonance 205
    I. M. Beterov, U. V. Brjzazovsky, V. L. Kurochkin, A. A. Chernenko, and V. B. Alman

Excitation Transfer and Radiative Decay in the n^2F and the $(n+3)^2S$ States of Rb 207
    B. Bieniak, M. Głódź, and J. Szonert

Time Behavior of the $7^3S_1 - 6^3P_2$ Transition Following Pulsed Laser Excitation of the 6^3P_1 State of Mercury 209
    K. Blagoev, D. Gruber, A. Dreischuh, A. Morozov, U. Reiter-Domiaty, and L. Windholz

The Effect of an Electromagnetically Induced Transparency on Laser-Assisted Collisional Ionization 211
    R. Buffa

High Contrast Measurement of a Laser-Induced Structure in the Ionization Continuum of Sodium by Photoelectron Spectrometry 213
    R. Eramo, S. Cavalieri, L. Fini, and M. Matera

Time Resolved Spectra of Emission Signals from Strontium Vapor after Switching Off Radiation Trapping with a Strong Laser Pulse 215
    L. Fini, M. Mazzoni, and R. Vaia

Experimental and Numerical Study of the Collision Induced Stimulated Effects in Alkali Vapors 217
    Z. Konefał

Theoretical Investigation of the CsHg Molecule 219
    R. Polly, D. Gruber, L. Windholz, and B. A. Hess

Collision Distribution of Particles Excited by the Laser Light 221
    S. Zielinska-Kaniasty

The LiHg $X^2\Sigma^+ \to 2^2\Pi_{3/2}$ Transition: Rotational Analysis and Potential Curves 223
    X. Li*, D. Gruber, P. Pircher, and L. Windholz

On the Validity of Different Intercombination Rules for the Parameters of the Ground State (X^1O^+) Potential 227
    G. D. Roston and M. S. Helmi

Interatomic Potentials for X^1O^+ and B^31 States of the Intercombination Cadmium Line 326.1 nm Broadened by Ar Pressure 229
    G. D. Roston, T. Grycuk, and M. S. Helmi

Van-der-Waals-Interaction Constant 231
    D. Neundorf

Radiation Spectroscopy of Heavy Rare-Gas Excimers and Their Mixtures 233
    N. A. Kryukov, P. A. Saveliev, and M. A. Tchaplyguine

DOPPLER-FREE AND ULTRA-FINE SPECTROSCOPY

High-Resolution Measurements and Multichannel Quantum Defect Analysis of Spectral Line Shapes of Autoionizing Rydberg Series 237
    K. Ueda**

Theory of Quantum Oscillations in Self-Broadening of 4^2S-n^2S Rydberg Transitions in Potassium 251
    R. M. Herman* and M. E. Henry

Line Shapes for a Pure Three Level System: Quantum Coherence Effects on Absorption and Birefringence of Helium 255
    F. S. Pavone, F. Bassani, G. Bianchini, P. Cancio, F. S. Cataliotti, T. W. Hänsch, and M. Inguscio*

Nonlinear Interference Effect in Ionic Zeeman Laser 259
    S. A. Babin, S. I. Kablukov, M. I. Kondratenko, and D. A. Shapiro

Non-Inverted Gain Line Shapes of the Cesium Resonance Transition at 894 nm 261
    F. S. Cataliotti, C. Fort, M. Prevedelli, T. W. Hänsch, and M. Inguscio

Quantitative Analysis of Second Derivative Absorption Line Shapes in Difficult Environments: Detection of Molecular Oxygen using a DFB-Laser at 761 nm 263
    C. Corsi, M. Gabrysch, M. Prevedelli, and M. Rosa-Clot

Diode Laser Spectroscopy in a Magnetron Discharge 265
    C. Csambal and V. Helbig

Spectroscopic Investigation of the Rubidium Resonance Line 267
    T. Rieper, T. Rose, V. Helbig, and D. Veza

Shape of D Line of Sodium by Single-Mode Laser Excitation 269
    J. H. Xu and R. M. Celli

Incomplete Optical Shielding in Cold Sodium Atom Traps 271
    V. Yurovsky and A. Ben-Reuven*

Structured Continua of the Intermediate Long-Range Cs_2 Molecules 275
    D. Veža, R. Beuc, S. Milošević, and G. Pichler

Two Step Resonances in a Rb Magneto-Optical Trap 277
    C. Gabbanini, A. Evangelista, S. Gozzini, and A. Lucchesini

Investigation of Radiation Trapping in a Cs Magneto-Optical Trap 279
    A. Fioretti, J. H. Müller, M. Colla, A. Molisch, and M. Allegrini

NEUTRALS: ATOMS AND MOLECULES

Purely Quasi-Molecular Radiation in the Hg-Inert Gas Atom Collisions 283
    A. Z. Devdariani**, M. G. Lednev, A. L. Zagrebin, and M. A. Tchaplyguine

Action Spectra in Collisions Between Neutral Atoms and Molecules 297
    K. M. Sando** and Y. M. Hung

Stark Effect in Some Lines of Br I 307
    A. Bacławski, A. Goly, I. Ksiażek, and T. Wujec

Light Absorption during a Resonant or Near-Resonant Collision: Study of the Cross Section in the Far Wing 309
    S. Cavalieri and M. Celli

Speed-Dependent Galatry Profile 311
    R. Ciuryło and J. Szudy

Speed-Dependent Correlations between Pressure and Doppler Broadening of the 748.8 nm Neon Line 313
    R. Ciuryło, A. Bielski, J. Szudy, and R. S. Trawiński

On the Temperature Dependence of the Stark Broadening Parameters of Ar I 425.9 nm Line 315
    S. Djurović, Z. Mijatović, R. Kobilarov, and N. Konjević

Observation of Collision-Time Asymmetry of the ^{114}Cd $\lambda=326.1$ nm Line Perturbed by Krypton 317
    J. Domysławska and R. S. Trawiński

Quasistatic Non-Binary Self-Broadening and Radiationless Transfer of Excitation in a System of Classical Oscillators 319
    Ya. A. Korennoy, A. B. Kukushkin, and A. G. Zhidkov

Quantum Width and Shift of Spectral Lines in the Non-Impact Regime 321
    W. C. Kreye and J. F. Kielkopf

Stark Effect in Some Lines of Neutral Argon Emitted from a Ferroelectric Plasma Source 323
    J. Kusz and D. Mazur

Pressure Broadening of $2p^53s-2p^53p$ Neon Lines 325
    P. J. Leo, D. F. T. Mullamphy, G. Peach, V. Venturi, and I. B. Whittingham

Regularities in the Behavior of Stark Widths of Argon Spectral Lines 327
    E. V. Sarandaev and M. Kh. Salakhov

Quantum Defect Theory for Momentum Space Wave Functions 329
    H. Van Regemorter and D. Hoang Binh

High Resolution Studies of Infrared and Raman Line Shapes at the University of Toronto 331
    R. Berman, J. R. Drummond, P. Duggan, S. Hamid Fakhr-Eslam, J. W. Forsman, A. D. May**, G. D. Sheldon, and P. M. Sinclair

Collisional Coupling Between Components of Molecular Spectral Lines 343
    O. Tarrini**, S. Belli, and G. Buffa

Velocity Selective Coherent Transient Study of Depolarizing Collisions in Molecular Gas 357
    L. S. Vasilenko, N. N. Rubtsova*, and E. B. Khvorostov

Self and Foreign Broadening and Shift Versus Temperature of Ammonia Lines 361
    G. Baldacchini, G. Buffa, F. D'Amato, M. DeRosa, F. Pelagalli, and O. Tarrini

Rotational Contribution to Raman Line Shape of Liquid Carbonyl Sulphide: MD Simulation 363
    Z. Gburski, H. Stassen, and A. Kachel

Collisional Self-Broadening, Shifting, and Line-Coupling of Rotational Transitions of CH_3F in Presence of Stark Fields 365
    V. Lemaire, L. Dore, G. Cazzoli, G. Buffa, O. Tarrini, and S. Belli

Collisional Broadening and Shift of Acetylene and Oxygen Overtones 367
    A. Lucchesini, M. DeRosa, A. Ciucci, C. Gabbanini, and S. Gozzini

Vibrational Line Width and Line Shift in (Binary) Systems
at High Pressures 369
    J. A. Schouten** and J. P. J. Michels

Vibrational Raman Spectrum of O_2 in Ar Matrix at Room
Temperature and Pressure to 60 GPa 383
    F. Gorelli, L. Ulivi, and M. Zoppi

Highlights and Open Problems on Line Shapes 385
    Round Table Introduction by J. Szudy

APPLICATIONS: ASTROPHYSICS, ATMOSPHERE, AND ENVIRONMENT

Line Profiles as Diagnostic and Actor: Abundance Anomalies in Stars 397
    G. Michaud** and J. Richer

Databases for Diagnostic of High Temperature Astrophysical Plasmas 411
    Round Table Introduction by M. Landini

Density Diagnostic of Astrophysical Plasmas 419
    M. Landini and E. Landi

The XUV Spectral Code of Arcetri 421
    M. Landini, E. Landi, and B. Monsignori Fossi

Effect of the Continuum and Selective Absorption Ratio on Calculated
Synthetic Solar Spectral Line Profiles 423
    A. Galal, N. Yousef, M. Behery, and R. Hamid

Laboratory Studies of Infrared Features Observed in the Atmospheres
of the Outer Planets 425
    P. Varanasi* and V. Nemtchinov

MD Simulations of the RT CIA of CO_2 for the Atmosphere of Venus 429
    M. Gruszka and A. Borysow

Millimeter Wave Farwings of Water Vapor in N_2 and CO_2 Atmospheres 431
    A. Bauer, M. Godon, J. Carlier, and R. R. Gamache

Raman Scattering as a Probe for Remote Sensing Measurement
of Water Temperature 433
    G. Cecchi, P. Mazzinghi, V. Raimondi, and M. Zoppi

Implementation of the Voigt Line-Shape Calculation in the Forward
Model for Operational MIPAS Retrievals 435
    M. Ridolfi, M. Höpfner, and P. Raspollini

Airborne Infrared Diode Laser Spectrometer for *in Situ* Measurements
of Stratospheric Trace Gas Concentration 437
    G. Toci, P. Mazzinghi, and M. Vannini

COLLISION INDUCED SPECTROSCOPY: SYMPOSIUM IN HONOR OF GEORGE BIRNBAUM

George Birnbaum and the Collision-Induced Raman and Infrared
Spectroscopies 441
    L. Frommhold*

Collision-Induced Light Scattering in Metal Vapors: The Mercury Spectra 453
    F. Barocchi**, M. Sampoli, and L. Ulivi

Collision-Induced Emission and Sonoluminescence 471
    L. Frommhold** and W. Meyer

Bull's Eye Model for Collision-Induced Light Scattering 485
    U. Balucani and R. Vallauri*

The Wings of the Rototranslational Raman Spectrum of Nitrogen at Low Density 489
    A. Borysow, Y. Fu, and M. Moraldi*

Non-Resonant Microwave Absorption and Collisional Coupling in Infrared Q Bands 493
    G. Buffa* and O. Tarrini

Collision-Induced Absorption Intensity Redistribution and the Atomic Pair Polarizabilities 497
    M. O. Bulanin*

Quantum Analysis of Experimental Anisotropic and Isotropic CILS Spectra from Ar Diatoms in the Frequency Band 0-400 cm^{-1} 501
    M. Chrysos*, O. Gaye, and Y. LeDuff

Measurements and Analysis of the He-Xe Absorption Spectrum in the Far Infrared 505
    P. Dore*, L. Frommhold, A. Nucara, and P. Postorino

Collisional Interference in the Vibration-Rotation Spectrum of HD-He 507
    B. McQuarrie and G. C. Tabisz*

The Far Infrared Continuum Spectrum of the Milky Way Explained by a Dust and Gas Model 511
    J. Schäefer*

Frequency-Dependent Quenching in the Emission Spectra $3P\Lambda$, $3D\Lambda \rightarrow 2P\Lambda$ of Lithium-Rare Gas Collision Molecules 515
    W. Behemenburg, J. Bonsmann, A. Kaiser, A. Makonnen, and W. Meyer

Molecular Dynamics Studies of Light Scattering from Ar_{13} Cluster: A Two-Body Correlations 517
    A. Dawid and Z. Gburski

Contributions of Multipolar Polarizabilities to the Anisotropic and Isotropic Light Scattering Induced by Molecular Interactions in Gaseous CF_4 519
    A. Elliasmine, J.-L. Godet, Y. LeDuff, and T. Bancewicz

Triple Transitions $3Q_1$ in Compressesd Hydrogen 521
    L. Frommhold and M. Moraldi

Improved Analysis of the Spectral Moments of CIA in N_2 and CO_2 Gases 523
    M. Gruszka and A. Borysow

Collision-Induced Absorption Near the 227 nm Hg Line in the Hg+Ar Mixture 525
    T. Grycuk, N. A. Kryukov, and M. G. Lednev

Bound Dimer Contribution in the Collision-Induced Pair Spectrum of Mercury Vapor..527
    N. Meinander, L. Ulivi, G. Pratesi, G. Cirnigliaro, and F. Barocchi

On the Nature of Collision-Induced Intensity and Bandshapes in the Region of N_2 and O_2 Fundamentals.................................529
    A. A. Vigasin

First Quantum Mechanical Computations of Collision-Induced Absorption Spectra of H_2 Pairs in the Second Overtone Band...............531
    C. Zheng, Y. Fu, and A. Borysow

Double Rotational Transitions of Molecular Hydrogen in the Condensed Phase...533
    M. Zoppi, L. Ulivi, M. Santoro, M. Moraldi, and F. Barocchi

Double Rotovibrational Transitions in Solid Hydrogen....................535
    M. Zoppi, L. Ulivi, M. Santoro, M. Moraldi, and F. Barocchi

Interaction-Induced Dipoles and Polarizabilities in Diverse Phenomena.......537
    G. Birnbaum* and B. Guillot

APPENDIX

Minutes of the Meeting of the International Committee....................551
List of Participants (with E_Mail)..552
Author Index..556
Subject Index...561

*Oral contribution speaker
**Invited speaker

PREFACE

The 13th International Conference on Spectral Line Shapes was held in Firenze (Italy) from June 16–21, 1996. It was attended by 160 participants from 21 countries. The spectral line shape conferences are held every other year, alternately in Europe and North America. The Proceedings of the past eight conferences (Berlin 1980, Boulder 1982, Aussois 1984, Williamsburg 1986, Torun 1988, Austin 1990, Marseille 1992, and Toronto 1994) have been published under the title **Spectral Line Shapes** by de Gruyter, Berlin (volumes 1–3); A. Deepak Publishing, Hampton, Virginia (volume 4); Ossolineum Publishing, Warsaw, Poland (volume 5); American Institute of Physics Press, New York (volumes 6 and 8); and Nova Science Publishers, Commack, New York (volume 7).

The conference program covered a wide range of subjects emphasizing the physical processes associated with the formation of line profiles. This included 17 invited lectures, 16 oral contributions, 103 contributed papers presented in three poster sessions, as well as three round table discussions. A special session was dedicated to a symposium in honor of our dear colleague George Birnbaum and consisted of 12 oral contributions in the field of collision induced spectroscopy. The program of invited speakers was suggested primarily by the International Committee, with particular emphasis on newly emerging results and techniques. Some of the contributed papers where chosen for an oral presentation on the same basis. All contributions to the conference, with few exceptions, have been included in the present volume. The minutes of the meeting of the International Committee are included in the Appendix.

The quantitative determination of the line profiles that are observed in absorption, emission, or scattering of electromagnetic radiation by plasmas and neutral fluids is a powerful tool for studying the fundamental physics of atomic and molecular interactions in any of the phases of matter. Line shape analysis is also a powerful diagnostic tool for many media in extreme conditions of temperature (gas discharges, flames, plasmas) or pressure (planetary atmospheres, systems under extremely high pressure conditions). Atmospheric, environmental, and astrophysical applications are also considered. These proceedings are intended to record the latest advances in the field and to stimulate new collaborations among scientists of different countries for novel and interesting applications of the line shape analysis technique.

Since the vast number of topics considered in this book would have suggested an excessive number of pages, we have decided to impose a page limit on each contribution thereby increasing, we think, the readability of the book. Following in the tradition of other volumes, the first two sections are dedicated to high and low density plasmas. Then, we include the section on atoms and molecules in strong laser fields (part 3), followed by that on Doppler-free and ultra-fine spectroscopy (part 4). Part 5 deals with the line shapes generated by the interaction of neutrals, atoms and molecules, where the relevant quantities are the single particle properties. This is a large section that ends with the introductory talk of the Round Table: *Highlights and Open Problems on Line Shapes*. The astrophysical, atmospheric, and environmental applications are the arguments of part 6. Finally, part 7 is dedicated to the interaction-induced spectroscopy and includes the symposium in honor of George Birnbaum.

This book is intended to record the latest advances in this interdisciplinary field and to encourage all scientists interested in line shapes to build up and maintain international contacts and collaborations. For this reason, we have included in the appendix a list of participants including their e-mail addresses (when available). We hope that this information may help future collaborations.

We would like to take this opportunity to thank all those individuals and organizations who invested time and effort to make this meeting possible. We acknowledge the precious organizational support from the *Area della Ricerca di Firenze* (CNR) and from the staff of the *Istituto di Elettronica Quantistica* (CNR). We are especially grateful to Carla Pardini for her skill and patience in the organization of this conference.

We gratefully acknowledge the friendly hospitality of Banca Toscana, who hosted the conference in its own auditorium, and the financial support provided by the Italian National Council of Research (CNR), the University of Florence, the Carlo Marchi Foundation, and the International Science Foundation. We would like to thank Prof. E. Garaci, President of CNR, Prof. M. Fontanesi, President of the Physical Sciences National Committee of CNR, and to Prof. S. Califano, Director of the European Laboratory for Nonlinear Spectroscopy (LENS), for their interest and help in the organization of the conference.

Last, but not least, we are extremely grateful to all the participants for the top-level science presented at the conference and for the live discussions during the various sessions. Finally, we were happy to observe that the participation level remained very high, up to the last concluding talk, in spite of the attractiveness of the city and the beautiful weather.

The 14th International Conference on Spectral Line Shapes will be hosted by Professor Roger Herman at Pennsylvania State University, USA, in the summer of 1998. We express our best wishes to him and his colleagues for a successful conference.

Marco Zoppi
Lorenzo Ulivi

International Committee

K. Burnett	Clarendon Laboratory	Oxford, U.K.
A. Z. Devdariani	Dept. Optics and Spectroscopy	St. Petersburg, Russia
N. Feutrier	Observatoire Paris-Meudon	Meudon, France
C. Iglesias	Lawrence Livermore Nat. Lab.	Livermore, CA, USA
J. Kielkopf	Department of Physics	Louisville, KY, USA
N. Konjevic	Faculty of Physics	Belgrade, Yugoslavia
D. May	Department of Physics	Toronto, Ont., Canada
H. Nguyen	Université P. et M. Curie	Paris, France
E. Oks	Department of Physics	Auburn, AL, USA
G. Pichler	Institute of Physics	Zagreb, Croatia
J. Seidel	Physik. Tech. Bundesanstalt	Berlin, Germany
R. Stamm	Université de Provence	Marseille, France
G. C. Tabisz	Department of Physics	Winnipeg, Man., Canada
R. Tipping	Dept. Physics and Astronomy	Tuscaloosa, AL, USA
E. Yukov	Lebedev Physical Institute	Moscow, Russia
M. Zoppi	National Council of Research	Firenze, Italy

Local Committee

B. Carli	CNR-Istituto di Ricerca Onde Elettromagnetiche, Firenze
S. Cavalieri	Laboratorio Europeo di Spettroscopia Non-lineare, Firenze
A. Giulietti	CNR-Istituto di Fisica Atomica e Molecolare, Pisa
M. Matera	CNR-Istituto di Elettronica Quantistica, Firenze
M. Mazzoni	CNR-Istituto di Elettronica Quantistica, Firenze
M. Moraldi	Università di Firenze, Dipartimento di Fisica, Firenze
F. Palla	Osservatorio Astrofisico di Arcetri, Firenze
C. Pardini	CNR-Area della Ricerca di Firenze
A. Tronconi	CNR-Area della Ricerca di Firenze
L. Ulivi	CNR-Istituto di Elettronica Quantistica, Firenze

Sponsors

The 13th International Conference on Spectral Line Shapes was sponsored by the *Consiglio Nazionale delle Ricerche*, the *Banca Toscana*, the *Fondazione Carlo Marchi*, the *Università di Firenze*, the *Comune di Firenze*, the *Azienda di Promozione Turistica di Firenze*, the *International Science Foundation*, the *International Union of Pure and Applied Physics*. In accordance with the IUPAP rules, no bona fide scientist was excluded from participation on the ground of national origin or political considerations unrelated to science.

HIGH DENSITY PLASMAS

Theoretical and Experimental Advances in Spectroscopy of Strong and/or Dynamic Electric Fields in Plasmas

E. Oks

Physics Department, Auburn University, Auburn, AL 36849-5311, USA

Abstract. The paper consists of 3 parts: 1) new type of shift of hydrogen and hydrogen-like lines; 2) resolution of a long-standing argument on inelastic interference terms; 3) the first measurements ever of a polarization of x-ray line profiles in a laboratory plasma.

NEW TYPE OF SHIFT OF HYDROGEN AND H-LIKE LINES

1. It is well known that there are two primary sources of the Center of Gravity Shift (CGS) of spectral lines in plasmas: the electron shift $\Delta\lambda_e$ and the ion shift $\Delta\lambda_i$. Here and below we consider only the CGS of Hydrogen or Hydrogen-Like Lines (HHLL).

The electron shift (usually, red) consists of two terms - due to virtual transitions between the states of different principal quantum numbers ($\Delta n \neq 0$) and between the states of the same n ($\Delta n = 0$). The first term $\Delta\lambda_e(\Delta n \neq 0)$ was calculated semiclassically by Griem [1].

As for the second term $\Delta\lambda_e(\Delta n=0)$, the semiclassical calculation of the CGS yields zero, despite generally non-zero contributions to the shifts of individual Stark components of HHLL. This can be understood considering a pair of lateral Stark components k and -k corresponding to radiative transitions $(n,q) \rightarrow (n',q')$ and $(n,-q) \rightarrow (n',-q')$, where $q=n_1-n_2$ and $q'=n_1'-n_2'$ are expressed via electric quantum numbers. The semiclassical electron ($\Delta n=0$)-shifts $\Delta_{\pm k}$ of both components in the pair are of an equal magnitude and of the opposite signs [2]. Since in the dipole approximation with respect to the interaction with quasistatic ions, both components have equal intensities $I_k^{(0)}$, the CGS of the pair remains unshifted. The summation of the CGS over all such pairs of lateral Stark components of a HHLL still yields zero. An allowance for central Stark components does not change the result since their semiclassical electron ($\Delta n=0$)-shift is zero. Only quantum calculations of the term $\Delta\lambda_e(\Delta n=0)$ yielded non-zero results [3,4].

The ion shift (usually, blue) is primarily caused by the non-uniformity of the ion field as had been first pointed out by Sholin [5] who calculated ion quadrupole corrections to energies and wave functions. Then from this approach Demura and Sholin deduced formulas for the ionic CGS for a binary microfield [6] and for the Holtsmark microfield [7]. Later their formulas were extended to the case of the Baranger-Mozer microfield [8]. The analytical results from [7] and [8] were used in numerical calculations [1,9] and [4], respectively.

2. A new result we report here is a <u>coupled ionic-electronic shift.</u> The gist of this new shift is the following. Consider again a pair of lateral Stark components k and -k corresponding to radiative transitions $(n,q) \rightarrow (n',q')$ and $(n,-q) \rightarrow (n',-q')$. The semiclassical electron ($\Delta n=0$)-shift Δ_k of a component in the pair is given by the formula [2]

$$\Delta_k = (2m_e^2)^{-1} 3\pi^2 \hbar^2 N_e [2m_e/(\pi T_e)]^{1/2} (n^2q - n'^2q')/Z_r, \quad (1)$$

where N_e and T_e are an electron density and a temperature respectively, Z_r is a radiator charge. Formula (1) shows that the components of the pair are shifted by electrons symmetrically: usually the blue component in the pair (characterized by $(nq-n'q') \equiv k > 0$) shifts to the blue side by Δ_k and the red component - to the red side by $\Delta_{-k} = -\Delta_k$ (see Fig. 1).

However, ion quadrupole corrections to intensities break down the symmetry. Indeed, intensities of the components $I_{\pm k}$ cease to be equal to each other:

$$I_{\pm k} = I_k^{(0)}(1 \pm B_k), \quad B_k \equiv \varepsilon_k^{(1)} a_0 (F/e)^{1/2}, \quad (2)$$

where a_0 is the Bohr radius, F is an ion field strength (see Fig. 1). Quantities ε_k (tabulated in [5] for neutral hydrogen) are of the order of n^2/Z_r, so that coefficients B_k are of the first order in terms of the small parameter of the nonuniformity μ defined as

$$\mu \equiv (n^2 a_0/Z_r)(\beta F_0/e)^{1/2}, \quad \beta \equiv F/F_0, \quad F_0 \approx 2.603 Z_i e N_i^{2/3}. \quad (3)$$

With the allowance for the asymmetry of the intensities, the CGS of the pair differs from zero, being $\delta_k = B_k(\beta)\Delta_k$. Therefore the CGS of the entire spectral line at a fixed value of the reduced ion field strength β becomes

$$\delta(\beta) = [\sum_{k>0} I_k^{(0)} B_k(\beta) \Delta_k] / [I_0^{(0)}/2 + \sum_{k>0} I_k^{(0)}] \equiv C_{nn} \cdot \beta^{1/2}, \qquad (4)$$

where $I_0^{(0)}$ is the intensity of the central Stark components.

3. To obtain a final expression for the CGS of the spectral line, we integrate (4) over the distribution $W(\beta)d\beta$ of the ion field strength. An important feature of the new shift is that this integration <u>converges.</u> Indeed,

$$\delta \equiv <\delta(\beta)> = C_{nn} \cdot <\beta^{1/2}>, \qquad (5)$$

where $<\beta^{1/2}>$ exists for field distributions corresponding to both neutral and charged points. In particular, for a neutral point we find for the binary and the Holtsmark distributions $<\beta^{1/2}> \approx 1.354$ and $<\beta^{1/2}> \approx 1.62$, respectively.

The convergence of the new shift is its primary distinction to all previously known ion shifts of neutral hydrogen lines. For example, the ion quadrupole shift of a Stark component Q_k, that up to now was considered as the leading ion contribution to the shift, has the following dependence on β:

$$Q_k(\beta) \propto \mu^3 \propto \beta^{3/2}. \qquad (6)$$

However, calculations of $<\beta^{3/2}>$ diverge at large β for field distributions at a neutral point. Therefore, in order to obtain a finite expression for $<Q_k>$, Demura and Sholin used upper cutoffs β_k [5]. Then by a subsequent summation over all Stark components they obtained a zero CGS for the H_β line and a non-zero CGS for the H_α and H_γ lines.

The convergence or divergence are important both from the mathematical and the physical points of view. An improper treatment of diverging integrals may lead to erroneous physical conclusions. In particular, we would like to point out that a <u>consistent calculation of the usual ion quadrupole CGS yields zero</u> in distinction to both the Demura-Sholin's paper [5] and to those Griem's results [1,9] that were based on [5].

To prove our point, we note that Demura and Sholin performed first the integration of $Q_k(\beta)$ over β and then the summation over k (i.e., over the Stark components with an allowance for quadrupole corrections (2) to their intensities). Obviously, these two operations commute. Therefore they could be also done in the opposite order: first the summation over k at fixed β and then the integration over β. However, we have found that the summation of $Q_k(\beta)$ over k yields zero regardless of the value of β. Then, naturally, the subsequent integration over β does not change the result: the CGS is zero.

This mathematical result has a deep physical meaning. The ion quadrupole CGS of a spectral line should be invariant with respect to the inversion of the sign of the ion field gradient. Therefore, the terms in the perturbation expansion that are linear with respect to the ion field gradient (and therefore linear with respect to Q_k) cannot contribute to the CGS. In other words, the first non-vanishing contribution to the CGS from the ion quadrupole interaction term V_{iQ} of the Hamiltonian could originate only from the perturbation expansion term of the second order with respect to V_{iQ}. [This is similar to the well-known fact that the ion dipole interaction term V_{iD} of the Hamiltonian does not contribute to the CGS in the first order of the perturbation theory and that the first non-vanishing contribution from V_{iD} to the CGS arises only in the second order of the perturbation theory. That fact resemble the invariance of the ion dipole CGS with respect to the inversion of the sign of the ion field.]

We also note that in the Demura-Sholin's calculational scheme there were used the same upper cutoff $β_k$ for the lateral components but a different cutoff for the central components. That is why they arrived at the correct zero CGS for the $H_β$ line but at the erroneous non-zero result for the CGS of the lines $H_α$ and $H_γ$ lines.

So the bottom line is that the usual ion quadrupole CGS is zero and that <u>the new coupled ionic-electronic shift is the leading ionic contribution to the CGS.</u>

4. Let us compare our new ionic CGS $Δλ_i = δλ_0^2/(2πc) = 1.62 C_{nn'} λ_0^2/(2πc)$, where $C_{nn'}$ is defined in (4), with experimental and previous theoretical results using as an example the H I $L_β$ line. The experimental CGS measured by Grützmacher and Wende [10] at the density $N_e = 2 \times 10^{17} cm^{-3}$ and the temperature T=16000 K is $Δλ_{exp} = (10±5)$ mÅ (red shift). At the same N_e and T we find that our ionic CGS is $Δλ_i ≈ -1.2$ mÅ (blue shift), so that its relative contribution is $|Δλ_i/Δλ_{exp}| ≈ (8 - 24)$%. Thus our new ionic CGS is a significant part of the total CGS.

As for the previous theoretical results, the most extensive calculations of the shift due to both electrons and ions were performed by Könies and Günter [4]. Their results show a strong, seemingly divergent dependence of the total CGS on the interval Δs of the integration over Δλ (see, e.g., Figs. 3,4 from [4]).

In our view, the primary source of the divergence of their results is their allowance for the ion quadratic shift $Δλ_{iD}^{(2)}$, that is the contribution of the ion dipole interaction term V_{iD} of the Hamiltonian in the second order of the perturbation theory. Indeed, since

$$\Delta\lambda_{iD}^{(2)}(\beta) \sim \mu^4\lambda_0 \propto \beta^2, \qquad (7)$$

then the integration over β diverges at large β ($<\beta^2>$ does not exist).

Our most important comment on the Könies-Günter paper [4] is the following. They did not allow for the ion octupole interaction V_{io} and they did not include the contribution of the ion quadrupole interaction V_{iQ} in the second order of the perturbation theory. However, the shift $\Delta\lambda_{iQ}^{(2)}$ due to the latter contribution as well as the shift $\Delta\lambda_{io}^{(1)}$ due to the contribution of the ion octupole interaction in the first order of the perturbation theory are of the same order of magnitude as the ion quadratic shift (as Sholin pointed out 27 years ago [5]):

$$\Delta\lambda_{iQ}^{(2)} \sim \Delta\lambda_{io}^{(1)} \sim \Delta\lambda_{iD}^{(2)} \sim \mu^4\lambda_0. \qquad (8)$$

It does not make any sense to allow for one term and to neglect two other terms of the same order of magnitude. The same mistake, that actually negates the results, was also made by Günter and Könies in their previous calculations of both shifts and widths [11].

Thus the most extensive calculations of the total shift $\Delta\lambda_{theor}$ [4] are, regrettably intrinsically inconsistent from the physical point of view. Therefore the quantitative comparison of the total CGS $\Delta\lambda_{theor}$ and $\Delta\lambda_{exp}$ has yet to be done. However, qualitatively it is already clear that at least for neutral hydrogen lines, while the purely electronic contribution $\Delta\lambda_e$ to the total CGS predominates, the new coupled ionic-electronic shift $\Delta\lambda_i$ is the leading of the remaining terms.

5. The new shift $\Delta\lambda_i$ should play even a more important role at higher densities. Indeed, it has the following scaling

$$\Delta\lambda_i \propto N_e^{4/3}n^5/(Z_r^3 Z_i^{1/3} T_e^{1/2}), \qquad (9)$$

demonstrating <u>the nonlinear dependence on the density N_e</u>. The electron shift scales linearly with the density $\Delta\lambda_e \propto N_e n^4/(Z_r^2 T_e^{1/2})$, so that <u>the relative contribution of the new shift increases with the density:</u>

$$|\Delta\lambda_i/\Delta\lambda_e| \propto N_e^{1/3} n/(Z_r Z_i^{1/3}). \qquad (10)$$

Therefore we could make two suggestions on possible experiments where the role of the new shift should be the most important. The first suggestion concerns neutral hydrogen lines, e.g., the L_β line. Given that at $N_e \sim 10^{17}\text{cm}^{-3}$ the ratio is $|\Delta\lambda_i/\Delta\lambda_e| \sim 0.1$, then at $N_e \sim 10^{19}\text{cm}^{-3}$ the ratio should be $|\Delta\lambda_i/\Delta\lambda_e| \sim (0.4 - 0.5)$. Observations of hydrogen lines at these densities are possible, as demonstrated by two experimental groups [12,13].

The second suggestion concerns H-like lines employed for diagnostics in laser fusion experiments at the densities $N_e \sim 10^{24} cm^{-3}$. For example, for the Ar XVIII L_β line in a hydrogen or deuterium plasma we would expect the ratio $|\Delta\lambda_i/\Delta\lambda_e| \sim 1$. However, in high temperature plasmas there occurs a decrease of the relative role of the electron shift compared to the plasma polarization shift (see, e.g., [14]), so that the latter may dominate the CGS.

RESOLUTION OF A LONG-STANDING ARGUMENT ON INELASTIC INTERFERENCE TERMS

1. It is well known that in the electron impact broadening operator $\Phi \propto (\mathbf{r}_a - \mathbf{r}_b^*)^2$, the Interference Terms are $\Phi_{IT} \propto (-2\mathbf{r}_a\mathbf{r}_b^*) = -2(x_a x_b^* + y_a y_b^* + z_a z_b^*)$. If one considers, for example HHLL in the parabolic quantization with the axis Oz along the quasistatic ion field $\mathbf{F}$, then due to the selection rules, the Elastic Contribution to the Interference Terms (ECIT) is $\Phi_{ECIT} \propto (-2z_a z_b^*)$, while the Inelastic Contribution to the Interference Terms (ICIT) is $\Phi_{ICIT} \propto (-2x_a x_b^* - 2y_a y_b^*)$.

The argument whether or not to include the ICIT has a long history. The ICIT were included both in the impact formalism of the Griem's book of 1964 [15] and in the unified theory [16] (the fact that in [15] and in his earlier papers Griem wrongfully omitted the complex conjugation sign in the interference terms is of a minor importance for our discussion). However, as early as in 1958, Baranger in his fundamental paper [17] clearly formulated that "the elastic parts of the scattering subtract in a coherent manner" but "the effects of inelastic collisions cannot be expected to add coherently". This seemed to contradict to the inclusion of the ICIT on the equal footing with the ECIT, as noted by Sholin et al [2].

To resolve the alleged contradiction, in 1973 Sholin et al introduced in [2] an additional averaging over instants of the closest approach of perturbers. In this scheme they demonstrated that in the limit of isolated components, nondiagonal elements of Φ (to which ICIT belong) "automatically" acquired a small factor of the order of $\gamma^2/(n\omega_F)^2$, where γ is the electron impact width and $n\omega_F$ is the quasistatic splitting of the upper multiplet by ions:

$$\omega_F \equiv 3\hbar F/(2Z_r m_e e). \qquad (11)$$

However, the averaging over instants of the closest approach had been already contained in the standard impact formalism, as could be seen, perhaps most clearly from the paper by Sahal-Brechot [18]. Since there is no need to perform the same averaging twice, the above result by Sholin et al is physically incorrect.

In 1973 Griem published comments [19] where he concluded that in the line core the ICIT should not be included. He also forecasted that "a proper theory ... would leave the ECIT intact, but would reduce the ICIT substantially". Then in his book of 1974 [20] Griem dropped the ICIT out of the impact broadening operator Φ (see, e.g., formula (110) in [20]). He also supplemented the no-ICIT formula for Φ by a comment that amounts to the following recipe: in the case of overlapping components (i.e., $\gamma \gg n\omega_F$ for HHLL) the ICIT should be retained; however, in the case of isolated components (i.e., $\gamma \ll n\omega_F$ for HHLL) the ICIT should be artificially dropped out.

In 1975 Hey and Griem [21] reiterated Griem's view from [19,20]. In 1976 Voslamber [22] sharply criticized this Griem's view concluding that the ICIT should be always retained. Griem and Hey [23] made an unconvincing attempt to refute the arguments by Voslamber (e.g., by referring again to some "future calculations").

All of the above, as well as private communications with the leading researchers in this field, shows that up to now there is no commonly accepted view on the ICIT. An additional illustration that the question still remains open is the following. In the paper [12] the experimental widths of the H_α line, upon discussions with Griem, were compared with calculations provided by Kepple based on Kepple-Griem paper of 1968 [24] (with the correction on the aforementioned complex conjugation [25]). The latter were presented in two versions - without and with the ICIT - without any comment on the preference. We also note that the ratio of the two version results was almost a factor of two, so that the question on the ICIT is of both the theoretical and the practical interest.

Therefore we have analyzed this problem. Below we present a proof that in the case of isolated components of HHLL, in the line core the ICIT automatically acquires a small factor of the order of $\gamma/(n\omega_F)$ compared to the ECIT. Therefore <u>the ICIT should be always retained</u> and there is no need in supplementing the impact formalism by the Griem's recipe [20] that required to artificially omit the ICIT in the case of isolated components.

2. In the impact formalism the lineshape is given by

$$I(\Delta\omega) = -\int_0^\infty dFW(F) \, \text{Re} \sum_{\alpha\alpha'\beta\beta'} \mathbf{d}_{\beta\alpha} \mathbf{d}_{\alpha'\beta'} g^{\alpha\beta}{}_{\alpha'\beta'} / \pi, \quad g \equiv G^{-1} = \{i[\Delta\omega - \omega_{ab}(F)] + \Phi\}^{-1} \quad (12)$$

We break down the operator G into diagonal and nondiagonal parts

$$G=G'+G'', \quad G'=i[\Delta\omega-\omega_{ab}(F)]+\Phi', \quad G''=\Phi'' \quad (13)$$

(diagonal and nondiagonal parts of operators are denoted by ' and '', respectively). Then we introduce a well-known formal expansion

$$g=g_0+g_1+\ldots, \quad g_0=G'^{-1}, \quad g_1=G'^{-1}G''G'^{-1}, \quad \ldots \quad (14)$$

and calculate the following ratio of matrix elements:

$$\varepsilon \equiv |g_1{}^{\alpha\beta}{}_{\alpha'\beta'}/g_0{}^{\alpha\beta}{}_{\alpha\beta}| = |(\Phi'')^{\alpha\beta}{}_{\alpha'\beta'}/\{i[\Delta\omega-\omega_{\alpha'\beta'}(F)]+(\Phi')^{\alpha'\beta'}{}_{\alpha'\beta'}\}|. \quad (15)$$

In the parabolic quantization with the $Oz \parallel \mathbf{F}$, we consider the case where components (α,β), (α',β'), ... of a HHLL are isolated: $\gamma \ll n\omega_F$. Then the core of the profile of the component (α,β) corresponds to detunings

$$\Delta\omega = \omega_{\alpha\beta} + \delta, \quad |\delta| \ll n'\omega_F. \quad (16)$$

Substituting (16) into (15) and using the selection rules for $(\Phi'')^{\alpha\beta}{}_{\alpha'\beta'}$ (see, e.g., [2]), we find

$$\varepsilon = |(\Phi'')^{\alpha\beta}{}_{\alpha'\beta'}| / \{[pn\omega_F+p'n'\omega_F+\delta+\mathrm{Im}(\Phi')^{\alpha'\beta'}{}_{\alpha'\beta'}]^2+[\mathrm{Re}(\Phi')^{\alpha'\beta'}{}_{\alpha'\beta'}]^2\}^{1/2}, \quad (17)$$

where generally

$$p, p'=-2,-1,0,1,2; \quad p^2+p'^2>0 \quad (18)$$

(particularly, for the ICIT it is $p,p'=-1,0,1$; $p^2+p'^2>0$). Introducing $\gamma \equiv \max|\mathrm{Re}(\Phi)^{\alpha\beta}{}_{\alpha'\beta'}|$, $d \equiv \max|\mathrm{Im}(\Phi)^{\alpha\beta}{}_{\alpha'\beta'}|$ and using the fact that $d < \gamma \ll n'\omega_F$, we finally obtain:

$$\varepsilon \approx \gamma/(|pn + p'n'|\omega_F) \ll 1. \quad (19)$$

Thus for the line core of isolated components of HHLL we arrive at the following conclusions: 1) expansion (14) is valid; 2) the contribution of any nondiagonal terms of the impact operator automatically acquires a small factor of the order of $\gamma/(n\omega_F)$ compared to the contribution of its diagonal terms; 3) since the ICIT and the ECIT belong to nondiagonal and diagonal terms, respectively, the ICIT automatically acquires the same small factor compared to the ECIT. This proves that <u>the ICIT should be always retained and there is no need in supplementing the impact formalism by the Griem's recipe [20]</u>.

As an illustration of these conclusions, we point out

that for the experiment [12], by taking Kepple's calculations with the ICIT (no ion dynamics) and allowing for the ion dynamics in frames of our new universal model of ion broadening [26], we achieve a good agreement with the experiment [12] (see Table 1 in [26]).

THE FIRST MEASUREMENTS EVER OF A POLARIZATION OF X-RAY LINE PROFILES IN A LABORATORY PLASMA[1]

The experimental idea how to perform X-ray polarization measurements is the following: if one observes a spectral line employing a crystal at the Bragg angle, and also observes the same line after rotating the crystal through 90°, the resulting spectra will correspond to two orthogonal linear polarizations. This could be achieved by using two spectrometers, so positioned with respect to each other that the plasma line source is in the plane of the spectrometer in one case and perpendicular to it in the other case.

We performed experiments at a powerful Z-pinch PHOENIX, where the current reaches peak values of several MegaAmperes during several tens of nanoseconds. We used two high-resolution, x-ray crystal spectrometers to register time-integrated profiles of the L_β and L_γ line of Al XIII in two orthogonal linear polarizations: one polarization - parallel to the discharge current, another - perpendicular to it.

It turns out that our experimental profiles of the L_β line become pretty close to straight lines being plotted as $\log[I(\Delta\lambda)/I(0)]$ v. $(\Delta\lambda)$, as demonstrated in Fig. 2. This "log-quasilinear" shape of HHLL is well known for Balmer lines observed in solar flares (see, e.g., [27,28]) and is considered as the manifestation of non-thermal processes in the solar plasma and of the Anomalous Electric Fields (AEFs) that accompany those processes.

While a variety of AEFs had been spectroscopically diagnosed in plasmas of electron densities $N_e \leq 10^{20} cm^{-3}$ (see, e.g. [29]), for a super high-density plasma of $N_e >> 10^{20} cm^{-3}$ expected at the final compression stage at the PHOENIX, all the types of AEFs, except those associated with the Langmuir waves at the plasma electron frequency ω_{pe}, would be strongly damped due to a high collision frequency γ_e. Indeed, using the standard formulas of

---
[1] This Part 3, performed in collaboration with E.J. Clothiaux, A. Schulz, V. Svidzinski, and J. Weinheimer, is sponsored by the Defense Nuclear Agency. The views expressed in Part 3 are those of the authors and do not reflect the official policy or position of the Department of Defense or the U.S. Government. Thanks are due to A.V. Demura who participated at the preliminary stage of the Z-pinch project. Also Part 1 of this paper was discussed with him.

plasma physics, it is easy to estimate that for those super high-density plasmas, out of all the types of AEFs only the Langmuir waves have the frequency significantly higher than the collisional damping rate γ_e.

The formation of log-quasilinear profiles of HHLL due to the AEFs at the frequency ω_{pe} can be explained as follows. It is well-known that under the action of the multi-mode AEF at the frequency ω_{pe}, the profile $I_k(\Delta\omega)$ of a lateral component number k splits up into equidistant satellites at frequencies $\Delta\omega = \pm p\omega_{pe}$, where p= 0, 1, 2, ... [29]. The envelope of the satellites of one lateral component is a Gaussian [29]. The width of the Gaussian is $(\Delta\omega)_{1/2} = c_k E_0$, where c_k is a generalized Stark constant of the component number k, E_0 is the amplitude of the AEF. However, each hydrogenic line (except for the L_α) has several lateral components (both in the blue and in the red sides of the line). The contribution of each component to the summary profile of the line is a Gaussian of a width that differs from one component to another. Therefore the resulting profile of the spectral line, being the sum of Gaussians of different widths, should fall off slower than any of the constituent Gaussians.

It turns out that for a relatively broad range of plasma and AEF parameters, the resulting theoretical profile of the L_β line demonstrates the log-quasilinear behavior. Figure 2 shows a typical experimental profile of the L_β line plotted in Log-v.-$\Delta\lambda$ coordinates (dotted line) and a theoretical fit calculated as described above (solid line). [The Doppler, opacity, and normal Stark effects are also incorporated in the calculations as the secondary broadening mechanisms.] It is seen that the theoretical profile broadened primarily by the AEFs is indeed log-quasilinear and fits very well the experimental profile.

For the polarization analysis the L_γ line is more appropriate than the L_β line since the former is more sensitive to the electric fields than the latter. Figure 3 shows experimental profiles of the L_γ line registered in two orthogonal linear polarizations (parallel and perpendicular to the discharge current) demonstrating a <u>significant polarization in the wings</u>.

We also used a flat-crystal spectrometer that registered unpolarized profiles of the L_γ, L_δ, and L_ϵ lines of Al XIII on one film in one shot. Our detailed analysis of all profiles of the L_β, L_γ, L_δ, and L_ϵ lines of Al XIII obtained during our experimental campaign at the PHOENIX, including the analysis of the anomalous intensity trend in the wings and the polarization analysis, has led to the following conclusions. In the compressed PHOENIX plasma of the density $N_e \approx 3 \times 10^{21} cm^{-3}$ there have been developed highly suprathermal Langmuir waves of the electric field amplitude $E_0 \approx 7$ GV/cm. Their angular distribution represents a circular ellipsoid, which is prolate along the discharge current, so that the ratio of its axes is approximately 3:1. This shape of the angular distribution of the Langmuir waves is consistent with the mechanism of their generation by electron beams (run-away electrons) travelling parallel to the bulk plasma current.

REFERENCES

1. Griem, H.R., *Phys. Rev. A* **28**, 1596-1601 (1983).
2. Sholin, G.V., Demura, A.V., and Lisitsa, V.S., *Sov. Phys. JETP* **37**, 1057-1065 (1973).
3. Boercker, D.B., and Iglesias, C.A., *Phys. Rev. A* **30**, 2771 (1984).
4. Könies, A., and Günter, S., *J. Quant. Spectr. Rad. Transfer* **52**, 825-830 (1994).
5. Sholin, G.V., *Opt. Spectrosc.* **26**, 275-282 (1969).
6. Demura, A.V., and Sholin, G.V., *Opt. Spectrosc.* **36**, 711 (1973).
7. Demura, A.V., and Sholin, G.V., *J. Quant. Spectr. Rad. Transfer* **15**, 881-899 (1975).
8. Halenka, J., *Z. Phys. D* **16**, 1-8 (1990).
9. Griem, H.R., *Phys. Rev. A* **38**, 2943-2952 (1988).
10. Grützmacher, K., and Wende, B., *Phys. Rev. A* **18**, 2140 (1978).
11. Günter, S., and Könies, A., "Hydrogenic Ion Lines Emitted from Dense Plasmas", in *Proc. Int. Conf "Physics of Strongly Coupled Plasmas"*, Singapore: World Scientific, 1996, pp.282-285.
12. Böddeker, St., Günter, S., Könies, A., Hitzschke, L., and Kunze, H.-J., *Phys. Rev. E* **47**, 2785-2791 (1993).
13. Parigger, C., Lewis, J.W.L., and Plemmons, D., *J. Quant. Spectr. Rad. Transfer* **53**, 249-255 (1995).
14. Nguen, H., Koenig, M., Benredjem, D., Caby.,M., and Couland, G., *Phys. Rev. A* **33**, 1279-1290 (1986).
15. Griem, H.R., *Plasma Spectroscopy*, New York: McGraw-Hill, 1962.
16. Vidal, C.R., Cooper, J., and Smith, E.W., *Astrophys. J. Suppl.* **25**, 37-136 (1973).
17. Baranger, M., *Phys. Rev.* **112**, 855-865 (1958).
18. Sahal-Brechot, S., *Astron. Astrophys.* **1**, 91 (1969).
19. Griem, H.R., *Comments At. Mol. Phys.* **4**, 75-83 (1973).
20. Griem, H.R., *Spectral Line Broadening by Plasmas*, New York: Acad. Press, 1974.
21. Hey, J.D., and Griem, H.R., *Phys. Rev. A* **12**, 169-185 (1975).
22. Voslamber, D., *Phys. Rev. A* **14**, 1903-1905 (1976).
23. Griem, H.R., and Hey, J.D., *Phys. Rev. A* **14**, 1906 (1976).
24. Kepple, P., and Griem, H.R., *Phys. Rev.* **173**, 317-325 (1968).
25. Kepple, P., private communication (1996).
26. Derevianko, A., and Oks, E., "A New Multi-Particle Model of Ion Broadening Applicable for both High and Low Densities", in *These Proceedings*.
27. Svestka, Z., *Adv. Astron. Astrophys.* **3**, 119-239 (1965).
28. Koval, A.N., and Oks, E., *Bull. Crimean Astrophys. Obs.* **67**, 78-89 (1983).
29. Oks, E., *Plasma Spectroscopy: The Influence of Microwave and Laser Fields"*, Springer Series on Atoms and Plasmas, vol. 9, New York: Springer, 1995.

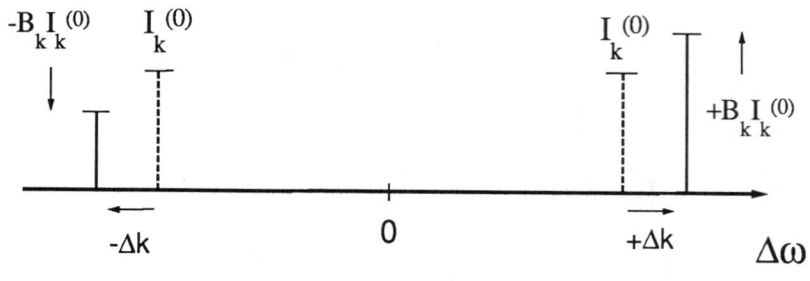

Fig. 1

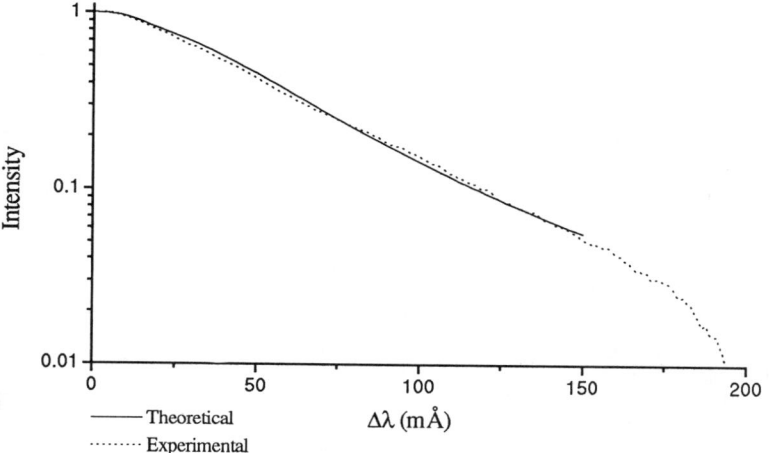

Fig. 2

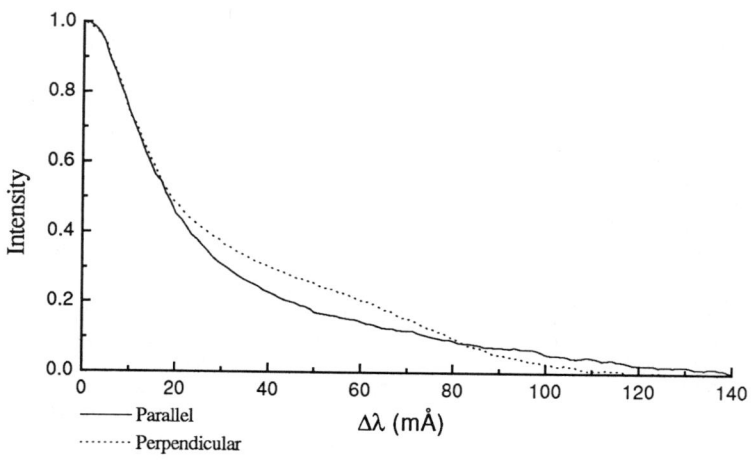

Fig. 3

A New Multi-Particle Model of Ion Broadening Applicable for both High and Low Densities.

A. Derevianko and E. Oks

Physics Department, Auburn University, Auburn, AL 36849-5311, USA

1. There are two groups of models for the ion dynamical broadening: Simulation Models (SM) and Analytical Models (AM) - see, e.g, [1] and references therein. SM are further subdivided into fully numerical models (where both the plasmic and the atomic parts of the problem are treated numerically) and semianalytical models (where only the first part is calculated numerically). SM are slow and expensive. Compared to them, AM, such as the Model Microfield Method (MMM) and the Frequency Fluctuation Model (FFM), are faster and less costly. However, all the models, while working relatively well at high densities, become progressively inaccurate at low densities. The FFM fails to reproduce the ion impact regime; the MMM reproduces this limit only at densities several orders of magnitude below the real "entrance" into the ion impact regime.

We have developed a new AM that is free from this shortcoming and is also simpler than most of its predecessors. Our model is based on <u>three</u> stochastic characteristics of the multi-particle ion microfield (in distinction to the MMM that employs only two of them). These three characteristics are used for separating the entire ensemble of microfields into two parts that act differently on radiating atoms: a dynamic part, that is then treated in the impact approximation, and the remaining part, treated in the quasistatic approximation. It turns out that our model, being simpler and faster than the MMM, is much more accurate than the MMM. In fact, for high and medium densities our model yields practically the same results as the MD and the experiments. For lower densities, in distinction to all the previous ion-dynamical models, our model reproduces the ion impact broadening limit [2] at and below its real entrance density and yields results in a better agreement with experiments than any of the predecessors.

2. At the first stage the goal is to divide the

ensemble of microfields into the dynamic and quasistatic subensembles and to calculate effective densities N_d and N_q of ions in each subensemble. For doing this we employ a crucial characteristic of the multi-particle ion microfield (neglected in the MMM) which is the frequency Ω of the variation of any field component at a fixed value of the fieldstrength:

$$\Omega \equiv [<(dF_\parallel/dt)^2>_F + <(dF_\perp/dt)^2>_F]^{1/2}/3^{1/2}F. \qquad (1)$$

Using the corresponding results by Chandrasekhar and von Neumann [3], this frequency can be represented in the form

$$\Omega(\beta) = (15/8)^{1/2}\Omega_0/\beta, \quad \beta \leq \beta_m; \quad \Omega(\beta) = \Omega_0(\beta)^{1/2}, \quad \beta \geq \beta_m; \qquad (2)$$
$$\Omega_0 = (2.603)^{1/2} N_i^{1/3} (2T_i/M_i)^{1/2}; \quad \beta_m = (15/8)^{1/3} \approx 1.233;$$

$\beta = F/F_0$ being the reduced fieldstrength, $F_0 \approx 2.603 Z_i e N_i^{2/3}$.

The separation procedure combines two well-known criteria: the modulation-type and damping-type determinations of the boundary between quasistatic and dynamic fields (see, e.g., [4]). In other words, it is based on comparing $\Omega(\beta)$ (Fig. 1, solid line) with both the instantaneous splitting of a hydrogen line $n \to n'$

$$\omega_\beta \equiv [(n^2-n'^2)\hbar F_0/m_e e]\beta \qquad (3)$$

(Fig. 1, dotted line) and the summary damping constant

$$\gamma = 2(\gamma_i + \gamma_e) \qquad (4)$$

(Fig.1, dashed line), where γ_i and γ_e are the ion and electron impact HWHM of the line, γ_i being calculated below. After finding the root β_b of the equation

$$\Omega(\beta) = \max(\omega_\beta, \gamma), \qquad (5)$$

we calculate the densities of the dynamic N_d and quasistatic N_q ions as follows:

$$N_d = pN_i, \quad N_q = (1-p)N_i, \quad p \equiv \int_0^{\beta_b} d\beta\, H(\beta), \qquad (6)$$

where $H(\beta)$ is the Holtsmark distribution function.

3. At the second stage the goal is to calculate the ion impact width due to the dynamic subensemble of the ion density N_d. It turns out that for this subensemble the number of ions in the ion Weisskopf sphere is smaller than one not only for the case where the initial density N_i is low but also for high densities N_i. This is due to the fact that in the latter case we have $N_d \ll N_i$. Therefore the ion impact broadening operator Φ_i can be calculated using the standard, binary ion impact theory from [2] by substituting N_i by N_d.

We note that the resulting ion impact HWHM γ_i is determined by the value of N_d, which in its turn was calculated at the stage 1 using the value of γ_i. Clearly, this algorithm is an iterative procedure that, fortunately, converges very rapidly.

4. At the third stage we calculate the final line profile by the formula

$$I(\Delta\omega) = -\int_0^\infty d\beta_q H(\beta_q) \pi^{-1} \mathrm{Re} \sum_{\alpha\alpha'\beta\beta'} d^\beta_\alpha d^{\alpha'}_{\beta'} \{[i\Delta\omega(\beta_q) + (\Phi_i(N_d) + \Phi_e)]^{-1}\}^{\alpha\beta}_{\alpha'\beta'}, \quad (7)$$

where

$$\beta_q = F/F_q, \quad F_q = 2.603 Z_i e N_q^{2/3}. \tag{8}$$

The above formalism contains both the ion quasistatic limit at high densities and the ion impact limit at low densities.

5. Using this model, we calculated Stark profiles of the H_α line and HWHM of these profiles for a broad range of densities from $10^{14} \mathrm{cm}^{-3}$ to $10^{19} \mathrm{cm}^{-3}$. The comparison of our results with experimental and previous theoretical results is presented in Table 1, where the last column shows the relative inaccuracy of our results compared to the experiments. The conclusions are the following.

1. Our model is in the excellent agreement with experiments [5,6] (1% to 6% accuracy) over the range of densities from $10^{15} \mathrm{cm}^{-3}$ to $10^{19} \mathrm{cm}^{-3}$. In this range it yields practically the same accuracy as the SM [7] and a much higher accuracy than the MMM [8], despite it is simpler and faster than the MMM.

3. At densities from $10^{14} \mathrm{cm}^{-3}$ to $10^{15} \mathrm{cm}^{-3}$, the MMM and even the SM [7] fail to reproduce the ion impact regime [2], while this regime is reproduced by our model at these densities.

4. At densities of the order of $10^{14} \mathrm{cm}^{-3}$, our model is still in a good agreement with the experiment [9] (within 15% accuracy), while the SM [7] underestimates the broadening by a factor of 2 or 3. We note that at this density the fine structure contribution (subtracted from the experimental HWHM) is $\approx 15\%$, so that one should not expect better than 15% agreement between the calculations based on Coulomb-radiator wave functions and the experimental Stark HWHM.

Thus our analytical model is not only simpler, faster, and more accurate than the MMM, but is also universal: it works well over five orders of magnitude density range.

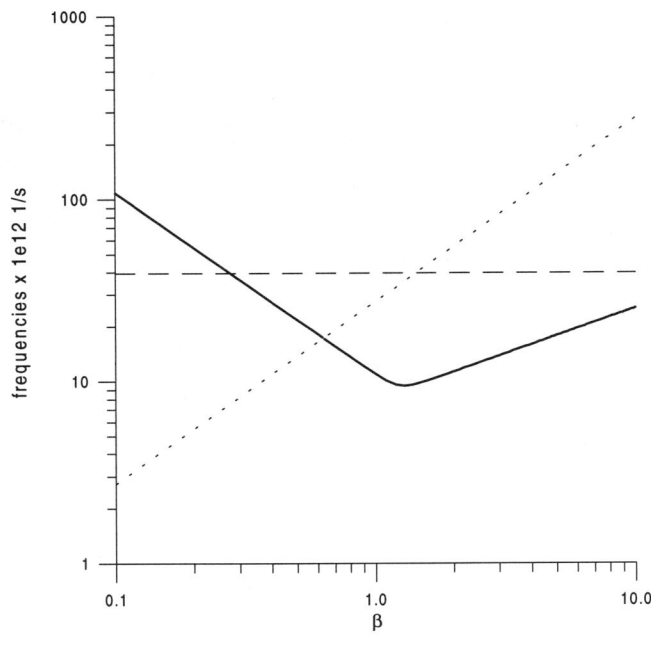

Figure 1

N_e	T, K	MMM [8]	Impact [2]	SM [7]	our	Experiment	error, %
9.27e18	11600				118	123 [5]	3.7
1.00e17	10000	4.3		4.6	4.7	4.6 [6]	2.0
1.00e16	10000	1.00		1.15	1.16	1.2 [6]	0.9
1.00e15	10000	0.22		0.28	0.30	0.28 [6]	6.5
1.00e14	10000		0.089	0.051	0.122		
1.35e14	13500		0.088		0.119	0.14 [9]	15

Table 1

REFERENCES

1. Alexiou, S., *J. Quant. Spectr. Rad. Transfer* **54**, 1-26 (1995).
2. Stehle, C., Mazure, A., Nollez, G., and Feautrier, N., *Astron. Astrophys.* **127**, 263-266 (1983); Stehle, C., and Feautrier, N., *J. Phys. B* **17**, 1477-1489 (1984); *Ann. Phys. Fr.* **9**, 697-704 (1984).
3. Chandrasekhar, S., and von Neumann, J., *Astrophys. J.* **95**, 489 (1942); **97**, 1 (1943).
4. Lisitsa, V.S., *Sov. Phys. Usp.* **20**, 603-630 (1977).
5. Böddeker, St., Günter, S., Könies, A., Hitzschke, L.,and Kunze, H.-J., *Phys.Rev.E* **47**, 2785 (1993).
6. Kelleher, D.E., Konjevic, N., and Wiese, W.L., *Phys. Rev. A* **20**, 1195 (1979); Ehrich, H., Kelleher, D.E., *ibid.* **21**, 319 (1980); Wiese, W.L., Kelleher, D.E., and Paquette, D.R., *ibid.* **6**, 1132 (1972).
7. Oza, D.H., Greene, R.L., and Kelleher, D.E., *Phys. Rev. A* **37**, 531-536 (1988).
8. Seidel, J., *Z. Naturforsch.* **32a**, 1207-1214 (1977).
9. Weber, E.W., Frankenberger, R., and Schilling, M., *Appl. Phys. B* **32**, 63-73 (1983).

Novel Diagnostic for nonthermal Electric Fields in Plasmas

S. Alexiou, A. Weingarten, Y. Maron, M. Sarfaty and Ya. E. Krasik

Weizmann Institute of Science, Rehovot 76100, Israel

Two new methods are proposed and illustrated for spectroscopic diagnostics of nonthermal electric fields in plasmas. These methods were used to analyse H_α and H_β profiles from a Plasma Opening Switch experiment.

In thermal plasmas, Stark Broadening has been often used to diagnose the plasma parameters. Can Stark broadening be also used to diagnose nonthermal plasmas? For dynamic Stark broadening, the essential difference is that the time variation of the nonthermal microfield is not known, unlike the thermal case. Thus, it would be very attractive to use a quasistatic approach, which does not invoke the *unknown* microfield dynamics. There are, however, two problems: First, the quasistatic approximation is known to be often quite bad and unreliable. Second, even if the quasistatic approximation were valid, the turbulent microfield distribution is not known. Past treatments have always, based on central limit theorem arguments for fully developed turbulence, assumed a 1 or more-dimensional Gaussian distribution. Besides the practical difficulty in telling experimentally whether one is dealing with a fully developped turbulence or not, the modern theory of turbulence does not support the Gaussian assumption over the entire electric field axis E. For example, in the self-similar range, the microfield distribution varies as E^{-5}.

The new method[1] consists of the following ideas: First, to ensure the validity of the quasistatic approximation, we limit ourselves to the short-time behavior of the dipole autocorrelation function C(t) (the Fourier transform of the lineshape). It is shown that if we neglect or deconvolve thermal broadening, that is, if the main broadening mechanisms are turbulent broadening and the Doppler effect, the short-time behavior of C(t) is quadratic in time, with the proportionality constant a linear function of the average of the square of the turbulent field. Thus plotting the quantity $\frac{1-C(t)}{t^2}$ vs. t for short times, it should tend, as t goes to 0 to a constant value that is linear in the average of the square of the turbulent field $\langle E^2 \rangle$. If we have C(t) from two lines, we can self-consistently determine this average and the Doppler temperature. The method assumes that turbulent broadening dominates thermal broadening (otherwise we have no real interest in a turbulent field determination). The key advantage of the method is that it makes *no* assumption that the nonthermal fields are quasistatic, nor on their distribution functions.

A complementary analysis can determine the "typical" field amplitude and frequency. This is to be understood in the following sense: Consider an oscillatory electric field with amplitude E and oscillation frequency ω. This is a *single* electric field, and no ensemble is considered here. The (unaveraged) contribution to the autocorrelation function C(t) from this field is compared directly with the experimental C(t). A "typical" field is defined so that these two functions are "close" for short and intermediate times(for example until C(t) has decayed to the 1/e level). In short, the decay rate of this single configuration C(t) is compared to the experimental. The effects of the field amplitude and frequency on the decay rate is that a larger amplitude results in a faster decay, while a larger frequency results in a slower decay; this means that there is an amplitude-frequency curve for each line that gives the amplitude and frequency pairs that result in decay rates matching the experimental. The intersection of 2 such curves from 2 different lines gives us an indication of the "best typical" amplitude and frequency, though quoting a single value is not what is important here, but rather the estimation of the "typical" field.

Clearly this approach completely neglects the statistics and is useful in the sense that a distribution of frequencies and amplitudes too much displaced and in a single direction from the typical field parameters will not be able to match the experimental C(t). We should also point out that this type of analysis can *not* be done on the frequency domain, since a line profile implies *averaging*.

An application of these methods to the data from a Plasma Opening Switch experiment using the observed H_α and H_β profiles obtained very consistent results, which were in excellent agreement, as far as the Doppler temperature was concerned with independent collisional-radiative calculations.

REFERENCES

1.S.Alexiou, A.Weingarten, Y.Maron, M.Sarfaty and Ya.E.Krasik, Phys.Rev.Lett. **75**, 3126(1995).

SCREENING POTENTIAL IN HIGH DENSITY PLASMAS

M. Amari, J.P. Arranz, J. Butaux, and H. Nguyen

Spectronomie des Gaz et des Plasmas - DRP Université Pierre et Marie Curie
4, place Jussieu Tour 22 - 75252 Paris cedex 05 - France

Abstract. On the basis of a two-ion center model, an accurate closed form of the screening potential is suggested for intermediate and high density plasmas.

By means of a two-ion-center model (1) we have shown that the screening potential in high density plasmas can be written in the form:

$$V_S = \begin{cases} \dfrac{(Ze)^2}{d}\left[1+\xi-\left(\theta-\dfrac{1}{2-\xi}\right)\xi^2\right], & \xi = 1 - R/d \geq 0, \\ \dfrac{(Ze)^2}{R}, & R > d \end{cases} \quad (1)$$

In Eq.(1), R, θ and d denote the interionic distance, the effective electron restoring force factor and the correlation length respectively. As shown in the following table, d/a varies very slowly with $\Gamma = (Ze)^2/akT$ (a is ion-sphere radius) and takes its minimum value $\cong 1.60$ for $\Gamma \cong 10$.

Γ	1	2	3	4	10	20	40	100	∞
d/a	1.874	1.715	1.656	1.628	1.598	1.609	1.630	1.651	1.699

For $\Gamma \in [1, 160]$, Eq.(1) agrees within 1% with the most accurate Monte Carlo data (2). The linear behaviour (3) of V_S for $R \in [0.4a, d]$ is explained in terms of antagonistic forces due to electrons localized in the neighbourhood of the two reacting ions. Indeed, we have shown that the relative contribution of the quadratic term in Eq.(1): $Q(\xi) = \xi^2[\theta - 1/(2-\xi)]/(1+\xi)$ is less than 0.04 for $\xi \in (0, 0.8)$.

For the sake of comparison at the extreme high density limit ($\Gamma \gg 1$) the screening potential in a body-centered-cubic crystal has been calculated in details and expressed as a series expansion including the hexadecapole term for small displacements from the equilibrium configuration and as a closed form fitting numerical results for the large ones. This result is in agreement with the above-mentioned two-ion-center model provided that the nearest neighbour distance in lattices has the meaning of correlation length and that surrounding ion sites are properly relaxed when very short distances between reacting ions are considered.

Finally, by replacing the Coulomb limit by a suitable screened asymptotic form, Eq.(1) has been generalized to plasmas with lower densities ($\Gamma \leq 1$) and compared with the solutions of the non-linear Poisson-Boltzman equations.

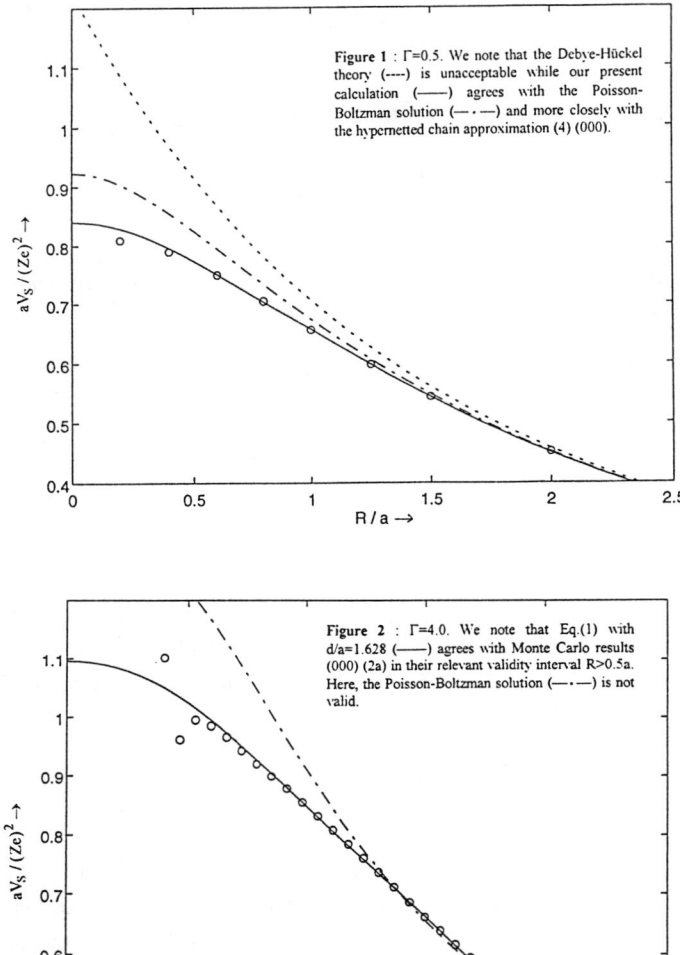

Figure 1 : Γ=0.5. We note that the Debye-Hückel theory (----) is unacceptable while our present calculation (——) agrees with the Poisson-Boltzman solution (—·—) and more closely with the hypernetted chain approximation (4) (000).

Figure 2 : Γ=4.0. We note that Eq.(1) with d/a=1.628 (——) agrees with Monte Carlo results (000) (2a) in their relevant validity interval R>0.5a. Here, the Poisson-Boltzman solution (—·—) is not valid.

REFERENCES

1. a/ Malnoult Ph., d'Etat B. and NGUYEN H., Phys. Rev. **40**, 1983 (1989) ; b/ Rosenfeld Y., Phys. Rev. **53**, 2000 (1996).
2. a/ Hansen J.P., Phys. Rev. **8**, 3098 (1973) ; b/ Ogota S., Phys. Rev. **53**, 1094 (1996).
3. De Witt H., Graboske H. and Cooper M.S., Astrophys. J. **181**, 439 (1972).
4. Springer J.F., Polkrand M.A. and Stevens Jr F.A., J. Chem. Phys. **58**, 4863 (1973).

Redistribution Function for Resonance Radiation in a Hot Dense Plasma

A. E. Bulyshev,[†] A. V. Demura,[‡] V. S. Lisitsa,[‡]
A. N. Starostin,[†] A. E. Suvorov,[†] and I. I. Yakunin[†]

[†]*State Research Center Troitsk Institute for Innovation and Thermonuclear Research, 142092 Troitsk, Russia*
[‡]*Russian Research Center "Kurchatov Institute", 123182 Moscow, Russia*

Abstract

The redistribution function for resonance radiation in the L_α spectral line of hydrogenic ions in a dense hot plasma is calculated. The calculation is based on a self-consistent solution of the equations for the populations of the excited ionic sublevels and for the polarizations of the transitions considered. Nonlinear interference effects due to mixing of atomic states in both static and dynamic ionic fields are thereby taken into account. Molecular dynamics methods are used to account for the evolution of the multiparticle ionic field resulting from thermal motion of the ions. We calculate the L_α line of the hydrogen-like argon ion in a plasma with electron temperature 1 keV and electron density $N_e = 10^{22} \div 10^{24}$ cm^{-3}. The rescattering function is compared with the approximation provided by complete frequency redistribution. The results demonstrate the limited usefulness of the latter approximation for a plasma consisting of multiply-charged ions.

A fundamental characteristic of the theory of resonance radiative transfer in an optically dense medium is the photon frequency redistribution function $R(\omega, \omega')$, which is the joint probability that a photon absorbed at frequency ω' is re-emitted at frequency ω. For a resonance spectral line, $R(\omega, \omega')$ ordinarily contains two types of contributions, corresponding to coherent and incoherent photon scattering. The ratio of these contributions depends on the line broadening mechanisms operating in the medium. For example, coherent scattering dominates when a probability of radiative decay A from an excited state is much greater then a rate of line-broadening collisions with pertubing particles ϕ. In opposite case the scattering becomes completely incoherent.

An important aspect of calculations of the function $R(\omega, \omega')$ in a dense plasma is to account for the plasma ionic microfield, which leads to Stark broadening of the ionic emission lines. In a dense plasma the photon frequency redistribution in the rest frame of the emitter results from fast "shaking" of the atomic states by electrons and slow shifts of the states by the ions.

For a static ionic field **F** the problem can be solved analytically for a model three-level system. In so doing it has been demonstrated that nonlinear interference effects play a role in the formation of the emission spectra and the photon frequency redistribution function. These effects are due to the interference of the atomic states in an external field, which in a plasma with multiply-charged ions leads to a strongly nonequilibrium distribution of the

© 1997 American Institute of Physics

population over the atomic sublevels. The rescattering function $R(\omega,\omega')$ is most sensitive to these effects. Here, first of all, if nonlinear interference effects are neglected, $R(\omega,\omega')$ is no longer positive definite and secondly, even for $A/\phi \ll 1$, when the contribution of the coherent component is negligibly small, $R(\omega,\omega')$ does not reduce to a product of independent profiles (absence of total frequency redistribution). The latter circumstance does not reduce simply to a trivial discrepancy between the microfield averaged product of the profiles and the product of the averages, rather, it reflects the interference of atomic states that results from nonlinear interference effects.

The next step in the calculation of $R(\omega,\omega')$ is to take into account the nonstationary ionic field generated by the thermal motion of the ions (ion dynamics). We make the first attempt to calculate the rescattering function $R(\omega,\omega')$ for the L_α line of hydrogenic ions, taking into account both non-linear interference effects and ion dynamics [1].

The rescattering of resonance radiation was described using the compaund density matrix method. This method is based on the equations for density matrix for the "atom (ion) + spontaneouse electromagnetic fields" compound system. It allows us to find self-consistently the populations of excited ionic sublevels and the polarizations of the transitions considered.

Numerical calculations of the function $R(\omega,\omega')$ both in static ion microfield and with taking ion dynamic into account for the L_α line were performed for the hydrogenic Ar ion in a hydrogen plasma with a temperature of 1 keV and electron density $10^{22} \div 10^{24}$ cm^{-3}. The ion microfield was modeled by the molecular-dynamics method.

We note that this range of hydrogen plasma density (with a small admixture of argon) is the main object of study in the compression dynamics of laser fusion targets. For these conditions the function $R(\omega,\omega')$ obviously differs from the total frequency redistribution approximation because of nonlinear interference effects, although it is ion dynamics (rather than static) that makes this approximation possible.

Let us now summarize the basic results obtained in the present work.

We have, for the first time, systematically calculated the rescattering function for resonance radiation in the real L_α spectral line of a multiply-charged ion in hot dense plasma, taking nonlinear interference effects and ion dynamics into account.

We have shown that the rescattering function differs from the total frequency redistribution approximation even at high densities, which may have an important influence on the interpretation of diagnostics involving radiation from impurities in fusion targets.

We have shown that ion dynamics strongly influences (in contrast to the emission profiles) the ratio of coherent and incoherent components of rescattering, even at comparatively low plasma densities.

REFERENCES

1. Bulyshev, A. E., Demura, A. V., Lisitsa, V. S., Starostin, A. N., Suvorov, A. E., and Yakunin, I. I., *Sov. Phys. JETP* **81**, 113–121(1995).

Hydrogen Lines with Central Stark Components as a Sensitive Magnetic Probe

Alexander V. Demura* and Eugene Oks**

*Hydrogen Energy and Plasma Technology Institute,
RRC "Kurchatov Institute", Moscow, 123182, Russia
**Department of Physics, 206 Allison Lab, Auburn University
Auburn, AL 36849-5311, USA

Abstract

Profiles of spectral lines in dense plasmas experience a significant Stark broadening. If there is also a magnetic field in a plasma, the analysis of the resulting line profiles becomes very complicated. Polarization difference contours obtained by subtracting two profiles of two orthogonal polarizations seem to be the most suitable diagnostic tool: plasma and field parameters can be more easily deduced from pronounced structures in the difference contours. It is shown that the difference contour of the central Stark components of hydrogen spectral lines can be used for a highly accurate measurements of the magnetic field, in particularly, in laser fusion plasmas, in the solar photosphere and in the chromospheres of flaring stars.

Consider a hydrogenlike radiator in a plasma in a uniform magnetic field $\vec{B}$, directed along the OZ axis. Its Hamiltonian $\hat{H}$ incorporates also the interaction with an ion produced electric microfield $\vec{F}$

$$\hat{H} = \hat{H}_o - \hat{\vec{d}}\vec{F} - \frac{e\hbar}{2mc}\hat{L}_z B + V_{er},$$

where $\hat{H}_o$, $\hat{\vec{d}}$, $\hat{\vec{L}}$ are the unperturbed Hamiltonian of the radiator, its dipole moment and angular momentum operators, respectively; e, m are the electron charge and mass; c is the speed of light; V_{er} is the radiator interaction with plasma electrons. Here it is assumed that the magnetic field is sufficiently large to break the LS coupling, but the diamagnetic interaction is neglected as well as the splitting of the energy levels due to the electron spin. We consider the case, when the Stark splitting of the energy levels ω_S is much greater than both the fine structure splitting and the Zeeman splitting ω_Z in a magnetic field [1]. To match the first condition outlined above one needs for the radiator of the charge of order ten to have the density of the field ions of the charge Z=1 of order $10^{23} - 10^{24}$ cm^{-3}. The ratio of the Zeeman splitting to the Stark one $\epsilon \equiv \frac{\omega_Z}{\omega_S} \ll 1$ represents the small parameter ϵ of the perturbation in the problem under consideration. The first condition $\omega_s \gg \Delta W_{fs}$ is important for a possibility to choose as a basis the parabolic wave functions with the quantization axis $\vec{OZ'}$ directed along the ion microfield $\vec{F}$. The wave functions and the energy levels thus can be found by the standard perturbation theory. We emphasize that the orientation of the "primed" reference frame, generally speaking, varies from one radiator to another. Polarized profiles $I_{\parallel}(\Delta\omega)$, $I_{\perp}(\Delta\omega)$ are observed in the "unprimed" reference frame with $\vec{OZ} \parallel \vec{B}$. We choose the direction of the observation towards the direction of the $\vec{OY}$ axis and two linear crossed polarizations along the $\vec{OZ} \parallel \vec{B}$ and $\vec{OX} \perp \vec{B}$ respectively. Then the polarization difference contour is $P(\Delta\omega) = I_{\parallel}(\Delta\omega) - I_{\perp}(\Delta\omega)$, where $\Delta\omega \equiv \omega - \omega_o$; ω_o is the unperturbed frequency due to a radiative transition between initial i and final f energy levels.

A starting formula for the line profile of the radiative transition between the upper set of sublevels $\{|\alpha><\alpha'|\} \in i$ and the lower set of sublevels $\{|\beta><\beta'|\} \in f$ defined in the primed reference frame contains lateral and central Stark components. It is very important that the central Stark components are not perturbed by the electric ion microfield and therefore affected by Zeeman splitting in a magnetic field in zero order of the perturbation. Thus, profiles of their Zeeman components are determined by the simultaneous action of the electron impact and Doppler broadening. A measured degree of the polarization may yield the information on the value and the direction of the magnetic field. The polarization difference profiles of a central Stark component exhibit the following characteristic structures: two lateral extremums (one in each wing) and a central extremum of the opposite sign. These features may be unambiguously used for the determination of the magnetic field.

Previously for the magnetic field determination it was suggested to use the contribution of the lateral Stark components into the polarization difference profile. However, in that case the polarization maximum is proportional to the second order of the small parameter $\epsilon \ll 1$ and thus is much less than for the central Stark components. Moreover, since the intensity of the lateral Stark components is distributed over the detunings from the line center of the order of ω_S, the total gain of the sensitivity in the current method compared to the previous one is of the order of $\epsilon^{-3} \gg 1$. However, the spectral resolution should be larger to detect the Zeeman caused polarization in the center of the line.

REFERENCES

1. Demura, A. V., Lisitsa, V. S., *Sov. Phys. JETP* **35**, 1130-1134 (1972).

The Influence of Charge-Exchange on Ly-Radiation

A.Devdariani*, E.Leboucher-Dalimier**, P.Angelo**,
P.Gauthier**, P.Sauvan**

*Department of Optics and Spectroscopy, University of St-Petersburg,
St-Petersburg, 198904, Russia
**Physique Atomique dans les Plasmas Dences, LULI/Université Paris VI - Ecole Polytechnique.
4 place Jussieu, 75252 Paris cedex 05, France

A dynamical approach to the description of the spectral line shapes generated by (Ze)* - Z collisions has been developed. It includes explicitly the interaction exchange between initial diabatic states as well as final states. On the basis of this approach a formula has been obtained which describes uniformly the central part as well as the far wings of the spectral lines for the case of Lyalfa radiation. It reproduces also the oscillating structure of the red satellite due to charge exchange between ions.

The experiments in hot and dense plasma give some evidence of the existence of quasimolecular emission in the spectral range of Lybeta line (1). Although the quasimolecular emission is influenced by plasma electrons, it is still reasonable to start from the basic problem of the (Ze)* - Z radiation without interaction with plasma electrons as a first approach because this is the unique example of system for which the energy terms and dipole moments are well known.

The aim of this work is to develop a dynamical approach to the description of the spectral line shapes generated by the (Ze)* - Z collisions, which includes the influence of nonadiabatic transitions on radiation. As a first case we deal with radiation which covers Lyα spectral range.

The solution of the problem must take into account six diabatic states produced by the interaction between ions and generated by the single ion state n=2. These states are connected by ion-ion interactions and interact with six continua which are bound to two ground diabatic terms produced by the ionic state n=1. The main nonadiabatic process merging from ion collisions and connecting diabatic states is charge exchange. The spectral line shapes which include the influence of nonadiabatic transitions can be calculated on the basis of the approach (2) generalized for the case of interacting bounderies of continua as reported in (3). In order to study the influence of charge exchange on radiation, one can follow a simplified picture. There are quasimolecular optical transitions between only two upper diabatic states and two ground states which describe an electron bound to the first or second nucleus.

The calculation of the dipole moment Fourier transform between excited and ground states leads to the following expression for the probability of Lyα radiation

$$J_1 = \frac{1}{\sqrt{2\pi\alpha}} \left(\frac{A}{\alpha}\right)^{-i\frac{\Delta\omega}{2\alpha}} \Gamma\left(i\frac{\Delta\omega}{\alpha}\right) D_{-i\frac{\Delta\omega}{\alpha}}\left(\frac{C}{\sqrt{\alpha A}} e^{-\frac{3i\pi}{4}}\right) exp\left(\frac{\pi\Delta\omega}{4\alpha} + i\frac{C^2}{4\alpha A}\right)$$

where $\alpha = Z/2$, $A = 3Z^2/2$, $C = 3Z^2/8$. The formula leads to the Lorentzian distribution for the central part of the line and to the correct quasistatic expression for the wings. It reproduces the oscillating structure of the red satellite due to charge exchange between ions. So, the approach proposed can be regarded as a basis for a proper consideration of the nonadiabatic transition influence on the spectral line shapes of lines produced in collisions of multicharge ions.

A.D. likes to gratefully acknowedge the support of LULI (PECO program) during his stay in Paris and the hospitality of Prof. E.Leboucher-Dalimier.

REFERENCES

1. E.Leboucher-Dalimier, A.Poquerusse, and P.Angelo, *Phys.Rev.E* **47**, R1467 (1993).
   P.Gauthier, P.Sauvan, P.Angelo, S.Alexiou, E.Leboucher-Dalimier, A.Calisti, and B.Talin. 13th ICSLS, invited paper, Firenze (1996)
   P.Angelo, H.Derfoul, P.Gauthier, P.Sauvan, A.Poquerusse, T.Ceccoti, E.Leboucher-Dalimier, T.Shepard, C.Back, M.Vollbrecht, I.Uschmann, and E.Förster, 13th ICSLS, Firenze (1996)
2. A.Z.Devdariani, Yu.N.Sebyakin, and V.N.Ostrovskii, *Sov.Phys.JETP* **49**, 266 (1979).
3. D.Crothers, A.Devdariani, and Yu.Sebyakin, 5th EPC *Conference on Atomic and Molecular Physics*, Edinburgh, Contributed Papers, Part I, 387 (1995).

On Importance of the Divergence of Ionic Microfields in Plasmas for the Shift and Asymmetry of Hydrogen Spectral Lines

J. Halenka, W. Olchawa, and B. Grabowski

Institute of Physics, Opole University, Oleska 48, PL-45-052 Opole, Poland

Abstract. It is shown that the contribution of the term proportional to the divergence of the ionic microfield to the total ionic shift and asymmetry of hydrogen lines is small compared to the others.

The Hamiltonian of hydrogen atom in the ionic microfield **E** of a plasma can be - taking unhomogeneities of this microfield into account - described as follows[1]

$$H = H_0 - \mathbf{d}\cdot\mathbf{E} - \frac{1}{6}\sum_{ij} Q_{ij} E_{ij} + \frac{1}{6}e_0 r^2 \nabla\cdot\mathbf{E}, \qquad (1)$$

where $\nabla\cdot\mathbf{E}$ is the divergence of the ionic microfield. When the microfield is described within the Debye-Hückel model, the divergence differs from zero. In Ref. 1 the term proportional to $\nabla\cdot\mathbf{E}$ has been neglected as being insignificant. In the present paper the importance of that term from the point of view of shift and asymmetry of H-lines has been examined in detail.

Taking into account the screening and ion-ion correlation effects we calculated the mean value of the divergence of the ionic microfield:

$$\frac{1}{6} < \{\nabla\cdot\mathbf{E}(\mathbf{R})\}_{\mathbf{R}\to 0} >_\beta = \frac{5}{(32\pi)^{1/2}} \frac{E_0}{R_0} b_\rho(\beta). \qquad (2)$$

For a neutral point in single ionised plasma the function $b_\rho(\beta)$ reveals features ilustrated in Fig. 1. The function fulfils the condition of quasi-neutrality of the plasma:

$$\int_0^\infty b_\rho(\beta) W_\rho(\beta) d\beta = 0. \qquad (3)$$

In order to estimate the influence of the examined term on shift and asymmetry of H-line profiles, the calculations have been executed within the quasi-static approximation for ions and the impact approximation for electrons. The impact operator Φ has been calculated, using computer simulation techniques, according to the expression

$$\Phi = -\hbar^{-2} \int_0^\infty \{\tilde{V}_e(0)\tilde{V}_e(t)\}_{av} dt . \qquad (4)$$

All matrix elements of the operators H, Φ, and $\mathbf{d}$ have been calculated with an accuracy to terms of the order $\propto (a_0 / R_0)^4$.

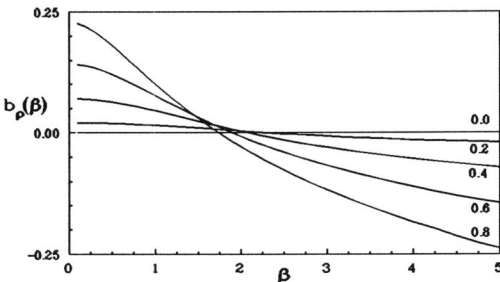

FIGURE 1. The function $b_\rho(\beta)$ defined by Eq. (2) at a neutral point for several values of $\rho = R_0/D$.

We found that the contribution of the examined divergence term to the total ionic shift amounts about 5 per cent only. The contribution of that term to the asymmetry of the intensity peaks of H_β is somewhat greater and equals to about 15 per cent. The calculated total shifts resulting from interactions with ions and electrons, being defined as the so-called ELC-shift, agree with experimental data.

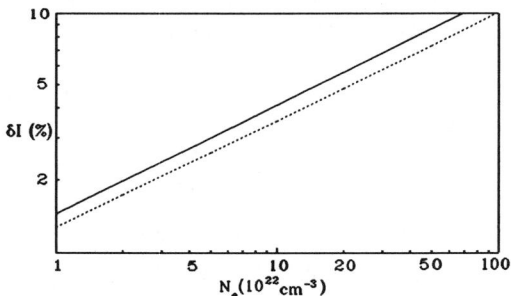

FIGURE 2. The asymmetry parameter of the H_β peaks $\delta I = (I_B - I_R)/I_B$ as a function of N_e. Solid line - the last term in Eq. (1) is included, dotted line - the term is neglected.

REFERENCES

1. J. Halenka, Z. Phys. D-Atoms, Molecules and Clusters **16**, 1 (1990).

The Frequency Fluctuation Model applied to the hydrogen-like helium Paschen-α line

T. Meftah, S. Alexiou, A. Calisti, L. Godbert,
R. Stamm and B. Talin

PIIM, centre Saint Jérôme, case 232, F13397 Marseille Cedex 20

Up to now the experiments with hot and dense plasmas rarely provide accurate line shapes corresponding to very well defined density and temperature conditions. This situation is due to the transient and inhomogeneous nature of laser produced plasmas. As a consequence experimental observations of lines in hot and dense plasmas do not generally provide a severe test for the theoretical models, and we must rely on other plasma sources to perform these. Various kind of plasmas created by discharges with electron densities between 10^{16} and 10^{18} cm^{-3} allow the observation of the Balmer and Paschen lines of hydrogenic helium. These lines are as many benchmark case for the modeling of Stark broadening for charged emitters. In this communication, we report on new calculations of the Paschen-α line shape of He$^+$ which are in agreement with recent precise and reliable experiments. We have used an accurate and fast line shape code [1] for obtaining these profiles by taking into account all the known broadening mechanisms. The atomic structure has been calculated by retaining the fine structure of the levels. Electron broadening is included by using an electron impact operator taking into account the effect of the energy separation of the perturbing levels, and the effect of the trajectory of the perturbing electrons [2]. An accurate description of electron broadening is essential since it contributes always significantly for the selected plasma conditions. In the present work we use the operator of Ref.2. This operator is straightforward to implement in our model for the diagonal matrix elements. However, for the Paschen-α, the interference terms are also very important and we had to make a decision on how to extend this operator to the interference terms. Briefly, this operator involves two energy differences, which for the diagonal elements are opposite, while for the interference terms there need be no definite relation between them. Thus, we used the same operator, taking as energy difference the root of the sum of the squares of the energy differences.

In our calculations, we use the Frequency Fluctuation Model (FFM) [1], which treats the ion dynamic effect by assuming an exchange process between the static radiative channels [1].

For a density domain of 2×10^{16} cm^{-3} to 4×10^{18} cm^{-3}, we have reported on Fig.1 the Full Width at Half Maximum (FWHM) obtained by our model (Crosses) for the plasmas conditions of three different experiments (temperatures between 4 and 7,5 eV, He+ or H+ perturbers [3-5]). These results demonstrate that a model retaining simultaneously all the broadening mechanisms allows an accurate density diagnostic based on the profile of weakly ionized emitters. The role of ion dynamics is confirmed for the benchmark case of He+.

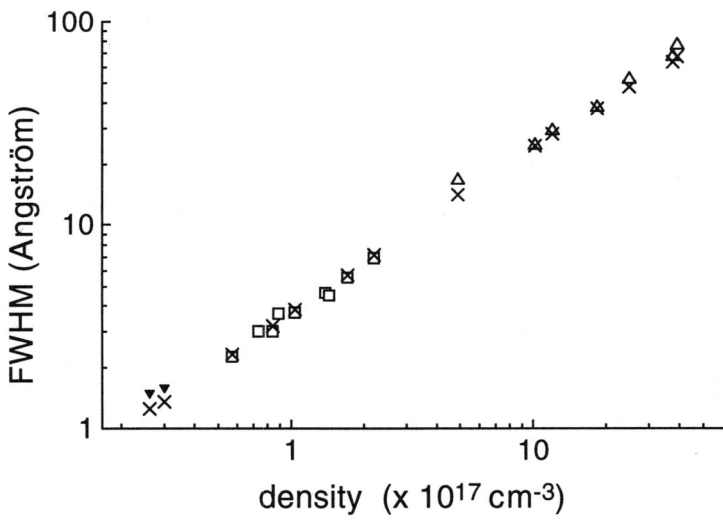

Fig.1 Comparison of the FWHM of the experiments of Stefanovic et al. (▼: [3] , T. Pittman et al. (: 4]), S. Büscher et al. (Δ : [5]), and our calculations (✗).

REFERENCES

[1] B. Talin et al., Phys. Rev. A**51**, 1918 (1995).
[2] S. Alexiou, Phys. Rev. A**49**, 106 (1994).
[3] I. Stefanovic, M. Ivkovic and N. Konjevic, Physica Scripta **52**, 178 (1995).
[4] T. Pittman et al., Phys. Rev. Lett. **45**, 723 (1980), Phys. Rev. A**33**, 1292 (1986).
[5] S. Büscher et al. , to appear in J. Phys. B.

Redistribution function of resonance radiation for atoms with degenerate structure

V. I. Savchenko*, N. J. Fisch*, A. A. Panteleev[+], A. N. Starostin[+]

*Princeton University, Plasma Physics Laboratory;
[+]Moscow Institute of Physics and Technology

Abstract

An analytical expression for the redistribution function of resonance radiation in plasma is found, using an atom + field density matrix approach to describe interaction with the quantized radiation fields. The relaxation of the atoms with degenerate structure of the levels is modeled using relaxation constants. The polarization of the emitted and absorbed photons are thus taken into account easily, with the scattering event being described by the redistribution tensor.

The problem of resonance radiation transfer has been studied intensively during the past years in connection with X-ray lasers and plasma diagnostics. The quantity of fundamental interest in this problem is the photon frequency redistribution function $R(\omega, \omega')$. To study polarization effects, we consider $J = 1 \Rightarrow J = 0$ atomic transition, taking into account the degenerate structure of the upper level. In this approach, the redistribution function becomes a tensor, with upper and lower indices corresponding to absorbed and emitted radiation respectively. In a following paper we analyze more realistic situation to include $J = 0$ sub-level of the upper level and take into account static and dynamic microfields.

The equation of motion for the "atom + spontaneous field" density matrix can be written in the form

$$i\frac{\partial \rho}{\partial t} = [H_0 + V + U, \rho] + i\Gamma(\rho) \qquad (1)$$

Here, H_0 is the unperturbed Hamiltonian of the atom; V and U represent the interaction of the atom with incident and scattered fields respectively; and $\Gamma(\rho)$ is a matrix of relaxation constants. Expanding the above equation in the perturbations U and V leads[1] to an expression for $R_{\sigma\sigma'}^{\sigma_1\sigma_2}(\omega, \omega')$.

The frequency dependence of the redistribution function $R_{\sigma\sigma'}^{\sigma_1\sigma_2}(\omega, \omega')$ is completely described by the pair of Lorentzians

$$q_\sigma(\omega) + i\phi_\sigma(\omega) = Re\frac{1}{(\omega - \omega_{0\sigma}) + i\Gamma} - iIm\frac{1}{(\omega - \omega_{0\sigma}) + i\Gamma}. \qquad (2)$$

For convenience, define
$$Q_{\sigma\sigma'}(\omega') + i\Phi_{\sigma\sigma'}(\omega') = -(q_\sigma(\omega') - q_{\sigma'}(\omega')) + i(\phi_\sigma(\omega') + \phi_{\sigma'}(\omega')). \tag{3}$$

An explicit by-component expression for the redistribution tensor is:
$$R^{\sigma\sigma}_{\sigma\sigma} = \frac{1}{2}\frac{\gamma}{\Gamma}(b-1)\phi_\sigma(\omega)\phi_\sigma(\omega') + \frac{\gamma}{\Gamma}\phi_\sigma(\omega')\delta(\omega-\omega'); \quad R^{\sigma'\sigma'}_{\sigma\sigma} = \frac{a\gamma}{2\Gamma}\phi_\sigma(\omega)\phi_\sigma(\omega'). \tag{4}$$

$$R^{\sigma\sigma'}_{\sigma\sigma'} = \frac{\gamma}{2}\frac{1}{\delta^2_{\sigma\sigma'}+\bar{\Gamma}^2-\bar{g}^2}\left[(\delta_{\sigma\sigma'}q_\sigma(\omega) - \bar{\Gamma}\phi_\sigma(\omega))\Phi_{\sigma\sigma'}(\omega') + (\bar{\Gamma}q_\sigma(\omega) - \delta_{\sigma\sigma'}\phi_\sigma(\omega))\Phi_{\sigma\sigma'}(\omega')\right]$$
$$+\pi\gamma(q_\sigma(\omega)q_{\sigma'}(\omega) + \phi_\sigma(\omega)\phi_{\sigma'}(\omega))\delta(\omega-\omega'). \tag{5}$$

$$R^{-10}_{10} = \frac{\gamma}{2}\frac{\bar{g}}{\delta^2+\bar{\Gamma}^2-\bar{g}^2}(\phi_1(\omega)\Phi_{-10}(\omega') - q_1(\omega)Q_{-10}(\omega')). \tag{6}$$

$$R^{10}_{-10} = \frac{\gamma}{2}\frac{\bar{g}}{\delta^2+\bar{\Gamma}^2-\bar{g}^2}(\phi_{-1}(\omega)\Phi_{10}(\omega') - q_{-1}(\omega)Q_{10}(\omega')). \tag{7}$$

The physical meaning of these tensor components is that they give the probability for the atom to absorb the photon, (which is described by the polarization matrix $\rho^\gamma_{\sigma_1\sigma_2}$), propagating in Z direction and to emit the photon $\rho^\gamma_{\sigma\sigma'}$ also in Z direction. To take into account angular dependence of the incident radiation, we have to multiply each tensor component by two Wigner matrices,
$$\sum_{\sigma_3\sigma_4} D^*_{\sigma\sigma_3}(\mathbf{n})D_{\sigma'\sigma_4}(\mathbf{n}), \tag{8}$$
and sum over indices σ_3 and σ_4. The angular dependence of the scattered light can be calculated similarly, with the summation being performed over the lower indices of the redistribution tensor.

We observe that this tensor is symmetric with respect to the interchange between upper and lower indices, which corresponds to time-reversal symmetry. Also, it is worthwhile to notice that components like $R^{\sigma_1\sigma_2}_{\sigma\sigma}$ are zero, since they describe the amplitude of the transformation of unpolarized light into polarized light, which can not take place (this may be changed in the presence of a laser field). As usual, one can see that there are coherent $\sim \delta(\omega-\omega')$, and incoherent contributions to $R(\omega,\omega')$. But in our case they do not appear in all components of the tensor. Scattering events like $\sigma'\sigma' \Rightarrow \sigma\sigma$ or $\sigma_1\sigma_2 \Rightarrow \sigma\sigma'$ correspond to completely incoherent processes, therefore $R^{\sigma\sigma'}_{\sigma\sigma'}, R^{\sigma_1\sigma_2}_{\sigma\sigma'}$ do not have coherent components.

This work was supported by the U. S. Department of Energy under Contract No. DE-AC02-CHO-3073.

REFERENCES

[1] A. N. Starostin et al, JETP 98, 1304 (1990).

Experiments on Hot and Dense Laser-Produced Plasmas

C. A. Back, N. C. Woolsey, A. Asfaw, S. H. Glenzer, B. A. Hammel,
C. J. Keane, R. W. Lee, D. Liedahl, J. C. Moreno, J. K. Nash,
A. L. Osterheld, A. Calisti*, R. Stamm*, B. Talin*, L. Godbert*,
C. Mossé*, S. Ferri*, and L. Klein*

Lawrence Livermore National Laboratory, P.O. Box 808, Livermore, CA 94551

* PIIM, Centre St Jerome, Université de Provence, Marseille, France

Plasmas generated by irradiating targets with ~ 20 kJ of laser energy are routinely created in inertial confinement fusion research. X-ray spectroscopy provides one of the few methods for diagnosing the electron temperature and electron density. For example, electron densities approaching 10^{24} cm^{-3} have been diagnosed by spectral linewidths. However, the accuracy of the spectroscopic diagnostics depends the population kinetics, the radiative transfer, and the line shape calculations. Analysis for the complex line transitions has recently been improved and accelerated by the use of a database where detailed calculations can be accessed rapidly and interactively. Examples of data from Xe and Ar doped targets demonstrate the current analytic methods. First we will illustrate complications that arise from the presence of a multitude of underlying spectral lines. Then, we will consider the Ar He-like $1s^2$ ($^1S_0$) - 1s3p ($^1P_0$) transition where ion dynamic effects may affect the profile. Here, the plasma conditions are such that the static ion microfield approximation is no longer valid; therefore in addition to the width, the details of the line shape can be used to provide additional information. We will compare the data to simulations and discuss the possible pitfalls involved in demonstrating the effect of ion dynamics on lineshapes.

Plasmas generated by irradiating targets with ~ 20 kJ of laser energy are routinely created in inertial confinement fusion research. In laser-produced plasmas, a single observed lineshape may be a convolution over several transitions instead of the result of a single transition. Furthermore, individual spectral lines can be affected by radiative transfer or ion dynamic effects. The experiments we discuss are motivated by both theory and applications. For instance the study of line formation can test basic physics models such as Stark broadening theory. However, spectroscopy can also be used to diagnose plasmas relevant to laser fusion, astrophysics, and atomic phyiscs in the laboratory. Some examples are experiments that explore beam propagation, ionization physics, and opacity (1-3).

In these studies, emission spectra provide non-perturbative diagnostics of the laser-produced plasmas. Spectroscopic diagnostics have been used to diagnose the electron temperature, T_e, by the ratio of line intensities, electron density, N_e, by

Stark broadening or line intensity ratios, and ion temperature, T_i, by Doppler broadening(4). The analysis of spectra often requires information about the atomic kinetics which involves vast amounts of atomic data and rate equation models. It also requires additional information about the hydrodynamic evolution of the plasma, and, if the plasma is rapidly evolving, time-dependent analysis may be required.

Figure 1 shows photographs of the two types of targets used to study the physics of hot, dense matter, gasbags and hohlraums (5,6). The gasbags are formed by inflating CH membranes which are glued to both sides of a 2.75 mm diameter washer. The gas-filled hohlraums are Au cylinders 2.5 mm long and 1.6 mm in diameter. Thin CH membranes are placed over the endcaps of the cylinder to prevent the gas from escaping. These targets are irradiated by the Nova laser, a Nd:glass laser which can deliver a total of ~ 25 kJ of laser light at 0.35 μm in 1 ns. The targets are irradiated by ten laser beams. Gasbags are irradiated in a uniform manner since the lasers are incident in two cones of beams that are 50 ° from the axis of the gasbag washer. Hohlraums are irradiated on the inside wall by laser beams which enter through the ends of the cylinder which create a plasma inside which is bathed in a radiation field created by the plasma formed at the Au wall.

The spectroscopic measurements are designed to diagnose laser-produced plasmas having electron densities of $n_e > 10^{21}$ cm^{-3} and temperatures T_e from 0.5 to 5 keV. The gasbags are open geometry targets that enable tests of atomic models, atomic kinetics, and benchmarks of hydrodynamic codes. Hohlraums, on the other hand, are closed geometry targets that enable studies of plasmas in extreme conditions of N_e, T_e, and T_r that cannot otherwise be achieved in laboratory plasmas. These closed geometry targets produce confined plasmas that may be subject to intense radiation fields produced by the conversion of laser light into x-rays at the inside wall.

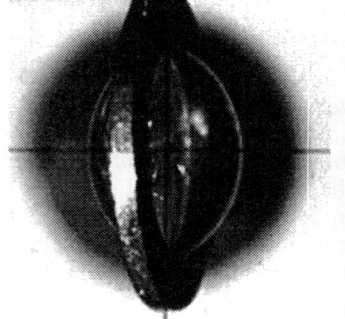
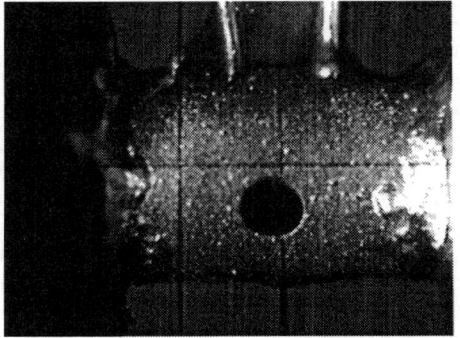

GASBAG   HOHLRAUM

FIGURE 1. The gasbag is formed by confining gas between two thin CH membranes. The plasma is formed by overlapping five laser beams on each membrane. The gas-filled hohlraum is an Au cylinder in which the laser beams enter through the endcaps which are covered by a thin CH membrane to enclose the gas.

Both targets produce mm-sized plasmas in approximately 300 ps, which is the time required to ablate the thin CH membranes and ionize the gas-fill of the target. Precision fabrication of these targets allows us to use different elements or gas fills to systematically vary the density, temperature, Z, or other plasma parameters of interest. The experiments are diagnosed with time-resolved Bragg crystal spectrometers and x-ray pinhole cameras.

Line shapes can be affected by practical constraints on the experiments as well as by physical processes. In this paper, three examples will be given from different experiments on the Nova laser: Ar spectra from gasbags, Xe spectra, also from gasbags, and Ar implosion spectra from hohlraums. In each of these examples, we will illustrate an aspect of the experimental analysis and for the last case, the Ar implosions, we will show detailed calculations that require sophisticated models to fit the entire lineshape.

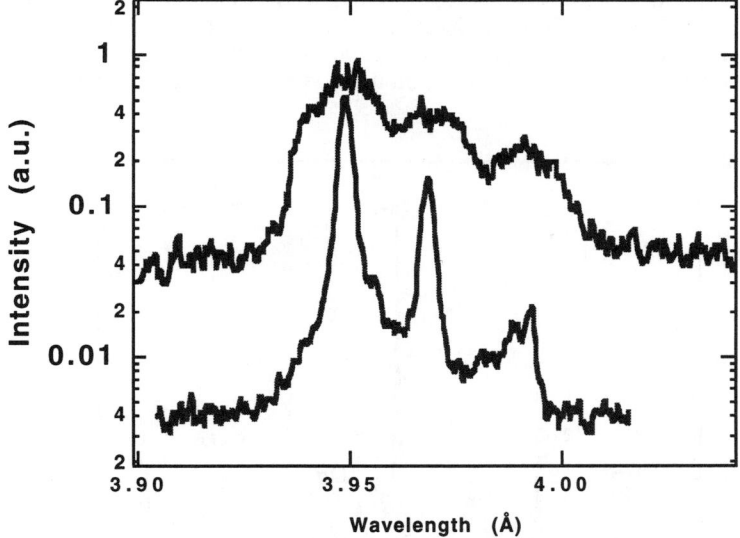

FIGURE 2. X-ray spectra of the ArXVII n=1-2 resonance line (3.949 Å) and its satellites (3.990 Å, 3.994Å) from an Ar doped gasbag. The upper spectrum is from a ~ 2 mm diameter source which broadens the spectral features. The lower spectrum is the same source viewed through a 250 μm slit that reduces the effective source size seen by the spectrometer.

Ar spectra are obtained from gasbag targets that are filled with neopentane gas, C_5H_{12}, which is doped with 1 % Ar. The Ar dopant concentration was chosen to remain optically thin for the plasma T_e and N_e of interest. These plasmas are nearly spherical with an average diameter of ~ 2.5 mm. Because the plasma is large, the experimental linewidth in this case is dominated by source broadening. The K-shell emission from these types of targets usually exhibit resonance lines and their associated satellites which have been used for temperature and density sensitive diagnostics. In the case illustrated in Figure 2 above, T_e is determined from the ratio of the n=2-1 resonance line of He-like Ar (He-α) to the Li-like satellites, labeled jkl in Gabriel notation (7). However, when the source size dominates, these individual features are washed out and cannot be resolved. We have employed 250 µm slits mounted 8 mm from the center of the targer to reduce the effective source size from the target that is observed by the spectrometer. Figure 2 shows an example of the gasbag spectra having a T_e of 3 keV with and without a source limiting slit. As can be observed, the detailed lineshapes cannot be analyzed if the plasma source size dominates the spectral line shape.

For the case of Xe spectra, another complication in interpreting the line shapes arises due to the presence of multiple ionization stages in the plasma. Xe spectra from Nova plasmas emit L-shell spectra that can be used for plasma diagnostics. For instance the n=4-2 lines can be used to diagnose T_e and the n=3-2 lines can be used to diagnose N_e (8). Unfortunately, L-shell spectra are extremely difficult to model because of the multitude of transitions present in the spectra. Gasbag plasmas at densities of 10^{21} cm^{-3} can exhibit emission from F-like to Mg-like ions

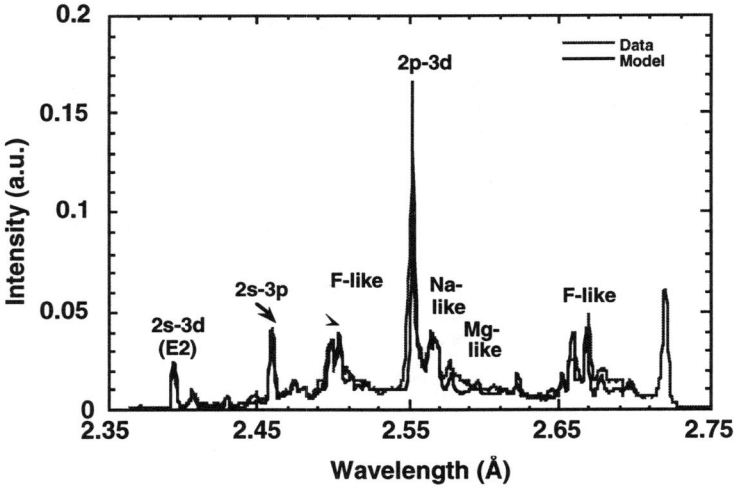

FIGURE 3. Spectra of the n=4-3 transitions of an Xe-filled gasbag. The plasma conditions measured by spectroscopy were $T_e = 3.5$ keV and $N_e = 2 \times 10^{21}$ cm^{-3}. To model this spectrum over 2000 transitions from a Xe atomic model were included.

when temperature approach 3 keV. For a realistic model of these transitions, over 2000 transitions are included in the atomic model used to fit the spectrum. The detail is necessary to isolate emission features that are sensitive to plasma conditions and that can be used to produce line intensity ratio diagnostics. In figure 3, we show the n=3-2 lines of Xe. Here the quadrupole line, identified as E2, relative to the F-like 2s-3p line has been found to be a sensitive measure of the density. Even with high resolution spectrometers, overlapping transitions in the L-shell spectra are unavoidable and make it imperative to have modeling capabilities to assist in finding good spectroscopic signatures.

From the experimental point of view, gasbag targets are a good testbed of atomic physics calculations important to line shapes. This point can be illustrated by a comparison of gasbag K- and L-shell spectra. Gasbags can be filled with a mix of gases, as already discussed, therefore, when more than one gas dopant is present in the gas mixture, the spectroscopic diagnostic from each can be cross calibrated for the same plasma conditions. For instance, we have performed experiments on gasbags filled with 1% Ar, 2% Xe, and 97% neopentane. The K-shell spectra from Ar are primarily determined by the Doppler and instrumental width. As before, the ratio of Ar He-like He-α to the jkl satellites can be used as a reliable temperature diagnostic in this case. On the other hand, line emission ratios from the L-shell spectral features of Xe, are not yet well-established diagnostics. Since the spectrum is a sum over lines from many ionization stages, the atomic model as well as the electron temperature may have an important effect on the intensity and width of the calculated spectral features. Thus, the comparison of L-shell line intensity ratios with the more reliable K-shell ratios allows us to validate the Xe spectroscopic diagnostics.

The modeling tools necessary to produce plasma diagnostics generally consist of three types of codes. The hydrodynamic behavior of the target is calculated by codes such as LASNEX (9). After generating temperature and density temporal histories, kinetics codes solve a system of rate equations to determine the populations of various levels and ionization states (10). These kinetics codes depend on data from large-scale atomic codes to calculate energy levels of the ion (11). Once the populations are calculated, the synthetic spectra can be generated. Given the large amount of data that can be accumulated during laser-produced plasma experiments, it is of great advantage that the line shape information is stored in an easily accessible database for use in real-time analysis.

With the experimental techniques and modeling capability in hand, we now consider an experiment that draws on both for a careful study of line shapes. For this case, we will look at the line shapes from an imploding capsule inside of a hohlraum. These experiments are performed with a hohlraum that is 2.5 mm in length and 1.6 mm in diameter. The hohlraum is heated by ten 0.35µm wavelength laser beams in 1 ns, producing x-rays which ablatively implode a 270 µm radius CH capsule that is filled with 50 atm of deuterium, D_2, and 0.1 atm of Ar. This target is diagnosed with time-resolved x-ray spectrometers that record the emission of the Ar XVIII $1s^2$ ($^1S_0$) - $1s3p$ ($^1P_0$) transition. Time-resolved x-ray pinhole cameras record 2-dimensional images of the imploding core, directly measuring the size of the imploding capsule and gas. Satellite lines on the long wavelength side of the transition, primarily due to contributions from Li-like Ar ions, cause the line to be asymmetric, but they provide spectral features which can be used to determine the T_e. In addition, at the high densities produced in these plasmas, the transition is strongly Stark broadened and the width provides an electron density diagnostic. Hence, the full analysis of this line shape can provide T_e and N_e as a function of time, as well as reveal other physical processes occuring in the implosion.

The Ar concentration is optimized by considering two criteria. The first is that the emission from the expected K-shell lines must be observable. Second, spectroscopic diagnostics are more sensitive when the emission is optically thin. In addition, for systematic studies, it is advantagous if the concentration is low enough to insure that the presence of the Ar does not significantly perturb the evolution of the deuterium plasma.

The plasma conditions in the imploding core are expected to produce a line shape that has a central intensity dip. Roughly speaking the dip is caused by splitting of the Stark components that comprise that transition due to ion microfields. In addition, each of these components is broadened by electron collisions. Calculations of the line shape, integrated over the predicted capsule conditions with spatial gradients in N_e and T_e, show a clear dip at the line center, see Figure 4.

Simulations of the temperature and density of the imploding capsule evolve on a ns timescale. Both T_e and N_e peak at approximately 1.8 ns after the start of the 1 ns square laser pulse when the capsule attains its minimum radius. At peak density, the temperature is gradient is expected to vary from 0.8 - 1.0 keV and the density is expected to compress to $1.3 - 1.5 \times 10^{24}$ cm^{-3}.

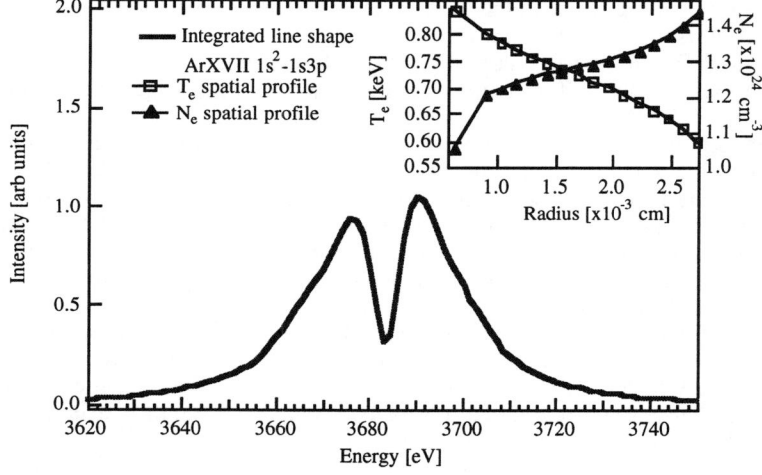

FIGURE 4. Calculated lineshape for the Ar XVII n=1-3 resonance line for the plasma conditions of an imploding capsule in a hohlraum. Temperature and density gradients are included in the calculation and they are shown as a function of radius in the inset box. The line shows a central dip due to the strong Stark effect at these high densities.

Spectroscopic data of very high quality have been obtained from imploded capsules inside of hohlraums. One spectrometer provides a survey of the spectral region from the He-like Ar $1s^2$ - $1s2p$ transition to the H-like Ar $1s$-$3p$ transition. The spectral resolution is ~ 500, which is sufficient to use the features of the lines for T_e diagnostics. A high resolution spectrometer, with resolving power of ~ 1800, records the He-like Ar $1s^2$ - $1s3p$ transition and is able to resolve the dip in the profile (12). The data and fits from the calculations show dramatic changes in the line shape as a function of time from the high resolution spectrometer. This can be observed in figure 5 where the lineouts at different times are shown for comparison. In these data, the Li-like satellites transition intensities increase late in time as the capsule cools.

In the analysis, the calculated spectra is iteratively fit to the data until the relative intensity of the Li-like satellites to the He-like resonance line were well matched. In the 10^{24} cm^{-3} electron density regime, this diagnostic is most useful from 400 to 900 eV. At temperatures outside this range, the satellites are not pronounced enough, or the contribution is too large to reliably determine the resonance lines width. When the high and low resolution Ar XVII spectra are analyzed we find that the temporal evolution of the core plasma is close to that predicted by hydrodynamic simulation. This indicates that the at the onset of the transitions of interest, the temperature is approximately 400 eV, while at the peak, the T_e has risen to 800 eV.

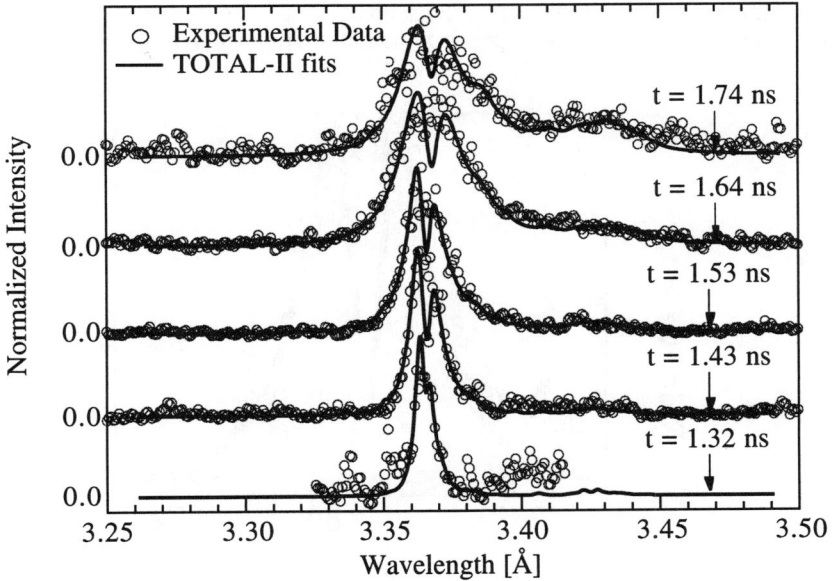

FIGURE 5. The data and calculations compared for the Ar XVII n=1-3 lineshape shown in Fig. 4. The data is obtained on an x-ray streak camera and lineouts of intensity vs wavelength are integrated over 100 ps and taken the times indicated in the figure.

The density is measured on the same experiments by using the full-width-at-half-maximum (FWHM) of the He-like Ar $1s^2$ - $1s3p$ transition. Since the density varies roughly as the 2/3 power of the FWHM it is a very sensitive measure of the plasma conditions, especially since the width of the line profile is relatively insensitive to changes in the electron temperature (13). Further, we have found excellent reproducibility of the data over a series of ~ 10 shots. The Doppler contribution to the line shape is negligible for this case.

Since this plasma is fairly well characterized in temperature and density, it is meaningful to look deeper at the details of the line shape. Figure 6 shows the calculated detailed line shape compared to the data. As mentioned before, the n=3-1 line should have a central dip for quasi-static line broadening calculations. However, figure 6 shows that there is no dip observed. The high quality of the data suggest that the absence of the dip is real and may be due to physical effects not yet included in the model. The reasons for the lack of the central dip could be one or more of the following. First, the satellites to the higher n levels, which for nln'l' n'≥4 fall under the transiton, are not included in the atomic model and the presence of emission from these transitions may fill in the dip. The n' =4 satellites have recently been observed by P. Beiersdorfer on the electron beam ion trap (EBIT) and the wavelengths are near the line center position (14). However, the dip in the line shape is absent even early in time when the n=3 satellites do not make a strong contribution ot the line shape. Therefore, this does not seem to be the mechanism that fills in the dip. The second possibility is that there are stronger temperature gradients in the experiment than those predicted by the simulations. In

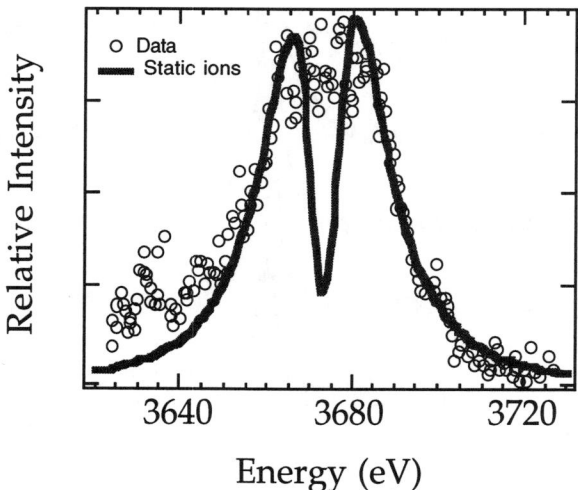

FIGURE 6. Calculated lineshape for the Ar XVII n=1-3 resonance line for the plasma conditions of an imploding capsule in a hohlraum. Temperature and density gradients are included in the calculation and they are shown as a function of radius in the inset box. The line shows a central dip due to the strong Stark effect at these high densities.

this case, the central dip is not as pronounced because emission from lower temperature Ar in the core will fill in the central portion of the line shape. This possibility is difficult to eliminate without further experiments and if experimentally confirmed, it would imply that the hydrodynamic simulations are not correctly modeling the plasma.

The most interesting explanation for the lack of a central dip is that ion dynamics of the D_2 gas is important. For most plasmas, the ion microfield is assumed to be slowly varying and the ion-emitter interaction is assumed to be quasi-static. However, for the implosion in which the ions are highly charged, the density is high, and the ionic perturbers are of relatively low mass, the ion microfield fluctuations may not be negligible. In this case, ion dynamics would fill in the dip by smearing it out (15). Fortunately for this n = 1-3 line shape, this effect would not significantly affect the FWHM of the line profile. Evidence of ion dynamics has been observed on low density gas-liner pinch experiments (16), but it has not yet been conclusively shown on high density plasmas(17, 18). Future experiments are planned to vary the different fill gases and thus change the ion dynamics(19). The challenge is that changing the gas fill can also affect the hydrodynamic evolution of the capsule because of differences in the radiative cooling of the gas. To conclusively demonstrate ion dynamics, it is desirable that N_e and T_e peak at the same time. Therefore experiments need to be carefully designed.

CONCLUSIONS

Experimental techniques and simulation capabilities for complex ions have been advanced enough to enable tests of complex line shape calculations. High-powered lasers can create mm-sized plasmas and multi-ionized species that can create special difficulties for line shape measurements. However, well characterized plasmas, such as those in hohlraum capsule implosions, can be produced. These extremely hot and dense plasma conditions are difficult to achieve in other plasma devices and the study of detailed line shapes can be fruitful. Experiments in the laboratory also afford the possibility of varying targets to explore effects such as ion dynamics in controlled experiments. For real-time analysis of the data, line shape calculations for these plasmas must either run quickly or rely on large databases.

ACKNOWLEDGEMENTS

This work was performed under the auspices of the U.S. Department of Energy by the Lawrence Livermore National Laboratory under Contract No. W-7405-ENG-48.

REFERENCES

1. Koch, J. A., et al. , *Phys. of Plasmas* **2**, 3820-3831 (1995).
2. Klisnick, A, et al. , *Phys. Rev. E* **53**, 5315-5322 (1996).
3. Perry, T. S., et al., *Phys. Rev. Lett.* **67**, 3784-3787 (1991).

4. Kauffman, R. L., "X-ray Radiation from Laser Plasma" in *Handbook of Plasma Physics*, vol. 3, eds. Rubenchik and Witkowski, pp. 111-162 (Elsevier Science, North-Holland, 1991) and references therein.
5. Lindl, J. D. *Phys. Plasmas* **2**, 3933-4024 (1995), and references therein.
6. MacGowan, B. J, et al *Phys. Plasmas* **3**, 2029-2040 (1996), and references therein.
7. Gabriel, A. , *Mon. Not. Roy. Astron. Soc.* **160**, 99-1009 (1972).
8. Keane, C. J., Hammel, B. A., Osterheld, A. L., Kania, D. R., *Phys. Rev. Lett.* **72**, 3029-3032 (1994).
9. Zimmerman, G. and Kruer, W., *Comments Plasma Phys. Controlled Fusion* **2**, 85 (1975).
10. Lee, R. W., and J. T. Larsen, *J. Quant. Spectrosc. Radiat. Transfer* to be published in (1996); also see the FLY users manual available from Cascade Applied Sciences.
11. Osterheld, A. L. et al., Proceedings of the 3rd International Colloquium on X-ray Lasers, pp. 309-314 (Institute of Physics, Briston, 1992).
12. Hammel, B. A. et al., *Rev. Sci. Instrum.* **61**, 2774-2777 (1990).
13. Griem, H. , *Spectral Line Broadening by Plasmas* (Academic, New York, 1974).
14. private communication, P. Beiersdorfer 1996.
15. Godbert, L., Calisti, A., Stamm, R., Talin, B., Lee, R., and Klein, L., *Phys. Rev. E* **49**, 5644-5651 (1994).
16. Glenzer, S. , Wrubel, T., Büscher, S. , Kunze, H.-J. , Godbert, L. , Calisti, A. , Stamm, R., Talin, B. , Nash, J. , Lee, R. W., and Klein, L. J., Phys. B **27**, 5507-5515 (1994).
17. Hammel, et al., *Phys. Rev. Lett.* **70**, 1263-1266 (1993).
18. Haynes, D. A. et al., *Phys. Rev. E* **53**, 1042-1050 (1996).
19. Woolsey, N. C. , et al. *Phys. Rev. E* **53**, 6396-6402 (1996).

SPECTROSCOPIC DIAGNOSTICS OF 2D EFFECTS IN DENSE PLASMAS. COMPARISON WITH 2D SIMULATIONS

P.Angelo*, H.Derfoul*, P.Gauthier*, P.Sauvan*, A.Poquerusse*,
T.Ceccotti*, E.Leboucher-Dalimier*, T.Shepard**, C.Back**,
M.Vollbrecht***, I.Uschmann***, and E.Förster***.

*Physique Atomique dans les Plasmas Denses, LULI/Université Paris VI - Ecole Polytechnique.
4 place Jussieu, 75252 Paris cedex 05, France
** Lawrence Livermore National Laboratory, University of California
L-473, P.O. Box 808, Livermore, CA94550 USA
*** X-ray Optics Group, Max-Planck-Gesellschaft,
Friedrich Schiller University Jena, 1 Max Wien Platz, D-07743 Jena, Germany

We create hot (T_e>200ev) and dense (N_e>10^{23}cm^{-3}) moderate Z (Al or F) plasmas by colliding foil and massive target experiments [1,2] with the laser beams of the LULI facilities operating at 0.263 µm (energy 25J, pulse duration 500ps). The colliding foil experiments are driven by the two-dimensional LASNEX code which predicts the choice of the foil thickness and initial distance for different focal spots and laser intensities in order to get the densest emissive plasma in the impact region (fig1). The simulations of these experiments are difficult because the foils are distorted during the interaction with the inhomogeneous laser beams on one hand (fig2) [3] and because the laser energy absorbed by the foils is not totally used to their motion on the other hand. Thus two-dimensional effects (lateral expansion, transverse thermal conduction) [4,5] are important and it is obvious that these effects are deleterious for the creation of a hot dense plasma. Concerning massive target experiments, 2D effects are also important due to the hot spots in the focal plane. For all these experiments, beam smoothing techniques have been recently implemented (RPP and PZP plates) in order to reduce the 2D effects and then enhance the overdense plasma X-Ray emission.

The two-dimensional effects have been exhibited and measured thanks to two independent diagnostics : a spectrography with a space resolution along the transverse direction (i.e. perpendicularly to the laser beam(s) axis (fig3)) and a monochromatic imaging of Lyβ emission (fig4 a,b,c).

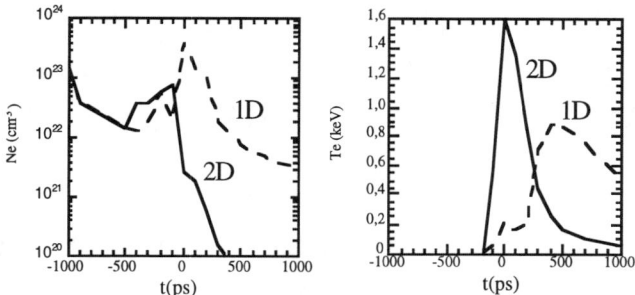

fig.1 Predictions from mono and two-dimensional codes (LASNEX 1D & 2D) versions for the rear face of two CF2 foils (initial thickness 3μm, initial separation 100μm) irradiated by two laser beams (0.26μm, 16J, 500ps, focal spot diameter 80 μm). 0 ps refers to the top of laser pulse.

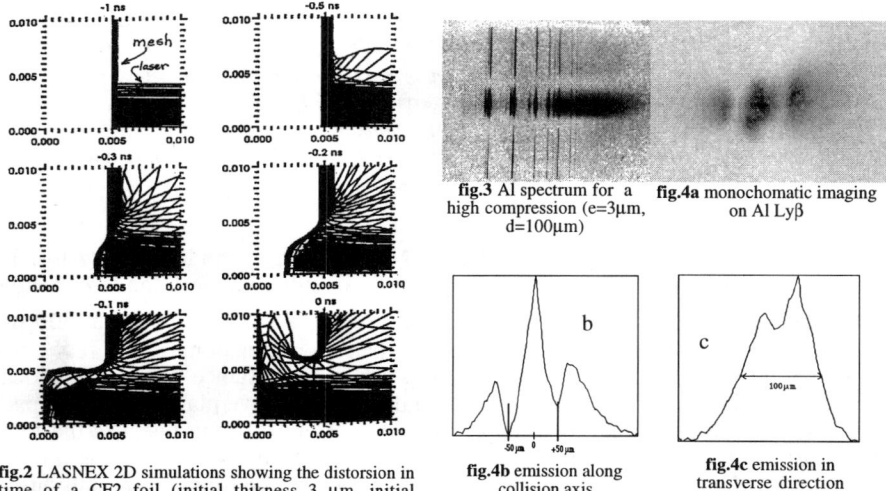

fig.2 LASNEX 2D simulations showing the distorsion in time of a CF2 foil (initial thikness 3 μm, initial separation 100 μm) irradiated by two laser beams (0.26μm, 16J, 500ps) focused onto 80μm focal spot.s One can also see lateral plasma expansion. The curves correspond to isodensities. 0 ns refers to the top of laser pulse; distances are in cm.

fig.3 Al spectrum for a high compression (e=3μm, d=100μm)

fig.4a monochomatic imaging on Al Lyβ

fig.4b emission along collision axis

fig.4c emission in transverse direction

REFERENCES

1. E.Leboucher-Dalimier, A.Poquerusse, and P.Angelo, *Phys.Rev.E* **47**, R1467 (1993).
2. E.Leboucher-Dalimier, P.Angelo, P.Gauthier,.Poquérusse, AIP Conference Proceedings 328, Spectral Line Shapes **8** 12th ICSLS, Toronto Canada 1994, ed. by A.D. May, J.R. Drummond, E. Oks
3. B. Faral, *Phys. Fluids* B**2**, 371 (1990)
4. M.H. Key, *Phys. Fluids* **26**, 2011 (1983)
5. F. Cottet et al, LULI report, 133 (1984)

Redistribution of Resonance Radiation in Hot and Dense Plasmas

C. Mossé*, S. Ferri*, A. Calisti*, B. Talin*, R. Lee**
and L. Klein***

(*)Laboratoire de Physique des Interactions Ioniques et Moléculaires, Centre Saint Jérôme,
13397 Marseille, France
(**)Lawrence Livermore National Laboratory, P.O. Box 808, L-447, Livermore, CA 94550, USA
(***)Department of Physics, Howard University, Washington, DC 20059, USA

Abstract. The development of a numerical procedure for the study of the resonant two photon plasma spectral properties is presented. The formalism is based on an extension of the Frequency Fluctuation Model (F.F.M.) that permits the calculation of the spectral line shape of Stark broadened lines emitted by multi-electrons ions and takes into account ion dynamics effect. We present a application for the study of resonant scattering by a calculation of the radiative redistribution function for the 3d-2p transition of MgIV.

The one photon emission or absorption spectroscopy model (FFM)[1] is extended to describe two photon processes by continuing the expansion of the linear response function to higher order. These higher order response functions describe either redistribution, when the power spectrum of the fluorescence induced by monochromatic pumping radiation is studied, or pump-probe phenomena when the change of the linear radiative response due to a monochromatic pump field is considered. The application described here is related to the induced fluorescence case.

The F.F.M. provides a practical numerical means to obtain the characteristics of the spectral inhomogeneities for arbitrary lines. Moreover, it gives a system of Stark dressed two-level states that permits to model radiative properties accounting for ion dynamics, even for complicated systems. This, allows to build a model designed to calculate radiative redistribution function $R(\omega_1,\omega_2)$[2], i.e., the probability that a photon with frequency ω_1 is absorbed from a radiation field and a photon is subsequently re-emitted at frequency ω_2 in the same line.

In order to investigate a tentative experiment, we have calculated the spectrum of radiation scattered by fluorine-like magnesium (Mg IV) in a dense plasma. The plasma is supposedly driven by a monochromatic X-ray laser field with a frequency laying near one of the $2p^43d$-$2p^5$ resonance [3]. The figure corresponds to the redistribution function $R(\omega_1,\omega_2)$ for the MgIV 3d-2p spectral line in a plasma with temperature $T_e=100$eV and electron density $N_e=10^{21}$ cm^{-3}. The curve in the rear part of the graph, is the absorption line shape for same plasma conditions.

For the plasma conditions above, the ion Stark splitting contribution to the line broadening is weak. Thus, as long as isolated resonances are considered the fluorescence pattern shows mainly complete redistribution over the investigated spectral domain.

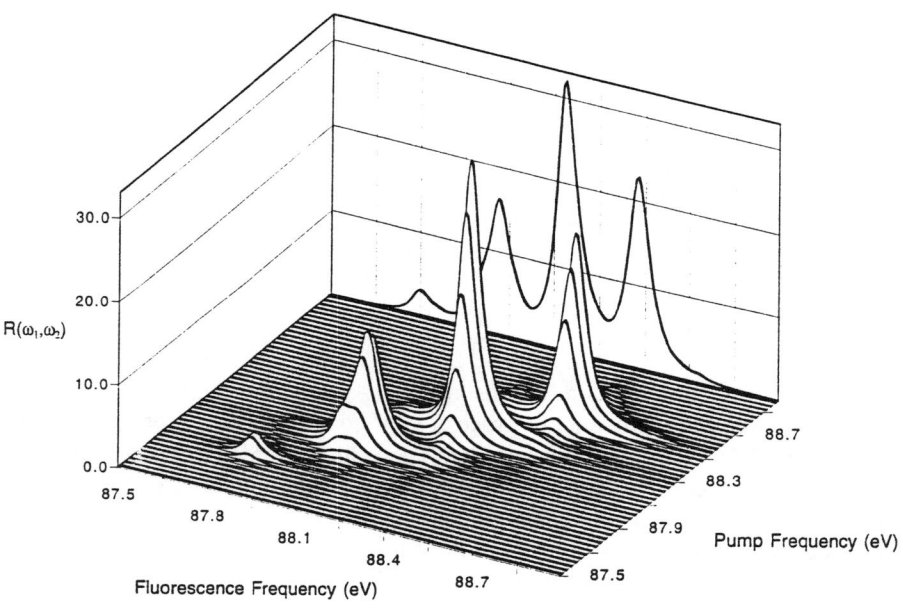

The redistribution radiative function of the 3d-2p transition of fluorine-like magnesium at $N_e=10^{21}$ cm^{-3} and $T_e=100$ eV

REFERENCES

[1] B. Talin, A. Calisti, L. Godbert, R. Stamm, R.W. Lee and L. Klein, Phys.Rev.A51, 1918 (1995)
[2] B. Talin and L. Klein, Phys. Rev. A26, 2717 (1982)
[3] C. Mossé, A. Calisti, M. Koubiti, R. Stamm, B. Talin, R.W. Lee and L. Klein, to be published

X-ray Spectral Line Shapes for the Excimer-Laser-Produced High Density Plasma Diagnostics

A. Magunov[†], A. Faenov[†], I. Skobelev[†], T. Pikuz[†], D. Batani[‡*], M. Milani[‡*], A. Conti[‡*], A. Masini[‡], M. Costato[¶], A. Pozzi[¶], E.Turcu[§], R. Allot[§], N. Lisi[§], M. Koenig[||], A. Benuzzi[||], F. Flora[a], T. Letardi[a], L. Palladino[1] and A. Reale[1]

[†] Multicharged Ions Spectra Data Center, VNIIFTRI, Mendeleevo, 141570, Russia
[‡] Dipartimento di Fisica, Universita' di Milano, Italy
[¶] Dipartimento di Fisica, Universita' di Modena, Italy
[§] Rutherford Appleton Laboratory, U.K.
[||] LULI, Ecole Polytechnique, France
[a] Dipartimento Innovazione, Settore INN-FIS, CRE ENEA, Frascati, Italy
[1] Dipartimento di Fisica, Universita' dell' Aquila, e INFN G.Coll.LNGS, Assergi, Italy
[*] Also Istituto Nazionale di Fisica Nucleare, Sezione. di Milano, Italy

The time and space-integrated emission spectra measurements have been performed in plasma produced by 308 nm wavelength XeCl laser radiation ($I_L = (4\text{-}10) \cdot 10^{12}$ W/cm^2, $\tau=10$ ns) and by 248 nm wavelength KrF laser pulse train radiation ($I_L = 5 \cdot 10^{15}$ W/cm^2, $\tau=7$ ps, 16 pulses in train) on (CF$_2$)$_n$ plane target. The lines' shapes and intensities modeling of Lyman series and He-like ion resonance series of fluorine up to $n = 7$ by fitting experimental data shows the considerable difference of plasma formation features for these two sets of the laser pulse parameters.

The X-ray spectra analysis based on the line profiles studying is the most informative method for the determination of laser-produced plasma parameters (see [1,2] and references inside) even without spatial and temporal resolution.

In the present work we consider the spectra of resonance line series in H-like and He-like fluorine ions in the region from 748 eV to 1245 eV obtained from the plasma produced by excimer XeCl laser radiation of "Hercules" facility in Frascati [3] and by KrF laser radiation at Rutherford Appleton Laboratory [4]. The geometry of experiments with plane crystal Bragg spectrometer and the axial symmetry of plasma expansion provides the spectral emission probability to be approximated by finite sum over the uniform plasma layers in the form

$$P(E) = \sum_p w_p Q_E(N_{ep}, T_{ep}) / \tau_E^p (1 - e^{-\tau_E^p}) e^{-\sum_{p'>p} \tau_E^{p'}} (1 + e^{-\tau_E^p} e^{-2\sum_{p'<p} \tau_E^{p'}}) \quad (1)$$

where $\tau_E^p = k_E(N_{ep}, T_{ep}) R_p$ is the optical depth of p layer with constant N_{ep} and T_{ep}, R_p is the time and space averaged size of uniform plasma layer, and w_p gives the weight for corresponding plasma parameters. We asumed $N_{ep} < N_{ep'}$ for $p > p'$. Equation (1) accounts for the absorption in internal and external plasma regions and by layer itself. The emissivity Q_E and absorption coefficient k_E are defined by the sum going on Lyman and He-like ion resonant line series terms and by photo-recombination continuum. Each line term contributes according the radiative transition rate, the level population, and the normalized spectral line profile function.

The line profiles for H-like F IX accounting for ionic field Stark broadening, electron elastic collision and Doppler broadening were taken in the form

$$S_{n1}(E) = \sum_\alpha \frac{\gamma_{n\alpha}}{\Delta E_n^D \sqrt{\pi}} \int_0^\infty W\left(\frac{E - E_{n1} - \Delta E_{n\alpha}^S \beta}{\Delta E_n^D}, \frac{\Delta E_{n\alpha}^c}{\Delta E_n^D}\right) P_a(Z_i, \beta) d\beta$$

where $\gamma_{n\alpha}$ is the relative transition rate for the sublevel $\alpha \equiv (n_1, n_2, m)$, $W(x,y)$ is the Voigt function for collision width $\Delta E_{n\alpha}^c(N_e, T_e)$ [5] and Doppler width $\Delta E_n^D(T_i)$ with the effective ion temperature T_i, $\Delta E_{n\alpha}^S$ is the linear Stark shift of sublevel in the mean ionic field $F_o = Z_i e/r_o$ with $r_o = 0.62 N_i^{-1/3}$, Z_i and N_i are the mean charge and density of ions, $P_a(Z_i,\beta)$ is the field distribution function with $a = r_o/r_D$ (r_D is the Debye radius) [6]. We used these approximation for the resonance transition in He-like fluorine, since the electron correlations are small compared with the perturbation by ionic field. For $n \leq 3$ levels the Coulomb approximation is worse. However, in optically thick plasma the line widths for these transitions are defined by Stark wings while the central part of line profiles are absorbed.

The values w_p and R_p were fitted in (1) to the experimental data and the results for w_p shown in Table strongly differed at precritical densities in two cases.

N_{ep} cm^{-3}	$5 \cdot 10^{20}$	10^{21}	$2 \cdot 10^{21}$	$5 \cdot 10^{21}$	10^{22}	$2 \cdot 10^{22}$
XeCl	$5 \cdot 10^6$	$1.7 \cdot 10^6$	$2 \cdot 10^4$	$5 \cdot 10^3$	$7 \cdot 10^3$	$7 \cdot 10^3$
KrF	-	$6 \cdot 10^3$	$2 \cdot 10^4$	$3.8 \cdot 10^4$	$6 \cdot 10^3$	$6 \cdot 10^3$

REFERENCES

[1] Yaakobi B., Steel D., Torsos E., Hauer A., Perry B. *Phys. Rev. Lett.* **39**, 1526 (1977)
[2] d'Etat B., Grumberg J., Leboucher E., Nguyen H., Poquerusse A. *J. Phys. B* **20**, 1733 (1987)
[3] Bollanti S., Di Lazzaro P., Flora F., et. al. *Physica Scripta* **51** 326 (1995)
[4] Turcu E., Ross I., et. al. 1994 *Applications of Laser Plasma Radiation* SPIE vol 2015 p 243
[5] Griem H. R. 1974 *Spectral line broadening by plasmas* (New-York, Academic)
[6] Tighe R. J. and Hooper C. F. Jr *Phys. Rev. A* **14**, 1514 (1976)

Refractive fringe and Stark diagnostics of laser induced plasmas on solid targets

S. Siano, G. Pacini, R. Pini, and R. Salimbeni
Quantum Electronic Institute, National Research Council, Firenze, Italy

INTRODUCTION

Refractive Fringe Diagnostics (RFD) is a relatively novel technique[1] to derive the electron density distribution of undercritical spherical plasmas induced by high power laser-target interaction. This technique is based on the deconvolution of interferograms obtained by pump-and-probe shadowgraphy scheme (fig. 1) and could represent an alternative to Stark line broadening and shift analysis that is more often employed for plasma diagnostics. Considering plasmas formation in air, during the early phase of expansion the presence of a strong continuum associated to *bremsstrahlung* and recombination emissions can limit spectroscopy analysis, not allowing the detection of line structure. On the other hand, the application of RFD is limited at longer times, whenever a shock wave detaches from the plasma front complicating the interference pattern. Hence, as other single wavelength interferometry techniques, RFD can be regarded as complementary to the Stark diagnostics.

In the present work we applied RFD to investigate the plasma mediated excimer laser ablation of alumina, silicon and steel sheets. Furthermore we compared the results of RFD with the ones obtained by means of time and space resolved spectral line broadening in the narrow time interval when both techniques were applicable.

RFD THEORY

The refractivity of a laser produced plasma on a target surface during the early phase of its expansion can be attributed to the electrons (neglecting heavy particle refractivity[2]). Then the refractive index is approximated by:

$$n = \sqrt{1 - n_e / n_c} \quad (1)$$

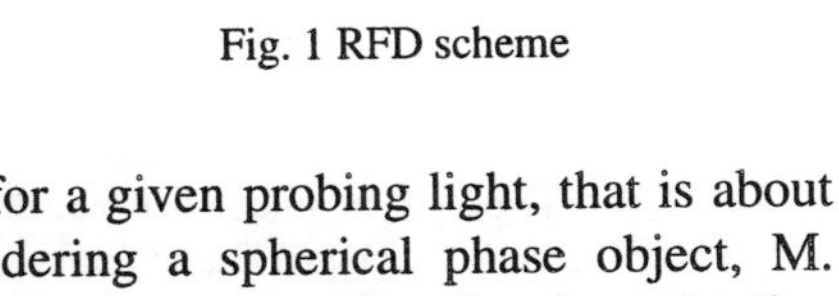

Fig. 1 RFD scheme

where n_c is the critical electron density for a given probing light, that is about 10^{22} cm^{-3} for UV wavelengths. Considering a spherical phase object, M. Michaelis and O. Willi[1] obtained the following expression for the refractive index at radius r_i:

$$n(r_i) = \frac{\cos\frac{\alpha_i}{2}\sin\beta_i}{\sin\beta_i + \frac{1}{2}\sin\frac{\alpha_i}{2}\cos(\frac{\alpha_i}{2}+\beta_i)}, \qquad (2)$$

where α_i and β_i are indicated in fig. 1 together with the schematic fringe structure is expected for a monotonically decreasing electron distribution toward the plasma boundary. The angles α_i and β_i are related to the plasma radius R_0 and to the fringes width d_i. Thus the electron density distribution can be obtained from the defocused shadowgraphy using eq. (1), by measuring directly the two last parameters.

EXPERIMENTS AND RESULTS

The pump laser was a long pulse XeCl laser (308 nm, 100 ns) and as the probe for RFD we employed a Nitrogen laser (337 nm, 0.5 ns). Plasma spectroscopy was done by collecting the emission with an optical fiber coupled with a monochromator (0.5 m, 1200 lines/mm). The spectra were acquired by an OMA III synchronized with the pump laser. Spatial resolution was achieved by imaging the plume on a screen (magnification ratio of 4) and scanning along the pump beam optical axis with the input end of the optical fiber.

RFD allowed to derive electron distributions up to 150 ns after the onset of the pump pulse, whereas broadening and shift effects were well detectable for neutral and ionized atomic lines at time delays greater than 150-200 ns. For the alumina target a comparison between the two techniques could be performed at a time delay of 150 ns. As shown in Fig. 2, the agreement is reasonably good, considering the complete independence of the two plasma diagnostics.

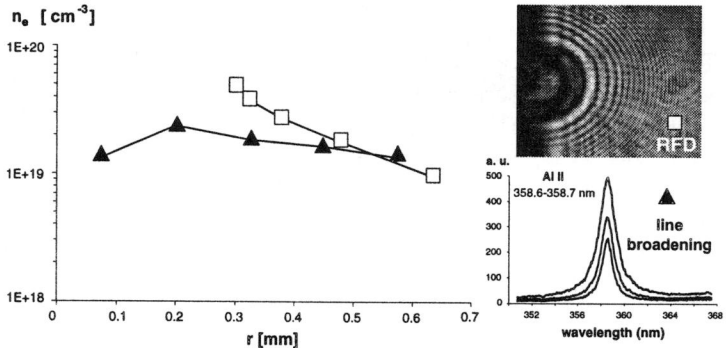

Fig. 2. Comparison between electron density distributions derived by RFD and Stark spectroscopy, respectively.

[1]   M. M. Michaelis and O, Willi, Opt. Commun. **36**, 153 (1981).
[2]   A. Giulietti, D. Giulietti, M. Lucchesi and M. Vaselli, Opt. Commun., **47**, 131 (1983).

Time-Resolved X-Ray Spectroscopy of Laser Heated Plasmas

A.Giulietti, D.Giulietti^, L.A.Gizzi* and O.Willi#

*Istituto di Fisica Atomica e Molecolare del CNR,
via del Giardino, 7   56127 Pisa, Italy*

^also at Dipartimento di Fisica, Università di Pisa
*presently at Istituto TESRE-CNR, Bologna
#The Blackett Laboratory, ICSTM, London

Abstract. Recent experimental results are presented on time-resolved spectroscopy of laser produced plasmas. In particular, K-shell emission of H-like and He-like Aluminium plasmas was investigated. The electron temperature was measured taking into account the opacity of the plasma. Re-heating effect due to the interaction of a delayed laser pulse with the preformed plsma was observed and studied. The results have been compared with the prediction of a steady state atomic physics numerical code.

X-RAY EMISSION AND TEMPERATURE MEASUREMENTS

The interaction physics of intense laser radiation with coronal plasmas of interest for inertial confinement fusion is being currently studied in many laboratories worldwide. Relevant results were obtained in a recent experiment supported by the EC Programme for the Access to the Large Facilities and performed at the Rutherford Appleton Laboratory. Test plasmas were produced by laser irradiation and consequent explosion of thin foil Al targets. The preformed plasma was then interacted after a suitable delay with a focused 600 ps laser pulse at an irradiance of 10^{15} W/cm^2. The test plasma was fully characterised[1,2] in terms of its electron density distribution and temperature evolution. The temperature measurements were performed by means of time resolved X-ray spectroscopy, using an X-ray streak camera coupled with a TlAP crystal spectrometer, shown in

Fig1. K-shell emission lines of He-like and H-like Al plasma were investigated. Line intensity profiles were found to be well fitted with a Gussian profile. The Ly-γ (1s-4p) to the He-γ (1s^2-1s4p) intensity ratio was measured at different times before and after the high irradiance delayed interaction.

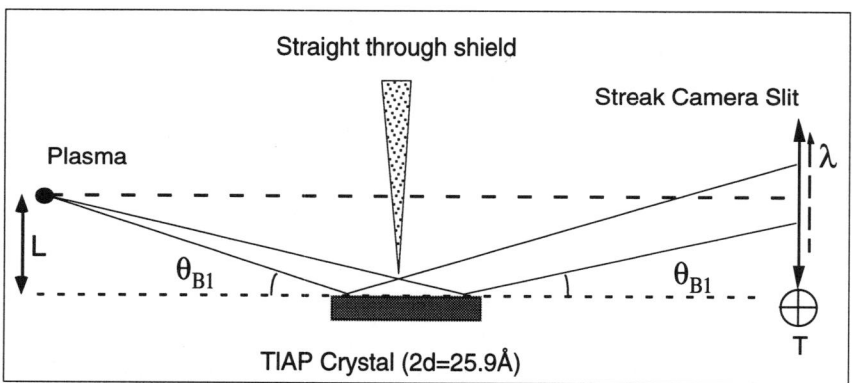

FIGURE 1. Crystal set up for time resolved X-ray spectroscopy.

The electon temperature was then obtained taking into account the opacity of the plasma. A significant re-heating of the preformed plasma was observed during the delayed interaction. The results have been compared with the prediction of the steady state atomic physics numerical code RATION[3].

Time-Resolved X-Ray Spectra

A typical time-resolved spectrum is shown in Fig.2a, while Fig.2b shows a 1-D spectral lineout taken 500 ps after the peak of the Heβ line emission and integrated over 50 ps, which is also the temporal resolution of the spectrum. Emission lines from the Heβ to the Lyδ are clearly visible with the Lyγ and Lyδ emerging from the He-like continuum. Line intensity profiles were found to be well fitted by a Gaussian profile with $\Delta\lambda_{FWHM} = 40$ mÅ $\pm 10\%$. All the lines, except the Heε and Lyβ, are well resolved allowing a direct evaluation of the line intensity. With the available spectral resolution (set by the source size), the Lyβ line is only partially resolved being merged with the Heε and higher quantum number He-like lines and with the He-like continuum edge.

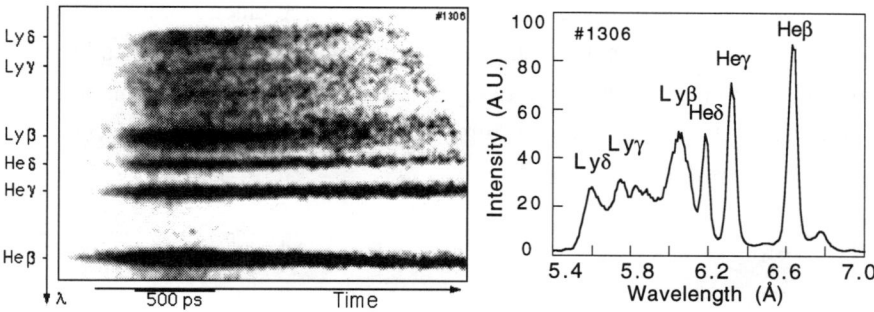

FIGURE 2. a) Time-resolved spectrum of K- shell Al plasma emission. Irradiance on target: 3.2 10^{13} W/cm^2. b) Lineout taken 500 ps after the peak of emission.

A systematic study of such kind of spectra are presently in progress both for basic studies of atomic and plasma physics[4] and for optimisation of X-ray sources for applications[5].

Temporal Evolution of Electron Temperature

The evolution of the electron temperature is obtained from time-resolved X-ray spectra by measuring, at the different times, the intensity ratios of suitable pairs of lines, provided that additional information on the physical plasma condition is available. In fact, simple analytical or semi-analytical models of plasma equilibrium (local thermodynamic or coronal equilibrium) cannot be applied over the whole plasma, due to the wide range of temperatures and densities typical of such inhomogeneous plasmas. Therefore, numerical models are necessary in order to perform detailed balance of all available ionic levels in a mixed collisional-radiative equilibrium. The steady state RATION code was used in our case since, in our conditions, atomic physics processes are typically faster than plasma hydrodynamic processes. Electron density was measured independently by optical interferometry on the plasma[1,2].

Finally, plasma opacity effect on the emission lines were also taken into account for temperature measurements. Calculations showed that, typically 1.5 ns after the peak of the plasma heating pulse, plasma opacity effects can be neglected in our case. Consequently, accurate temperature measurement could be obtained at the time of the delayed intaraction, that is, 2.5 ns after the peak of the plasma heating pulse.

In Fig. 3a a time resolved spectrum is shown, as obtained from the plasma emission in the two subsequent stages of i) plasma formation and heating ii) delayed interaction and re-heating. In Fig 3b the evolution of the electron plasma temperature is plotted, as obtained from mesurements of the Ly-γ to He-γ line ratio at different times.

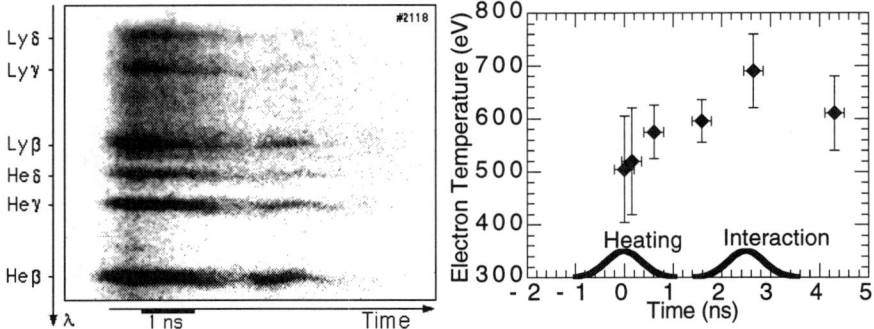

FIGURE 3. a) Time-resolved spectrum of K- shell Al plasma emission including the delayed interaction re-heating. Irradiance at plasma formation: 2.8×10^{13} W/cm^2. Irradiance at delayed interaction: 3.0×10^{14} W/cm^2. Peak plasma density during delayed interaction: 10^{20} cm^{-3}.   b) History of electron temperature.

AKNOWLEDGEMENTS

Athors are grateful to the Central Laser Facility staff at RAL. The eperiments were supported by EC funding in the framework of the Large Facility Acess scheme of the Human Capital and Mobility Programme.

REFERENCES

[1] L.A.Gizzi *et al.*, Phys. Rev. E **49**, 5628 (1994)
   L.A.Gizzi *et al.*, Phys. Rev. E **50**, 4266 (1994)
[2] M. Borghesi *et al.*, submitted to Phys. Rev. E (1996)
[3] R.W.Lee *et al.*, J. Quant. Spectr. Radiat. Transfer **32**, 91 (1984)
[4] A.Macchi *et al.*, Il Nuovo Cimento D **18**, 727 (1996).
[5] D.Giulietti *et al.*, Il Nuovo Cimento D **17**, 401 (1995).

High Precision X-ray Measurements of Polarization Shifts in Dense Aluminum Plasma

Oldřich Renner,[1] David Salzmann,[2] Peter Sondhauß,[3]
Eckhart Förster,[3] Abdeslem Djaoui,[4] and Eduard Krouský[1]

[1] *Inst. of Physics, Academy of Sciences CR, 18040 Prague, Czech Republic*
[2] *Soreq Nuclear Research Centre, 81800 Yavneh, Israel*
[3] *Research Unit X-ray Optics of the Max-Planck-Gesellschaft, 07743 Jena, Germany*
[4] *Rutherford Appleton Laboratory, Didcot, Oxon OX11 0QX, United Kingdom*

Abstract. The spatially resolved profiles and positions of the spectral lines Al Ly$_{\gamma-\varepsilon}$ emitted from a laser-produced plasma were recorded using a vertical-geometry Johann spectrometer. The observed red shifts were related to plasma densities via numerical simulation of the experiment. The experimental results agree well with the plasma polarization shifts predicted by the ion sphere model.

The plasma polarization shift (PPS), i.e. the shift of the spectral line in hot dense plasma relative to the same bound-bound transition in isolated ions, due to perturbers in vicinity of the radiators, is essentially a many-body problem [1]. Except for H and He$^+$ ions [2, 3], where the agreement between theory and experiment is satisfactory, there is no generally accepted theoretical approach to this fundamentally important phenomenon. Furthermore the results of some experiments performed for higher-Z ions are inconsistent in claiming the existence [4] or non-existence [5] of the blue/red shifts. The validity of the previous results was limited mainly by instrumental and opacity/plasma gradient effects. The aim of our experiments was to obtain high-precision spectroscopic data for Lyman series of aluminum which may allow a reliable identification of the PPS and could be used as a benchmark for different theoretical models of the line shifts.

The magnitude of the PPS expected for ions immersed in a dense laser-produced plasma can be estimated using an ion sphere model [6]. Within the framework of this model, the formula for the line shift of the spectral lines is

$$\Delta h\nu = \frac{2\pi}{3}\bar{Z}e^2 n_i (r_f^2 - r_i^2), \quad (1)$$

where $\bar{Z}$ is the average ion charge, n_i is the ion density, and $r_{i(f)}^2 = \langle i(f)|r^2|i(f)\rangle$ is the mean square radius of the initial (final) electronic states. For hydrogenic states $r_{nl}^2 = a_0^2 n^2/2Z^2 \cdot [5n^2 + 1 - 3l(l+1)]$, where a_0 is the Bohr radius. The resulting formula for the Lyman $np \to 1s$ transitions in hydrogen-like ions is given by

$$\Delta h\nu(np \to 1s) = 4.223 \cdot 10^{-3} \text{eV} \frac{\bar{Z}}{Z^2} \left(\frac{n_i}{10^{21}\text{cm}^{-3}}\right) \left\{5n^2(n^2-1) - 6\right\}. \quad (2)$$

This formula suggests that at ion densities $(1-2) \cdot 10^{21}$ cm^{-3} which correspond to the conditions of our experiment, the measurable line shifts can be expected for Al Lyman series members γ and higher.

The experiment was carried out at the MPQ iodine laser system ASTERIX. The laser beam containing energy 170 J in a 0.4 ns pulse at $\lambda = 0.44$ μm irradiated aluminum dot target with dimensions $100 \times 100 \times 1$ μm^3 deposited on plastic 10 μm thick. The Al plasma created in a focal spot of diameter 250 μm was confined in a constrained flow by the plasma from the surrounding substrate. Its radiative characteristics were studied by a novel x-ray instrument, the vertical dispersion version of the Johann spectrometer [7]. A combination of its high collection efficiency, dispersion (150 mm/Å), spectral (3500-7000) and 1-D spatial resolution (10 μm), allowed well resolved emission profiles of the higher Al Lyman series members to be measured as a function of the distance from the target surface, i.e. as a function of the plasma density.

The profiles and positions of the Ly$_\gamma$ ($\lambda = 5.739$ Å), Ly$_\delta$ ($\lambda = 5.604$ Å), and Ly$_\varepsilon$ ($\lambda = 5.534$ Å) lines were observed at angle of emission 2° to the target surface. The spectra recorded in single shots on Agfa Structurix D7 x-ray film were converted to an intensity and wavelength scale [8] and corrected for motional Doppler shifts due to variable angle of observance and the plasma blow-off velocity along the plasma column. An example of reconstructed spectra is shown in Fig. 1. The equidistant spatially resolved lineouts of the spectral lines were fitted with the symmetric Gaussian profiles and related to ion densities via the numerical simulation of the experiment described further.

In a recent version of the 1D hydrodynamics code MEDUSA [9], the hydrodynamic equations are supplemented by a time-dependent non local thermodynamic equilibrium average-atom model. This allows to calculate the populations of the ground and excited ion states. The relevant plasma parameters (density, temperature, ion charge and velocity) corresponding to emission at a given distance from the target surface were averaged over all timesteps by using the emission rates of individual spectral lines as weighing factor.

The experimental results obtained for members γ, δ and ε of the Al Lyman series are shown in Fig. 2. The energy shifts for each transition are related to line positions defined at the distance 200 μm from the target surface; here the tabulated unperturbed energies (i.e. the transition energies of the isolated ion) were corrected for the PPS and the Doppler shift expected at plasma parameters inferred from the simulations. The overall line position uncertainty is indicated by error bars. The line energies shift to the red with the growing ion density and the upper principal quantum number of the relevant transition, as foreseen by the theory of the plasma polarization shift.

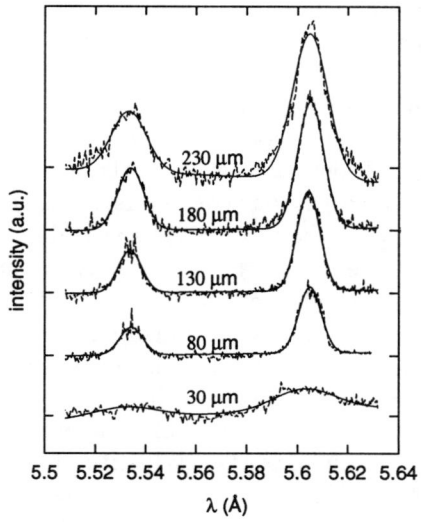

FIGURE 1: Profiles of the spectral lines Al Ly$_\delta$ and Ly$_\varepsilon$ recorded at distances $30-230$ μm from the target surface.

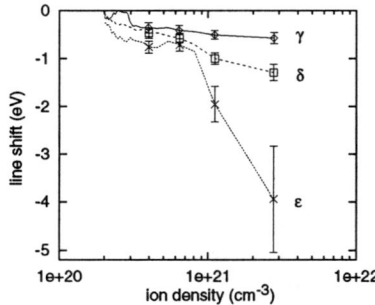

FIGURE 2: The experimentally observed energy shifts of the Ly$_\gamma$ - Ly$_\varepsilon$ lines as a function of the plasma ion density.

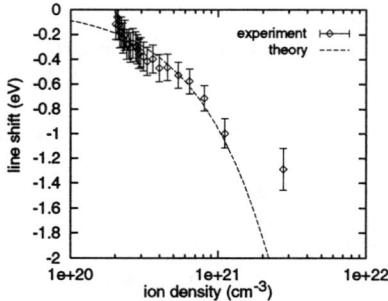

FIGURE 3: The calculated (dashed line) and observed (points) energy shifts of the spectral line Al Ly$_\delta$.

The quantitative comparison of the line shifts predicted by the ion sphere model and experimental results obtained for Lyman δ transition is provided in Fig. 3, where the density dependent line positions are related to unperturbed transition energy 2212.4 eV. The measured shifts are generally small, approximately 8-10% of the spectral line width (FWHM), but well above the experimental errors. The red shift observed for the highest ion density is somewhat lower than expected, but in view of the high accuracy required for these measurements, the agreement between the theoretical values and experimental data is very good.

In conclusion, we report the first reliable measurements of the PPS for high-Z ions. This effect, although predicted by plasma theory, has never been observed without controversy. By using the novel high-dispersion x-ray spectrometer, the shifts of the spectral lines Al Ly$_{\gamma-\varepsilon}$ were identified and correlated with the plasma densities inferred from numerical simulation of the experiment. The behaviour of the observed red shifts, i.e. their increase with the growing ion density and with transitions from the higher excited states, indicate that we have observed the real plasma polarization shifts, and not phenomena induced by combined effects of the plasma gradients and opacities.

The authors gratefully acknowledge the help of the scientific and technical staff from Max-Planck-Institut für Quantenoptik in performing the experiments. This work was supported by Grant Agencies of the Czech Republic and the Academy of Sciences under Grants No. 202/94/0710 and 110427, and by EEC Large Facility Program under contract ERB-CH-CT92-0006.

REFERENCES

[1] H.R. Griem, *Spectral Line Broadening by Plasmas* (Academic, New York and London, 1974).
[2] T.L. Pittman and C. Fleurier, *Phys. Rev. A* **33**, 1291 (1986).
[3] St. Bödecker, S. Günter, A. Könies, L. Hitzschke and H.-J. Kunze, *Phys. Rev. E* **47**, 2785 (1993).
[4] K.G.H. Baldwin, J.R. Liu, J.D. Kilkenny and D.D. Burgess, *J. Phys. B* **19**, L179 (1986).
[5] S. Goldsmith, H.R. Griem and L. Cohen, *Phys. Rev. A* **30**, 2775 (1984).
[6] D. Salzmann and H. Szichman, *Phys. Rev. A* **35**, 807 (1987).
[7] O. Renner, T. Mißalla and E. Förster, *Proc. SPIE* **2523**, 155 (1995).
[8] O. Renner, T. Mißalla, P. Sondhauß, E. Krouský, E. Förster, C. Chenais-Popovics and O. Rancu, *J. Appl. Phys.* (submitted).
[9] A. Djaoui and S.J. Rose, *J. Phys. B* **25**, 2745 (1992).

Quasimolecular Features in Hot Dense Plasma Emission

P. Gauthier, P. Sauvan, S. Alexiou, P. Angelo, E. Leboucher-Dalimier

Physique Atomique dans les Plasmas Denses, LULI, CNRS
Université Paris VI, 4 place Jussieu, 75252 Paris Cedex 5, France
and Ecole Polytechnique, 91128 Palaiseau, France

A. Calisti, B. Talin

PIIM, Université de Provence, Centre de St Jérôme, 13397 Marseille, France

In this work we present new experimental spectra from thin foil collision experiments showing quasimolecular satellite-like features and a theoretical analysis demonstrating the observability of these features under specific experimental conditions.

In moderately coupled plasmas ($\Gamma_{i-i} \approx 1$), the mean interionic distance can be of the order of the typical spatial extention of bound states and the possibility of generation of N-center emissive structures can be no longer ignored. In fact it may be seen from Molecular Dynamics simulations that it is sufficient, under some conditions, to describe the bound states in terms of 2-center wavefunctions.

Considering an instantaneous ionic configuration around a given ion of interest 0, a necessary condition for the usefulness of a 2-center approach is that the Next Nearest Neighbor (NNN) be significantly farther away from 0 than is its Nearest Neighbor (NN). To address this question, we have computed the NN and NNN probability densities as a function of their distance r from the emitter 0, assumed to be at the origin. Fig.1 shows the NN and NNN conditional probability densities $P_{NN}(r)$ and $P_{NNN}(r)$, for three values of the ion-ion coupling parameter Γ_{i-i}, both subject to the constraint that the ion at the origin and the NN are also mutual NNs. In this way, for the selected configurations, the NNN is excluded from a large volume delimited by the association of 2 r_{NN}-truncated spheres and as a consequence it is expected to have only a weak influence on the dicenter system. The solid and dashed lines in fig.1 correspond to $\Gamma_{i-i} = 1.8$ and 0.7 respectively. In each case, the NN distribution is the one at the left. It may be seen that for a sizeable proportion of the total number of configurations, about 30 %, the ion at the origin and the NN are mutual NNs. In addition, for each value of the ionic coupling parameter we observe that the NN and NNN probability densities are rather well separated. So, at the parameters considered, we note that, for a large fraction of ionic configurations, the influence of the ionic component of the plasma on a given ion of interest can be modelized, in a good approximation, in terms of the only NN interaction. To treat *exactly* this NN interaction and then to calculate the bound states of the corresponding dicenter, we used a molecular basis. In this way, the multipolar interaction is considered at all orders. The bound states were calculated in the framework of the Born-Oppenheimer approximation which implies that all the quantities of interest depend on the internuclear separation as a parameter. In addition, in order to account for the lowering of the ionization potential, we adopted a cylindrically symmetric confining volume defined from the particular ionic equipotential which satisfies the electrical neutrality condition. Such a

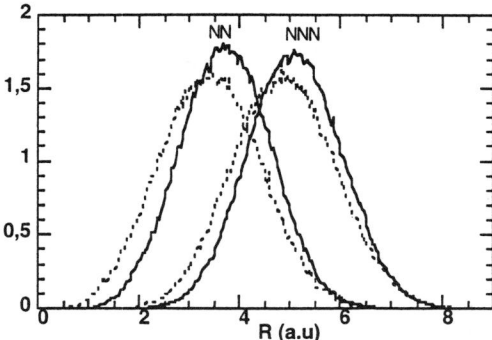

Fig.1 : Nearest neighbor and next nearest neighbor probability densities as a function of the internuclear separation for 2 different values of the ion-ion coupling parameter $\Gamma_{i\text{-}i}$: $\Gamma_{i\text{-}i}$ =1.8 (i.e. Ne = 2.10^{23} cm^{-3}, Te = 300 eV) (solid line); $\Gamma_{i\text{-}i}$ = 0.7 (i.e. Ne = 2.10^{23} cm^{-3}, Te = 800 eV) (dashed line).

molecular confining volume was first introduced by P. Malnoult et al [1] who calculated the bound states of a one electron molecular ion embedded in a uniform electron gas. In the present work, in order to take into account of finite temperature effects, we solved the Schrödinger and Poisson equations in elliptical coordinates according to a *selfconsistent field routine*. This procedure was applied for calculating the adiabatic electronic energies of a F^{8+}/F^{9+} molecular ion embedded in a fluorine plasma for various (N_e,T_e) plasma conditions. Results for $N_e = 2.10^{23}$ cm^{-3} and $T_e = 300$ eV are shown in fig.2. We have represented the only transition energies belonging to the spectral range of the FLyβ line, which is for many pratical reasons the line under study in our colliding foil experiments (see below). It may be seen that two particular electronic transitions (namely, the transitions (1) 541_g-210_u and (2) 650_u-100_g) of the embedded-in-plasma F^{8+}/F^{9+} molecular ion can be detected on experimental spectra implying electronic densities of the order of 10^{23} cm^{-3}. As known from neutral pressure broadening, the satellite-like features occur mainly near frequencies corresponding to extrema in the difference of two potential curves of the optical transition under consideration [2]. We have also checked that these transitions, first, have some minima at internuclear separations for which the NN probability density is high (fig.3), and second, have appreciable dipolar matrix elements for a large range of internuclear separations (fig.4).

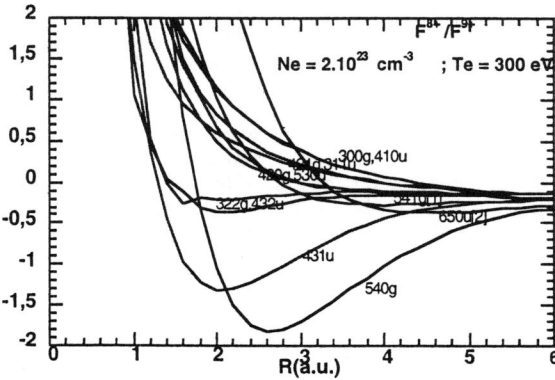

Fig.2 : Transition energies (in atomic unities) belonging to the FLyβ spectral range for an embedded-in-plasma F^{8+}/F^{9+} molecular ion as a function of the internuclear separation. The energies are labelled by the spherical quantum numbers nlm associated to the upper level of the transition.

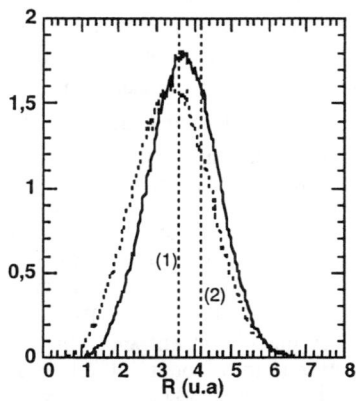
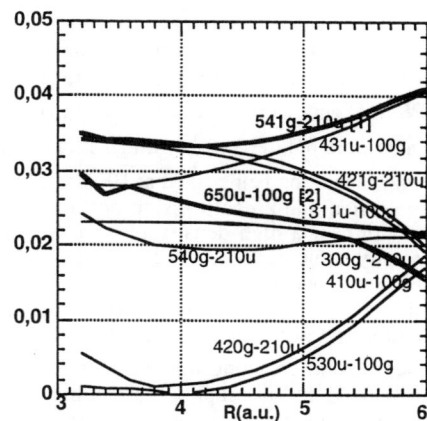

Fig.3 : NN probability densities in the case of a fluorine plasma (Z=9) characterized by an electron density equal to 2.10^{23} cm^{-3} and 2 temperature conditions : Te = 300 eV (solid line) and Te = 800 eV (dashed line). The vertical bars indicate the internuclear separations for which the transition energies (1) and (2) exhibit their flat minima.

Fig.4 : Dipolar matrix elements (in a.u.) of the 12 transitions belonging to the spectral range of the FLyβ for a F^{8+}/F^{9+} molecular ion embedded in a plasma whose mean electron density is equal to 2.10^{23} cm^{-3} and the electron temperature is Te = 300 eV. Each level implied in the transition is labelled by the spherical quantum numbers nlm.

The evolution of the spectral positions of transitions (1) and (2) for different mean electron densities and temperatures is presented in table 1. The key point is that these spectral positions are *insensitive* to variations of the plasma parameters and in fact, pratically identical to the pure quasimolecular results without any plasma electrons. As a consequence these molecular satellites can be expected to be detected even if the experimental observation technics imply a time and space integration over various (but of course dense) plasma conditions. Moreover, since plasma polarization shift predictions are usually an overestimate of the actual shift [3], and since in our case they turned out to be pratically zero for any reasonable choice of plasma parameters, *we may take this as an indication that the dynamic shifts will also not change the satellite positions.*

N_e(cm^{-3})/ T_e (eV)	9.10^{22}	1.10^{23}	2.10^{23}
300	[1] 12.72(2) Å	[1] 12.72(3) Å	[1] 12.73(2) Å
	[2] 12.77(2) Å	[2] 12.77(1) Å	[2] 12.77(0) Å
600	[1] 12.72(2) Å	[1] 12.72(4) Å	[1] 12.72(9) Å
	[2] 12.77(2) Å	[2] 12.77(2) Å	[2] 12.76(4) Å
800	[1] 12.71(8) Å	[1] 12.72(1) Å	[1] 12.72(6) Å
	[2] 12.77(1) Å	[2] 12.77(0) Å	[2] 12.76(3) Å

Table 1 : Spectral positions of the molecular transitions (1) and (2) for various (dense) plasma conditions.

In order to exhibit experimentally the two previous molecular transitions, we created hot and dense plasmas by the colliding foil technics. The experiments were carried out on the LULI facilities and employed intense (> 10^{14} W/cm^2) laser pulses of 500ps duration (FWHM) and operating at a wavelength of 0.26 µm (after frequency quadrupling). Two beams were focused on very thin Teflon foils (thickness = 5 µm) facing each other, their initial separation being fixed at 100 µm. Due to the laser-matter interaction, the foils are accelerated and a compression occurs in the plane where they collide leading to the formation of a high density plasma. Simulations performed by T. Shepard with the hydrodynamic code LASNEX [4] have shown that it is possible to create, from the collision of two Teflon foils, a fluorine plasma characterized by an electron density of the order of 2.10^{23} cm^{-3} and an electron temperature greater than 300 eV for such irradiation conditions and for the quoted values of the thickness and the initial foil separation. A space resolution along the collision axis ensured by the 10 µm entrance slit of the PABURCE spectrograph [5] and a high magnification ratio of about 120 allowed an accurate exhibition from the compressed central zone. Reproducible satellite-like features have been obtained in the spectra of the H-like Fluorine Lyβ line. Fig.5 shows typical FLyβ line profiles from two different shots showing a reproducible satellite feature at the same spectral location (12.72 Å). Such a satellite line can be fully interpreted, according to the previous theoretical analysis, in terms of the transition coupling the molecular states 541_g and 210_u. The second feature at 12.76 Å in fig.5b) (dashed vertical bars) corresponds to a possible second molecular satellite but we must emphasize that this latter is hardly spectrally resolved from dielectronic satellites belonging to the same wavelength range. As a consequence, the transition (2) cannot be definitively exhibited for our experimental conditions. In addition, *unlike the satellite at 12.72 Å*, the satellite line at 12.76 Å has not been detected in all spectra.

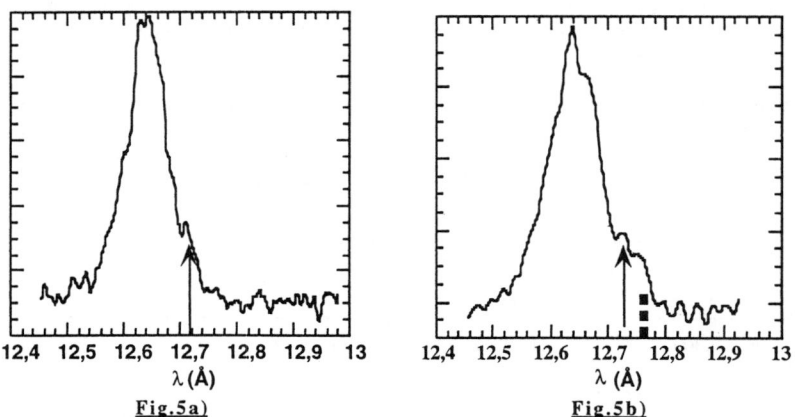

Fig.5a)   Fig.5b)

Profiles of the Lyβ line of fluorine emitted by the densest part of the plasma resulting from the collision of two thin Teflon foils. These two profiles correspond to two different laser shots. The arrows show the molecular satellite at 12.72 Å.

REFERENCES

[1] P. Malnoult, B. D'Etat and H. Nguyen, Phys. Rev. A**40**, 1983 (1989)
[2] S. Y. Chen and M. Takeo, Rev. Mod. Phys. **29**, 29 (1957)
[3] C. A. Iglesias, J. Quant. Spectrosc. Rad. Transf **51**, 1918 (1995)
[4] T. D. Shepard, Private Communication.
[5] A. Poquérusse, Opt. Com. **58**(2), 108 (1986)

QUASIMOLECULES AND SPECTRAL LINE BROADENING IN DENSE PLASMAS

J.P. Arranz, J. Butaux, H. Nguyen, and A. Reggadi

Spectronomie des Gaz et des Plasmas - DRP Université Pierre et Marie Curie
4, place Jussieu Tour 22 - 75252 Paris cedex 05 - France

Abstract - The energy levels and transition probabilities for transient diatomic molecules have been obtained by means of a self-consistent field method and used to compute the static and dynamical NeX Lyman α and Lyman β line profiles emitted from dense plasmas. The typical excitation of molecular satellite lines is pointed out.

In astrophysical and inertial confinement plasmas, electron densities $N_e \approx 10^{23-24}$ cm^{-3} and temperatures $T_e \approx 10^{2-3}$ eV are usual. Under such conditions the correlation parameter lies in the range $0.5 < \Gamma < 10$ and the radiator is strongly coupled to the neighboring ions and electrons. Indeed, the most important ion effects are mainly due to the nearest neighbor. For that reason, we have studied in details the electronic structure on the basis of transient diatomic molecules which takes accurately into account the screening effects due to electrons and level shift effect due to the nearest ion (two-ion center model).

For given N_e, T_e and each interionic distance the self-consistent field method (1,2,3) has been used to obtain the bound electron energies, radiative probabilities and also the local electron density. The numerical results show that the Stark effect is drastically reduced and many potential extrema appear at short distances.

These data and spatial distribution functions (4) are used to compute the spectral profiles of NeX Lyman α and NeX Lyman β both with and without the quasistatic approximation.

It is worth to emphasize that the quasistatic profiles depend on the temperature, since the energies do so (via the electron screening). The quasistatic profiles exhibit singularities which originate from the above-mentioned potential extrema.

The dynamical profiles have been computed, assuming ion classical trajectories in high density plasmas. We note that, except the intensest one, most of the satellites are strongly smoothed in the dynamical profile.

Figure 1 shows the dynamical NeX Lyman β line profile for $N_e = 2 \times 10^{23}$ cm^{-3} and T_e=150eV and T_e=450eV We note that two molecular satellites originating from the minimum of the transition energy for the component 101g → 000u and 200u→ 000g appear clearly on the Lyman β red wing. The relative positions of satellites and line peaks can be explained by recalling that the energy level (101g) is nearly free from plasmas conditions while the other ones decrease for decreasing temperature and increasing density.

Finally, the half width at half maximum of the global profile of Lyman α is compared to previous results on other ions under different conditions, computed

© 1997 American Institute of Physics

either by Molecular Dynamic Methods (7) or Model Microfield Methods. We have shown that these results can all be fairly represented by the empiric formula

$$W_{1/2}(eV) = 1.83 \left(\frac{kT_e(eV)}{500(Z-1)^2} \right)^{0.41} \left(\frac{N_e(cm^{-3})}{10^{22}(Z-1).} \right)^{0.507}$$

As shown in the table, our results fit very well this expression. This is a priori surprising, since we use a quasimolecular model instead of an atomic one as the previous works. This can be understood by the fact that the spectral widths are mainly governed by the weak interactions which are well taken into account in both models.

Z-1	N_e (cm^{-3})	kT_e (eV)	Formula	M.D.	M.M.M	Present work
12	4×10^{21}	233	0.031	0.031	0.033	
12	4×10^{21}	862	0.053	0.053	0.040	
17	1.5×10^{23}	862	0.210	0.21	0.21	
9	2×10^{23}	900	0.576			0.58
9	2×10^{23}	150	0.276			0.31

Figure 1: DYNAMICAL NE X LYMAN β LINE PROFILE
($N_e = 2 \times 10^{23}$ cm^{-3}, T_e = 150, 450 eV)

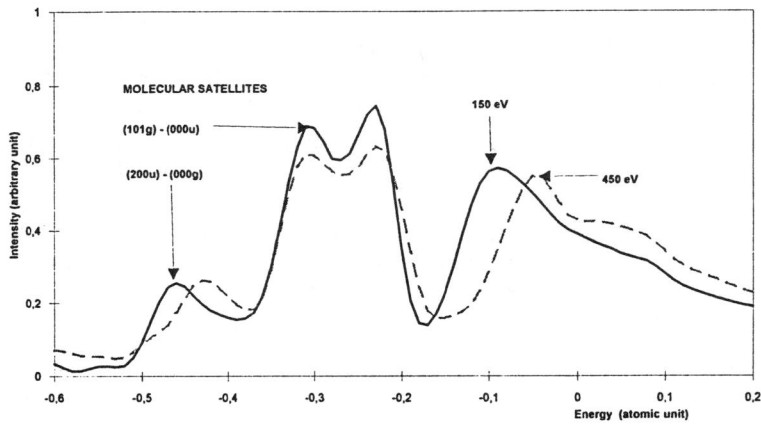

REFERENCES

(1) H. Nguyen, M. Koenig, D. Benredjem, M. Caby, G. Coulaud, Phys. Rev.A **33**, 1279 (1986)
(2) B. d'Etat, Thèse Paris (1987); P. Malnoult, B. d'Etat and H. Nguyen, Phys. Rev.40,1983 (1989)
(3) A Reggadi, Thèse Paris (1995)
(4) M. Amari, J.P. Arranz, J. Butaux, H. Nguyen (this conference)
(5) B. d'Etat, J. Grumberg, E. Leboucher, H. Nguyen, A. Poquérusse, J. Phys. B **20**, 1733 (1987)
(6) Ph. Malnoult and H. Nguyen, "Spectral Line Shapes", Vol.6 (American Institute of Physics, New York, 1990) p.142, Ph. malnoult, Thèse Paris (1989)
(7) R. Stamm, B. Talin, E.L. Pollock, C.A. Iglesias, Phys. Rev. A **34**, 4144 (1986)
(8) D. Gilles, C. Stehlé, J. Phys. II France 5,75 (1995)

Effects of Atomic States Interference on Spectral Line Shapes in MMM-Theory

I.N.Kosarev[◊], C.Stehlé[+], N.Feautrier[+], A.V.Demura[*] and V.S.Lisitsa[*]

[◊]Institute of Physics & Power Engineering, 249020, Obninsk, Russia
[+]Observatoire de Paris, Section de Meudon, F-92195, Meudon Cedex, France
[*]Russian Research Center "Kurchatov Institute", 123182, Moscow, Russia

Abstract. The influence of nonlinear interference effects (NIEF) on line shape is investigated. Ion dynamics is taken into account within the frame of Model Microfield Method. The present research is performed within atomic density matrix formalism. Numerical calculations of both lifetimes of radiating states and line shapes are performed for the spectral doublet $(1s-2s)2^1S-(1s-4p)4^1P$, $(1s-2s)2^1S-(1s-4d)4^1D$ of helium-like multicharged ions in hot dense plasmas. It is found that ion microfield essentially influences on the difference of populations of radiating $4^1P, 4^1D$ states. NIEF contribution in both allowed and forbidden components is calculated at various plasma conditions. Results demonstrate that account of NIEF is essential in the calculation of line shapes of multicharged ions.

The standard theory of spectral line broadening is based on the calculation of only spectral functions of radiating states, while the calculation of populations of radiating states is considered as a separate problem in atomic kinetics. However it has been demonstrated that the line broadening and the kinetic problems must be taken into account selfconsistently[1-4]. In such calculation the effect of atomic states interference (so called nonlinear interference effect - NIEF) due to action of plasma ion electric field is of importance. However the problem of account for ion dynamic effect is essential. In present consideration the effects of ions motion are considered within the model microfield method (MMM) which has been efficiently applied to the description of atomic spectral line shapes in plasmas[5-8]. Analytical calculations of metastable state lifetimes in plasma performed within MMM[9] have demonstrated the possibility of MMM-theory to reproduce all known analytical solutions for these lifetimes. This possibility of MMM is important for calculation of spectra with account of NIEF[1-4].

Radiating atom or ion is modelled by the three-level system. The transition 3-1 is assumed to be dipole forbidden, whereas the transition 2-1 is dipole allowed (levels 2 and 3 are separated in energy by an interval $\hbar\omega_{32}$). The states 3 and 2 are mixed both by electron inelastic collisions and ion microfield. This scheme is the simplest one for investigation of NIEF[1-4].

The equations for density matrix of upper radiating states ρ^0 in ion field $\vec{F}$ are following [2]

$$\dot\rho^0_{22} = -\Gamma_2\rho^0_{22} - 2\Phi(\rho^0_{22} - \rho^0_{33}) - iV_{23}\rho^0_{32} + iV_{32}\rho^0_{23} + Q_2,$$

$$\dot\rho^0_{33} = -\Gamma_3\rho^0_{33} + 2\Phi(\rho^0_{22} - \rho^0_{33}) + iV_{23}\rho^0_{32} - iV_{32}\rho^0_{23} + Q_3,$$

$$\dot\rho^0_{32} = -(\Gamma_2/2 + \Gamma_3/2 + i\omega_{32})\rho^0_{32} - iV_{32}(\rho^0_{22} - \rho^0_{33}),$$

$$\rho^0_{23} = (\rho^0_{32})^*, \qquad (1)$$

where Γ_2, Γ_3 are radiative decay rates of the states 2 and 3 respectively, Q_2, Q_3 are pumping rates, 2Φ is the matrix element of electron impact broadening operator, $V = \vec{d}\vec{F}$ is interaction potential between the radiator and plasma ion microfield.

For the matrix ρ describing a polarization at the radiating transition to the level 1 the following equations are valid at the first order in the interaction with a spontaneous radiating field G:

$$\dot\rho_{21} = -(i\omega_{21} + \Gamma_2/2 + \Gamma_1/2 + \Phi)\rho_{21} - iV_{23}\rho_{31} + i\rho^0_{22}Ge^{-i\omega t},$$

$$\dot\rho_{31} = -(i\omega_{31} + \Gamma_3/2 + \Gamma_1/2 + \Phi)\rho_{31} - iV_{23}\rho_{21} + i\rho^0_{32}Ge^{-i\omega t}, \qquad (2)$$

where $\omega_{ij} = \omega_i - \omega_j$ (i,j=1,2,3), Γ_1 is radiation decay rate of the state 1.

The selfconsistent solution of two systems of equations (1,2) with application of MMM procedure of averaging leads to the following expression for the line shape $I(\omega)$:

$$I(\omega) = -\left(\pi\{\rho^0_{22}\}_{MMM}\right)^{-1} Re(i(0,0,0,0,1,0)\{\hat{T}^{(1)}\}_{MMM}\vec{Q}), \qquad (3)$$

$$\{\hat{T}^{(1)}\}_{MMM} = \{\Delta\hat{T}\} + \{v\hat{T}'\}\{\hat{I} - v^2\hat{T}'\}^{-1}\{v\Delta\hat{T}\} +$$

$$+ \{v\Delta\hat{T}\}\{\hat{I} - v^2\hat{T}'\}^{-1}\{v\hat{T}'\} + \{\hat{T}'\}\{\hat{I} - v^2\hat{T}'\}^{-1}\{v^2\Delta\hat{T}\}\{\hat{I} - v^2\hat{T}'\}^{-1}\{v\hat{T}'\},$$

$$\hat{T}' = \begin{bmatrix} \hat{T}^0 & \hat{0} \\ \hat{0} & \hat{T}^\mu \end{bmatrix}, \quad \Delta\hat{T} = \hat{T}'\hat{G}\hat{T}', \qquad (4)$$

where $\hat{T}^0$ is Fourier transform (at zero frequency) of the static evolution operator for "populations" (1) and $\hat{T}^\mu$ is Fourier transform (at zero frequency) of the static evolution operator for polarizations (2).

Numerical calculations for line shapes of multicharged helium-like ions have been performed either with Baranger-Mozer (at low densities) formulation or with Monte-Carlo simulation (at high densities) of field distribution function[10].

MMM calculations have been performed for populations of 4^1P, 4^1D states of helium-like ion Al^{+11} at plasma temperature T=350 eV.

TABLE 1. Populations of metastable state (normalized to ρ_{22}^0).

N_e, cm^{-3}	MMM, numerical	radiative-collisional model	static ion microfield
10^{20}	1.27	1.39	1.33
10^{19}	3.45	4.94	4.499
10^{18}	22.7	40.4	37.5
10^{17}	199.	395.	385.

As one can see from this table ion dynamical microfield essentially changes difference of populations of radiating states.

The account of both the ion motion and NIEF for line shapes of helium-like multicharged ions of Al^{+11} and Ni^{+26} are presented in the Fig. It is seen that NIEF can strongly modify the line shapes both allowed and forbidden. There is an increase of intensities in the nearest wings of both components for relatevly low plasma densities.

The calculations performed demonstrates that selfconsistent account of atomic level population kinetics and polarization at the observed transition under the influence of dynamical ion microfield is of importance for correct description of spectral line shapes of ions in hot dense plasmas.

REFERENCES

1. V.S.Lisitsa, *Atoms in Plasmas*, Springer, 1994
2. A.V.Anufrienko, A.L.Godunov et al., Sov. Phys. JETP **71**, 728 (1990)
3. A.V.Anufrienko, A.E.Bulyshev et al., Sov. Phys. JETP **76**, 219 (1993)
4. I.N.Kosarev and V.S.Lisitsa, Sov. Phys. JETP **79**, 64 (1994)
5. A.Brissaud and U.Frisch, JQSRT **11**, 1767 (1971)
6. C.Stehlé, A.Mazure, G.Nollez and N.Feautrier, Ast. Astroph., **127**, 263 (1983)
7. C.Stehlé, Astron. Astrophys., **292**, 699 (1994)
8. B.Talin, R.Stamm, V.P.Kaftanjian and L.Klein, Astrophys.J. **322**, 804 (1987)
9. I.N.Kosarev and V.S.Lisitsa, J. Phys. B **29**, 1 (1996)
10. A.V.Demura, D.Gilles and C.Stehlé, JQSRT **54** 123 (1995)

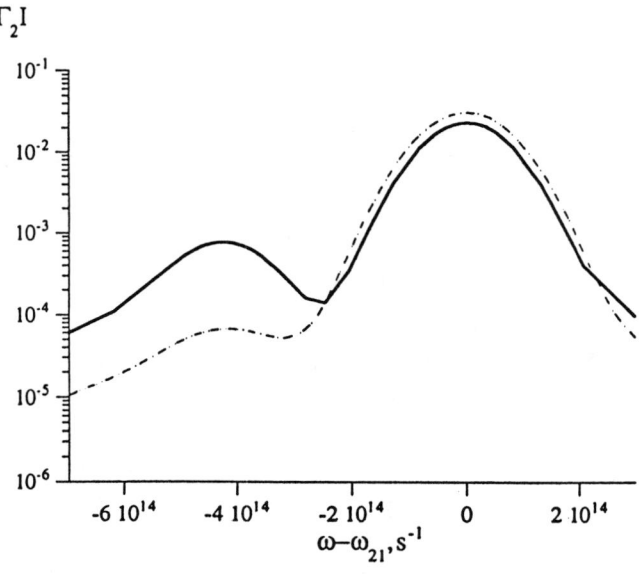
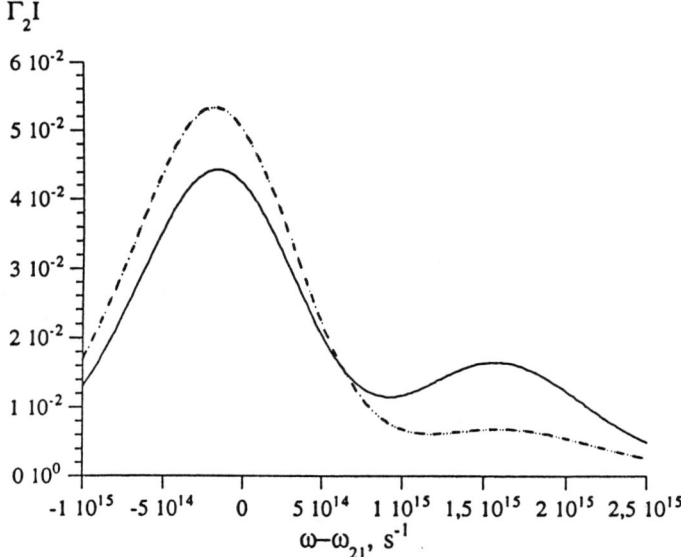

FIGURE. The line shapes $\Gamma_2 I$ with (solid curves) and without (dashed curves) NIEF versys frequency shift $\omega - \omega_{21}$ for He-like ions in dense plasmas: upper curves - Al^{+11}, $N_e=10^{18} cm^{-3}$, T=350 eV; lower curves - Ni^{+26}, $N_e=10^{21} cm^{-3}$, T=1 keV.

Line Profile Measurements in a Gas-Liner Pinch Discharge

Th. Wrubel[1], S. Glenzer[2], S. Büscher[1] and H.-J. Kunze[1]

[1] Institut für Experimentalphysik V, Ruhr-Universität, D-44780 Bochum, Germany
[2] Lawrence Livermore National Laboratory, Livermore, CA 94550, USA

We report on recent line shape measurements performed in a gas-liner pinch discharge. The 2s3s-2s3p singlet and triplet transitions in NeVII were measured which are of interest for astrophysical line formation calculations. The measured widths of the Lorentzian line shapes are in good agreement with semi-classical calculations which include electron quadrupole broadening. The Stark widths of the 3s ^{2}S – 3p ^{2}P^o transitions in lithiumlike NV and FVII were measured optically thin. At high plasma densities ($\leq \times 10^{19}$ cm^{-3}) the widths which are dominated by electron collisional broadening show deviations from linear scaling. Furthermore, we investigated spectral line shapes of transitions between levels of high principal quantum numbers of lithiumlike ions. The measurement of the n=4 to n=5 transition in FVII, where forbidden transitions and various (n-1,l)-(n,l') fine structure components overlap, show excellent agreement with a suitable line shape code. Finally, we present measurements of the line shapes of P$_\alpha$ and P$_\beta$ of hydrogenlike helium. New scalings are presented as well as a detailed analysis of the dip of the P$_\beta$ line which is predicted to be sensitive to ion dynamics.

INTRODUCTION Spectral line profiles and Stark widths are an important diagnostic tool applied to several kinds of plasmas, e.g. inertial confinement fusion plasmas, high-density laser-produced plasmas or astrophysical plasmas. It is of utmost importance to find experimental scalings in order to test approximations used in theoretical calculations. We studied spectral line shapes of both lowly and multiply ionized species with a gas-liner pinch where the plasma was diagnosed independently by Thomson scattering.

Isolated spectral lines of nonhydrogenic ions were studied where the line shape is a Lorentzian and the width is mostly determined by electron collisional broadening [1]. However, our last contribution at the 12th ICSLS [2] showed that the line profiles of electron-collisional-broadened spectral lines are not adequately described by several theoretical approximations especially for emitters with large spectroscopic charge number [3]. While detailed calculations in [4] show good agreement for all lines of the light species the calculated widths seem to be systematically lower for FVII and NeVIII. Regarding the seemingly excess broadening in NeVIII recent calculations in [5] show good agreement with our experimental results [6].

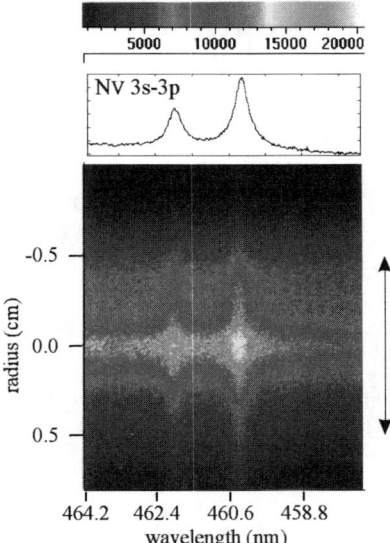

Figure 1. CCD-spectrum of the NV3s-3p douplet transition spatially resolved over the radius of the plasma column. The plasma diameter is about 2 cm. The emitter are located in the center of the hydrogen plasma with a diameter smaller than 3 mm where the plasma is homogeneous in respect to density and temperature.

Even in the transition region of linear to quadratic Stark effect there is a lack of a proper broadening theory which is capable of calculating line broadening data, e.g. for astrophysical line formation calculations [7]. Approaching this theoretical region where broadening by fields of disturbing ions begins to play the dominant role, we investigated spectral line shapes of transitions between levels of high quantum number of lithiumlike fluorine FVII which is an extension of earlier work on CIV, NV and OVI [8]. In order to describe such line profiles we used a suitable line shape code [9] where satisfactory agreement could only be achieved when taking into account ion dynamic effects.

Finally, we measured line shapes of hydrogenlike helium, the widths and shifts of the P_α and P_β lines [10]. An accurate measurement of the dip in the center of the HeII P_β shows the predicted behaviour of ion dynamical effects, i.e. growth of the dip with increasing density and decreasing temperature, but existing calculations cannot properly describe our experimental profiles.

EXPERIMENTAL SETUP Since the gas-liner was described at the last ICSLS-conference [2] only a short summary is given here. It is basically a z-pinch equipped with a special gas inlet system. Triggering the main capacitor bank (11.2 μF, 27-40 kV) the injected hydrogen is compressed onto the axis giving a plasma of 1-2 cm in diameter. An on-axis valve injects the so-called test gas seeding the imploding plasma with atoms emitting the lines of spectroscopic interest. These atoms are located in a small, homogeneous part of the hydrogen plasma (see figure 1) resulting in the outstanding advantages of the gas-liner pinch for the goal of line profile measurements [11]: cold boundary

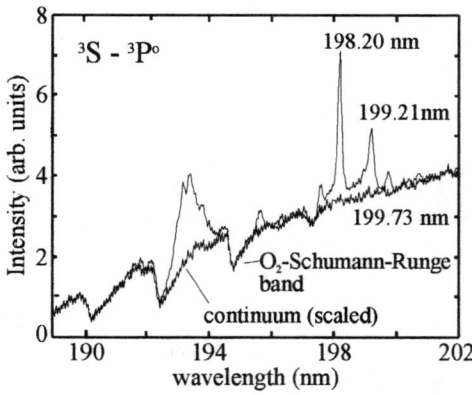

Figure 2. The multiplet components of the 2s3s-2s3p triplet line of berylliumlike neon. There is no radiation of other impurity ions. The absorption lines belong to the Schumann-Runge band of O_2 of air.

layers absorbing the radiation are practically absent. There is no necessity for an Abel-inversion and the optical depth can be controlled by reducing the amount of the injected test-gas. Furthermore, the underlying continuum radiation or line radiation of impurity ions can be measured by switching off the on-axis valve.

The electron density varies between $(0.5\text{-}10)\times 10^{18}$ cm^{-3} and the temperature ranges from 3 eV to 50 eV depending on the discharge conditions. The plasma parameters are determined independently by Thomson scattering [12] simultaneously with spectroscopic measurements in the visible and vacuum ultraviolet spectral range. Compared to previous spectroscopic measurements in the visible spectral range, the simultaneous diagnostic using a second visible spectrograph (Czerny Turner, Spex M 1000, 1200 l/mm, 750 nm blazed) equipped with an OMA-system (OMA III, EG&G model 1456-990G) offers, in addition to higher precision, the possibility of taking into account shots of higher density which occur stochastically. The measurement of the FVII transition in the vuv spectral range was performed with a 1m vuv spectrometer (McPherson model 225) having a 1200 lines/mm grating blazed at 120 nm which was equipped with an OMA-system giving an reciprocal dispersion of 0.0207 nm/pixel.

NeVII 2s3s-2s3p TRANSITIONS The line profiles of the singlet and triplet transitions are important for astrophysical line formation calculations in order to determine plasma parameters of, e.g., the atmospheres of hot stars. Figure 2 shows the NeVII triplet line at λ=198.20/199.21/199.73 nm together with a spectrum with hydrogen bremsstrahlung radiation in first order where the line profile is essentially determined by the apparatus profile. We succeeded in measuring the multiplet components in fifths order with a reciprocal linear

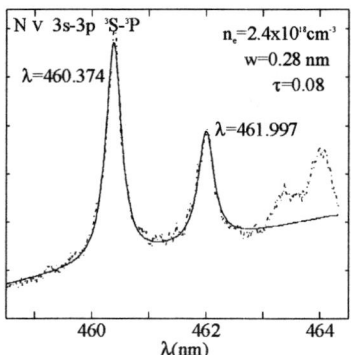 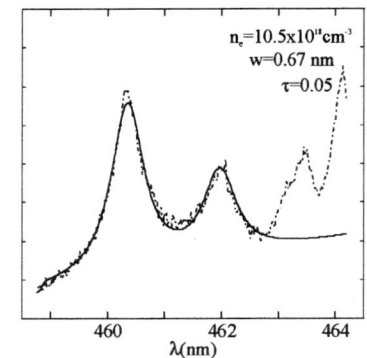

Figure 3. The 3s-3p douplet in NV along with fitted Voigt profiles taking into account radiational transport. The density was measured simultaneously with Thomson scattering.

dispersion of 0.0036 nm/pixel applying an interference filter (Schott, UV-KMD 12-1, $\lambda_{max} = 196.5$ nm, 25% transmission) and taking the mean of 10 spectra. We calculated a profile consisting of a Lorentzian which is convoluted with both a Doppler profile and the measured apparatus profile. By varying the width of the Lorentzian this profile is well fitted to the experimental spectrum by a least-squares method. The $^1S - ^1P^o$ singlet line at $\lambda = 364.36$ nm was measured in second order (0.0099 nm/pixel). The fitted Stark widths and the measured plasma parameters applying Thomson scattering along with error bars are compiled in table 1 of Ref. [13] where we give more detailed information about the measurements.

The experimental widths are compared with theoretical data according to semi-classical and semi-empirical calculations. The semi-empirical calculations according to Ref. [14] clearly underestimate the width, especially in the case of the triplet transition. In contrast to this approximation the semi-classical method of Griem [1] results in widths which are comparable with the experimental ones. The semi-classical approximation according to [15] yields values which are in good agreement with the experimental data. Best agreement is found with non-perturbative calculations performed by Alexiou [5] showing the importance of electron quadrupole contributions to the width. Furthermore, this calculations show consistent agreement with other earlier studies on highly ionized ions [6].

3s-3p TRANSITION IN NV AND FVII
We succeeded in measuring the Stark width of $^2S - ^2P^o$ doublet transitions at high densities ((1-$10)\times 10^{18}$cm^{-3}). The emission was checked to be optically thin by determining the optical depth from the intensity ratio of the doublet components. For lower densities we confirm the results of earlier measurements at the gas-liner pinch

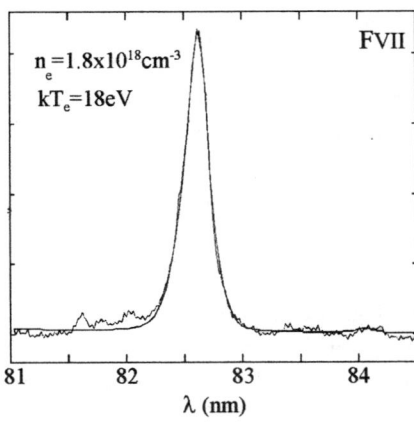

Figure 4. Recorded line spectra of n=4 - n=5 transitions in FVII. The fitted profile which was calculated using the Frequency Fluctuation method taking into account the apparatus profile is in excellent agreement with the measured one. The density and temperature were determined by Thomson scattering.

[6] whereas for higher electron densities the Stark width of the transition in NV shows deviation from linear scaling. This deviation could be explained by the increasing overlap of the multiplet components and, therefore, at high densities the transitions cannot be treated as isolated lines (see figure 3). In the case of the FVII line the components are still isolated from each other and we found a slight deviation from linear scaling which might be due to dynamical Debye shielding effects [3]. The temperature varies over a range of 10 eV for all data points and local inhomogeneities in the plasma center occurring at the highest densities and, therefore, an increase of the error bars of the density determination cannot be ruled out.

n=4 - n=5 TRANSITION IN FVII Specific discharge conditions using unusual high test-gas concentrations were connected with observed instabilities where we found population inversion, e.g. the intensity of 4f-3d transition was enhanced over the intensity of the 4d-3p line in OVI and FVII [16]. In order to investigate these inhomogeneities and its density fluctuations along the z-axis we measured the n=4 - n=5 vuv-transitions using a CCD-camera. We employed homogeneous discharge conditions for calibration and measured the line profile shown in figure 4 at a density of $n_e = 1.5 \times 10^{18}$ cm^{-3}. Extending our earlier work on CIV, NV and OVI [8] where the importance of ion dynamic effects was verified [9] we found excellent agreement of the profile with spectral line shape calculations. In this kinetic theory model the profile obtained from the quasistatic approximation is divided into radiative channels and exchanges between these frequency groups are included. Taking into account the Doppler and apparatus contribution which gives about 50% to the width we found excellent agreement between calculated and measured line profile.

HeII P_α AND P_β TRANSITIONS The line shapes of HeII P_α and P_β

which are often used in plasma diagnostics were measured for optically thin plasma conditions at densities up to 4×10^{18} cm^{-3} and $n_e = 0.9 \times 10^{18}$ cm^{-3}, respectively, and temperatures of 3.5 – 7.5 eV [10]. In the case of the HeII P_α line new empirical scalings can be given which improves density diagnostics at high densities. While the width of this transition shows significant discrepancies with some modern line shape theories, very recent calculations with improved electron broadening operator show good agreement [17]. In the case of the HeII P_β line profile we found good agreement with several calculations. The dip which presents a critical test for line shape calculations including ion dynamics was measured excluding disturbing mechanisms, and it shows qualitatively the expected density and temperature dependence, i.e. the dip decreases with increasing temperature and decreasing density. The present data agree with a scaling law to within a factor of 1.6 [10].

REFERENCES

[1] H. R. Griem, *Spectral Line Broadening by Plasmas*, Academic Press, New York (1974)
[2] S. Glenzer, in *Spectral Line Shapes Vol. 8*, 12 th ICSLS, Toronto, Canada, June 1994, eds. A. D. May, J. R. Drummond and E. Oks, AIP Conf. Proc. 328, AIP Press, New York, 134 (1994)
[3] H. R. Griem, *Principles of Plasma Spectroscopy*, in print, Draft (1996)
[4] S. Alexiou, Phys. Ref. A **49**, 106 (1994)
[5] S. Alexiou, Phys. Ref. Lett. **75**, 3406 (1995)
[6] S. Glenzer, N. I. Uzelac, and H.-J. Kunze, Phys. Ref. A **45**, 8795 (1992)
[7] K. Werner, U. Heber and K. Hunger, Astron. Astrophys. **244**, 437 (1991)
[8] S. Glenzer, Th. Wrubel, S. Büscher, and H.-J. Kunze, R. Stamm, B. Talin, J. Nash, R. W. Lee and L. Klein, J. Phys. B: At. Mol. Opt. Phys. **27**, 5507 (1994)
[9] A. Calisti, L. Godbert, R. Stamm and B. Talin, J. Quant. Spectrosc. Radiat. Transfer. **51**, 59 (1994)
[10] S. Büscher, S. Glenzer, Th. Wrubel and H.-J. Kunze, J. Phys. B: At. Mol. Opt. Phys., in print
[11] H.-J. Kunze, in *Spectral Line shapes*, edited by R. J. Exton (Deepak, Hampton), Vol 4 (1987)
[12] Th. Wrubel, S. Glenzer, S. Büscher and H.-J. Kunze, J. Atmos. Terr. Phys. **58**, 1077 (1996)
[13] Th. Wrubel, S. Glenzer, S. Büscher and H.-J. Kunze, Astron. Astrophys. **307**, 1023 (1996)
[14] M. S. Dimitrijević and N. Konjević, J. Quant. Spectrosc. Radiat. Transfer **24**, 451 (1980)
[15] J. D. Hey and P. Breger, J. Quant. Spectrosc. Radiat. Transfer **24**, 427 (1980)
[16] S. Glenzer and H.-J. Kunze, Phys. Rev. E **49**, 1586 (1994)
[17] S. Alexiou, these proceedings

**LOW AND MODERATELY
DENSE PLASMAS**

Closing in on Stark Broadening for weakly coupled plasmas

Spiros Alexiou

PAPD, Universite Paris VI/LULI Ecole Polytechnique, France

Recent major advances in Stark broadening in weakly coupled plasmas have resulted in the subject achieving a very satisfactory standing. Some of these advances are discussed in detail, particularly the progress in electron impact broadening and ion dynamics and an overview of the status and remaining fronts in Stark broadening is given.

1. INTRODUCTION

Let me start this talk with a provocative statement on the significant advances made recently on the subject of spectral line broadening in weakly coupled plasmas: On virtually every front, we are closing in. The choice of this word is not accidental; I mean to say more than progress is being made: I mean to tell you that the subject of Spectral Line Broadening in weakly coupled plasmas (where this includes from the atmospheric plasmas to some plasmas of ICF) is close to becoming a completed subject for the great majority of the existing cases known. This closing in process has been the outcome of precision experiments as well as theoretical work. I will divide this talk into two parts, dealing with electron broadening and ion dynamics; though advances in the subject of nonthermal effects and diagnostics are also being made[1,2], some aspects of this subject are discussed elsewhere in this book[1].

2. STANDARD APPROXIMATIONS AND TIMESCALES

This is hardly the place to present the standard Stark-broadening theory, i.e., the impact approximation for electrons and the quasistatic approximation for ions. I will thus limit myself to a brief look at this old subject from a different angle. Line broadening is the result of the fact that, in the presence of a stochastic interaction, the time evolution operator matrix elements (a linear combination of which is the autocorrelation function C(t), i.e., the Fourier transform of the line profile) have essentially random phases after a characteristic time; hence, C(t) is zero after such a characteristic time[3]. This, in turn, yields a line profile with a finite width. It is this characteristic time, hereafter called "memory loss time" or "Time of Interest"(TOI) and its relation to the fluctuation rate(FR) of the plasma-emitter interaction that determines the appropriate dynamical regime and thus the approximations, if any, that may be used.

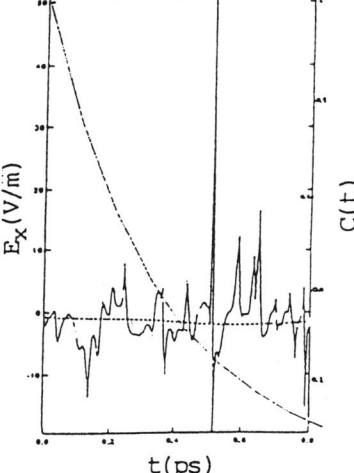

Fig.1a: Typical Impact TOI vs. FR

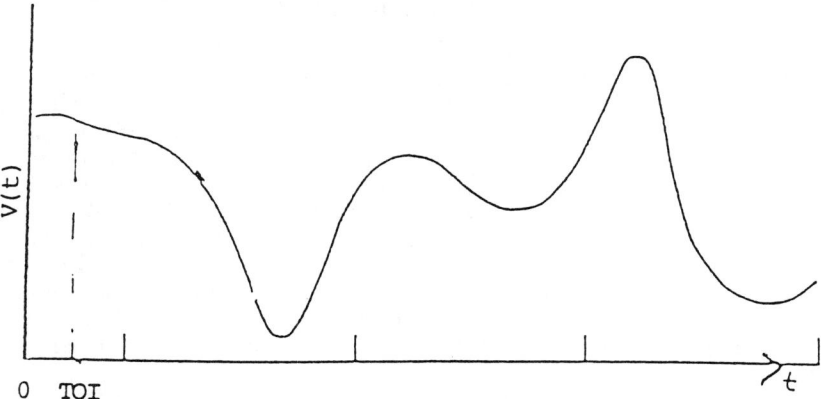

Fig.1b:Typical quasistatic TOI vs. FR

If TOI is much larger than 1/FR (i.e. HWHM≪ FR), the impact approximation is valid over almost all of the profile: The exception being the very far wings. If TOI is much less than 1/FR (i.e. HWHM≫ FR), the quasistatic approximation is essentially valid over all of the line profile. However, we note that a factor of 10 may not be sufficient to satisfy the ≫ in this context. The intermediate case is the ion dynamical regime. The nonlinear aspect of this problem is that TOI is determined by the emitter-plasma interaction(hence also FR) itself.

Fig.1a shows a typical microfield time history and C(t) for an "impact" case: Field fluctuations occur on a very short time scale compared to the decay of C(t). We note the isolated nonsmooth features in the microfield.

Fig.1b shows a typical microfield time history and C(t) for a quasistatic case. C(t) has decayed before the field can change appreciably. Had C(t) decayed on the time scale of the entire horizontal axis shown in Fig.1b, or on the time scale of one or a few peaks in Fig.1a, we would have been in the ion-dynamical regime.

3. ELECTRON BROADENING (IMPACT THEORY)

The major factor behind these advances on electron impact broadening have been the precision experiments by the Bochum group[4-8]. It is these experiments that are enabling us to finally give *definitive* answers to a number of problems in Stark broadening, *without* the uncertainities found in the previous literature. This has allowed us to develop predictive capabilities that satisfy the needs of the lineshape "users" community. In this talk, I will limit myself to the experiments with isolated lines, for a good reason: These lines are usually unaffected by ion effects, whether static or dynamic. Consequently, by *isolating* the electron broadening effects, they provide an excellent testing ground for our understanding of electron broadening. For over 30 years, electron broadening was thought to be well understood in terms of the impact theory[9]. This theory defines a collision operator, which is essentially a self-energy[3], and predicts a Lorentzian lineshape. But for practical applications, the question arises *which* impact theory to use. By this I mean that there have been a number of different calculations, ranging from semiempirical formulas[10-12] to close coupling (CC) calculations[13-14], which are not always in agreement with each other. The Bochum experiments were initially compared with the predictions of three such semiclassical calculations[10,12,15] and important differences by factors as large as 2 were found. Discrepancies by factors of 2 to more than 3 were also obtained with Seaton's[13] CC calculations. For all cases agreement was obtained (within the error bars) by more careful calculations based on the standard perturbative theory with dipole[16] and dipole plus quadrupole[17] terms that used the exact dipole and quadrupole functions without approximations. The only exception to this agreement was the Li-like Neon 3s-3p line, where agreement was marginal; this case provides much insight, so we discuss it more in detail later. These calculations quoted a weak collision contribution that was presumably being accurately computed by the standard perturbative treatment for impact parameters larger than $\rho_{min}(v)$ (which was self-consistently determined) and an *estimated error bar* for the contribution of the strong ($\rho \leq \rho_{min}(v)$) collisions. At this stage, we define strong collisions as collisions with impact parameter less than $\rho_{min}(v)$, for which the standard semiclassical perturbative theory is not valid.

Because of these results, there was somewhat of a cloud hanging over the Li- like Neon results, with indirect suggestions that this was an experimental problem. Indeed, even for the best agreement achieved[17], it would have been necessary to assume the maximum possible strong collision contribution, which, in addition, would need to be almost constant with impact parameter and velocity, in order to have agreement within the error bounds. Clearly there was a persistent disagreement here: The purpose of giving error bars is that this is the honest perturbative semiclassical answer. It is not to be able to hide behind them. There are indications that strong collisions are very important for Neon; But, these should have been correctly bounded by the error bars. This was, therefore, an indication that the error estimates themselves were inaccurate.

At this point one should mention that the standard impact calculations divide the phase space into two parts: One part, corresponding to large impact parameters, which gives rise to weak collisions and one, corresponding to small impact parameters, giving rise to strong collisions; the latter can be (and have in general always been) a problem for standard impact calculations. However, we must make a distinction between collisions that belong to the strong collision phase space part because they cannot be treated semiclassically and those which can be treated semiclassically, but not perturbatively. Historically, for a semiclassical collision to be treated perturbatively, it had been considered a necessary and sufficient condition that the first nonvanishing term after unity in the perturbation expansion of the S-matrix be $\ll 1$. It turns out that for many lines considered in the Bochum experiments, the contribution of the latter type of strong collisions, i.e. collisions that may be treated semiclassically, but not perturbatively, is substantial. Therefore it made a lot of sense to try to improve the calculations [16,17] by including a nonperturbative calculation, in order to refine the error estimates.

The final answer to the problems with the standard theory came during the testing of the new version of a code that was designed to handle the phase space part of the semiclassical, but strong collisions. The problem was that the usual unitarity check[18], which for over 30 years had been considered necessary and sufficient for a perturbative treatment, is not sufficient. The check can fail for small impact parameters[19]. Let me show you an example of a very simple line, Be-like Neon 2s3s-2s3p singlet.

As a reminder, the half width(HWHM) in the impact approximation is written as

$$HWHM = 2\pi n \int \rho d\rho \int dv v f(v) Re\{I - S_a(\rho,v)S_b^{-1}(\rho,v)\} \quad (1)$$

where n is the electron density and $f(v)$ is the Maxwellian velocity distribution. Hence the width is proportional, in the impact approximation, to the real part of $\{S_a(\rho,v)S_b^*(\rho,v) - I\}$, where S_a and S_b are the S-matrices for the upper and lower levels respectively and $\{...\}$ denotes an average over angles, while I is the unit matrix. The S-matrix is the limit of the time-evolution operator $U(-\infty,t)$ as $t \to \infty$. We further note that perturbation theory for the above angular average has always been expected to be accurate as long as $\{S_a(\rho,v)S_b^*(\rho,v) - I\}$ is small compared to unity.

In Fig.2, we plot the real part of the angular average $\{...\}$ of the expression $\{U_a(t;\rho,v)U_b^*(t;\rho,v) - I\}$, which, when averaged over the impact parameter ρ

and velocity v, and taken at $t \to \infty$, gives us the width. u is the hyperbolic trajectory parametrization variable related to time t by:

$t = \frac{Z_e m e^2}{4\pi\epsilon_0 m v^3}(\epsilon \sinh[u] - u)$ For both graphs, the velocity is $v = 2 \times 10^6$ m/sec and both are calculated numerically, allowing only dipole interactions. For the solid line, the impact parameter is $\rho = 5\text{Å}$, while for the dashed line $\rho = 0.68\text{Å}$. This figure shows an important point for a simple line and with only dipole interactions considered: For the larger of the two impact parameters considered, the evolution of the unitarity expression, $\{U_a(t;\rho,v)U_b^*(t;\rho,v) - I\}$ is as expected, namely we have very weak collisions that slowly destroy some of the coherence and $\{S_a(\rho,v)S_b^*(\rho,v) - I\}$ is small compared to unity. Things are completely different for the smaller impact parameter. $\{S_a(\rho,v)S_b^*(\rho,v) - I\}$ is indeed small compared to unity, but $\{U_a(t;\rho,v)U_b^*(t;\rho,v) - I\}$ is not small compared to unity for all times, as perturbation theory would require. This is why the exact result has no resemblance to the perturbative result, although both are small compared to unity. In other words, the usual unitarity criterion failed and a perturbative calculation was grossly in error, even though the perturbative expression for $\{S_a(\rho,v)S_b^*(\rho,v) - I\}$ was indeed much less than unity. Practically speaking, this means that previous calculations, even those with self-consistent minimum impact parameters, did not always correctly discriminate between the weak and strong collisions, particularly for situations where small impact parameters (eccentricities) were involved. This was precisely the situation for the Li-like NeVIII 3s-3p results of the Bochum experiments, where the new, nonperturbative calculations give semiclassical collision contributions to the FWHM of 0.9 and 0.95Å, for the two plasma parameters considered, compared to 0.6Å of the older self-consistent calculations. Thus the Li-like NeVIII result was *not* an experimental problem!

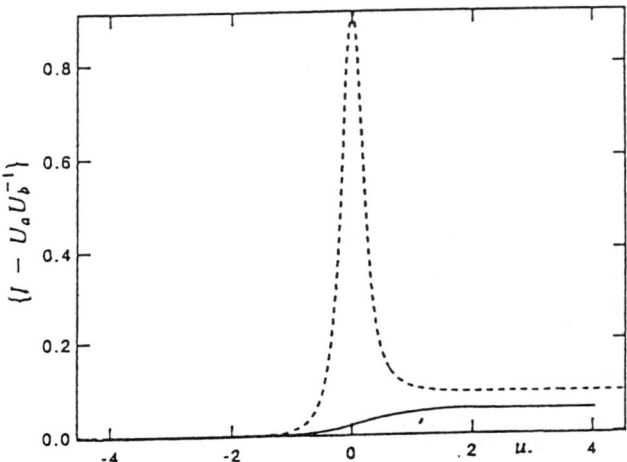

Fig.2: Illustration of the failure of the usual unitarity criterion

Given the inadequacy of the old unitarity criterion, a new one had to be found. The new criterion takes account of the fact that the second order term in the perturbation expansion of U(t) is not necessarily a maximum at $t = \infty$ (which is what the usual way of checking unitarity was assuming) and one needs to use the *maximum value* of this term, rather than the value at $t = \infty$. Thanks primarily to the ingenuity of A.Poquérusse, this was recently accomplished in a very efficient manner[20] for the dipole case and it is noteworthy that the new unitarity test function A is even simpler to use than the old one!

Theory vs. Experiment: I: Li-like 3s-3p

Species	$T_e(eV)$	$n_e(10^{18}e/cm^3)$	Exp(Å)	DK	HB	G	CC	PR	NP
CIV	7.0	1.5	6.7±0.4	0.84	0.84	1.11	0.71	1-1.44	
CIV	8.6	2.4	9.7±0.5	0.87	0.88	1.14	0.78	1.04-1.5	
NV	14.9	1.2	2.2±0.1	0.66	0.73	0.85	-	0.73-1.14	0.92-1.03
NV	18.7	1.6	2.7±0.1	0.71	0.76	0.89	-	0.74-1.13	0.91-1.045
NV	21.8	2.0	3.4±0.2	0.72	0.79	0.91	-	0.7-1.07	0.853-0.992
NV	23.9	2.3	3.8±0.2	0.7	0.78	0.89	-	0.705-1.07	0.85-1
OVI	8.3	1.0	1.0±0.1	0.7	0.78	0.83	-	0.86-1.3	1.16-1.24
OVI	11.5	1.3	1.4±0.1	0.61	0.69	0.72	0.55	0.707-1.05	0.944-1.03
OVI	15.6	2.1	1.8±0.3	0.73	0.83	0.86	0.7	0.8-1.18	1-1.13
OVI	17.5	2.4	2.1±0.2	0.66	0.76	0.79	0.66	0.75-1.11	0.956-1.07
FVII	14.4	1.57	0.87±0.1				-	0.705-1.225	0.96-1.06
FVII	16.6	2.1	1.11±0.13				-	0.702-1.21	0.955-1.06
FVII	18.5	2.92	1.49±0.18				-	0.7-1.2	0.94-1.05
NeVIII	29.7	2.8	1.2±0.1	0.39	0.55	0.45	0.35	0.52-0.76	0.8-0.93
NeVIII	42.5	3.2	1.2±0.1	0.41	0.56	0.48	0.3	0.52-0.77	0.77-0.97

Theory vs. Experiment: II: BIII 2s-2p

$T_e(eV)$	$n_e(10^{18}e/cm^3)$	Exp(Å)	DK	HB	G	CC	PR	NP
10.6	1.81	0.22±0.03	0.47	0.8	0.84	0.55	0.53-1.19	0.85-1.02

Theory vs. Experiment: III: Be-like Ne 2s3s-2s3p

Line	$T_e(eV)$	$n_e(10^{18}e/cm^3)$	Exp(Å)	DK	HB	G	PR	NP
Singlet	19.0	3.5	1.762	0.636	0.87	0.78	0.63-1.03	0.95-1.096
Triplet	20.5	3.0	0.453		0.52	0.77	0.65 0.713-1.13	0.983-1.13

Theory vs. Experiment: IV: OIV 3s-3p

$T_e(eV)$	$n_e(10^{18}e/cm^3)$	Exp(Å)	DK	HB	G	IS	PR	NP
4.7	0.56	0.62±0.08	0.55	0.55	0.73	1.11	0.36-1.21	0.93-1.
7.5	1.03	0.96± 0.096	0.55	0.61	0.73	1.15	0.41-1.19	0.96-1.05
8.5	0.99	0.98± 0.09	0.49	0.57	0.7	1.07	0.38-1.06	0.86-0.96
10.3	1.63	1.28± 0.13	0.58	0.7	0.85	1.32	0.47-1.25	0.99-1.13

Table 1.: Theory vs. Experiment for isolated ion lines.

In Table 1, theoretical and high-quality experimental results are compared. The experimental FWHM is listed, together with the *ratio* of the theoretical to the experimental width. The theoretical width is calculated according to the semiempirical formula of Dimitrijevic and Konjevic(DK)[10], the simplified semiclassical(SC) theory of Hey and Breger(HB)[15], Griem's formula (G)[9], the CC calculations of Seaton[13-14], the SC perturbative calculations by Alexiou and Ralchenko(PR)[17] and the new nonperturbative calculations(NP)[19]. The last two collumns quote a minimum FWHM ratio on the left, corresponding in the PR case to the perturbative SC contribution and in the NP case to the entire SC contribution and the sum of the previous number and a strong collision estimate, obtained from Eq.(1) by replacing $\{I - S_a(\rho,v)S_b^{-1}(\rho,v)\}$ by 1. For the OIV 3s-3p line, IS are the predictions of code ISOLINE discussed later on(with the same strong collision convention).

One should note the excellent results of the new nonperturbative calculations. Further, it is of special interest to note the unsatisfactory results of the CC calculations, which appear to have a larger discrepancy with experiment than *any* semiclassical calculation. We make the point about the CC calculations as there are many who may feel that semiclassical calculations have had their day and CC calculations should be the state of the art. Whatever one may wish to believe, CC calculations *systematically underestimate* the electron impact broadening, even for the cases where they should have been the most reliable(low-lying lines and low temperature)[7], but the reasons for this are not clear. If the failure of CC for the 3s-3p discrepancies is attributable to too high a principal quantum number and/or temperature, the BIII 2s-2p results are much less so. This is a highly applied field, and a theory that is exact in principle, but will sometimes fail, *without warning, i.e. without being able to draw a borderline between the cases it can and those it cannot handle correctly,* is not too useful. Having made these remarks about CC calculations, I'd like to make two things clear: First, this is *not* a declaration of war on CC, whose potential cannot be underestimated: rather, it is a call for more work. It is important to pinpoint the exact place where CC and SC NP calculations disagree. Second, as is made aboundantly clear from the comparison with experiment in Table 1, SC is by *no means* speaking from a position of weakness.

4. DEVELOPMENTS IN ION DYNAMICS

4a. Brief Introduction to Ion Dynamics

Ion Dynamics is a subject that in my opinion is obscured by much of the early literature. What should be made aboundantly clear is that the subject is one of *time scales*(as is, indeed, all of the subject of line broadening). It is not that ions are fast or slow that matters; in fact there is a 0 probability density for 0 ionic velocities. The issue is the relative timescales of the ionic field fluctuations versus the inverse of the HWHM of the line. This HWHM is to be calculated by taking into account *all* broadening mechanisms, *including* dynamic ions. Hence whether ions are static or dynamic is determined by the *electrons* as well as other broadening mechanisms[16], such as natural broadening and of course ion broadening itself. As we move to high densities, the electronic(and ionic, but a quasistatic ionic perturbation is ineffective for lines with unshifted components) perturbation becomes strong enough to produce memory loss on shorter time scales, and eventually the timescale is short enough to be shorter than the time scale of ion field fluctuations[3]. It is only in this limit that ions may be treated as quasistatic, because the time of interest is short. Of course there are lines, such as the Lyman β line, which may be computed *as if* ions were quasistatic, except near the line center.

4b. What methods are available

I have recently given a semianalytical solution by means of a collective coordinates approach[3], which is effectively as exact as simulations. This development is important in that it represents a practically exact semianalytic solution to a problem thought unsolvable. The key idea is that the microfield in the ion dynamical case is smooth over the relevant time scale. Hence, knowledge of the microfield(say the electric field for a dipole interaction) at a few time points completely determines the interaction, and, via the Schrödinger equation, the U-matrices and line profile. The idea is *not* equivalent to a polynomial fit to the interaction, since the polynomial fit encounters the time-ordering problem. The results of the collective coordinates method are practically as exact as the exact solution, *configuration for configuration*, for configurations that are indeed smooth. For nonsmooth configurations the method still works, but needs more points and loses its computational efficiency. This is overcome by the FST(Frequency Separation Technique) to be discussed below.

Nevertheless, for todays highly complex experiments, there is a very strong emphasis on *fast and reliable* methods for calculating Stark profiles and in this respect stochastic methods, such as the BID method of Boercker, Iglesias and Dufty[21] and especially the Frequency Fluctuation Model (FFM)[22,23] developed by the group of Marseille, with collaboration from Lee and Klein, is particularly promising. The key point is that these methods -especially the FFM are extremely fast and practical methods, so that it is unlikely that the collective coordinates method will be able to compete with them speedwise. The problem with the BID and FFM is that their theoretical basis is not as solid as the simulation[24-27] or collective coordinate methods and that results derived from these were not always as good. Nevertheless, the FFM's record in practical application is impressive and I cannot help mentioning recent FFM calculations of the He^+ Paschen-α line[28]. These calculations, which are also in excellent agreement with benchmark simulation[29] calculations for this line, reproduce the experimental results[30] with excellent accuracy. This is particularly important for such a complex line as the P_α and this comparison must be considered a triumph for the FFM on the one hand and a serious failure of the Model Microfield Method(MMM)[31] on the other, since MMM[32] calculations underestimate the width by a factor of 2.

4c. The MMMs burial ceremony

Based on experience with hydrogen lines, as far back as 1994, cases where the MMM would fail by more than 30% were unknown to the author[3,33]. On the other hand, MMM calculations for charged emitters had not been attempted.(In fact, there is *no sound scientific reason* for using the MMM for ion emitters, since the BID is a much better MMM variant). Today, such calculations, based on an unsophisticated extension to charged emitters[32,34] have been performed. By comparing these results to benchmark calculations[35], 30% errors in the line widths have been found for the MMM. These benchmark calculations have required overcoming[36] a technical problem, namely that each perturber has its own "time" for a hyperbolic parametrization by new inversion formulas. The Relaxation Theory[37] has also been shown to have problems towards the impact limit. MMM discrepancies of factors of 2 have also been found with respect to benchmark calculations[25] for the L_α line of H-like Argon and the He^+ P_α line mentioned before with respect to both benchmark calculations and experiment. The *key* factor which weighs heavily against the MMM as a viable method for calculating lineshapes from charged emitters is that in the last two comparisons these discrepancies were found quite far from the impact regime, which means that the FST idea cannot cure these problems completely. The MMM had been fairly successful for hydrogen lines in the early days of the ion dynamic research, when no other method was available. For neutral hydrogen, it may still be fairly reliable; however, the present cases demonstrate clearly its limitations as a method for charged emitters. It may be interesting to note that in all three cases considered, the MMM underestimated the width. Problems with the Relaxation Theory[37](which overestimates the width at low densities) are not similar, as these problems had to do with the transition to the impact regime, and may thus be solved by the new Frequency Separation Technique(FST)[38].

As discussed in detail later on, the problems of *all* methods designed for ion dynamics, except the MMM, which also has problems in other regimes, have to do with the transition to the impact limit and the new FST[38] is designed to take care of exactly this problem.

4d. FST[38]

The FST (Frequency Separation Technique) concept comes from the following observation: The line profile is determined(via the Schrödinger equation) by the plasma-emitter interaction, V(t). V(t) is due to *both* electrons and ions, yet in line broadening calculations one almost always calculates their contributions separately. It must be emphasized that *there is no mathematical justification for this step*: The usual *intuitive* explanation is that for the vast majority of practical cases, either electrons or ions dominate the broadening, so that this separation should not introduce a large error[39]. In fact, several tests of the electron-ion separation in a parameter range where electron or ion broadening was dominant have given excellent practical results. There remains the question of what happens when electron and ion broadening are comparable. To illustrate the separation hypothesis for comparable electron and ion broadening effects we show in Fig.3 a calculation of the H_α line emitted from a plasma with electron density of $4\times10^{17}cm^{-3}$ and temperature of 1 eV, parameters chosen to give a comparable electronic and ionic contribution. Fig.3a presents C(t), resulting from a calculation of the emitting atom perturbed by a)electrons only(dashed-dotted), b)protons only(dotted) and c) electrons and protons(solid line). Fig.3b shows the corresponding line profiles. The electron and ion broadening contribute almost equally to the line shape as is evidenced by their individual C(t). The product of these individual correlation functions is represented by the dashed line, and it can be seen that it is in good agreement with the correlation function calculated with both electrons and ions. This calculation indicates that the electron-ion separation is a valid assumption *even* in those rare cases where ions and electrons have comparable contributions, and where the separation would seem to be least justified.

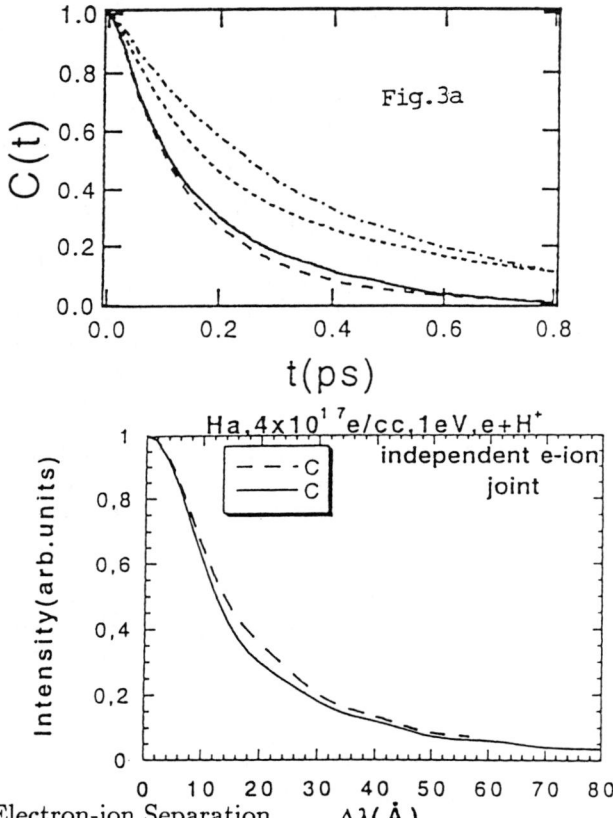

Fig.3: Electron-ion Separation

The main difference between the ionic and electronic fields is their fluctuation frequency. In this regard, the electron-ion separation is, in effect, a frequency separation of the total field into fast and slow(not necessarily quasistatic) components and we have just seen that the separation appears to be valid for all regimes.

There are two implications in the electron-ion separation: First, *to the extent that* electrons may be treated in the impact approximation(IA), and to the extent that the IA is a solved problem, their part is *solved* and we only need a convolution of the ionic and electronic profiles.

The second implication of the electron-ion separation is that if we were allowed to decompose the total electron-ion field into a fast electron part and a slow ionic part, can we do the same with the ion field alone? In other words, if $C_{if}(t)$ and $C_{is}(t)$ denote the autocorrelation functions(AF) due to the interaction of the emitter with the fast and slow parts of the ionic field respectively(we will specify below what constitutes a fast or a slow component), and $C_i(t)$ is the ionic AF, can one write

$$C_i(t) \approx C_{if}(t)C_{is}(t)? \qquad (2)$$

It can be shown analytically[38] by means of a *Gedankencalculation* that this separation is valid in two cases: 1) the case where either the fast or slow component is dominant and also 2)the case of short times, where C(t) is close to unity and the relative error is small, in agreement with the results of the numerical simulations. Of course, for very long times, the slow component is also well described by the impact approximation and hence the separation is also valid.

This fast-slow separation is in fact quite general and applies both to the electron-ion separation and to the fast-slow separation of the ion component. In a practical test[3] of the frequency separation on the *ionic component alone*, excellent results were obtained; however, this test was done for a parameter range where the slow component was dominant. There remains the question of what happens when the fast and slow components are comparable, and this is a more important question in the context of the fast-slow ion field separation, because there is no neutrality constraint relating the fast and slow components, as in the electron-ion separation. That is, it is clear that the FST is correct at short times and when either the fast or slow component is dominant; however, it is less clear when these components are comparable. Although smoothness considerations imply that the error introduced by the FST should be small, this proposition must be confirmed by a calculation.

Fig.4 illustrates the accuracy of the FST applied to the ion broadening in a parameter range where the fast and slow ion components are comparable, i.e. Lyman α transition from a plasma of singly ionized argon perturbers in a 1.38 eV, $1.2 \times 10^{16} cm^{-3}$ plasma. Using the same methods as for the calculation in Fig.3, we have calculated in Fig. 4a the autocorrelation function C(t) due to a) the fast ionic field component only(dotted line), b) the slow component only(dashed) and c) both fast and slow component(i.e. the net ionic) (solid line). Also shown is the product of a) and b)(dash-dotted line). Fig.4b shows the corresponding line profiles. We note once again that the frequency separation seems to work very well: there are *no* logarithmic scales in *any* of these comparisons.

Thus far we have said nothing on what constitutes a "fast" or "slow" component. The key point is that the fast component must be chosen so that it *is* correctly treated by the IA. The IA is valid[9] for perturbers with $v/\rho \gg HWHM$, where v and ρ are the particle velocities and impact parameters respectively. The way to "chop off" the fast component is to determine a $v/\rho = \Omega \geq HWHM$; perturbers involving higher frequency components(i.e. higher v/ρ) belong then to the fast component. Technically speaking, the rigorous way to do this is as follows: Assuming that we have two fast and reliable routines, one to predict the width contribution of the slow components with frequencies $\leq \Omega$ (such a method would be the FFM) as a function of Ω and one to predict the width contribution of the fast components with frequencies $\geq \Omega$ as a function of Ω(i.e. a modified impact routine), it only takes a couple of iterations to achieve a good Ω determination. However, simplifications may be possible in a number of cases.

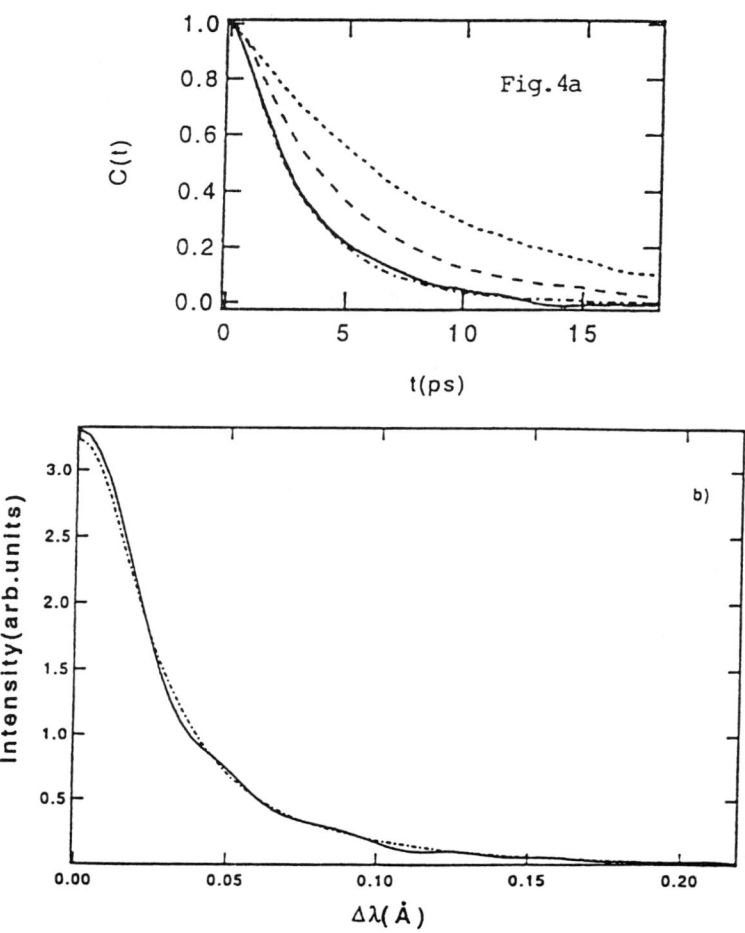

Fig.4: Ionic fast-slow separation

It should be stressed that we *chose* to separate out the impact fast component, not only because the impact limit is a *known* limit, but also because slower, non-impact components are correlated and may *not* be frequency-separated(otherwise one could continue by separating the slow component into a faster and slower part): Only components varying rapidly on the memory loss time scale may be separated out. An important implication of this fact is that at high densities a nonnegligible part of the electrons is no longer impact and becomes correlated with the ionic field.

4e. Explanation of the convergent theory results

At this point I would like to make a connection to a recent important paper by Ispolatov and Oks[40], which has justly made some waves in the field. They showed analytically that by treating exactly the component of the electronic electric field that is parallel to the quasistatic ion field, they obtain significantly different results from the usual standard theory at high densities. This was a shock to the community, which had at least since 1962[41] accepted that the impact electrons do not couple to the quasistatic ion field. With hindsight, there should have been no shock at all(just like ion dynamics should have been obvious from the very beginning!): As we just said, at high densities, a nonnegligible part of the electron distribution is not treatable as impact on the memory loss time scale and thus becomes correlated with the ion field.

4f. What does one gain with the FST?

We now come to the implications of the frequency separation for line broadening calculations: With the FST the ionic calculation consists of computing the impact contribution(fast part) and the slow part contribution. It is clear that in this way the ion impact limit is recovered *exactly and in as sophisticated a form as desired*: For example, one may use the perturbative IA[9], the *exact* semiclassical dipole results[42] (for hydrogenic emitters) or fully non-perturbative results including higher multipole effects[19]. Further, this separation of the ionic field into fast(impact) and slow components improves the performance and/or reliability of *every* method capable of dealing with the intermediate(ion dynamical) regime, so that the method has only to compute the contribution of the *slow* part: Simulations[24-27] have problems concerning their efficiency toward the impact regime, both because a large number of particles needs to be simulated and because the Schrödinger solver has to integrate over sharp peaks. Such peaks are due to fast perturbers, which are correctly included in the impact part and do not appear in a simulation of the slow part. For the collective coordinate method[3], this separation is very important for the efficiency of the method, which would otherwise have to solve large linear systems. Finally, stochastic methods have problems in attaining the ion impact limit. For the FFM, the impact limit is not a limit of the formalism[23] and is not formally recovered, since this is not a simple component mixing effect. The BID[21] and MMM[31], can only recover the *dipole perturbative* impact limit(which may often be inaccurate[43]). Experience with the BID is very limited in this respect, but the MMM, even for hydrogen, where is works reasonably well, can only recover the impact limit with a "delay"(which may be related to the fact that it can achieve the perturbative impact limit rather than the time- ordered impact limit[42]).

In the FFM (and also BID) all the time variation of the microfield is modelled via the covariance of the electric field

$$\Gamma_{EE}(t) = \langle \mathbf{E}(t) \cdot \mathbf{E}(0) \rangle \tag{3}$$

With the frequency separation, it is only the slow part of this covariance that will appear in the FFM(or BID); the fast part *has been chosen* so that it is treatable by the IA. An explicit example[38] of such a truncated covariance, is the covariance computed in the Debye-shielded model of Brissaud et.al.[42]. (one may note that the slow covariance[38] does *not* reduce to the result of Ref.[44] as $\Omega \to \infty$, due to the contribution of more or less "head on" collisions, which are, however, correctly included in the fast part).

The most important *practical* application is that with the frequency separation coupled to the FFM we have an ultrafast robust method, valid for *all parameter ranges and all types of emitters*, a development of great importance for plasma spectroscopy. We note that the combination of the FFM and FST will provide a unified formulation for all emitters, neutral or charged for all plasma conditions. This now moves the focus of the outstanding problems to the computation of the strong interactions to the impact limit, which is a subject of much current interest. In other words, things have now changed dramatically, in that we have fast and reliable methods to calculate the slow part and it is the impact part for which we do not have comparably general, fast and reliable methods!

5. STANDING OF STARK BROADENING-DIRECTIONS FOR THE FUTURE
5a. Methods and Computer Codes for the Users Community

The needs of the plasma spectroscopy community(and with this I mean people working with experiments on ICF, short-pulse lasers, opacities, kinetics etc.) are variable. Often, e.g. for radiative transfer simulations, an estimate of Stark widths will be sufficient, while other times (mainly individual line diagnostics) more accurate widths will be required. To respond to these demands, the line-broadening community has and continues to produce methods to suit both needs. In doing so, we have often succeeded in combining convenience and speed with accuracy. Such examples are the FFM and the analytical formulas[45] for isolated ion widths, both of which are undergoing new improvements. At the present time, we have methods (and, equally importantly, computer programs) to suit every need: For fast, reliable width calculations within the dipole approximation, the code ISOLINE based on the analytical formulas of Alexiou and Maron(AM)[45] does very well, obtaining agreement with experiment that is usually better than 20%. When this fast code does not work, this is due to quadrupole effects. The fully nonperturbative code STARCODE gets good agreement. ISOLINE can be extended to include hydrogenic quadrupole channels, at essentially no extra cost to the calculation. For general calculations, the PIM/PAM/POUM code, based on the FFM for the ionic treatment and the AM broadening operator for the electron impact contribution, is a general code, capable of handling *any* line,

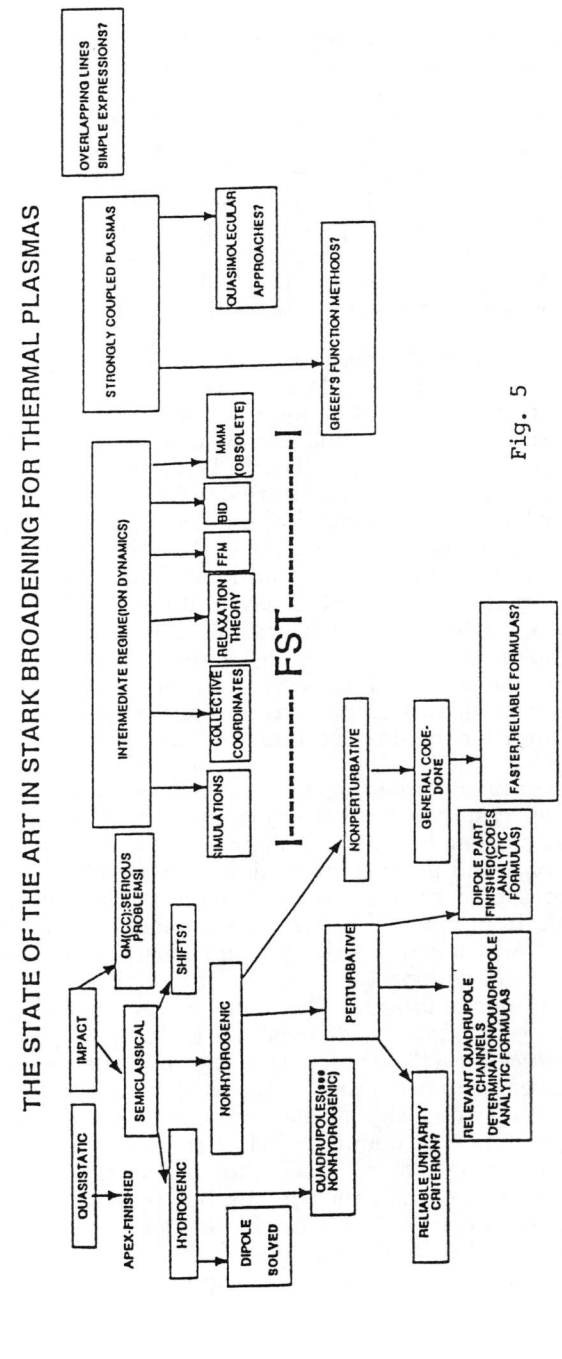

Fig. 5

in a very fast and reliable manner, and includes ion dynamics. In the future, the addition of the FST improvement should make it even more reliable. Finally, for benchmark calculations, a fully exact simulation code MARA, valid for *any* line, is available. This code employs a number of speedups, such as symmetry relations[46] to reduce the number of systems to be solved, and extrapolations, to reduce the computer time. MARA can perform independent particle simulations, or, it can accept electric field histories from a MD simulation. Simpler simulation codes(HSTRK and HSTRKZ) exist for hydrogen-like lines, which employ the Gigosos et.al.[47] group-theoretical solution. The advantage here is that these are more efficient for hydrogen-like species without fine structure and in the dipole approximation.

5b. The situation today: closed subjects and remaining Fronts

I would like to finish this talk with Fig.5 showing what, in my opinion, is the situation today in plasma spectroscopy of thermal plasmas and what the most important outstanding problems and frontiers are: Starting with the quasistatic theory[9], this is now a finished subject, at least practically, since the appearance of APEX[48]. With regard to the impact theory: For hydrogenic emitters, and within the context of a dipole interaction, the problem is solved in the semiclassical approximation. We note that CC calculations by Unnikrishnan and Callaway[49] have not settled any questions. Thus, for hydrogenic emitters calculations should take notice of the nonperturbative exact dipole semiclassical solutions[42], which make a difference compared to perturbative calculations. With regard to the quadrupole and higher multipoles, we find ourselves in the same calculational positions as for the nonhydrogenic lines, with the exception that quadrupole effects are more pronounced for nonhydrogenic emitters than for hydrogenic ones, since the collision operator for the dipole interaction decays with energy separation[16].

One of the controversial topics in the field is the issue of shifts. I do not believe that, at least at present they can in general be calculated reliably by any sort of an independent-particle approach(this includes independent particle simulations, and perhaps even Molecular Dynamics simulations) for a very simple reason: For widths we are looking at the difference of a quantity, possibly small, from unity. For shifts, we are looking for the differences of a small quantity, which can in principle change sign, from zero. Hence, shift calculations should address the question of sensitivity of the results to possible errors and uncertainities. The point is that $\mathrm{Im}\{I - S_a S_b^*\}$, the quantity that needs to be integrated over velocities and impact parameters to obtain the shift is a *highly oscillatory* function, except at large impact parameters and further a very significant part of the shift contribution comes from small impact parameters; very significant cancellation takes place and thus even the *sign* of the shift is in doubt. SC shift calculations can only be justified under the assumption that the oscillations at small impact parameters result in such cancellation that the contribution from large impact parameters dominates. Even if one could get $\mathrm{Im}\{I - S_a S_b^*\}$ right, a very slight deviation from a Maxwellian distribution could even change the sign of the shift. Fig.6 illustrates this point by showing the imaginary part of $\{I - S_a(\rho,v)S_b^{-1}(\rho,v)\}$

for the BIII 2s-2p line from Table 1, calculated by allowing only dipole interactions(unlike the results in Table 1). That said, I certainly find nothing strange in encountering shifts in plasmas and Green's function[50,51] approaches may be able to address these problems in a general fashion.

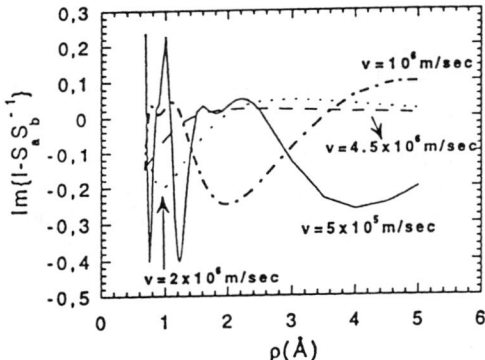

Fig.6: BIII 2s-2p shift integrand behavior

The subject of strongly coupled plasmas is only now starting to be explored. For a variety of reasons, there is strong interest in the radiative properties, including lineshapes, of strongly coupled plasmas today and experiments have started probing this new area. The difficulties in the field is that we do not have independent diagnostics, analogous to the Thomson scattering diagnostics of the Bochum group, and, moreover, we usually have highly transient, nonhomogeneous plasmas with various propagating shocks.

I believe that the two kinds of theoretical studies initiated, Green's functions approaches[50-53] and quasimolecular approaches[54-58] are important ways of dealing with this new subject. The Green's functions approaches are a complete microscopic theory and capable of treating not only quantum effects, but to go beyond the usual scattering theory and consider scattering in a disordered medium; the problem is, that their practical results are still sometimes unsatisfactory, mainly with regard to the treatment of strong overlapping collisions, as in the ionic case; more specifically, they often fail to recover well known results from the weak-coupling regime. For example, dissapointing results have been obtained in the recent paper by Seidel, Arndt and Kraeft[53], which predicts shifts larger than the widths and in marked disagreement with experiments. Similar problems have been seen in other papers[52]. It is hoped that future improvements will make this a reliable method.

A different approach, starting from experimental results, where sattelite-like structures have been found in strongly coupled plasmas, is being followed by E.Leboucher's group[56-58], where 2-center structures are considered, following previous theoretical work by Malnout, d'Etat and Nguyen[54] and by Salzmann, Stein and coworkers[55]. The idea behind this model is that by using a set of molecular rather than atomic states as basis, the static effects of the nearest neighbor interactions are treated correctly and automatically to all orders, with proper account of quenching and charge exchange[59]. Clearly this model is limited to a specific range of plasma parameters that will enable, for example, neglect of the next nearest neighbor and so on; A lot of work has been done on this recently[58].

6. CONCLUSIONS

The major recent advances in Stark broadening have given definitive answers to a number of formerly open questions. Reliable methods and computer codes have been developed to suit every need. Important unresolved problems in the field involve understanding the origin of the CC vs. experiment discrepancies and establishing more definitive results for the newly emerging field of lineshapes in strongly coupled plasmas. Experiments to probe this domain[56,60] and theoretical work will hopefully soon lead to a better understanding in this remaining big frontier.

Aknowledgements

The work described here was done during my stay at the Weizmann Institute, The Universite de Provence-Marseille and Universite Paris VI and Ecole Polytechnique and I would like to thank Y.Maron, R.Stamm, B.Talin and E.Leboucher for their hospitality and fruitful discussions and R.W.Lee for his support. It is a pleasure to aknowledge Yu.Ralchenko, who provided the atomic data for the electron impact calculations and Dr.R.D.Cowan for useful discussions on his atomic structure code.

REFERENCES

1. E.Oks, these proceedings.
2. S.Alexiou, A.Weingarten, Y.Maron, M.Sarfaty and Ya.E.Krasik, Phys.Rev.Lett.**75**, 3126(1995) and these proceedings.
3. S.Alexiou, J.Quant.Spectrosc.Radiat.Transf.**54**,1(1995).
4. S.Glenzer, N.I.Uzelac and H.J.Kunze, Phys.Rev. **A45**, 8795(1992).
5. N.I.Uzelac, S.Glenzer, N.Konjevic, J.D.Hey and H.J.Kunze, Phys. Rev. **A47**, 3623 (1993).
6. S.Glenzer, J.D.Hey and H.J.Kunze, J.Phys.B. **27**, 413(1994).
7. S.Glenzer and H.J.Kunze, Phys.Rev.A**53**, 2225(1996).
8. Th.Wrubel, S.Glenzer, S.Büscher, H.J.Kunze and S.Alexiou, Astron. Astrophys. **306**, 1023(1996); Th.Wrubel, these proceedings.
9. H.R.Griem, *Spectral Line Broadening In Plasmas* (Academic, New York, 1974).

10. M.Dimitrijevic and N.Konjevic, JQSRT **24**, 451 (1980).

11. J.Puric, M. Cuk, M.S.Dimitrijevic and A. Lesage, Ap.J.**382**, 353(1991).

12. H.R.Griem, Phys. Rev. **165**, 258(1968).

13. M.J.Seaton, J.Phys.B**21**,3033(1988).

14. M.J.Seaton, J.Phys.B**20**, 6363(1987).

15. J.D.Hey and P.Breger, S.Afr.J.Phys.**5**, 111(1982); J.D.Hey and P.Breger, JQSRT**23**, 311(1980); ibid.,**24**, 349(1980); J.D.Hey et.al., J.Phys.B**23**, 241(1990).

16. S.Alexiou, Phys.Rev. A**49**, 106(1994).

17. S.Alexiou and Yu.Ralchenko, Phys.Rev.A**49**, 3086(1994); ibid. **50**, 3552(1994)

18. H.R.Griem, M.Baranger, A.C.Kolb and G.Oertel, Phys.Rev.**116**, 4(1959).

19. S.Alexiou, Phys.Rev.Lett.**75**, 3406(1995).

20. A.Poquérusse, S.Alexiou and E.Leboucher, to appear in JQSRT.

21. D.B.Boercker, C.A.Iglesias and J.W.Dufty, Phys.Rev.A**36**, 2254(1987).

22. A.Calisti, L.Godbert, R.Stamm and B.Talin, JQSRT **51**, 59(1994).

23. B.Talin, A.Calisti, L.Godbert, R.Stamm, L.Klein and R.W.Lee, Phys.Rev. E **51**, 1918(1995).

24. R.Stamm, E.W.Smith and B.Talin, Phys.Rev.A**30**, 2039(1984).

25. R.Stamm, Y.Botzanowski, V.P.Kaftandjian, B.Talin and E.W.Smith, Phys. Rev. Lett. **52**, 2217 (1984); R.Stamm, B.Talin, E.Pollock and C.A. Iglesias, Phys. Rev.A **34**, 4144 (1986).

26. G.C.Hegerfeldt and V.Kesting, Phys.Rev.A**37**, 1488 (1988).

27. J.Seidel, in *Spectral Line Shapes*, Vol.6, edited by L.Frommhold and J.Keto, AIP Conf. Proc. No. 216 (AIP, New York, 1990).

28. T.Meftah, S.Alexiou, A.Calisti, L.Godbert, R.Stamm and B.Talin, "The FFM applied to the H-like P_α line", these proceedings.

29. Calculations done by Code MARA.

30. S.Büscher et.al., to appear in J.Phys.B

31. A.Brissaud and U.Frisch, JQSRT **11**, 1767(1971).

32. C.Stehlé, Astron.Astrophys.**292**, 699 (1994).

33. J.Seidel, in *Spectral Line Shapes*, editor B.Wende (de Gruyter, Berlin, 1981).

34. C.Stehlé, in *Radiative Properties of Hot Dense Matter*, W. Goldstein, C. Hooper, J.C.Gauthier and R.W.Lee, editors. World Scientific, Singapore(1990).

35. S.Alexiou et.al., to be submitted.

36. A.Poquérusse, S.Alexiou and E.Klodzh, JQSRT**56**, 153(1996).

37. R.L.Greene, Phys.Rev.A**14**, 1447(1976); JQSRT**30**, 409(1983); D. H. Oza, R. L. Greene and D.E.Kelleher, Phys.Rev.A**37**, 531(1988).

38. S.Alexiou, Phys.Rev.Lett.**76**, 1836(1996).
39. J.Seidel, private communication.
40. Y.Ispolatov and E.Oks, J. Quant. Spectrosc. Radiat. Transf.**51**,129 (1994).
41. M.Baranger, in Atomic and Molecular Processes, D.R.Bates, editor, Academic, 1962.
42. H.Pfennig, JQSRT**12**, 821 (1972); V.S.Lisitsa and G.V.Sholin, Sov. Phys. JETP **34**, 484(1972)).
43. V.Kesting, in *Spectral Line Shapes*, Vol.6, edited by L.Frommhold and J.Keto, AIP Conf. Proc. No. 216 (AIP, New York, 1990).
44. A.Brissaud et.al., J.Phys.B**9**, 1129(1976).
45. S.Alexiou and Y.Maron, JQSRT**53**, 109(1995).
46. S.Alexiou et.al., to be submitted.
47. M.A.Gigosos, J.Fraile and F.Torres, Phys.Rev.A**31**, 3509 (1985).
48. C.A.Iglesias, H.E.DeWitt, J.L.Lebowitz, D.McGowan and W. B. Hubbard, Phys. Rev. A**31**, 1698(1985).
49. K.Unniskrishnan and J.Callaway, Phys.Rev.A**43**, 3619(1991).
50. S.Günter, Phys.Rev.E**48**, 500(1993).
51. S.Günter, these proceedings.
52. M.W.C.Dharma-wardana et.al., Phys.Rev.A **21**, 379(1980).
53. J.Seidel, S.Arndt and W.D.Kraeft, Phys.Rev.E **52**, 5387(1995).
54. Ph.Malnout, B.d'Etat and Hoe Nguyen, Phys.Rev.A**40**, 1983 (1989).
55. J.Stein et.al., Phys.Rev.A**39**, 2078(1989); D.Salzmann et.al., Phys.Rev.A **44**, 1270 (1991).
56. E.Leboucher-Dalimier, A.Poquérusse and P.Angelo, Phys.Rev.E **47**, R1467 (1993).
57. E.Leboucher-Dalimier, 12th ICSLS, editors A.D.May, J.R.Drummond and E.A.Oks, AIP Conference Proceedings No. 328, New York, (1995).
58. P.Gauthier, P. Sauvan, P. Angelo, S. Alexiou, E. Leboucher-Dalimier, A. Calisti et B. Talin, "Quasimolecular features in hot dense plasma emission", these proceedings.
59. Le Quang Rang and D.Voslamber, J.Phys.B.**8**, 331(1975).
60. D.J..Heading, J.S.Wark, G.R.Bennett and R.W.Lee, JQSRT**54**, 167(1995).

Shifted and asymmetric profiles of hydrogen and hydrogenic ion lines

Sibylle Günter and Axel Könies*

Max-Planck-Institut für Plasmaphysik, Boltzmannstr. 2, 85748 Garching, Germany
* Teilinstitut Greifswald, Koitenhäger Landstraße, 17491 Greifswald, Germany

Abstract. A quantum statistical many-userdefs.texparticle theory has been applied to calculate the profiles of hydrogen and hydrogenic ion lines. Whereas the plasma electrons are treated quantum mechanically, for the dynamic ions, the well-known model microfield method has been used. The evaluated line profiles are shifted and asymmetric. Besides the electronic contributions to the line shift, the shift due to the inhomogeneities of the ionic microfield as well as that due to the quadratic Stark effect have been included. Calculating the profiles of hydrogenic ion lines, also fine structure splitting has been taken into account.

INTRODUCTION

As it is well known, due to the interaction between the radiators and the plasma environment, spectral lines emitted from dense plasmas are broadened, shifted and asymmetric. A lot of theories have been developed to describe the profiles of hydrogen lines. Including ion dynamics via computer simulations, an excellent agreement between theoretical and experimental line widths has been reached. The deviations for the line shifts and, especially, for the asymmetry parameters are mostly larger. The reasons for these discrepancies are twofold.
One of them is the treatment of strong electron-radiator collisions to the line shift. Line shift calculations mostly are based on a Born approximation for the radiator-perturber interaction. In order to avoid an overestimation of the strong collision contributions (or even divergences), usually, one applies a cutoff procedure [1, 2]. The resulting line shifts depend strongly on the choice of this parameter. Some years ago, line shift calculations based on the phase shifts for elastic scattering of the perturbing electrons at excited atomic states have been carried out by Unnikrishnan and Callaway [3, 4]. Since they

included explicitly only the states 1s, 2s, 2p, 3s, 3p, 3d in the corresponding close-coupling equations, the resulting shifts are too small, exept for the L_α-line.

The second reason for discrepancies is the definition of the line shift. Often the theoretical shift simply is regarded as a superposition of electronic and ionic contributions to the line shift [1,5]. For comparison with experiments, however, all contributions to shift and asymmetry should be included in the line profile calculations in such a way that shifted and asymmetric line profiles result.

The aim of this paper is to calculate such line profiles which allow to determine shifts and asymmetry parameters in the same manner as from experimental data. The theoretical approach applied here is based on a Green's function technique [6–10]. Thereby, the perturbing electrons are treated quantum mechanically. Contributions of strong collisions between radiator and perturbing electrons are calculated via partial summation of the corresponding T-matrix as described in [9,10].

The influence of the dynamic ionic microfields on the line profile is described via the well-known model microfield method [11–13]. In order to account for shift and asymmetry, however, the time development operator for a constant ionic microfield strength $U(\Delta\omega|\vec{E})$ has been modified. Thus, it contains the electronic shift, the shift due to inhomogeneities of the ionic microfield and the quadratic Stark effect. Furthermore, the trivial asymmetry and that because of the frequency dependence of electronic width and shift [14,15] are included.

THEORY

During the last years, a quantum mechanical many-particle approach to the optical properties of dense plasmas has been developed [6–10]. It has been shown that the shapes of spectral lines are related to the two-particle contribution of the polarization function. For the line profile one finds

$$I(\Delta\omega) \sim \sum_{ii'ff'} I_{if}^{i'f'}(\Delta\omega) \int \frac{d\vec{P}}{(2\pi)^3} e^{-\frac{P^2}{2Mk_BT}} \langle i|\langle f| [U(\Delta\omega)]^{-1} |f'\rangle|i'\rangle \qquad (1)$$

with

$$U(\Delta\omega) = \Delta\omega - \frac{\vec{P}\vec{k}}{M} - \frac{k^2}{2M} - \text{Re}\,\{\Sigma_i - \Sigma_f\} + i\text{Im}\,\{\Sigma_i + \Sigma_f\} + i\Gamma^v. \qquad (2)$$

M denotes here the radiator's mass, k_B is the Boltzmann constant, T the temperature of the heavy plasma particles, and k is the wave number of the emitted radiation. The sum has to be carried out over all transitions which contribute to the spectral line. The intensity

$$I_{if}^{i'f'}(\Delta\omega) = M_{if}^{(0)}(\vec{k})\ [M_{i'f'}^{(0)}(\vec{k})]^*\ I(\Delta\omega). \tag{3}$$

besides the transition matrix elements contains the frequency dependence of the dipole radiation and that resulting from the Boltzmann occupation of the radiator's states. It is responsible for the so-called "trivial asymmetry".

The influence of the surrounding plasma on the radiator is contained in the self energy Σ. Its real part corresponds to the shift of the concerning energy value, whereas its imaginary part gives the width.

For the plasma conditions which are of interest here, it is possible to decouple the ionic and the elecronic subsystems of the plasma. Assuming a static ionic microfield $\vec{E}$, one finds for the self energy

$$\Sigma = \Sigma^{ion}(E) + \Sigma^{el}(E, \Delta\omega). \tag{4}$$

Whereas the ionic part of the self energy describes the influence of the static ionic microfields on the radiator, the electronic contribution depends on both, the ionic microfield and the detuning $\Delta\omega$ from the line center. Since the main contributions to the self-energy arise from the upper level of the radiating transition, only the frequency dependence of that level has been taken into account.

We want to account for ion dynamics applying the well-known model microfield method. Therefore, instead of the time development operator (2) we use that of a Kangaroo-process, see [12]

$$\begin{aligned}\langle U(\Delta\omega)\rangle_{KP} &= \langle U(\Delta\omega|\vec{E})\rangle_s + \\ &+ \langle \Omega(E)\,U(\Delta\omega|\vec{E})\rangle_s [\langle \Omega(E)\rangle_s - \langle \Omega^2(E)\,U(\Delta\omega|\vec{E})\rangle_s]^{-1} \times \\ &\times \langle \Omega(E)\,U(\Delta\omega|\vec{E})\rangle_s.\end{aligned} \tag{5}$$

Since we are interested in shifted and asymmetric line profiles, the time development operator for a constant ionic microfield as given in [13], however, will be modified here to be

$$\begin{aligned}U(\Delta\omega|\vec{E}) &= [\,\Delta\omega - \frac{\vec{P}\vec{k}}{M} - \frac{k^2}{2M} - \mathrm{Re}\,\{\Sigma(\Delta\omega, E)\} + \\ &\quad i\Omega(E) + i\mathrm{Im}\,\{\Sigma(\Delta\omega, E)\} + i\Gamma^v\,]^{-1}.\end{aligned} \tag{6}$$

$\Omega(E)$ is the jumping frequency which is determined by the field autocorrelation function [16].

Now, we have to specify the self energy Σ. As mentioned above, the self energy may be given as the sum of the electronic and the ionic parts. Applying a quantum mechanical many-particle theory, the electronic contribution to the self energy has been evaluated in [10]. Restricting to a second order Born approximation one finds

$$\langle i|\Sigma(E_i^0 + \Delta\omega, \beta)|i\rangle = -\frac{1}{e^2}\int \frac{d\vec{q}}{(2\pi)^3} V(q) \sum_\alpha |M_{i\alpha}^{(0)}(\vec{q})|^2 \qquad (7)$$

$$\times \int_{-\infty}^{\infty} \frac{d\omega}{\pi}[1 + n_B(\omega)] \frac{\mathrm{Im}\,\varepsilon^{-1}(\vec{q}, \omega + i0)}{E_i^0 + \Delta\omega - E_\alpha(\beta) - (\omega + i0)}.$$

Here, $n_B(\omega)$ is the Bose function

$$n_B(\omega) = \frac{1}{e^{\frac{-\omega}{k_B T}} - 1}, \qquad (8)$$

and β denotes the Holtsmark normalized field strength.
The matrix elements $M_{n\alpha}^{(0)}$ are given by

$$M_{i\alpha}^{(0)}(\vec{q}) = \int \frac{d\vec{p}}{(2\pi)^3} [\Psi_i^{ei}(\vec{p})]^* [Z_0 e \Psi_\alpha^{ei}(\vec{p}) - e \Psi_\alpha^{ei}(\vec{p} + \vec{q})], \qquad (9)$$

where Ψ^{ei} are the radiator's wave functions, and Z_0 is the nuclear charge of the radiator.
In principle, the α-sum has to be carried out over all bound and scattering states of the radiator. We restrict ourselves here, however, to those states which principal quantum numbers n_α are within

$$n_i - 1 \leq n_\alpha \leq n_i + 2,$$

where n_i is the principal quantum number of the considered state. Many particle effects are contained in the inverse dielectric function ε which has been taken within the random-phase approximation.
Instead of using a Born approximation, via partial summation of the corresponding three particle T-matrix for the perturber–radiator–interaction strong collisions between the radiator and a perturbing electron can be included in a systematic manner [9, 10]. In principle, that leads to a cutoff of the integral (7) for large transition momentums q.
Taking into account the ionic contributions to the self energy, besides the well-known linear Stark effect also the quadrupole interaction between the radiator

and inhomogeneities of the ionic microfield has been considered. Within parabolic states, for the resulting ion quadrupole shift one finds

$$\Sigma^q_{ii'}(\beta) = \begin{cases} \frac{\pi}{3}a_0^2 e^2 n_e (n^i)^2 \sqrt{n_1^i(n^i - n_1^i)(n_2^i + 1)(n^i - n_2^i - 1)} B_\rho(\beta) & : \begin{array}{l} n_1^i = n_1^{i'} - 1 \\ n_2^i = n_2^{i'} + 1 \end{array} \\ \frac{\pi}{3}a_0^2 e^2 n_e (n^i)^2 \left((n^i)^2 - 1 - 6(n_1^i - n_2^i)^2\right) B_\rho(\beta) & : i = i' \\ \frac{\pi}{3}a_0^2 e^2 n_e (n^i)^2 \sqrt{(n_1^i + 1)(n^i - n_1^i - 1)n_2^i(n^i - n_2^i)} B_\rho(\beta) & : \begin{array}{l} n_1^i = n_1^{i'} + 1 \\ n_2^i = n_2^{i'} - 1 \end{array} \end{cases}$$
(10)

The function $B_\rho(\beta)$ has been given by Halenka [17]. It is a generalization of the Chandrasekhar- von Neumann function $B(\beta)$ [18] introduced by Demura and Sholin [19] into line broadening theory. In difference to that function, it takes into account the screening of the ionic microfield by the plasma electrons as well as ion pair correlations.

Further, the quadratic Stark effect has been included approximately using the well-known formula [20]

$$\Sigma^k_{ii}(\beta) = -\frac{1}{16}(n^i)^4 \left(17(n^i)^2 - 3(n_1^i - n_2^i)^2 - 9(m^i)^2 + 19\right)\beta^2 E_0^2.$$
(11)

RESULTS FOR HYDROGEN LINES

The theory outlined above has been applied to calculate shifted and asymmetric profiles of hydrogen lines (up to an upper principal quantum number of 5). This way we are able to employ the same definitions for the resulting shifts and asymmetry parameters as it has been done in corresponding experiments. In order to determine the shifts of spectral lines from experimental profiles, various definitions have been applied, among them the maximum shift (for lines with a central Stark component) and the dip shift (for lines without a central component), respectively. Wiese introduced the shift of the estimated line center (ELC shift) [21]. Further, sometimes the center of mass shift

$$\Delta = \int_{-\Delta s}^{\Delta s} P(\Delta\lambda) d(\Delta\lambda)$$
(12)

is applied. Due to the asymmetry of the spectral lines, the resulting shifts depend strongly on the applied shift definition.

Lyman Lines

Whereas for the Balmer lines theoretical and experimental shifts mostly are in reasonable agreement, for long time there was a large discrepancy for L_α and L_β. For example, the experimental shift of the L_α line was given by Grützmacher [22] to be (5.8 ± 0.6) mÅ for an electron density of about $2 \cdot 10^{17}$ cm^{-3}. The theoretical shift given by Griem, however, was a factor of 2 larger (12 mÅ). One of the reasons for this discrepancy is the different definition of both shifts. Whereas Griem gives the superposition of electronic and ionic shifts, the measured one is a center of mass shift. Our evaluated center of mass shift is given in Fig. 1. It is to be seen that this shift depends on the integration intervals Δs (see Eq. 12).

Figure 1: Center of mass shift Δ of the L_α line vs the integration interval Δs ($n_e = 2 \cdot 10^{17}$ cm^{-3}, T=13200K).

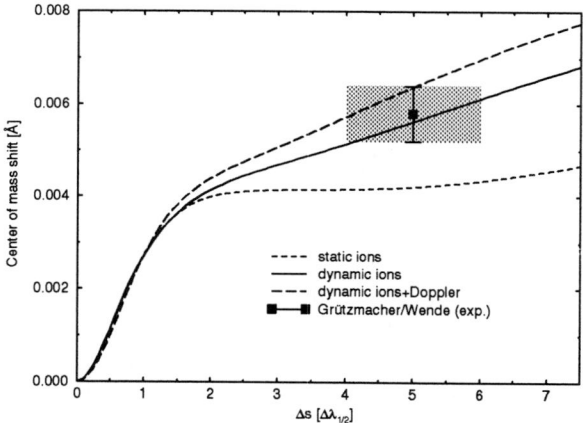

Δs is given in units of the FWHM.

Unfortunately, the experimental values for Δs are not well known. Nevertheless, it can be stated that the large discrepancy between theoretical and experimental shift of the L_α line has been removed. The same is true for the shift of the L_β line.

In Fig. 2 the frequency dependent asymmetry parameter

$$A(\Delta\lambda) = \frac{P^{red}(\Delta\lambda) - P^{blue}(\Delta\lambda)}{P^{red}(\Delta\lambda) + P^{blue}(\Delta\lambda)} \qquad (13)$$

of the L_β line is to be seen. There is a very good agreement in the line wings, where the plasma ions may be considered to be static. The larger difference

Figure 2: Asymmetry parameter A vs $\Delta\lambda$ for the L_β line ($n_e = 2 \cdot 10^{17}$ cm^{-3}, T=16000K).

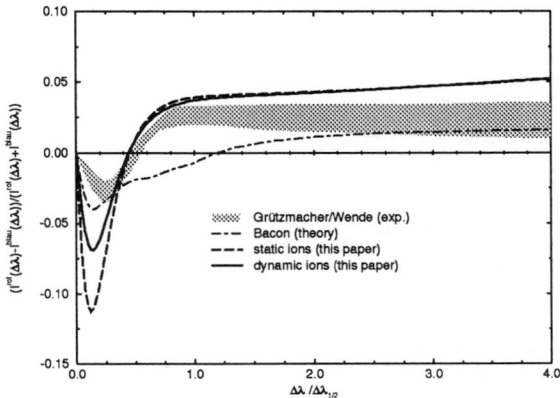

The experimental values are taken from [23], Bacon's data are published in [24].

in the line center is probably a result of an underestimation of ion dynamic effects within the applied MMM.

Balmer Lines

Fig. 3 shows the theoretical half width

$$\alpha_{1/2} = \frac{\Delta\lambda_{1/2}}{E_0}, \qquad (14)$$

of the H_β line compared to experimental results. We can state an excellent agreement between our theory and the experiment.
In Fig. 4, the theoretical asymmetry parameters

$$A = \frac{I^{blue} - I^{red}}{I^{blue}}, \qquad (15)$$

of the H_β-line are compared to experimental and former theoretical results. Of course, there are a lot of experimental data that could be used for comparison. For clarity, however, only a few results are included in Fig. 4. Our theoretical evaluations have been carried out for the same plasma conditions

Figure 3: Width of the H_β line in an argon plasma

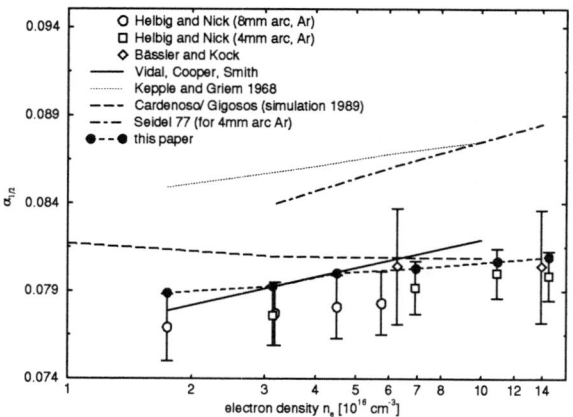

Figure 4: Peak asymmetry of the H_β line vs the electron density n_e

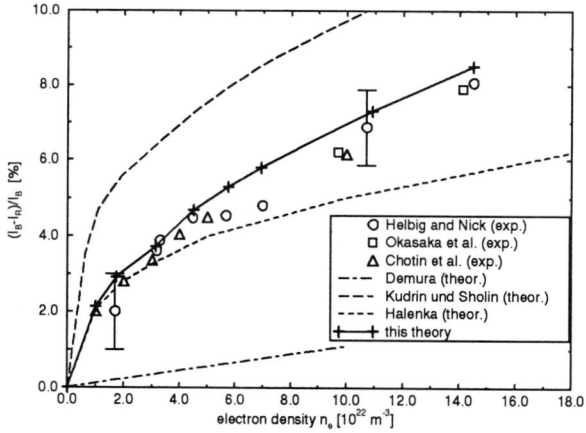

The calculations are carried out for the same temperatures as in the experiment [25]. Further experimental values are taken from [26] and [27]. For comparison theoretical values given by Demura [28], Kudrin/Sholin [29] and Halenka [17] are included.

as they where in the experiment of Helbig and Nick [25]. We find an excellent agreement between the theoretical and the corresponding experimental results. Former theoretical results by Kudrin and Sholin [29] and Demura [28], however, strongly differ from any experiments. For lower electron densities also the theoretical results given by Halenka [17] agree very well with the experimental data.

Figure 5: Shift of the H_β line according to different definitions

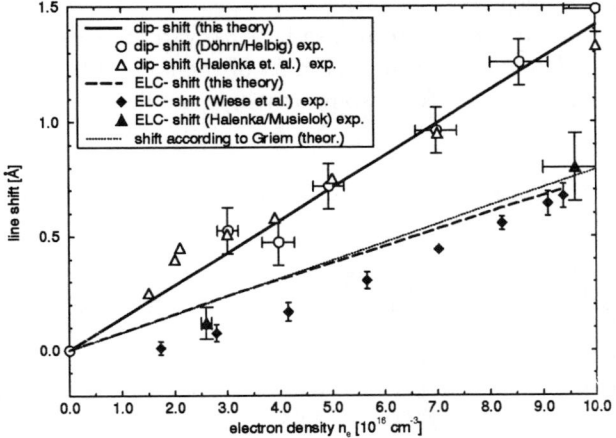

The experimental dip shifts are taken from [30, 31], the ELC shifts are given in [21, 32], and the theoretical shifts given by Griem are taken from [1].

In Fig. 5 the shift of the H_β-line according to various definitions is shown. It becomes evident that different shift definitions yield clearly distinct results in theory as well as in experiments. Since the dip shift, in principle, is given by the sum of the electronic shift contributions ($\propto n_e$) and the ion quadrupole shift ($\propto n_i$) of the two inner Stark components, a linear behaviour of the theoretical shift results. That linear behaviour is in good agreement with the experimental results. For the ELC-shift, however, the different shifts of the outer Stark components become important. They cause a non-linear behaviour with respect to the electron density.

Whereas the theoretical dip shift agrees excellently with corresponding experimental results, the theoretical ELC-shifts are larger than the experimental ones for lower electron densities. As it has to be expected, the theory gives no ELC-shift for vanishing electron density. Assuming, however, a nearly linear behaviour of the ELC-shift for low densities, from experimental values a negative shift for $n_e \to 0$ results.

The shift given by Griem [1,5] agrees very well with our theoretical ELC-shift. However, Griem did not calculate a shift from a complete shifted and asymmetric profile. He obtained his values by a superposition of the electronic contributions to the shift and the shift caused by the ionic microfield at a quarter of the maximum intensity. This purely theoretical shift definition considers the line asymmetry and appears to be a good approximation for the ELC-shift.

Due to the central component, the H_α line is much less asymmetric than the H_β line. Therefore, the influence of the various shift definitions is less important. In the table we compare our theoretical shifts with various experimental results. The agreement with the experimental shifts of the line center is excellent whereas the ELC shift appears to be somewhat too small.

Table. Shifts of the H_α-line for different experimental conditions and shift definitions

	$n_e(10^{16}cm^{-3})$	$T(K)$	$\Delta\lambda_{theo}(\text{Å})$	$\Delta\lambda_{exp}(\text{Å})$	
max. shift					
	9.0	12600	0.44	0.41 ± 0.06	[33]
	10.0	12000	0.48	0.43	[34]
	100.0	62700	5.35	5.33 ± 0.3	[35]
shift $\frac{1}{2}$ max					
	8.8	13000	0.38	0.38 ± 0.4	[36]
ELC- shift					
	9.0	12600	0.41	0.47 ± 0.08	[33]
	9.21	13000	0.42	0.52	[21]

Paschen lines

The Figs. 6 and 7 show the shift of the P_α and the P_β line, respectively. It becomes evident that many-particle effects are relevant for the P_α shift at higher electron densities. A good agreement with the experimental data only can be reached with the inclusion of dynamic screening.

Figure 6: Shift of the P_α line

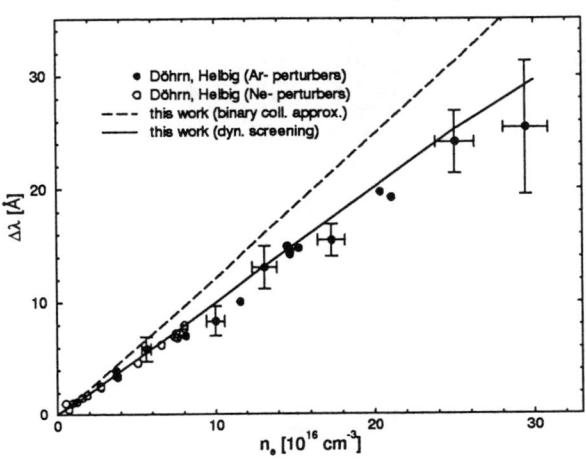

Figure 7: Shift of the P_β line

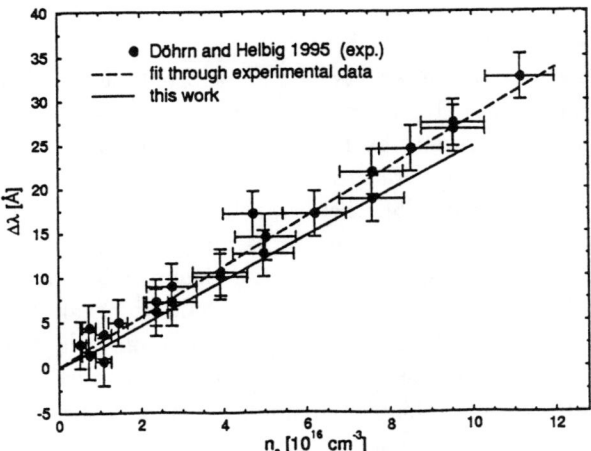

RESULTS FOR HYDROGENIC ION LINES

In order to evaluate hydrogenic ion lines, we changed the theory outlined above in two points. We accounted for fine structure splitting and used therefore the Pauli wave functions instead of the ordinary spherical wave functions. Further, applying the model microfield method, the field autocorrelation function has been changed for charged radiators (see [37]).

Figure 8: The influence of fine structure splitting, ion dynamics and Doppler effect on the profile of the HeII H_α line ($n_e = 5.5 \cdot 10^{16}$ cm^{-3}, $T = 44000$ K)

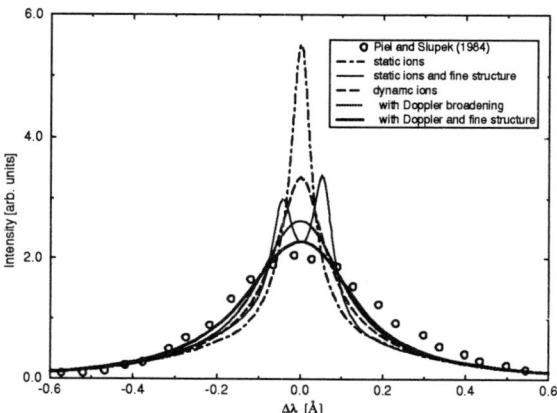

In Fig. 8 the influence of various effects on the line profile such as fine structure splitting, ion dynamics and Dopppler broadening are investigated. Analogous to the corresponding experiment by Piel and Slupek [38], an electron density of $5.5 \cdot 10^{16}$ cm^{-3} and a temperature of 44000 K have been chosen. As it has been already outlined in [38], there is a large discrepancy between the experimental and the theoretical line width assuming a static ionic microfield and neglecting fine structure splitting as well as Doppler broadening. Taking into account the Doppler broadening only, the half width already increases considerably. The same effect can be found including only the fine structure splitting. Considering simultaneously the fine structure splitting, the effects due to ion dynamics and the Doppler broadening, one reaches a much better agreement between experimental and theoretical results.

The influence of fine structure splitting is much more relevant for the higher charged hydrogenic ion lines. As it is to be seen in Fig. 9, fine structure splitting is much more important than ion dynamics for the H_α line of CVI.

Figure 9: Profile of the Balmer-α line of CVI (perturbers: C^{5+})

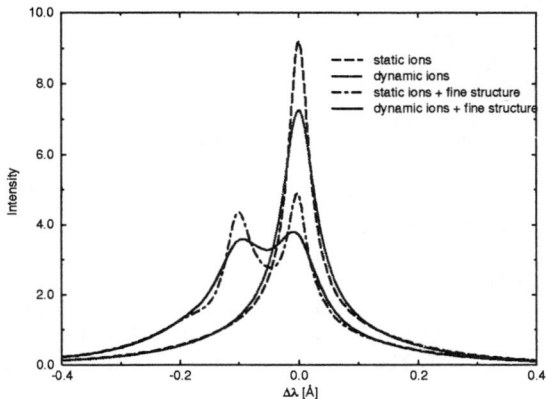

Therefore, discussing the influence of ion dynamics as in [39], one has to include fine structure splitting was not has been done there.

CONCLUSIONS

Asymmetric and shifted line profiles for hydrogen and hydrogenic ion lines have been calculated including ion dynamics. Besides the electronic shift, the ion quadrupole shift as well as the quadratic Stark effect have been accounted for. The frequency dependent electronic widths and shifts have been calculated using a Green's function technique, where the electrons are treated quantum mechanically and many-particle effects are included in a natural way. Strong collision contributions have been treated via partial summation of the corresponding perturbation series.

For comparison with experimental results, the asymmetry parameter and the line shift have been determined from theoretical profiles according to the same definitions as it has been done in the corresponding experiments. It has been shown that in such a way an excellent agreement between theoretical and experimental results for hydrogen lines has been reached.

Dealing with hydrogenic ion lines, fine structure splitting becomes more important and has to be taken into account. Even if fine structure is included the widths of the theoretical line profiles tend to be too small. The reason for this discrepancy may be an underestimation of ion dynanamic effects within the applied model microfield method.

REFERENCES

1. H.R. Griem. *Phys. Rev.*, A38:2943–2952, 1988.
2. H.R. Griem. *Spectral Line Broadening in Plasmas*. Academic Press, New York, 1974.
3. K. Unnikrishnan and J. Callaway. *Phys. Rev.*, A43:3619–3623, 1991.
4. K. Unnikrishnan and J. Callaway. *Phys. Rev.*, A44:3001–3006, 1991.
5. H.R. Griem. *Phys. Rev.*, A28:1596–1601, 1983.
6. G. Röpke, T. Seifert, and K. Kilimann. *Ann. Phys. (Leipzig)*, 38:381–395, 1981.
7. L. Hitzschke and G. Röpke. *Phys. Rev.*, A37:4991–4994, 1988.
8. S. Günter, L. Hitzschke, and G. Röpke. *Phys. Rev.*, A44:6834–6844, 1991.
9. S. Günter. *Phys. Rev.*, E48:500–505, 1993.
10. A. Könies and S. Günter. *Phys. Rev.*, E52:6658–6663, 1995.
11. A. Brissaud and U. Frisch. *J. Math. Phys.*, 15:524–534, 1974.
12. J. Seidel. *Zs. Naturforsch.* , 32a:1195–1206, 1977.
13. J. Seidel. *Z. Naturforsch.*, 35a:679–689, 1980.
14. S. Günter and A. Könies. *J. Quant. Spectrosc. Rad. Transfer*, 52:819–824, 1994.
15. A. Könies and S. Günter. *J. Quant. Spectrosc. Rad. Transfer*, 52:825–830, 1994.
16. A. Brissaud, C. Goldbach, L. Leorat, A. Mazure, and G. Nollez. *J. Phys. B: At. Mol. Phys.*, 9:1129–1146, 1976.
17. J. Halenka. *Z. Phys. D – Atoms, Molecules and Clusters*, 16:1–8, 1990.
18. S. Chandrasekhar and J. von Neumann. *Astrophys. J.*, 97:1–27, 1943.
19. A.V. Demura and G.V. Sholin. *J. Quant. Spectrosc. Rad. Transfer*, 15:881–899, 1975.
20. L.D. Landau and E.M. Lifschitz. *Lehrbuch der theoretischen Physik. Band 3*. Akademie Verlag, Berlin, 1986.
21. W.L. Wiese, D.E. Kelleher, and D.R. Paquette. *Phys. Rev.*, A6:1132–1153, 1972.
22. K. Grützmacher and B. Wende. *Phys. Rev.*, A16:243–246, 1977.
23. K. Grützmacher and B. Wende. *Phys. Rev.*, A18:2140–2149, 1978.
24. M.E. Bacon. *J. Quant. Spectrosc. Rad. Transfer*, 17:501–512, 1977.
25. V. Helbig and K.P. Nick. *J. Phys. B: At. Mol. Phys.*, 14:3573–3583, 1981.
26. R. Okasaka, M. Nagashima, and K. Fukuda. *J. Phys. Soc. Jpn.*, 42:1339–1347, 1977.
27. J.L. Chotin, J.L. Lemaire, J.P. Marque, and F. Rostas. *J. Phys. B: At. Mol. Phys.*, 11:371–383, 1978.
28. A.V. Demura, V.V. Pleschakov, and G.V. Sholin. Kurtschatov Institute Moscow, 1994. Preprint (russ.).
29. L.P. Kudrin and G.V. Sholin. *Sov. Phys. Dokl.*, 7:1015–1017, 1963.
30. A. Döhrn, V. Helbig, S. Günter, and A. Könies. In A. D. May, J.R. Drummond, and E. Oks, editors, *Spectral Line Shapes*, volume 8, pages 68–69, New York, 1994. American Institute of Physics.
31. J. Halenka. *J. Quant. Spectrosc. Rad. Transfer*, 42:571–573, 1989.
32. J. Halenka and J. Musielok. *J. Quant. Spectrosc. Rad. Transfer*, 36:233–237, 1986.
33. J. Halenka. In Bakes and Sörley, editors, *Proc. XVII ICPIG*, pages 993–995, Budapest, 1985.
34. A. Döhrn. PhD thesis, Universität Kiel, 1995.
35. St. Böddeker, S. Günter, A. Könies, L. Hitzschke, and H.-J. Kunze. *Phys. Rev.*, E47:2785–2791, 1993.
36. D.E. Kelleher, N. Konjevic, and W.L. Wiese. *Phys. Rev.*, A20:1195–1196, 1979.
37. M. Stobbe, A. Könies, S. Günter, and J. Halenka. submitted to Phys. Rev. E (1996).
38. A. Piel and J. Slupek. *Zs. Naturforsch.* , 39a:1041–1048, 1984.
39. C. Stehle. In A. D. May, J.R. Drummond, and E. Oks, editors, *Spectral Line Shapes*, volume 8, pages 36–57, New York, 1994. American Institute of Physics.

Profiles of High Principal Quantum Number Balmer and Paschen Lines from Alcator C-Mod Tokamak Plasmas

B. L. Welch, H. R. Griem, J. L. Weaver and J. U. Brill
Institute for Plasma Research, University of Maryland, College Park, MD 20742-3511

J. Terry, B. Lipschultz, D. Lumma and G. McCracken
Plasma Fusion Center, Massachusetts Institute of Technology, Cambridge, MA 02139-4307

S. Ferri, A. Calisti, R. Stamm and B. Talin
Université de Provence, Marseille

and R. W. Lee
University of California, Berkeley, CA 94720-7300

Atomic hydrogen and its isotopes contribute to line and continuum radiation especially from edge and divertor regions of high temperature tokamaks, and also from main chamber regions. Lower members of the line series exhibit Doppler broadening and Zeeman effects, whereas higher members are mostly Stark-broadened. Measurements of their line widths[1,2] can therefore be used to infer electron densities in the various regions by comparison with calculated line widths,[3] e.g., for the Balmer series up to $n = 12$. These calculations are updated here by using a code based on the frequency fluctuation model, radiative channel method[4,5] to include any ion-dynamical effects, and $\Delta n \neq 0$ perturbing levels. They are also extended to the Paschen series, whose Stark widths are larger by the square of the wavelength ratio.

We first use the "two-state" Liouville space version of the "standard" model of plasma line broadening, i.e., the quasistatic approximation to describe ion-microfield effects and the impact approximation to account for electron-atom collisions. For a given ion microfield this results in a very large set of generalized Lorentzians, with asymmetric dispersion terms. These profile components are characterized by two sets of complex numbers, the generalized relative intensities and frequency shifts. Effectively, the atoms are replaced by ion-field dressed atoms, and the line spectrum of any series is generated very efficiently. This allows us to include many more levels than is customary.

Because of the large number of Stark components, we use a renormalization procedure based on a statistical analysis to replace the actual set of components by a smaller set of representative "radiative channels", in which a given atom radiates until a field fluctuation transfers it to another channel. This combination of a frequency fluctuation model with the standard model results in line shapes

$$I(\omega) = \text{Re} \sum_{i,j} p_j \sqrt{1 + ic_i/a_i}(\omega - f + iW + i\phi)^{-1}\sqrt{1 + ic_j/a_j}, \quad (1)$$

with p_j being the probability of (dressed and renormalized) initial state j and c_i and a_i corresponding to the profile asymmetries and relative intensities. The quantities f, W and ϕ stand for microfield Stark shifts, channel-to-channel transition rates and the electron broadening operator. The transition rates are written in terms of the decay rate ν of the field-field autocorrelation function and the initial probabilities,

$$W_{i,j} = \nu p_i, \quad i \neq j; \quad W_{i,i} = -\nu(1 - p_i). \quad (2)$$

The characteristic frequency ν of the microfield fluctuations is obtained from molecular dynamics calculations.

We also studied, in the standard model, the effects of the $\Delta n \neq 0$ interactions. The major effect of including these interactions is a smoothing of central dips or peaks which are further smoothed by Doppler broadening. Including $|\Delta n| \leq 2$ interactions is more than sufficient, and there is very little increase of line widths, even from $|\Delta n| = 1$.

Ion-dynamical effects are not very important for lines which are sufficiently sensitive to Stark broadening to be useful for density determinations. Figure 1 shows comparisons of Stark profiles with and without ion-dynamical effects for the $n = 8$ Balmer line. Even for the high ion (and atom) temperature case, the differences are small enough not to be significant in the present context after allowance is made for a Doppler broadening of about 2 Å for deuterium. Possible exceptions are plasmas with very small atom-to-ion temperature ratios.

For density measurements, only the calculated Stark widths are required, corrected for Doppler and instrumental broadening according to experimental conditions and instrumentation. We therefore fitted our calculated Stark (PPP) widths to expressions corresponding either to the quasistatic $N_e^{2/3}$ scaling or to a weighted average of this scaling and a linear impact theory scaling. Figure 2 demonstrates such fits for Balmer $n = 8$ at $T_e = T_i = 4\,\text{eV}$. The weighted average of the two density scalings gives better fits to our calculations, indicating the relative importance of broadening by electrons. However, at high densities, e.g., $n = 12$ line widths exceed the width corresponding to the electron plasma frequency, thus invalidating the usual impact approximation.

Before correcting for this transition to a more quasistatic electron broadening, our line widths show a relatively strong dependence on the electron temperature, decreasing by factors of $1.5-2$ as the temperature increases from 1 to 10 eV. This complicates direct comparisons with the earlier calculations;[3] but the trend would be for a decrease of inferred electron densities by a factor ~ 2.

Use of the present calculations in fitting measured Balmer line spectra[1] from various regions of the tokamak plasmas would mostly indicate somewhat lower electron densities than those based on previous Stark broadening calculations.[3] However, no detailed calculations had been available for the Paschen lines. We therefore used the present calculations to fit these Paschen lines for $N_e \approx 4 \times 10^{19}\,\text{m}^{-3}$ and $N_e \approx 4 \times 10^{20}\,\text{m}^{-3}$ plasmas. These spectra are shown on Figs. 3a and b, respectively, indicating that Paschen lines can indeed be used for electron densities extending below $N_e = 10^{20}\,\text{m}^{-3}$.

It may be recalled that an earlier measurement[6] of the Paschen series at lower densities and temperatures could essentially be interpreted by assuming both electrons and ions to act quasistatically beyond the half widths of most of the lines exhibiting significant Stark broadening.

Our measurements and calculations indicate that high members of the Balmer and Paschen series have line widths which can be used to infer electron densities in tokamak experiments and other laboratory plasmas at relatively low electron densities. Our numerical procedure is sufficiently fast to synthesize entire synthetic spectra for fits to experimental data. Such synthetic spectra show, e.g., $n \geq 10$ lines having merged into an extended free-bound continuum as first discussed by Inglis and Teller.[7]

This research is partially supported by the U.S. Department of Energy and the National Science Foundation.

REFERENCES

1. B. Welch, H. Griem, J. Terry, C. Kurz, B. LaBombard, B. Lipschultz, E. Marmar and G. McCracken, Phys. Plasmas **2**, 4246 (1995).
2. B. L. Welch, H. R. Griem, J. L. Weaver, J. L. Terry, R. L. Boivin, B. Lipschultz, D. Lumina, E. S. Marmar, G. McCracken and J. C. Rost, Proc. 10th APS Topical Conf. on Atomic Processes in Plasmas (in press).
3. R. D. Bengtson, J. D. Tannich and P. Kepple, Phys. Rev. A **1**, 532 (1970).
4. A. Calisti, F. Khelfaoui, R. Stamm and R. W. Lee, Phys. Rev. A **42**, 5433 (1990).
5. L. Godbert, A. Calisti, R. Stamm, B. Talin, R. W. Lee and L. Klein, Phys. Rev. A **49**, 5644 (1994).
6. G. Himmel and F. Pinnekamp, J. Quant. Spectrosc. Radiat. Transfer **13**, 555 (1973).
7. D. R. Inglis and E. Teller, Astrosphys. J. **90**, 439 (1939).

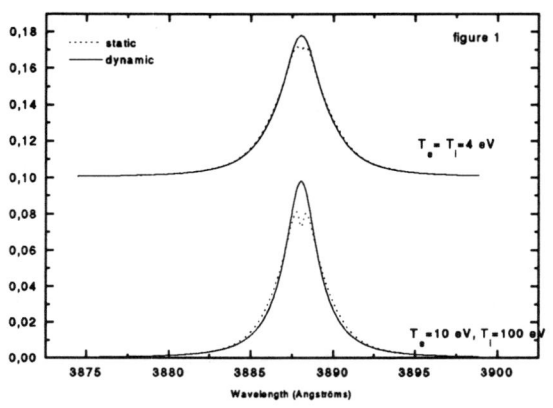

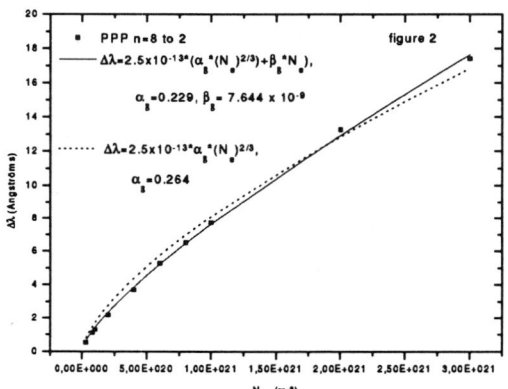

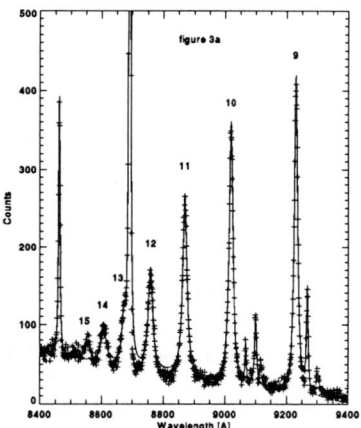
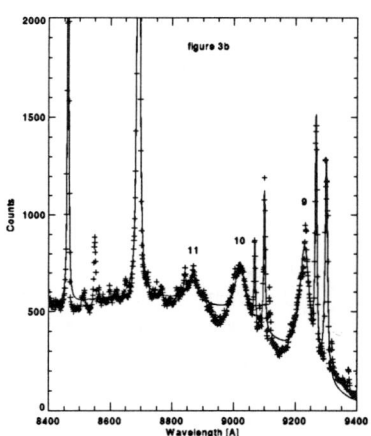

Semiclassical Stark Broadening Calculations of Neutral Oxygen Lines

Nébil Ben Nessib[*], Zohra Ben Lakhdar[*] and Sylvie Sahal-Bréchot[†]

[*]Laboratoire de Physique Atomique et Moléculaire, Département de Physique, Faculté des Sciences de Tunis, 1060 Campus Universitaire, Tunis, Tunisia and [†]Laboratoire Astrophysique, Atomes et Molécules, Unité associée au C.N.R.S. No 812, Département Atomes et Molécules en Astrophysique, Observatoire de Paris-Meudon, 92190 Meudon, France

Abstract. Using the semiclassical perturbation approach by Sahal-Bréchot (1969,1974,1991) updated by Dimitrijevic and Sahal-Bréchot (1984,1995), we have calculated electronic and ionic widths and shifts of several OI spectral lines. The TOPbase (Opacity Project) atomic structure has been used. We have compared our results with those of available experiments and with Griem's (1974) theoretical data.

By using the semiclassical perturbation Sahal-Bréchot approach (1-4) updated by Dimitrijevic and Sahal-Bréchot (5,6) and the TOPbase sophisticated atomic structure data, we have calculated electronic and ionic impact widths and shifts of several OI spectral lines. We have also calculated the quasistatic ion broadening parameter A with the TOPbase atomic structure (7-9).

For collisions with electrons, the impact approximation is always valid for all lines. For collisions with ions, the experimental values shows that for the two first lines (4368Å and 3947Å), the impact approximation can be applied, but for the two other lines (5436Å and 5330Å), the quasistatic approximation has to be applied.

By comparing our results with the experiments by Wiese and Murphy (10), Miller and Bengtson (11) Goly *et al.* (12,13), we find a better agreement with our computations than with Griem's formulation (14). The recent experimental results for the Stark widths of the 4368Å OI line by Mijatovic *et al.* (15), are very close to our calculated values (16) within seven percent. The Griem's results are roughly thirty percent higher.

On the whole, our results are in better agreement with experiments than the calculations by Griem, especially for the widths where the agreement can attain a few percent.

Table 1. Experimental and calculated total widths and shifts for OI lines. W_m and d_m are the experimental total widths and shifts. Ratios of measured and calculated widths and shifts are given : W_m/W_B and d_m/d_B for the present work (16) and W_m/W_G and d_m/d_G with Griem's formalism.

Line	N_e 10^{16} cm^{-3}	T_e K	W_m Å	W_m/W_G	W_m/W_B	d_m Å	d_m/d_G	d_m/d_B	Ref.
4368 Å	1.91	10100	0.39	1.27	1.03	-	-	-	12
	2.40	10600	0.51	1.29	1.07	0.10	1.17	0.69	15
	2.99	10900	0.58	1.15	0.97	-	-	-	12
	3.12	10980	0.67	1.28	1.07	0.13	1.17	0.69	15
	3.97	11580	0.78	1.14	0.97	0.12	0.85	0.51	13
	4.41	13570	0.82	1.02	0.88	-	-	-	13
	4.94	11900	1.07	1.24	1.06	-	-	-	12
	5.70	12080	1.08	1.10	0.93	0.20	0.87	0.59	10
	7.13	12700	1.57	1.25	1.07	-	-	-	12
	7.84	12500	1.44	1.04	0.90	0.27	0.97	0.51	13
3947 Å	3.97	11580	0.62	1.17	0.92	0.05	-	2.82	13
	4.41	13570	0.64	1.02	0.92	-	-	-	13
	5.70	12080	0.83	1.05	0.986	-0.13	11.92	4.78	10
	7.84	12500	1.20	1.12	0.895	0.09	-	2.27	13
5436 Å	3.17	11120	5.64	0.87	0.937	3.13	1.06	0.81	13
	7.04	12480	10.6	0.71	0.725	6.50	1.00	0.73	13
	10.0	11000	27.2	1.10	1.392	10.8	-	0.92	11
5330 Å	3.17	11120	15.8	0.61	0.681	2.8	0.48	0.29	13
	7.04	12480	33.8	0.67	0.66	8.6	0.69	0.40	13
	10.0	11000	33.2	0.81	1.012	21.6	-	0.72	11

REFERENCES

1. Sahal-Bréchot S., *Astron. Astrophys.* , **1**, 91 (1969)
2. Sahal-Bréchot S., *Astron. Astrophys.* , **2**, 322 (1969)
3. Sahal-Bréchot S., *Astron. Astrophys.* , **35**, 321 (1974)
4. Sahal-Bréchot S., *Astron. Astrophys.* , **245**, 322 (1991)
5. Dimitrijevic M.S. and Sahal-Bréchot S., *J.Q.S.R.T.*, **31**, 301 (1984)
6. Dimitrijevic M.S. and Sahal-Bréchot S., *Physica Scripta.*, **52**, 41 (1995)
7. "The Opacity Project", Vol. 1, compiled by the Opacity Project Team, Institute of Physics Publ., Bristol, U.K. (1995)
8. Cunto, W, Mendoza C., Ochsenbein, F. and Zeippen, C.J., *Astron. Astrophys.* , **275**, L5 (1993)
9. Zeippen C.J., *Physica Scripta.*, **T58**, 43 (1995)
10. Wiese W.L. and Murphy P.W., *Phys. Rev.*, **131**, 2108 (1963)
11. Miller M. H. and Bengtson R.D., *Phys. Rev.A*, **1**, 983 (1970)
12. Goly A, Rakotoarijimy D and Weniger, S, *J.Q.S.R.T.*, **30**, 417 (1983)
13. Goly A. and Weniger, S, *J.Q.S.R.T.*, **38**, 225 (1987)
14. Griem H.R. "Spectral Line Broadening by Plasmas", Academic Press Inc., New York (1974)
15. Mijatovic Z, Konjevic N., Kobilarov R. and Djurovic S. *Phys. Rev. E*, **51**, 613 (1995)
16. Ben Nessib N., Ben Lakhdar Z. and Sahal-Bréchot S., *Physica Scripta.*, in press (1996)

Line Asymmetries and Plasma Statistics

A. V. Demura[*], D. Gilles[**] and C. Stehlé[***]

[*]Hydrogen Energy and Plasma Technologies Institute, Russian Research Center "Kurchatov Institute", Moscow 123182, Russia, [**]C.E.A.-C.E.L.V., F-94195 Villeneuve St-Georges Cedex, France and [***] D.A.R.C. & UPR 176 du C.N.R.S., Observatoire de Paris, 5 Place Jules Janssen, F-92190, Meudon, France.

Abstract. Our purpose is to investigate the sensitivity of the line asymetry to the choice of the statistics included in the calculation of the field distribution function P(b) and the universal functions $B_D(b)$ and $B_{D0}(b)$ (Nearest Neighbour, Holtsmark or more realistic models). Ion dynamics effects are known to be important at low and intermediate densities. Their contribution to the line center is also checked.

The asymetries of the line of hydrogen-like system in opticaly thin media is attributed to different physical reasons. One of them is the fine structure which will not be considered here for simplicity.

We are mainly interested in this paper to the analysis of the quadrupolar interaction with the spatial gradient **G** of the plasma microfield, whose components are given by, $G_{ij} = \partial F_i / \partial x_j - (1/3 \sum \partial F_i / \partial x_i) \delta_{ij}$ and the monopolar interaction with the Laplacian $\Delta F = \sum \partial F_i / \partial x_i$.

This electric microfield $\vec{F}$ is the sum of all the elementary microfields due to the surrounding plasma ions, with arbitrary spatial locations and velocities, the gradient is also the sum of all the elementary gradients, and the interaction potential between the bound electron of the hydrogenic system and the plasma (dipole $\vec{d}$, quadrupole $\overline{\overline{Q}}$, and distance to the nucleus r) is equal to

$$V(F) = -\vec{d}\cdot\vec{F} + \sum Q_{ij} G_{ij} - \frac{1}{6} e r^2 \Delta F$$

Even if $\vec{F}$ is choosen along the quantization axis, the calculation of the line shape is very difficult to perform, because the gradient contains five independent stochastic variables $\{G_{ij}\}$, and one must average in fact over seven independent variables (F_z, $\{G_{ij}\}$ and ΔF). One escapes this difficulty by replacing the field gradient and its Laplacian by their conditional averages to a fixed value of the microfield $\vec{F}$. Choosing $\vec{F}$ along the z axis, the plasma-radiator interaction potential may be written as[1],

© 1997 American Institute of Physics

$$V(F) = -d_z F_0 \beta + Q_{zz} \frac{\pi N_e e}{3} B_D(\beta) - \frac{2\pi N_e e}{3} r^2 B_{D0}(\beta)$$

where $F_0 = e/r_0^2$ is the normal Hotsmark field value, r_0 is the mean interionic distance, defined by $4\pi N_e r_0^3/3 = 1$, $\beta = F/F_0$ and N_e is the electronic density.

The effects of the plasma statistics are illustrated on the figures below, which represent the Lyman α line of He^+ perturbed by He^+ ions at $10^{18} cm^{-3}$ and for a temperature of 10^4 °K. Doppler and monopolar effects are not included. Holtsmark and Baranger-Mozer are used for the calculation of $P(\beta)$ and $B_D(\beta)$. Ion dynamics effects are included by the Model Microfield Method.

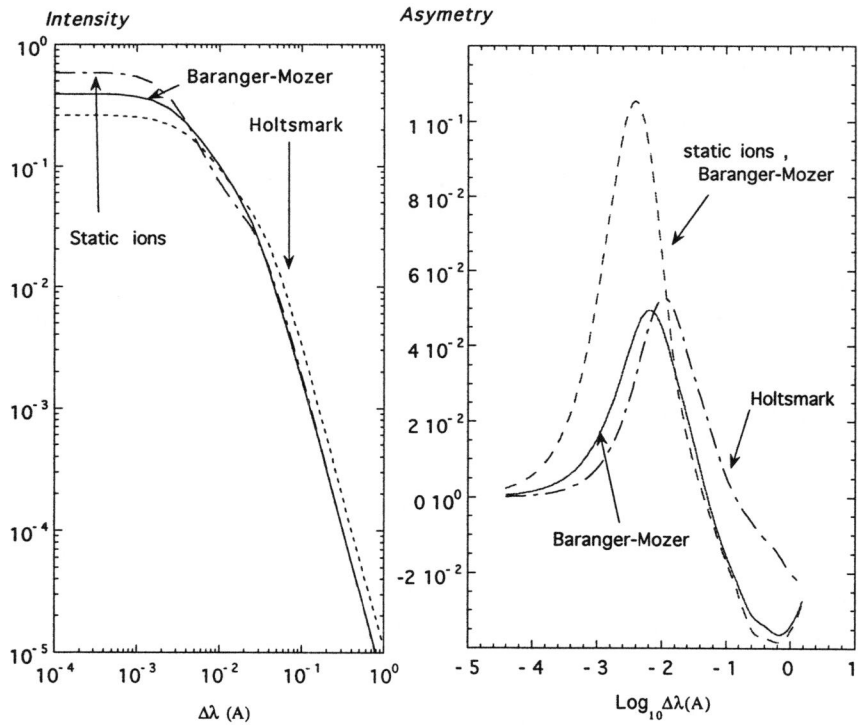

REFERENCES

1. Demura, A.V., Gilles, D. and Stehlé, C., J.Q.S.R.T. **54**, 123-136, (1995).
2. Stehlé, C., A.&A., **292**, 699, (1994).

The Shift of Stark Broadened H_γ Line

J. Halenka, W. Olchawa and T. Wujec

Institute of Physics, Opole University, Oleska 48, PL-45-052 Opole, Poland

Abstract. Line profile of H_γ have been measured at low electron densities and parameters of the line shift have been determined. The experimental results have been compared with calculated ones.

To date, measurements[1] of the pressure shift of H_γ line have been performed for plasmas of electron concentrations from the range $2 \cdot 10^{16}\,cm^{-3} - 10^{17}\,cm^{-3}$ only. In astrospectroscopy the pressure shift is very important for interpretation of redshifts in white-dwarf spectra.[2] In the pioneer's and succeeding papers on that subject mainly measurements have been carried out on the H_γ-line in its „astrophysical line center", i.e. in the region ± 20 Å in both sides from the line minimum. In the case of white-dwarf spectra this region is predominantly formed by absorption of radiation in plasma layer of electron concentrations of the order of $N_e \propto 10^{15}\,cm^{-3}$. The main aim of this paper is to investigate the pressure shift and asymmetry of the H_γ line, approximately at these physical conditions.

The helium plasma with small (~ 3%) hydrogen admixture has been generated in the electric arc. Using the OMA detector, the line profiles of H_β, H_γ, HeI3889, and HeI6878 have been recorded at a spectral resolution of about 0.01 Å. The electron concentration N_e has been determined from the FWHM of H_β line, taking into account the contribution from the ion-dynamic effects to the line width. The electron temperature has been determined on the basis of measured emission coefficients of HeI lines, assuming pLTE conditions. For plasmas of N_e from the range $2.5 \cdot 10^{15}\,cm^{-3} - 4 \cdot 10^{15}\,cm^{-3}$, and T_e from the range 14 kK - 17 kK, the measured shifts of H_γ amount for the line maximum from -0.08 Å to -0.10 Å (towards violet), and for the ELC from 0.03 Å to 0.05 Å (towards red).

Calculations of the H_γ shift for the same plasma conditions have been performed within the quasi-static approximation for ions and within the impact approximation for electrons. For the ELC-shift, the theoretical results agree with the experimental ones. The calculated shifts for the line-maximum are towards red, i.e. are inqualitative disagreement with the measurements. One possible reason for this discrepancy could be the contribution from very weak molecular H_2 in vicinity of the H_γ line core. Therefore the experimentally determined position of the H_γ line maximum is not a good measure of the line shift.

TABLE I. Comparison of measured values of the Experimental Line Center (ELC) with calculated ones.

N_e (10^{15} cm^{-3})	T (kK)	ELC (Å) expt. ±0.01	ELC (Å) this cal.	ELC (Å) cal.[3]
2.7	14.8	0.03	0.023	0.018
3.1	15.8	0.03	0.026	0.020
3.4	16.2	0.05	0.022	0.022
3.9	17.1	0.04	0.033	0.025

TABLE II. Comparison of measured values of the Shift of Line Maximum (SLM) with calculated ones.

N_e (10^{15} cm^{-3})	T (kK)	SLM (Å) expt. ±0.01	SLM (Å) this cal.	SLM (Å) cal.[3]
2.7	14.8	-0.09	0.054	0.046
3.1	15.8	-0.08	0.062	0.054
3.4	16.2	-0.09	0.059	0.059
3.9	17.1	-0.10	0.079	0.069

REFERENCES

1. W. Wiese, D. Kelleher, and D. Paquette, Phys. Rev. **A6**, 1132 (1972).
2. B. Grabowski, J. Madej, J. Halenka, Ap. J., **313**, 750 (1987).
3) S. Günter, private communication.

Determination of the Total Number of Absorbtion Atoms in a Plasma by Parameters of Self-reversed Spectral Line

G.G.Il'in, O.A.Konovalova, E.V.Sarandaev and M.Kh.Salakhov

*Department of Physics, Kazan State University,
420008, Kazan, Tatarstan, Russia*

Abstract. The problem of determining of the total number of absorbtion atoms in a plasma from the distance between maxima of strongly asymmetric self-reversed spectral line with the quadratic Stark-effect are considered under the conditions when the broadening by charge particles predominates.

Possibilities have been considered of determining of the total absorption atoms number $N_i = \int_0^{r_0} n_i(r) dr$ (r_0 is the radius of absorption zone) from measured distance Δ_s between maxima of strongly asymmetric self-reversed spectral line with the quadratic Stark-effect under the conditions when the broadening by charge particles predominates. A theoretical analysis of self-reversed profiles has been carried out on the basis of results of computer calculation of an emission transfer equation for different lines of sight of strongly ingomogeneous axially symmetric plasma described in [1]. The source function $\varphi(r)$ characterising the relative distribution of emission atoms relative to absorption atoms was set as $\varphi(r) = \varphi(0)[1-(r/R_0)^m]$ at $0 \le r \le R_0$ (R_0 is the radius of emission zone) and $\varphi = 0$ at $r > R_0$ and the radial concentration variation of absorption atoms on level i was set as $n_i(r) = n_i(0)[1+ag(r/R_0)^2]\exp[-g(r/R_0)^2]$. The broadening profile was represented in terms of the Stark profile $j_{\alpha,R}(x)$ [2] with the electron impact half-width $\delta_e(r) = \delta_0 \exp[-b(r/R_0)^2]$ at $0 \le r \le R_0$ and $\delta_e(r) = \delta_0 \exp(-b)$ at $r > R_0$ and electron impact shift $\Delta_e(r) = \eta \delta_e(r)$. For the computer calculation of the emission transfer equation the central part of the profile $j_{\alpha,R}(x)$ given in the tabulated form in [2] was presented in the form of two Student distributions with different parameters dependent on the ionic broadening parameter $\alpha(r) = \alpha_0[\delta_e(r)/\delta_0]^{1/4}$ as a first approximation and the Debay shielding parameter R. For the wings of the profile $j_{\alpha,R}(x)$ the approximate formulae [3] are used.

The calculations of asymmetric self- reserved profiles has been carried out for different lines of sight in the plasma cross-section at the different values of the absorption parameter $p_0 = (h\nu_0 B N_i)/\pi c \delta_0$ (B is the Einstein coefficient for absorption; ν_0 is the unperturbed spectral line frequency), $m, a, g, b, \eta, \alpha_0$ and R. The calculated data have permitted to investigate in detail for the central line of sight the theoretical dependences of the relative distance between the self- reversal maxima $2s = \Delta_s / \delta_0$ on p_0 which allow to determine p_0 and then N_i by the measured Δ_s and δ_0. The δ_0 value may be determined from the wings asymmetry to ν_0 of self-reversed line itself with the known η [2]. The δ_0 knowledge permits to determine the electron concentration n_{e0}, α_0 and R with using of tabulated data of [2]. It is found that for the N_i determination by Δ_s it is necessary to know a, g and b. The question of the a, g and b determination by the absorption spectra is considered in [1]. The calculated data permit to consider the question of the a, g and b determination by the asymmetric self- reversed profiles themselves. For the $\delta(r)$ and b determination the wings asymmetry of self- reversed profiles measured for different lines of sight may be used. The a and g determination may be carried out by the ratio of peak intensities I_{max1} / I_{max2} (I_{max1} is the intensity of the large peak of self-reversebility, I_{max2} is the intensity of the small peak of self-reversebility). The developed technique of the N_i determination by Δ_s was used to the Al I 396.1 nm self-reversed resonance spectral line under the conditions of a low-voltage impulsing charge [4]. The large peaks asymmetry ($I_{max1} / I_{max2} = 4$) shows on the large value of $a \approx 30$ and the small value of $g \approx 0.1$ which qualitatively well agree with the relative $n_i(r)$ distribution of aluminium atoms determined by the absorption spectra in [1]. At $\delta_0 = 0.55 \overset{0}{A}$, $b = 2$, and $2s = 1.8$ it was found that $N_i = 3.6 \cdot 10^{14} cm^{-2}$ with the most probable relative error (20-30)%.

REFERENCES

1. Fishman, I.S, Il'in, G.G.,and Salakhov, M.Kh., Spectochim.Acta Part B **50**, 947-959 (1995).; **50** 1165-1178 (1995).
2. Griem H.R., Spectral Line Broadening by Plasmas, New York: Academic Press,1974.
3. Il'in,G.G., Konovalova, O.A., Dep. VINITI N1553-B96 (1996).
4. Fishman,I.S., Salakhov,M.Kh, Sarandaev,E.V., and Semin,P.S., Opt. Spectrosc., **51,** 785-793 (1981).

The Polarized Spectrum of Hydrogen in the Presence of Electric and Magnetic Fields

R. Casini and E. Landi Degl'Innocenti

Dipartimento di Astronomia e Scienza dello Spazio, Università di Firenze
Largo E. Fermi 5, I-50125 Firenze, Italy

Abstract. We briefly comment on previous numerical and analytical work we have done on the quantum problem of the hydrogen atom in the presence of stationary electric and magnetic fields.

Detection and measurement of stationary electric and magnetic fields represent a main issue in the understanding of the equilibrium condition and dynamical evolution of solar plasmas. In fact, while the importance of solar magnetic fields has long been acknowledged, only in relatively recent times have electric fields started being investigated, their occurrence in solar plasmas being supported by present MHD models of solar structure (1).

As in the case of magnetic-field diagnostics, where the investigation of the Zeeman effect on spectral lines plays a dominant role, spectropolarimetry techniques are the most effective for the diagnostics of electric fields, through the investigation of the associated Stark effect. In particular, hydrogen (and hydrogen-like) spectral lines are most useful in this task, due to their sensitivity to the *linear* Stark effect. This is the reason why, in the last few years, we have undertaken the effort of a thorough study of the problem of the hydrogen atom subject simultaneously to electric and magnetic fields, particularly emphasizing the corresponding polarization properties of the hydrogen spectrum.

We wrote a numerical code for the calculation of the polarization profiles of hydrogen lines formed in the presence of stationary electric and magnetic fields of arbitrary orientation (2). In our approach we chose to neglect *configuration-mixing*. Fine-structure terms, which were usually neglected in previous work from other authors, are properly taken into account to ensure the correct behavior of the theoretical spectrum in the limit of low-intensity fields. Our code might then also be applied to Stark-broadening calculations of hydrogen (and hydrogen-like) lines at low plasma densities. In that case, in fact, the neglect of fine structure in the calculation of lower hydrogen transitions (e.g., Hα) is largely responsible for the observed discrepancies with the experimental line profiles (3).

The numerical calculations of polarization profiles can be very time consuming for relatively large values of the principal quantum numbers of the transition. For this reason, we pursued an analytical approach aimed at the determination of some relevant integral properties of the polarization profiles—namely, the first- and second-order moments, related to the centers of gravity and to the dispersions of the polarization profiles, respectively—which do not require a previous thorough calculation of such profiles. We were able to obtain, through an intense application of Racah's algebra, rather compact formulae for those moments, as functions of the principal quantum numbers of the transition and of the external fields (4,5). Taking the center of gravity of the intensity profile in the absence of external fields as the reference frequency for the calculation of the moments, we found

$$\langle\omega(i)\rangle = -\delta_{i3}\frac{\mu_0}{\hbar}B\cos\vartheta_B, \qquad i = 0, 1, 2, 3, \tag{1}$$

and

$$\langle\omega^2(0)\rangle = \langle\omega^2(0)\rangle_{\text{ff}} + \frac{e_0^2 a_0^2 E^2}{\hbar^2}\left[A_0(n,m) + \frac{1}{2}(1 - 3\cos^2\vartheta_E)A_2(n,m)\right]$$
$$+ \frac{1}{2}\frac{\mu_0^2 B^2}{\hbar^2}(1 + \cos^2\vartheta_B), \tag{2a}$$

$$\langle\omega^2(1)\rangle = \frac{3}{2}\frac{e_0^2 a_0^2 E^2}{\hbar^2}A_2(n,m)\sin^2\vartheta_E\cos 2\varphi_E$$
$$- \frac{1}{2}\frac{\mu_0^2 B^2}{\hbar^2}\sin^2\vartheta_B\cos 2\varphi_B, \tag{2b}$$

$$\langle\omega^2(2)\rangle = \langle\omega^2(1)\rangle\{\cos 2\varphi \to \sin 2\varphi\}, \tag{2c}$$

$$\langle\omega^2(3)\rangle = 0, \tag{2d}$$

where $i = 0, 1, 2, 3$ enumerates the four Stokes parameters I, Q, U, V. The field polar angles (ϑ) are measured from the line-of-sight, while the field azimuthal angles (φ) are measured from the chosen reference direction for positive Q, normal to the line-of-sight. The dimensionless coefficients $A_0(n,m)$ and $A_2(n,m)$, and the second-order moment of the intensity profile in the field-free case, $\langle\omega^2(0)\rangle_{\text{ff}}$, are tabulated for all the hydrogen transitions up to the level $n = 50$ (6).

From simple inspection of the above formulae, one can draw directly some important conclusions. (a) The electric field does not modify the center of gravity of the circular-polarization profile with respect to the value determined by the magnetic field. This is a rather important issue for the diagnostics of solar magnetic fields by means of longitudinal magnetographs. (b) While the electric-field contributions to the second-order moments depend on the transition (through the coefficients $A_0(n,m)$ and $A_2(n,m)$)—so that one can choose particularly Stark-sensitive lines for electric-field diagnostics (7)—the magnetic-field contributions, to both the first- and second-order moments, do not depend on the transition. This means that the *effective Landé factor* is equal to 1 for any hydrogen line, so that all hydrogen lines have the same sensitivity to the magnetic field.

REFERENCES

1. Foukal P., and Hinata S., *Solar Phys.* **132**, 307 (1991)
2. Casini R., and Landi Degl'Innocenti E., *Astron. Astrophys.* **276**, 289 (1993)
3. Ehrich D.L., and Kelleher D.E., *Phys. Rev. A* **17**, 1686 (1978)
4. Casini R., and Landi Degl'Innocenti E., *Astron. Astrophys.* **291**, 668 (1994)
5. Casini R., and Landi Degl'Innocenti E., *Astron. Astrophys.* **300**, 309 (1995)
6. Casini R., *Astron. Astrophys. Suppl.* **114**, 363 (1995)
7. Casini R., and Foukal P., *Solar Phys.* **163**, 65 (1996)

A New Source for the Experimental Study of Ions at Low Temperature

Frank C. De Lucia

*Department of Physics, The Ohio State University,
Columbus, OH 43210*

Abstract. A new source for the experimental study of ions at low temperatures (1 - 50 K) has been developed. It is based on a combination of two experimental methods which we have previously described: (1) the production of large concentrations of molecular ions in magnetically lengthened negative glow discharges, and (2) the collisional cooling of condensable spectroscopically active molecules to cryogenic temperatures. The study of molecular species in this regime is of both fundamental and astrophysical importance. Experimental details as well as scientific results will be discussed.

SCIENTIFIC MOTIVATION

Collisions between molecular ions and neutrals at low temperature are of both fundamental and practical interest. Much of this interest is driven by the basic role these species play in the chemistry of cold interstellar molecular clouds and by basic questions in chemical physics. As a consequence, we have combined collisional cooling methodology[1] with our magnetically enhanced molecular ion production technique[2] to provide a general environment for a wide range of scientific studies at very low temperature.[3]

The Role of Temperature in Collisions

In systems where all of the degrees of freedom can be defined by a single temperature, rotational spectroscopic studies of linear and polyatomic molecules ordinarily have been done in the regime $h\nu_r \ll kT$. As a result, for thermally populated levels $J_{max} \sim 10 - 100$ and the partition function $Q_r \sim 10^2 - 10^5$.

In this regime, there is in general spectral complexity, and for collisions a multitude of "open" channels, with "classical" collision properties which result from averaging over these many channels. Bohr stated this in more general and eloquent terms in his well known Correspondence Principle:

The predictions of the quantum theory for the behavior of any physical system must correspond to the prediction of classical physics in the limit in which the quantum numbers specifying the state of the system become very large.

As a result, the effect of a large reduction in temperature on molecular collisional processes can be significant, transforming a subject which is largely classical near room temperature into a subject with dramatic quantal features at low temperature. For example, at 1 K the orbital angular momentum of a He atom colliding with CO is L ~ 2 and the collisional channels associated with the J = 1 rotational state of CO are energetically marginally available. However, at 300 K, the orbital angular momentum associated with the collision is L ~ 30 and a multitude of collisional channels through J = 10 are open. The effect of the increase in open channels at higher temperature can be seen in Fig. 1, which shows the calculated

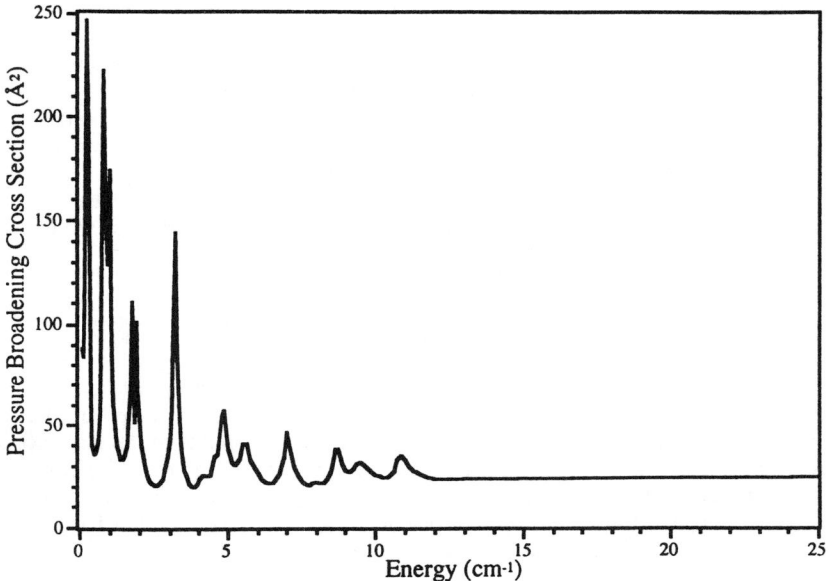

FIGURE 1. The effect of temperature and energy on the pressure broadening cross sections of the J = 1 - 0 transition of CO in He.

pressure broadening cross section for the J = 1 - 0 transition of CO in He at 115 GHz as a function of collision energy. At low temperature (2 cm^{-1} ~ 3 K) individual resonances, associated with the formation of quasi-bound states, are observable in the pressure broadening cross section, as well as in the state-to-state rates and the pressure shifts. At higher temperature the resonances slowly merge and disappear, and the cross section over a very wide temperature range is reduced to essentially that of the classical "size" of the molecules. In this higher energy regime semi-classical, Anderson-like theories[4] are appropriate, whereas at lower temperatures fully quantal, "exact" theories such as those embodied in scattering codes such as MOLSCAT[5] are required.

The Experimental Objective

Because of our desire to explore these phenomena, we have sought to develop a general, expandable methodology to study the regime for which $h\nu_r \geq kT$,

$J_{max} \sim 1 - 5$, and $Q_r \sim 1 - 10$. In this regime, there is not only significant spectral simplification, but also in collisions a much closer and more interesting relation between experimental observables and fundamental parameters. Additionally, in the millimeter and submillimeter (mm/submm) spectral region $h\nu_{exp} \sim h\nu_r \sim kT$, and this spectral region is especially advantageous for many scientific studies.

THE EXPERIMENTAL FOUNDATIONS

The work described in this paper rests on the foundations provided by earlier developments in our laboratory: (1) The production of large quantities of molecular ions in magnetically lengthened negative glow discharges, and (2) the collisional cooling of spectroscopically active molecules to cryogenic temperatures.

Ion Production Environments

The first step toward the overall goal was the development of a technique for ion spectroscopy that produces signals two orders of magnitude larger than previous techniques.[2] This was especially important because prior to this development, microwave ion spectroscopy was reputed to be extremely difficult and time consuming because of the long searches for weak lines. The physical basis of the method is the extension of the ion rich negative glow region of an anomalous glow discharge by means of an axial magnetic field. In addition to the very large gains in signal strength (~x100), the enhancement is also ion specific, thus providing a powerful discriminant. Figure 2 shows the observed signals as a function of applied magnetic field for two common molecular ions, HCO^+ and N_2H^+.

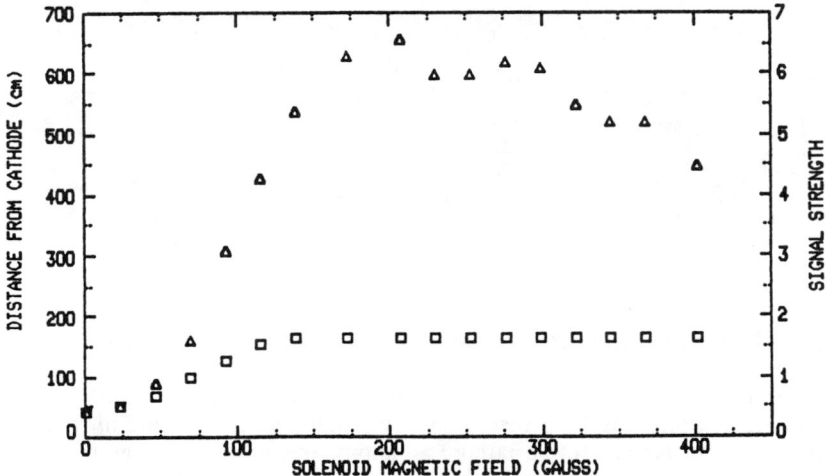

FIGURE 2. Signal strength of HN_2^+ (squares) and HCO^+ (triangles) as a function of magnetic field.

Figure 3 shows a particularly simple version of the magnetically lengthened negative glow cell. This cell is of all glass construction with Teflon end caps. The

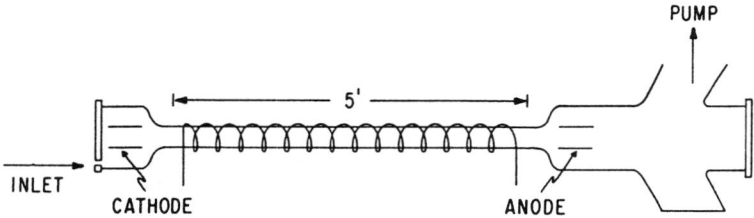

FIGURE 3. Cell for the study of molecular ions in a magnetically lengthened negative glow.

solenoid section is immersed in liquid nitrogen. This basic approach has been used by us[6-9] and a number of other workers to detect a wide variety of molecular ions.[10-12]

In our original work it was pointed out that the magnetically lengthened negative glow cell is functionally equivalent to an electron gun injecting magnetically confined electrons into a electric field free region. Based on this equivalence a new cell was developed to provide the first step in combing the two experimental techniques. This system is shown in Fig. 4.

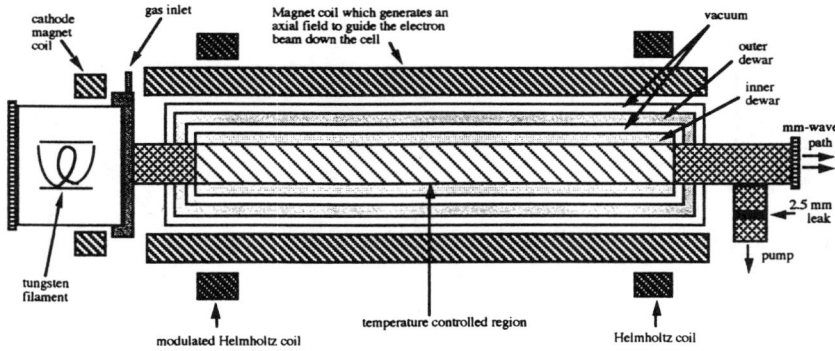

FIGURE 4. Cell with thermionic cathode for molecular ion study.

In this cell the electrons are produced by a thermionic cathode and guided through a metallic tube which can be cooled either by liquid helium or liquid nitrogen. Because the requirements for sustaining an anomalous glow discharge place severe constraints on pressure, gas mixture, voltage, and current, this new cell provides significantly greater flexibility for the optimization of the production of ions. Additionally, because the parameters of the systems are much better determined, it is possible to model the ion production. This makes possible more predictive ion production schemes as well as the quantitative measurement of the physical and chemical parameters involved.

The Collisional Cooling of Neutrals

The second step was the development of collisional cooling (CC) as a simple methodology for the study of gas phase neutral molecules at temperatures far below their freezing points.[1,13-16] This is accomplished by injecting small amounts of warm, spectroscopically active gas into cells filled with cold He or H_2. Because at typical spectroscopic pressures the warm gasses collisionally cool to the temperature of the background gas in far fewer collisions than required to reach a wall (where they condense), these collisionally cooled molecules are attractive subjects for a wide range of studies.

Figure 5 shows the layout of one of the collisional cooling systems. The entire apparatus is in a vacuum chamber maintained at a pressure below 10^{-5} Torr.

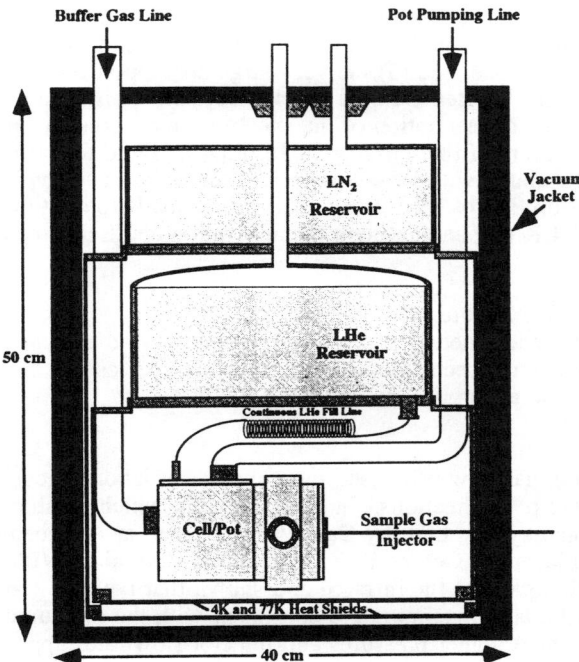

FIGURE 5. Collisional cooling system for the study of gas phase molecules at very low temperature.

Shields at 77 K (maintained by liquid nitrogen) and at 4 K (maintained by liquid helium) surround the region which contains the collisionally cooled cell. In this figure, the front window is shown as the open circle on the cell/pot. The cell is connected to a helium pot which is filled via a continuous fill capillary from the helium reservoir and pumped via an external pumping line. The cell is made of oxygen-free high conductivity copper and has 5 mil mylar windows. All removable flanges are sealed with indium. The temperature of the cell is varied by pumping on the pot through a vacuum regulator valve, which accurately controls the pressure and thus the temperature of the liquid helium. Since in this system the cell is isolated from the helium reservoir, temperatures above 4.3 K are also readily maintained. This technique allows the straightforward variation and control of the temperature, which is read directly with calibrated germanium resistance thermometers. The cell pressure is measured by a capacitance manometer at room temperature, with a correction for the effects of thermal transpiration.

At the side of the experimental chamber, warm spectroscopically active gas is injected via a vacuum insulated tube. Provisions exist for electrically heating this tube, but under normal conditions this has not been necessary for the relatively volatile gases used. The collisionally cooled cell is first filled with a static pressure of helium or hydrogen in equilibrium with the wall temperature. The spectroscopically active gas then cools as it collides with the cold buffer gas. The quantitative relations that make this a useful and general technique are: (1) the injected gas cools very rapidly as it collides with the buffer gas, requiring far fewer than 100 collisions to closely approach the temperature of the background; (2) at typical pressures, about 10 000 collisions are required for the gas to reach the walls and to condense; and (3) the very large absorption coefficients characteristic of very low temperature spectroscopy make possible large dilution ratios that ensure the concentration of the spectroscopic gas is so small as to not perturb the temperature of the buffer gas. In order to avoid the corrections to lineshape that are associated with large absorption, the flow rate is typically adjusted to produce an absorption of 1% - 10%. This corresponds to dilution ratios at temperatures around 4 K (depending upon the absorption coefficient of the gas) of $\sim 10^{-4} - 10^{-6}$.

Among the attractive features of collisional cooling are:
    (1) A well defined kinetic temperature continuously adjustable down to ~1 K.
    (2) Generality, any spectroscopically interesting species can be studied.
    (3) Macroscopic absorption path lengths.
    (4) Simplicity.

Collisional cooling has now been used by a number of laboratories in a rather wide variety of applications. In an experiment which combined a collisionally cooled system built by Dan Willey's group at Allegheny College with the infrared lasers of George Flynn's group at Columbia, Willey et al.[17] have extended the technique into the infrared and shown that both the translational temperature and rotational temperature are consistent with the measured wall temperature at the somewhat higher pressures and injector flow rates appropriate for infrared studies.

At Brookhaven, Trevor Sears' group has developed a multipass system for the infrared and in their initial experiments have studied N_2O, NO, and NO_2 near 5 K.[18] They were able to establish that both the rotational and translational temperatures were that of the cell walls, and were able to take advantage of the significantly reduced Doppler widths to resolve hyperfine structure. They also expect to be able to study the evolution of chemically reactive species following laser initiated photochemistry.

Miller's group at North Carolina has adapted the technique for the study of aerosols and found that they can vary the size distribution over a wide range by simple adjustment of cell pressure, gas composition and flow rate.[19]

In another collaborative experiment a collisionally cooled system built by Willey's group has been used in conjunction with a special high spectral purity diode laser system developed by Arlan Mantz at Franklin and Marshall to study CH_3F at 7.5 K. In contrast to the other experiments, this work determined rotational and translational temperatures about twice that of the cell temperature. This result underscores the fact that these are quasiequilibrium experiments and that if collisionally cooled systems are to be used in collisional studies which require a known and controlled kinetic collision temperature, then it is important to understand the thermodynamics of the system and to do careful diagnostics.

In the United Kingdom, this method has also been used in conjunction with an FTIR by Ballard and Newnham,[20] who are interested in upper atmospheric problems. Likewise, in related work we have developed liquid nitrogen cooled systems for mm/submm studies related to the atmosphere of the earth and other planets.[15] Finally, in an especially interesting experiment Willey's group has shown maser action in an collisionally cooled cell without any additional pump energy.[21] In this experiment differential rotational relaxation of the initially warm NH_3 in collision with cold He produces a significant population inversion on the (J,K = 4,3) transition.

THE PRODUCTION OF THERMAL IONS AT VERY LOW TEMPERATURE

For this work a novel apparatus was developed to produce large quantities of thermal molecular ions at temperatures well below the freezing points of their precursor neutrals. The challenge for our studies was the development of a system which produces an adequate number density of molecular ions at temperatures far below the freezing point of their neutral precursors while simultaneously allowing the pressure and temperature of the collisional environment to be controlled and characterized.

The Collisionally Cooled Ion Cell

Figure 6 shows a general purpose system for the production of ions and other transient species at very low temperature in an environment suitable for quantitative collisional studies. The central region, in which the observations are made, is a collisional cooling cell with open ends for the electron beam which is generated by a heated filament and accelerator region to the left. In the collisionally cooled region, H_2 is maintained between 10 - 100 K (or He between 1 - 100 K) by contact with the liquid/vapor He cooled copper walls, whose temperature is regulated by a feedback loop. In the middle of the central region, trace amounts of CO or other condensable precursor gas are continuously injected near ambient temperature into the cold H_2/He buffer gas, and cooled within 10 - 100 collisions to the measured cell wall temperature. The CO or other precursor is effectively cryo-

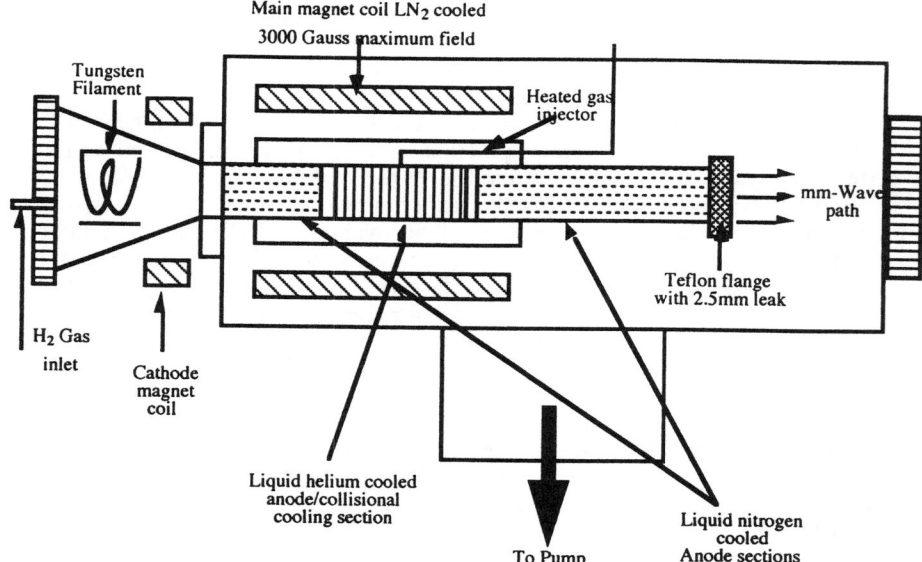

FIGURE 6. System for the study of macroscopic samples of molecular ions at very low temperature.

pumped by the helium temperature walls and has significant concentration only in the middle of the central region, as do reaction products, such as HCO^+. The central helium cooled region is surrounded by a 77 K heat shield and thermally isolated at both ends from the 77 K cylindrical sections by mylar cylinders sealed with indium. The H_2/He buffer gas inlet is at one end and a Teflon cap (which is transparent to the mm/submm probe) with a 1 mm diameter hole is used at the other to maintain a pressure differential between the experimental region and the guard vacuum region.

The electron gun consists of a heated tungsten filament located in a glass appendage to the main vacuum chamber for thermal and electrical isolation. A potential of ~1 keV is maintained between the filament and the entrance to the 77 K cylinder. In addition to the axial magnetic field provided by the main solenoid, a separately adjustable magnetic field in the region of the cathode is used to optimize the trajectories of the electrons through the experimental region. In operation, electrons are produced by the heated filament, accelerated by the electric field to ~1 keV, guided to and through the collisional cooling region by the axial magnetic field, and collected in the exit 77 K region with most of their energy intact.

The Diagnostics of the Collisionally Cooled Ion Cell

Because in collision experiments the kinetic temperatures of the collision partners must be well defined, careful diagnostics are required if reliable results are to be obtained. This is especially true at very low temperatures. In our earlier work on the development of the collisional cooling method, we reported diagnostic studies designed to measure and characterize the kinetic temperatures. For the study of cold ions additional factors, chiefly the open cell ends and the energy deposited by the electron beam, need to be considered. In order to understand and calibrate the kinetic temperature in the cold ion production region, we developed a quantitative model of heat flow.[3] This model shows that in the worst case the H_2 conduction between the 77 K and low temperature regions raises the temperature along the central axis a maximum of 0.5 K in the region populated by HCO^+. Additionally, because the electron beam deposits energy throughout the volume in proportion to current and gas density, it is possible to make comparisons between an analytic model and observations. These investigations identified large regions of pressure, temperature and current space where the electron beam has a negligible effect on measured collisional parameters.

Two examples of these comparisons are shown in Figs. 7 and 8. The range of pressures included in the fit was chosen to avoid the loss of collisional cooling efficiency at low pressure and electron beam heating at high pressure. The figures illustrate the different thermal effects nicely. Figure 7 clearly shows the beam heating which increases with hydrogen density and reduced hydrogen thermal conductivity at low temperatures. These density conductivity characteristics combined with the inverse temperature increase in the pressure broadening parameter gives the observed high pressure features in Fig. 8. The low pressure "hook" from loss of collisional cooling efficiency, a low density effect common in collisional cooling spectra, and the result of the partially cooled Doppler widths being attributed to pressure broadening, is also apparent in Fig 8. Additional details of these diagnostics are given in a recent paper.[3]

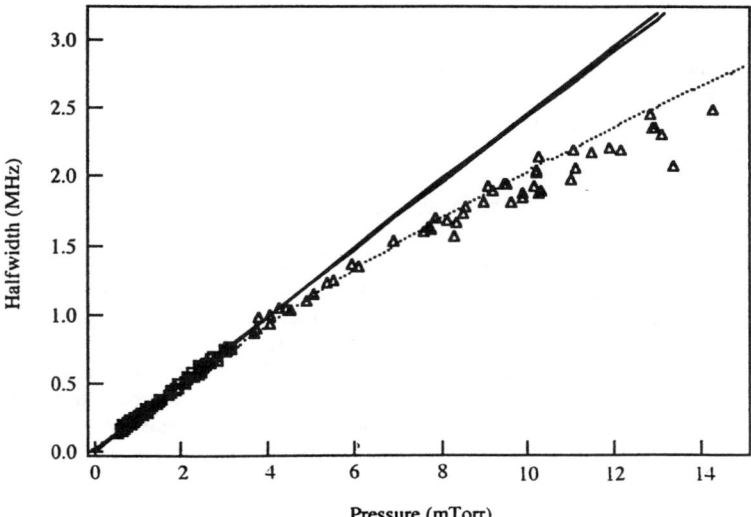

FIGURE 7. HCO^+ halfwidth as a function of pressure at 11.2 K and 8 ma beam current. The two solid lines represent fits to the linear portion of the data and the dashed line the prediction of the electron beam heating model.

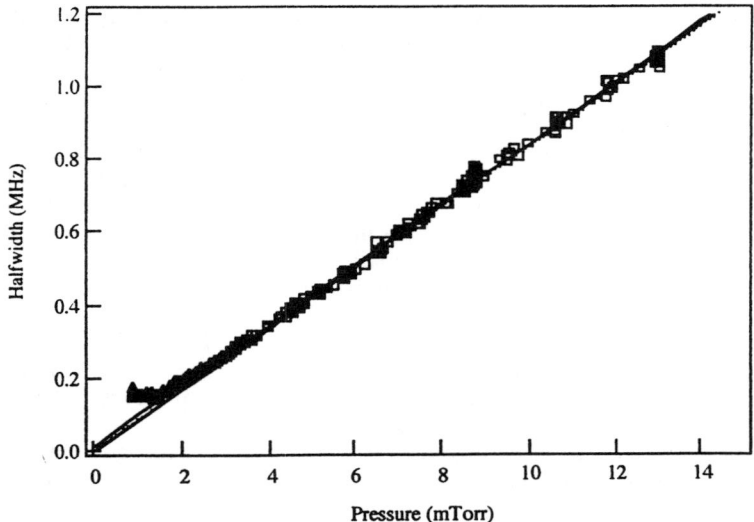

FIGURE 8. HCO$^+$ halfwidth as a function of pressure at 28 K and 10 ma beam current. The two solid lines represent fits to the linear portion of the data and the dashed line the prediction of the electron beam heating model. The low pressure data show the 'hook' from loss of cooling.

The Results

Because the principal purpose of this paper has been to describe the development of an experimental technique, only two representative results will be shown here. Additional information on the scientific results which have been obtained with these techniques can be found in a number of papers in the literature which have been cited above.

Figure 9 shows a comparison between the pressure broadening of the J = 3 - 2 transition of HCO$^+$ by H$_2$ as a function of temperature and a simple Langevin capture theory.[22] It is interesting to note that although these cross sections are perhaps five times larger than for neutrals at low temperature, they are still much lower than some of the initial estimates which would have precluded the observation of these species even at higher temperatures.[23] In contrast to the results for the neutral species CO - He shown below in Fig. 10, theory begins to under predict (by about 30%) the observed cross sections at the lowest temperatures. We are currently working to implement more exact quantal techniques using the MOLSCAT codes to see if a better agreement between experiment and theory can be obtained.

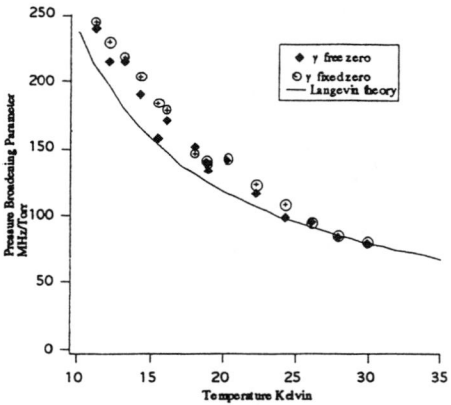

FIGURE 9. Pressure broadening parameters as a function of temperature for the J = 3 - 2 transition of HCO^+ broadened by H_2.

Collisions between neutrals at low temperature have been more thoroughly explored both experimentally and theoretically. Figure 10 shows a comparison between both the pressure broadening and pressure shifts for the J = 1 - 0 and J = 2 - 1 transitions of the CO - He system. In these figures calculations based both on an early intermolecular potential due to Thomas, Kramer, and Dierksen[24] and a more recent potential due to Chuaqui et al.[25] are shown. Overall the potential of Chuaqui et al. shows much better agreement (especially in the calculation of the pressure shift).

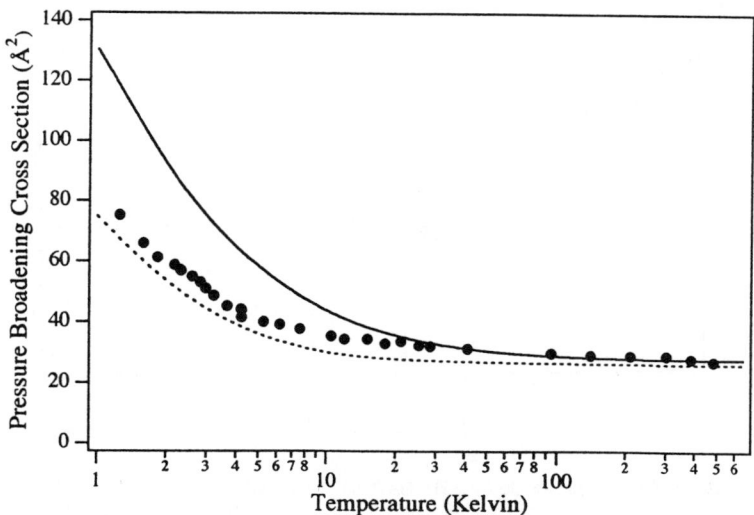

FIGURE 10. Comparison between experiment and theory for the pressure broadenings of the J = 1 - 0 transition of the CO - He system. The solid line is calculated from the potential of reference 27 and the dashed line from the potential of reference 26.

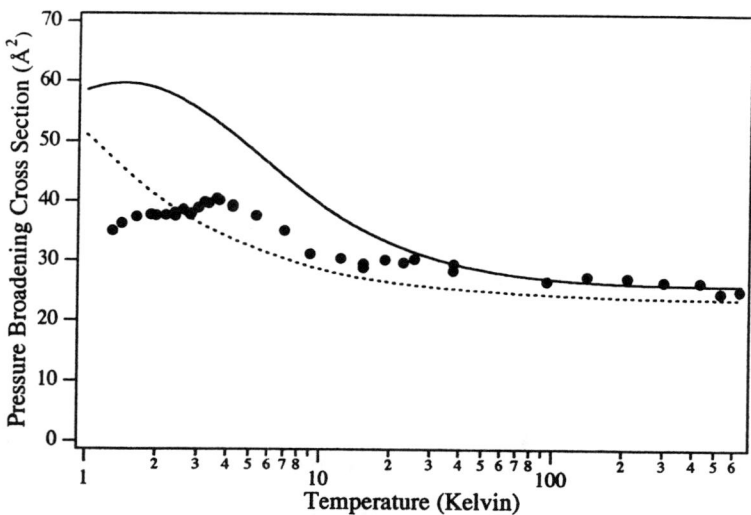

FIGURE 11. Comparison between experiment and theory for the pressure broadenings of the J = 2 - 1 transition of the CO - He system. The solid line is calculated from the potential of reference 27 and the dashed line from the potential of reference 26.

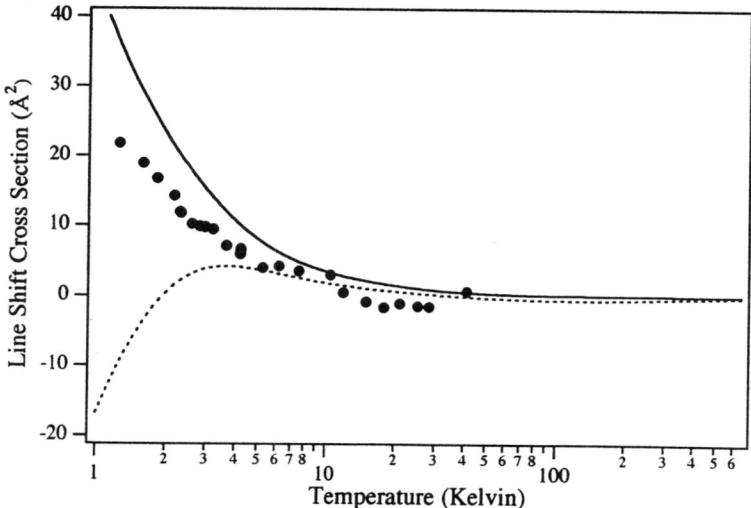

FIGURE 12. Comparison between experiment and theory for the pressure shifts of the J = 1 - 0 transition of the CO - He system. The solid line is calculated from the potential of reference 27 and the dashed line from the potential of reference 26.

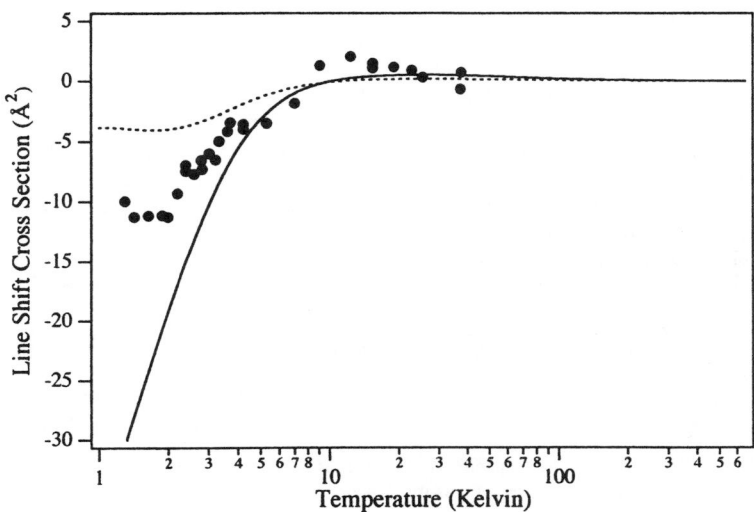

FIGURE 13. Comparison between experiment and theory for the pressure shifts of the J = 2 - 1 transition of the CO - He system. The solid line is calculated from the potential of reference 27 and the dashed line from the potential of reference 26.

ACKNOWLEDGMENTS

It is a pleasure to acknowledge the contributions of many students, post-docs and colleagues to the work described here. Among them are included Wayne C. Bowman, Grant M. Plummer, Daniel R. Willey, Daniel C. Flatin, John C. Pearson, Lee Oesterling, Matthew M. Beaky, Thomas M. Goyette, and Eric Herbst. This work has been supported by the National Science Foundation and the Army Research Office. We are also grateful to the Ohio Supercomputer Center for use of its Cray YMP-8.

REFERENCES

1. Messer, J. K., and De Lucia, F. C., " Measurement of Pressure-Broadening Parameters for the CO-He System at 4 K," Phys. Rev. Lett. **53**, 2555-2558 (1984).
2. De Lucia, F. C., Herbst, E., Plummer, G. M., and Blake, G. A., "The Production of Large Concentrations of Molecular Ions in the Lengthened Negative Glow Region of a Discharge," J. Chem. Phys. **78**, 2312-2316 (1983).
3. Pearson, J. C., Oesterling, L. C., Herbst, E., and De Lucia, F. C., "Pressure Broadening of Gas Phase Molecular Ions at Very Low Temperature," Phys. Rev. Lett. **75**, 2940-2943 (1995).
4. Anderson, P. W., "Pressure broadening in the microwave and infra red region," Phys. Rev. **76**, 647-661 (1949).
5. Green, S., and Mutso, "National Resource for Computation in Chemistry Software Catalogue," Lawrence Berkeley Laboratory, Berkeley, CA (1980).

6. Bowman, W. C., Plummer, G. M., Herbst, E., and De Lucia, F. C., "Measurement of the J = 0 → 1 Rotational Transitions of Three Isotopes of ArD$^+$," J. Chem. Phys. **79**, 2093-2095 (1983).
7. Plummer, G. M., Herbst, E., and De Lucia, F. C., "Laboratory Measurement of the J = 2 → 3 Rotational Transition Frequency of HC^{17}O$^+$," Ap. J. **270**, L99-L100 (1983).
8. G. M. Plummer, Herbst, E., and De Lucia, F. C., "Laboratory Measurement of the P(2,1) Submillimeter Transition Frequency of H$_3$O$^+$," J. Chem. Phys. **83**, 1428-1429 (1985).
9. The Millimeter and Submillimeter Spectrum of CF$^+$. Grant M. Plummer, Todd Anderson, Eric Herbst, and Frank C. De Lucia, J. Chem. Phys. 84, 2427-2428 (1986).
10. Warner, H. E., Conner, W. T., Petrmichl, R. H., and Woods, R. C., "Laboratory detection of the 1_{10} - 1_{11} submillimeter wave transition of the H$_2$D$^+$ ion," J. Chem. Phys. **81**, 2514 (1984).
11. Blake, G. A., Laughlin, K. B., Cohen, R. C., Busarow, K. L., and Saykally, R. J., "Laboratory measurement of the pure rotational spectrum of vibrationally excited HCO$^+$," Ap. J. **316**, L45 - L48 (1987).
12. Bogey, M., Demuynck, C., Denis, D., Destombes, J. L., and Lemoine, B., "Laboratory measurement of the 1_{10}- 1_{11} submillimeter line of H$_2$D$^+$," Astron. and Astrophys. **137**, L15 - 16 (1984).
13. Willey, D. R., Goyette, T. M., Ebenstein, W. L., Bittner, D. N., and De Lucia, F. C., "Collisionally Cooled Spectroscopy: Pressure Broadening below 5 K," J. Chem. Phys. **91**, 122-125 (1989).
14. Goyette, T. M., Ebenstein, W. L., and De Lucia, F. C., "Rotational and Vibrational Temperatures in a 77 K Collisionally Cooled Cell," J. Mol. Spectrosc. **140**, 311-321 (1990).
15. Goyette, T. M., and De Lucia, F. C., "Collisional Cooling as an Environment for Planetary Research," J. Geophys. Research **96**, 17455-17461 (1991).
16. Beaky, M. M., Flatin, D. C., Holton, J. J., Goyette, T. M., and De Lucia, F. C., "Hydrogen and Helium Pressure Broadening of CH$_3$F between 1 K and 600 K," J. Mol. Struct. **352/353**, 245-251 (1995).
17. Willey, D. R., Ross, K. A., Mullin, A. S., Schowen, S., Zheng, L., and Flynn, G., "Gas-Phase Infrared Spectroscopy of N$_2$O in an Equilibrium Cell at 10 and 5 K," J. Mol. Spectrosc. **169**, 66-72 (1995).
18. Jin, P., Wang, H.,Oatis, S., Hall, G., and Sears, T., "Very-Low-Temperature Infrared Laser Absorption Spectroscopy of N$_2$O, NO, and NO$_2$," J. Mol. Spectrosc. **173**, 442-451 (1995).
19. Dunder, T., and Miller, R. E., "Infrared spectroscopy and Mie scattering of acetylene aerosols formed in a low temperature diffusion cell," J. Chem. Phys. **93**, 3693-3703 (1990).
20. Ballard, J., and Newnham, D. A., OSA Topical Meeting, Santa Fe, NM, Feb. 9 - 11 (1995).
21. Willey, D. R., Southwick, R. J., Ramadas, K., Rapela, B. K., and Neff, W. A., "Laboratory Observation of Maser Action in NH$_3$ through Collisional Cooling," Phys. Rev. Lett. **74**, 5216-5219(1995).
22. Ausloos, P.,*Interaction Between Ions and Molecules,* Plenum Press, New York (1975).
23. Anderson, T. G., Gudeman, C. S., Dixon, T. A., and Woods, R. C., "Pressure broadening of the HCO$^+$ J = 0 - 1 transition by hydrogen," J. Chem. Phys. **72**, 1332-1336 (1980).
24. Thomas, L. D., Kraemer, W. P., and Diercksen, G. H. F. , "Rotational Excitation of CO by He Impact" Chem Phys. **51**, 131-139 (1980).
25. Chuaqui, C. E., Le Roy, R. J., and McKellar, A. R. W., "Infrared spectrum and potential energy surface of He-CO," J. Chem. Phys. **101,** 39-61 (1994).

Rotational Spectroscopy of Unstable Molecules Produced in a Low Density Plasma

Luca Dore, Claudio Degli Esposti, and Gabriele Cazzoli

*Dipartimento di Chimica "G. Ciamician", Università di Bologna,
Via Selmi 2, 40126 Bologna, Italy*

Abstract. Millimeter-wave spectroscopy has been exploited to study molecular ions produced in a d.c. glow discharge with magnetic confinement. A species of astrophysical interest like protonated formaldehyde (H_2COH^+) is being studied in order to determine its molecular structure. The analysis of a Coriolis-type interaction in the rotational spectra of the formyl ion (HCO^+) isotopomers allows an accurate determination of the equilibrium structure of this ion.

Laboratory spectra of molecular ions support the identification of species important in interstellar chemistry. The proton transfer reaction $H_3^+ + X \rightarrow HX^+ + H_2$ is the driving force of the interstellar chemistry. Formaldehyde was among the first polyatomic molecules discovered in space, and protonated formaldehyde is predicted to play a key role in both the formation and the depletion of H_2CO.

Protonated formaldehyde was first spectroscopically identified by Amano and Warner[1] by observing the ν_1 fundamental band at 2.9 μm. Later Chomiak et al.[2] succeeded in observing the millimeter-wave laboratory spectrum. Following this observation and a later investigation by Dore et al.,[3] we were able to assign the rotational spectrum of $H_2{}^{13}COH^+$.[4] As a further step toward the goal of determining the molecular structure, we have now observed and assigned the millimeter-wave spectrum of the fully deuterated species. The spectroscopic constants obtained by fitting the transition frequencies to an asymmetric rotor Hamiltonian using Watson's A-reduced I^r representation are reported in Table I.

TABLE I Ground state spectroscopic constants of D_2COD^+

A /MHz	102 145.142(14)	Δ_J/kHz	37.645(37)	δ_J/kHz	9.098(14)
B /MHz	27 062.9732(22)	Δ_{JK}/kHz	208.52(56)	δ_K/kHz	229.88(77)
C /MHz	21 328.1040(29)	Δ_K/kHz	1 645.9(43)		

In their study of the C-D stretch fundamental band of DCO^+, Kawaguchi et al.[5] found the vibration-rotation interaction constant α_1 to be smaller than the α_1 constants of the isoelectronic species DN_2^+ and DCN. They ascribed this anomaly to a Fermi-type interaction mixing the $v_1,v_2^l,v_3=10^00$ state with the excited bending state 04^00. However, the experimental data then available were not sufficient to give an unequivocal proof for this interpretation, even with a subsequent observation of the microwave spectra of Hirota and Endo.[6]

We have extended the observation of the rotational spectra to the 01^11, 03^10 and 04^00 excited vibrational states,[7] and we have shown that a Coriolis-type interaction perturbs the 01^11 and 10^00 states.

This Coriolis-type interaction affects also the rotational spectra of the normal species and of the other isotopomers, so that it has to be accounted for in computing the equilibrium structure. To this purpose, it should be more appropriate to use the following expression for the equilibrium rotational constant at the α level:

$$B_e = 1.5\, B_{00^00} - 0.5\, (B_{10^00} + 2B_{01^11}) + B_{01^10} + 0.5\, B_{00^01} - 0.5\, B_{02^00}$$

In the combination $B_{10^00} + 2B_{01^11}$, in fact, the Coriolis contributions to each constant cancel out. This equation has been used to derive the equilibrium structures for pairs of isotopomers reported in Table II employing, for the hydrogen containing isotopomers, the rotational constants obtained by Warner.[8]

The consistency between the different structures, along with the agreement with the theoretical results, confirms that this method is suitable to correct for the effect of the Coriolis interaction and it led to an accurate determination of the equilibrium structure for the formyl ion.

TABLE II Equilibrium structures for the formyl ion

	r_{CO} (Å)	r_{CH} (Å)
HCO^+ / $H^{13}CO^+$	1.105582	1.092026
HCO^+ / $HC^{18}O^+$	1.105595	1.091957
$HC^{18}O^+$ / $H^{13}CO^+$	1.105556	1.092162
DCO^+ / HCO^+	1.105580	1.092032
DCO^+ / $H^{13}CO^+$	1.105581	1.092032
DCO^+ / $HC^{18}O^+$	1.105581	1.092029
4 isotop. fit	1.1055808(4)	1.0920310(18)
theory Ref. 9	1.1055(1)	1.0919(3)
theory Ref. 10	1.1058(2)	1.0919(5)

REFERENCES

1. T. Amano and H.E. Warner, *Astrophys. J.* **342**, L99 (1989).
2. D. Chomiak, A. Taleb-Bendiab, S. Civiš, and T. Amano, *Can. J. Phys.* **72**, 1078 (1994).
3. L. Dore, G. Cazzoli, S. Civiš, and F. Scappini, *Chem. Phys. Lett.* **244**, 145 (1995).
4. L. Dore, G. Cazzoli, S. Civiš, and F. Scappini, presented at the XIV Colloquium on High Resolution Molecular Spectroscopy, Dijon, France, September 11-15, 1995.
5. K. Kawaguchi, A.R.W. McKellar, and E. Hirota, *J. Chem. Phys.* **84** (1986) 1146.
6. E. Hirota and Y. Endo, *J. Mol. Spectrosc.* **127** (1988) 527.
7. L. Dore and G. Cazzoli, *Chem. Phys. Lett.* in press.
8. H.W. Warner, *Ph.D. Thesis*, Wisconsin-Madison, 1988.
9. C. Puzzarini, R. Tarroni, P. Palmieri, S. Carter, and L. Dore, *Mol. Phys.* **87**, 879 (1996).
10. P. Botschwina and S. Schmatz, in *The Structure, Energetics and Dynamics of Organic Ions* (T. Baer, C.-Y. Ng and I. Powis Eds.), John Wiley & Sons Ltd, 1996.

Plasma Broadening and Shifting of Analogous Spectral Lines Along Isoelectronic Sequences

B.Blagojević, M.V.Popović, N.Konjević and M.S.Dimitrijević*

Institute of Physics, 11080 Belgrade, P.O.Box 68, Yugoslavia
*Astronomical Observatory, 11050 Belgrade, Volgina 7, Yugoslavia
e-mail: eblagojb@ubbg.etf.bg.ac.yu

Abstract. The Stark widths and shifts of the $3s^2S-3p^2P^0$ transitions along the isoelectronic sequences of lithium and boron, the $3s^3S-3p^3P^0$ transitions of beryllium sequence and the $3p^2P^0-3d^2D$ transitions of boron sequence have been studied theoretically using impact semiclassical method and experimentally observed in the plasma of a low pressure pulsed arc. The plasma electron densities were determined from the width of the HeII P_α line while the electron temperatures were measured from relative line intensities. To estimate the influence of different ions to the Stark width of lines, evaluation of the plasma composition data is performed and in conjunction with our theoretical results contribution of ion broadening estimated. Furthermore in our theoretical calculations for the first time we included the influence of perturbing levels with different parent therms to width and shift of the investigated OIV spectral lines.

THEORY

By using the semiclassical-perturbation formalism [1] we have calculated electron and all relevant perturbing ions impact broadening parameters for the $3s^2S-3p^2P^0$ transitions along the isoelectronic sequences of lithium and boron, the $3s^3S-3p^3P^0$ transitions of beryllium sequence and the $3p^2P^0-3d^2D$ transitions of boron sequence. Energy levels needed for these calculations have been taken from [2]. Oscillator strengths were calculated by using the method described in [3], see also [4]. The contribution of higher energy levels is estimated as in [5].

EXPERIMENT

The light source was a low pressure pulsed arc with a quartz discharge tube 10 mm internal diameter. The distance between aluminum electrodes was 161 mm, and 3 mm diameter holes were located at the center of both electrodes to allow end-on plasma observations. A 30 mm diaphragm placed in front of the focusing mirror ensures that light comes from the narrow cone about the arc axis. All plasma observations are performed with 1-m monochromator with inverse linear dispersion 0.833 nm/mm in the first order of the diffraction grating, equipped with the photomultiplier tube and stepping motor. The discharge was driven by: 15.2 μF low inductance capacitor charged to 3.0; 3.8 and 6.0 kV, critically damped current pulse duration τ = 7.7 μs, pressure of the gas mixtures p = 3 torr, The spectral line profiles were recorded with instrumental half widths of 0.0168 nm. The experimental apparatus and procedure are briefly described in [6].

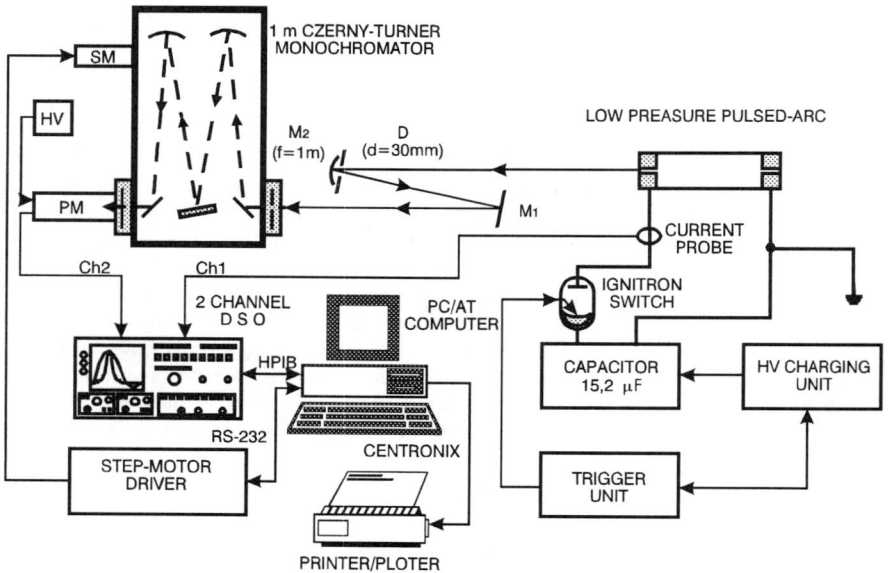

Figure 1. The experimental setup.

EXPERIMENTAL RESULTS AND DISCUSSION

The experimental results for Stark widths and comparisons with theoretical results for $3s^2S-3p^2P^0$ transitions along the isoelectronic sequences of lithium and the $3s^3S-3p^3P^0$ transitions of beryllium sequence are given in Figs.2-3 while the experimental results for the $3s^2S-3p^2P^0$ and the $3p^2P^0-3d^2D$ transitions of boron sequence are given in Fig.4. In order to evaluate contribution of ion impact widths it was necessary to compute plasma composition data for the conditions of width measurements. In the studied electron temperature range and within the estimated uncertainties the experimental Stark widths agree well with the results of our semiclassical electron impact widths. The only exception are BII lines which agree better with modified semiempirical formula [7] see Fig.2. For the conditions of the present experiment, estimated contribution of the ion broadening has never exceeded seven percents of the total line width. So within the precision of this experiment it was not possible to detect its contribution with certainty.

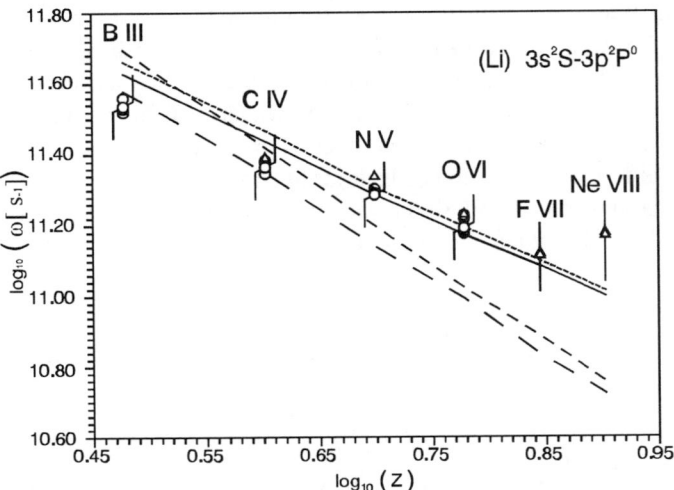

Figure 2. Stark widths Li-like spectral lines (in angular frequency units) as a function of $\log_{10} Z$ for $3s^2S$-$3p^2P^0$ multiplets. Theory: ········, semiclassical electrons + ions impact widths, ———, semiclassical electrons only; - - -, semiclassical approximation (Eq.(526) taken from [8]); — — —, modified semiempirical formula [7]. Experiment: O, our data; Δ, Glenzer et al [9,10].

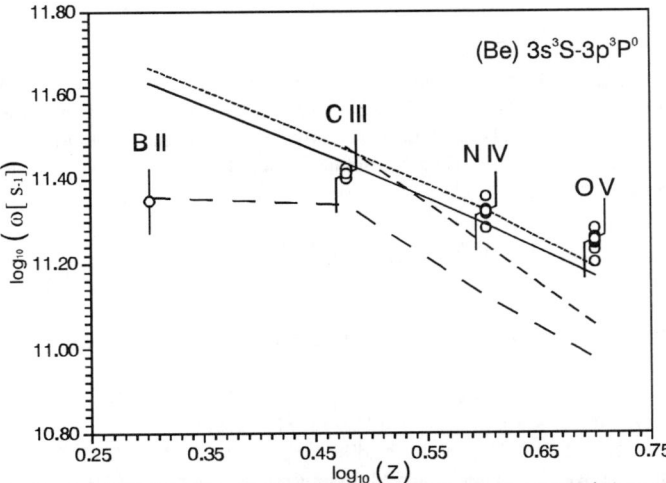

Figure 3. Stark widths Be-like spectral lines (in angular frequency units) as a function of $\log_{10} Z$ for $3s^3S$-$3p^3P^0$ multiplets. Theory: ········, semiclassical electrons + ions impact widths, ———, semiclassical electrons only; - - -, semiclassical approximation (Eq.(526) taken from [8]); — — —, modified semiempirical formula [7]. Experiment: O, our data.

The first calculation of our semiclassical electron impact shifts of the investigated spectral lines shows an agreement within estimated uncertainties with the experimental Stark shifts, except for the 3s-3p and the 3p-3d transitions of OIV and the 3s-3p transitions of NIV, NV and CIV which had the opposite sign in relation to the experimentally measured ones. For OIV, we included several transitions with different parent terms which change the sign of the shift [12]. The improved calculations for OIV shifts are in an good agreement with the experiment. Further calculations are in progress.

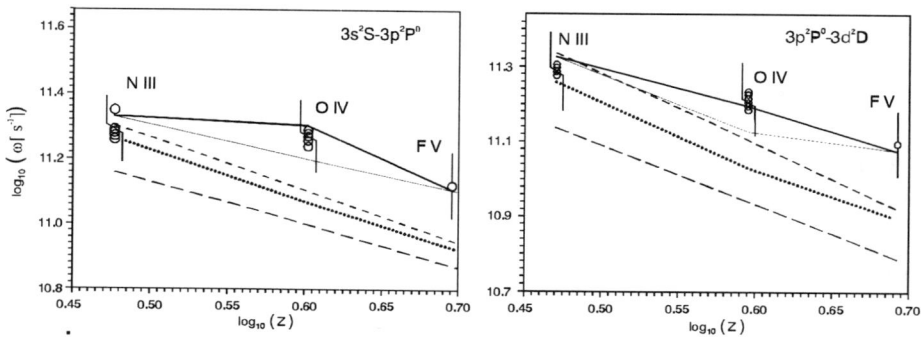

Figure 4. Stark widths B-like spectral lines (in angular frequency units) as a function of $\log_{10}Z$ for $3s^2S-3p^2P^0$ and $3p^2P^0-3d^2D$ multiplets. Theory: ········, semiclassical electrons + ions impact widths, ———, semiclassical electrons only; - - -, semiclassical approximation (Eq.(526) taken from [8]); — — —, modified semiempirical formula [7]. Experiment: O, our data; Δ, Glenzer et al [11].

REFERENCES

1. S.Sahal-Brechot, Astron.Astrophys. **1**, 91; **2**, 322 (1969).
2. S.Bashkin and J.J.Stoner,Jr., *Atomic Energy Levels and Grotrian Diagrams*, Vol. 1 (North Holland, Amsterdam, 1975).
3. D.R.Bates and A.Damgaard, Trans.Roy.Soc.London, Ser.A **242**, 101 (1949).
4. G.K.Oertel and L.P.Shomo, Astrophys.J.Suppl.Ser. **16**, 175 (1968).
5. H.Van Regemorter, Hoang Binh Dy, and M.Prud'homme, J.Phys.B **12**, 1073 (1979).
6. B.Blagojević, M.V.Popović, N.Konjević, and M.S.Dimitrijević, Phys Rev.E **50**, 2986 (1994).
7. M.S.Dimitrijević and N.Konjević, JQSRT **24**, 451 (1980).
8. H.R.Griem, *Spectral Line Broadening by Plasmas*, (Academic, New York, 1974).
9. S.Glenzer, N.I.Uzelac, H.-J.Kunze, Phys.Rev.A **45**, 8795 (1992).
10. S.Glenzer, N.I.Uzelac, H.-J.Kunze, *Spectral Line Shapes*, edited by R.Stamm and B.Talin (Nova Science, Commack, NY, 1993), Vol.7.
11. S.Glenzer, J.D.Hey, H.-J.Kunze, J.Phys.B **27**, 413 (1994).
12. B.Blagojević, M.V.Popović, N.Konjević, and M.S.Dimitrijević, Phys Rev.E, in press (1996).

On the Stark Broadening of B III Spectral Lines

Milan S. Dimitrijević [1] and Sylvie Sahal-Bréchot [2]

[1] *Astronomical Observatory, Volgina 7, 11050 Belgrade, Yugoslavia*
e-mail : mDimitrijevic@aob.aob.bg.ac.yu
and
[2] *Observatoire de Paris, 92195 Meudon Cedex, France*
e-mail : sahal@obspm.fr

Abstract. Using a semiclassical approach, we have calculated electron-, proton-, and ionized helium-impact line widths and shifts for 12 Be III and 27 B III multiplets. The obtained results have been compared with available experimental and theoretical data. The complete results will be published elsewhere. Here, we discuss the comparison of results for B III, with experimental and other theoretical data.

For studies as e.g. numerical modelling of stellar plasma or abundance determinations, data on B III lines may be of interest. Moreover, Stark broadening of B III lines is of interest for the investigation and diagnostic of laboratory and laser-produced plasma, as well as for the research of regularities and systematic trends. In (1)-(3) Stark widths of B III lines have been calculated within the semiempirical method (4), the modified semiempirical method (5), the simplified semiclassical method ((6), Eq. 526) and its modification (5). Moreover, Stark widths and shifts for B III 2s - 2p, 2s - 3p, 2p - 3s, 2p - 3d, 3s - 3p and 3p - 3d have been calculated by Seaton (7) within the quantum mechanical strong coupling method. Stark broadening parameters of B III have been also investigated experimentaly in two contributions. In Djeniže et al. (8), the results concerning the B III 4f ^{2}F° - 5g^{2}G 4497.6 Å line, measured in a pulsed linear arc plasma, have been reported. Srećković et al. (9) measured in a linear, low-pressure pulsed arc operating in O_2, the Stark widths of two lines within the B III $2s\ ^2$S - $2p\ ^2$P° multiplet.

The B III Stark broadening data are of interest and for studies of regularities and systematic trends within isoelectronic sequences. In our previous articles, Stark broadening data for Be II, C IV, N V, O VI, F VII, Ne VIII, Na IX, Al XI, Si XII all belonging to the lithium isoelectronic sequence have been calculated. Consequently, the results for B III will complete this set of data.

By using the semiclassical-perturbation formalism (10, 11), we have calculated electron-, proton-, and ionized helium-impact line widths and shifts for 27 B III

By using the semiclassical-perturbation formalism (10, 11), we have calculated electron-, proton-, and ionized helium-impact line widths and shifts for 27 B III multiplets, for perturber densities 10^{17} - 10^{21} cm^{-3} and temperatures T = 10,000 - 300,000K, in order to continue our research of multiply charged ion line Stark broadening parameters.

The unique experimental result convenient for comparison, is the Stark widths of the two lines within the B III $2s$ ^{2}S - $2p$ ^{2}P° multiplet, measured by Srećković et al. (9) in a linear, low-pressure pulsed arc operating in O_2. They found a large disagreement between their Stark widths and the results of (1), obtained within the modified semiempirical approach (5). For the B III 2065.77 Å line, they found that the ratio of measured to theoretical Stark width is 7.8 and for 2067.23 Å line 6.7 for the temperature of 48000 K at an electron density of 2.55 x 10^{17} cm^{-3}. Corresponding ratios with our results with ionized oxygen-impact broadening included, are 3.9 and 3.5 respectively, which is better but not satisfying.

We may compare available theoretical results for B III $2s$ ^{2}S - $2p$ ^{2}P° multiplet for the temperature of 160000 K at an electron density of 1 x 10^{17} cm^{-3}. Our full width at half maximum is W = 0.0103 Å, and the agreement is closest with calculations of Dimitrijević and Konjević (1) who used the simplified semiclassical approach of Griem ((6), Eq. 526), since they obtained W = 0.00892 Å. Within the modified semiempirical approach (5), same authors obtained W = 0.00449 Å, which is two times smaller. Within the close coupling quantum mechanical approach Seaton (7) obtained W = 0.00602 Å, which is also in disagreement with experiment and with our calculations. In order to clarify the situation, particularly since B III $2s$ ^{2}S - $2p$ ^{2}P° multiplet is important for the consideration of Stark broadening parameters within the lithium isoelectronic sequence, we recommend a new experimental determination of Stark broadening parameters particularly for this multiplet.

REFERENCES

1. Dimitrijević, M.S., and Konjević N., 1981, in *Spectral Line Shapes*, ed. B. Wende, W.de Gruyter and Co., New York, 521.
2. Dimitrijević, M.S., *Bul. Obs. Astron. Belgrade* **139**, 31 (1988).
3. Dimitrijević, M.S., *Astron. Astrophys. Suppl. Series* **76**, 53 (1988).
4. Griem, H.R., *Phys. Rev.* **165**, 268 (1968).
5. Dimitrijević, M.S., Konjević, N., *J. Quant. Spectrosc. Radiative Transfer* **24**, 451 (1980).
6. Griem, H.R., *Spectral Line Broadening by Plasmas*, Academic Press, New York, 1974.
7. Seaton, M.J.: 1988, *J.Phys.B* **21**, 3033 (1988).
8. Djeniže, S., Srećković, A., Labat, J., Plati\v sa, M., *Physica Scripta*, **45**, 320 (1992).
9. Srećković, A., Djenže, J., Platiša, M., in *XVI Int. Symp. Phys. Ioniz. Gases,* (Beograd), Contributed papers and Abstracts of Invited Lectures and Progress Reports, ed. M. Milosavljević , Institute of Nuclear Sciences "Vinca", Beograd, 1993, 201.
10. Sahal-Bréchot, S., *Astron. Astrophys.* **1**, 91 (1969).
11. Sahal-Bréchot, S., *Astron. Astrophys.* **2**, 322 (1969).

On the Stark Broadening of Mg II Spectral Lines

Sylvie Sahal-Bréchot [1] and Milan S. Dimitrijević [2]

[1] *Observatoire de Paris, 92195 Meudon Cedex, France*
e-mail : sahal@obspm.fr
and
[2] *Astronomical Observatory, Volgina 7, 11050 Belgrade, Yugoslavia*
e-mail : mDimitrijevic@aob.aob.bg.ac.yu

Abstract. Using a semiclassical approach, we have calculated electron-, proton-, and ionized helium-impact line widths and shifts for 67 Mg II multiplets. The obtained results have been compared with available experimental and theoretical data. The complete results will be published elsewhere. Here, we discuss the comparison of our results with experimental and other theoretical data.

The study Mg II lines does not only interest laboratory plasma research and plasma devices development. Due to the cosmical abundance of magnesium and its ionization potential value, Mg II lines are present in solar and stellar spectra and the corresponding Stark broadening data are important for stellar spectra analysis and synthesis, as well as for abundance determinations and stellar plasma modelling and research.

In order to provide the needed Stark broadening data, we have calculated within the semiclassical - perturbation formalism (1, 2) electron-, proton-, and ionized helium-impact line widths and shifts for 67 Mg II multiplets, for a perturber range of densities 10^{16} - 10^{19} cm^{-3} and temperatures T = 5,000 - 150,000 K. We discuss here the obtained results, together with a comparison with experimental and other theoretical data.

A detailed critical analysis of Mg II experimental data with the special emphasis on Mg II 3s - 3p resonance line has been performed in (3), and it was concluded that the results of Goldbach et al. (4) and Roberts and Barnard (5), which adequately account for the critical factors, provide the most reliable data for the Mg II resonance lines. These results are in accordance with the strong coupling quantum mechanical calculations (6,7) and about two times smaller than results of full semiclassical calculations (present results and results from (8, 9). One should be noted that the semi - classical method gives often results of lower accuracy for ionic resonance lines, especially at lower temperatures, since the full quantum mechanical approach is needed for appropriately including the various short range effects. We notice a good agreement between semiempirical calculations performed in (10) with the Griem's semiempirical (11) and the modified semiempirical approach (12), with the most reliable experimental results (Roberts and Barnard (5), Goldbach *et al.* (4) and with the quantum mechanical calculations (Bely and Griem (7) ; Barnes (6)). This is promising for the use of the much simpler modified semiempirical method (12) when there are no sophisticated results available, or when the use of the quantum close-coupling sophisticated method needs a considerably higher effort and does not promise a higher accuracy (e.g. lack of reliable atomic data or very high levels involved). For non - resonant lines there are much less data, which are additionaly of lower accuracy. Our results are in excellent agreement with the experimental result of Chapelle and Sahal - Bréchot (13) for the Mg II 3d - 4f transition, which has the best experiment accuracy for non-resonant lines according to critical reviews of experimental data (3, 14, 15).

One can see that there are less experimental data for the shift and that they are of lower accuracy. It is difficult to make a final conclusion since even the sign of experimental shifts are different. New and high precision measurements would be very useful.

REFERENCES

1. Sahal-Bréchot, S., *Astron. Astrophys.* 1, **91** (1969).
2. Sahal-Bréchot, S., *Astron. Astrophys.* **2**, 322 (1969).
3. Konjević, N., Dimitrijević, M.S., Wiese, W.L., *J.Phys.Chem.Ref.Data* **13**, 649 (1984).
4. Goldbach, C., Nollez, G., Plomdeur, P., Zimmermann, J.P., *Phys. Rev. A* **25**, 2596 (1982).
5. Roberts, D.E., Barnard, A.J., *J. Quant. Spectrosc. Radiative Transfer* **12**, 1205 (1972).
6. Barnes, K.S., *J.Phys. B* **4**, 1377 (1971).
7. Bely, O., Griem, H.R., Phys. Rev. A **1**, 97 (1970).
8. W. W. Jones, S. M. Benett and H. R. Griem, *Calculated Electron Impact Broadening Parameters for Isolated Spectral Lines from the Singly Charged Ions: Lithium through Calcium*, Tech. Rep. No 71-128, Univ. of Maryland, College Park, Maryland (1971)
9. Griem, H.R., *Spectral Line Broadening by Plasmas*, Academic Press, New York, 1974.
10. Dimitrijević, M.S., Konjević, N., *Astron. Astrophys.* **102**, 93 (1981).
11. Griem, H.R., 1968, *Phys. Rev.* **165**, 268 (1968).
12. Dimitrijević, M.S., Konjević, N., *J. Quant. Spectrosc. Radiative Transfer* **24**, 451 (1980).
13. Chapelle, J., Sahal-Bréchot, S., *Astron. Astrophys.* **6**, 415 (1970).
14. Konjević, N., Wiese, W.L., *J.Phys.Chem.Ref.Data* **5**, 259 (1976).
15. Konjević, N., Wiese, W.L., *J.Phys.Chem.Ref.Data* **19**, 1307 (1990).

Experimental Stark Widths, Shifts and Transition Probabilities of several ArII lines

J. A. Aparicio*, M. A. Gigosos[†], S. Mar[†] and V. R. González[†]

*Departamento de Física Aplicada, Universidad de Salamanca, Spain. [†]Departamento de Óptica y Física Aplicada, Universidad de Valladolid, 47071 Valladolid, Spain

Abstract. This paper is an extensive experimental contribution to the knowledge of ArII atomic parameters. This specie, which is very important for many astrophysical and industrial plasma diagnostics, has been extensively studied. However, there are still great differences in the experimental Stark widths and shifts coefficients, as well as a great lack of transition probability data, especially for lines coming from the very highly excited energy levels.

Keywords: Plasma spectroscopy, plasma diagnosis.

Interferometric and spectroscopic end-on measurements have been carried out on a plasma created by discharging a capacitor bank, charged up to 20 μF, on the electrodes of a lamp filled with a continuous flow of pure argon (in case of ArII A_{ki}-values determination) or helium-argon mixtures (for the Stark widths and shifts) at different pressures. Both measurements have been taken 2 mm from the lamp axis and from symmetrical positions referred to it.

One-wavelength interferometry allowed us to determine plasma refractivity changes and from them the electron density. It ranges usually between 2×10^{22} to 2×10^{23} m^{-3} with an uncertainty lower than 10%). Spectroscopic records of the ArII spectrum with exposures from 3 to 5 μs have been taken at different times of the plasma emission. This lasts approximately 500 μs. In all cases the excitation temperature (which ranges between 13000 and 26000 K, with an estimated error lower than 5%) has been calculated by means of a Boltzmann-plot[1] of the ArII lines, using the last compilation of transition probabilities of this specie[2]. In pure argon conditions, electron density has also been calculated from the Stark broadening of H_α

line, which appears in this plasma as an impurity. This line has been calibrated in this laboratory from simulation techniques[3,4]. Doppler and instrumental broadening were taken into account in the calculation of the Stark width from the total width.

The transition probabilities result from the intensities of the ArII lines along the plasma life and this temperature. Some of the A_{ki}-values obtained for the lines coming from the very highly excited ArII energy levels are listed in Table 1. In Table 2, the slopes B_ω and B_d of the linear fittings between the Stark widths and shifts, and the electron density for some ArII lines are shown.

Table 1. A_{ki}-values ($\times 10^8$ s^{-1})

λ (nm)	A_{ki}	λ (nm)	A_{ki}	λ (nm)	A_{ki}	λ (nm)	A_{ki}	λ (nm)	A_{ki}	λ (nm)	A_{ki}
362.21	0.338	379.66	0.148	395.84	0.028	420.34	0.221	454.78	0.148	495.51	0.154
363.70	0.151	379.94	0.213	397.94	1.042	421.87	0.348	456.10	0.021	509.05	0.134
363.98	0.480	380.32	0.876	398.82	0.047	422.26	0.860	456.38	0.448	516.27	0.186
365.09	0.091	380.95	0.339	401.12	0.055	422.99	0.282	456.44	0.321	516.58	0.137
365.53	0.292	381.90	0.166	401.98	0.360	425.56	0.033	457.29	0.021	521.68	0.147
365.61	0.070	382.57	0.474	403.14	0.074	427.52	0.232	461.12	0.130	581.27	0.076
366.04	0.613	382.68	0.322	403.38	0.775	429.80	0.220	461.41	0.134	597.33	0.0018
370.99	0.041	384.15	0.231	407.08	0.642	433.71	0.400	470.36	0.299	598.59	0.053
371.72	0.053	386.85	1.698	407.24	0.743	436.78	0.575	472.16	0.178	618.71	0.085
371.82	1.261	387.21	0.219	409.95	0.080	438.51	0.262	478.62	0.114	637.59	0.0065
372.04	0.140	388.03	0.202	411.64	0.075	440.49	0.187	479.21	0.053	647.53	0.018
372.45	0.279	390.06	0.071	412.97	0.448	443.38	1.076	486.59	0.190	648.01	0.021
373.79	1.307	391.16	0.074	415.61	0.285	444.58	0.081	486.76	0.113	661.44	0.117
376.35	0.215	392.60	0.749	416.90	0.058	449.85	0.469	488.23	0.696	662.10	0.054
376.53	0.697	393.25	1.028	417.93	0.105	451.00	0.022	491.43	0.022	669.63	0.024
377.05	0.322	394.61	0.863	418.97	0.067	453.55	0.094	494.29	0.134	705.50	0.056
378.08	0.797	395.27	0.181	419.99	0.147	453.76	0.176	494.94	0.016		

Table 2. Coefficients B_ω and B_d ($\times 10^{-23}$ pm/m^{-3}) obtained in this work.

λ (nm)	B_ω	B_d	λ (nm)	B_ω	B_d	λ (nm)	B_ω	B_d	λ (nm)	B_ω	B_d
458.99	29.4	-2.19	465.79	30.0	-4.06	497.22	27.6	-5.35	500.93	29.9	-5.61
460.96	27.5	-3.93	496.51	36.5	-5.57	484.78	28.5	-6.40	480.60	27.9	-5.02
457.93	28.7	-1.27	688.66	47.1	-3.28	506.20	29.9	-6.99			
476.49	30.1	-3.54	434.81	20.7	-3.57	493.32	30.5	-7.56			

The authors would like to acknowledge the financial support of the DGICYT under Contract No. PB94-0216.

REFERENCES

1. M. A. Gigosos, S. Mar, C. Pérez and I. de la Rosa, Phys. Rev. E, 49, 2, 1575 (1994).
2. V. Vjunović and W. L. Wiese, J. Phys. Chem. Ref. Data, 21, 919 (1992).
3. M. A. Gigosos and V. Cardeñoso, J. Phys. B: At. Mol. Opt. Phys. 22, 1743 (1987).
4. V. Cardeñoso and M. A. Gigosos, Phys. Rev. A39, 5258 (1989).

Stark broadening of several N II lines

C. Pérez, I. de la Rosa, M. A. Gigosos, J. A. Aparicio and S. Mar

Departamento de Óptica. Universidad de Valladolid. 47071 Valladolid (Spain).

Keywords : plasma spectroscopy, plasma diagnostics.

The Stark broadening of several NII lines have been measured in a pulsed discharge. These data have interest for different kinds of applications. As a probe of this, there are a great number of papers providing these experimental data [1–11].

The electron density, determined by two wavelengths interferometry, lies in the interval from 0.05 to 1.1×10^{23} m^{-3}. The excitation temperature, derived from the Boltzmann plot of several NII lines, varies from 5000 to 8000 K.

Results from this work and a review from previous experiments are shown in Table. All widths have been normalized at 10^{23} m^{-3}. In general, our results agree well with others. Eventhough we have worked with a very large number of data, the errors derived from measurements of the width of very narrow lines can be estimated to be, in same cases, even near 20%. Among the results new data is provided by this work. But as it is a work still in progress we hope to be able to provide more and more accurate data including shifts measurements.

[1] R.A. Day and H.R. Griem, Phys. Rev. A **140**, 1129 (1965).
[2] H.F. Berg, W. Ervens, and B. Furch, Z. Phys. **206**, 309 (1967).
[3] N.W. Jalufka and J.P. Craig, Phys. Rev. A **1**, 221 (1970).
[4] N. Konjevic, V.Mitrovic, Lj. Cirkovic, and J. Labat, J: Fizika. **2**, 129 (1970).
[5] M.V. Popovic, M. Platisa, and N. Konjevic, Astron. Astrophys. **41**, 463 (1975).
[6] E. Kllne, L.A. Jones, and A.J. Barnard, J. Q. S. R. Transfer **22**, 589 (1979).
[7] S.T. Purcell and A.J. Barnard, J. Q. S. R. Transfer **32**, 205 (1984).
[8] T.L. Pittman and N. Konjevic, J. Q. S. R. Transfer **36**, 289 (1986).
[9] J. Puric, A. Sreckovic, S. Djenize, and M. Platisa, Phys. Rev. A **36**, 3957 (1987).
[10] L. Istrefi, Rev. Roum. Phys. **33**, 667 (1988).
[11] S. Djenize, A. Sreckovic, and J. Labat, Astron. Astrophys. **253**, 632 (1992).

This work have been financed by Spanish D.G.I.C.Y.T under Contract Pb-94-0216.

Stark FWHM: Results from this experiment and review of previous data.

λ (Å)	Transition			FWHM (Å) at $10^{23}\,m^{-3}$	
				This work	Previous Experiments
3995.00	$3s^1\,P^0$	—	$3p^1\,D$	0.28	0.34 [2], 0.30 [3], 0.29 [4], 0.31 [5], 0.77 [6], 0.30 [7], 0.30 [10], 0.36 [11]
4041.31	$3d^3\,F^0$	—	$4fG[9/2]$	1.09	0.83 [5], 0.70 [8], 0.52 [9], 0.90 [11]
4447.03	$3p^3\,D$	—	$3d^3\,D^0$	0.35	0.45 [2], 0.28 [3], 0.31 [5], 0.58 [6], 0.32 [11]
4530.41	$3d^1\,F^0$	—	$4fg[9/2]$	2.19	1.60 [1], 2.20 [2], 2.2 [3], 1.28 [8]
4601.48	$3s^3\,P^0$	—	$3p^3\,P$	0.38	0.34 [7]
4607.16	$3s^3\,P^0$	—	$3p^3\,P$	0.35	0.3 [7]
4613.87	$3s^3\,P^0$	—	$3p^3\,P$	0.30	0.14 [1], 0.40 [2], 0.35 [3], 0.33 [5], 0.3 [7], 0.23 [11]
4621.39	$3s^3\,P^0$	—	$3p^3\,P$	0.35	0.31 [5], 0.33 [7]
4630.54	$3s^3\,P^0$	—	$3p^3\,P$	0.38	0.40 [2], 0.35 [3], 0.28 [4], 0.33 [5], 0.33 [7]
4643.09	$3s^3\,P^0$	—	$3p^3\,P$	0.35	0.35 [3], 0.28 [7]
4788.13	$3p^3\,D$	—	$3d^3\,D^0$	0.37	
4803.29	$3p^3\,D$	—	$3d^3\,D^0$	0.36	0.35 [5], 0.62 [6], 0.26 [11]
4994.36	$3p^3\,S$	—	$3d^3\,P^0$	0.45	
5005.15	$3s^5\,P$	—	$3p^5\,P^0$	0.37	
5045.10	$3s^3\,P^0$	—	$3p^3\,S$	0.32	0.29 [7]
5535.10	$3s^5\,P$	—	$3p^5\,D^0$	0.42	
5666.63	$3s^3\,P^0$	—	$3p^3\,D$	0.47	0.41 [11]
5676.02	$3s^3\,P^0$	—	$3p^3\,D$	0.51	0.41 [11]
5679.56	$3s^3\,P^0$	—	$3p^3\,D$	0.54	0.41 [11]
5686.21	$3s^3\,P^0$	—	$3p^3\,D$	0.49	0.38 [11]
5710.77	$3s^3\,P^0$	—	$3p^3\,D$	0.47	0.44 [7]
5927.81	$3p^3\,P$	—	$3d^3\,D^0$	0.57	
5931.78	$3p^3\,P$	—	$3d^3\,D^0$	0.56	
5940.24	$3p^3\,P$	—	$3d^3\,D^0$	0.60	
5941.65	$3p^3\,P$	—	$3d^3\,D^0$	0.50	0.83 [7]
5952.39	$3p^3\,P$	—	$3d^3\,D^0$	0.58	0.62 [7]
6481.73	$3s^1\,P^0$	—	$3p^1\,P$	0.66	
6482.05	$3s^1\,P^0$	—	$3p^1\,P$	0.63	
6610.56	$3p^1\,D$	—	$3d^1\,F^0$	0.79	

A New Deconvolution Method Applied to the Si II (1) Lines Profiles

A. Lesage[*], M. Depiesse[**], D. Meiners[***], J. Richou[**] and F. Wollschläger[***]

[*]*DASGAL/URA335, Observatoire de Paris-Meudon, 92195 Meudon CEDEX, France*
[**]*Laboratoire d'Opto-Electronique, Université de Toulon et du Var, 83957, La Garde, France*
[***]*Institut für Experimentalphysik, Heinrich-Heine-Universität, D 40225 Düsseldorf, Germany*

Abstract. Biraud's method of deconvolution is applied for the first time in spectroscopy to unfold the λ 3862, 3856, and 3854 Å lines profiles. The emitting plasma is generated in a diaphragm shock tube. The test gas is Neon plus a small admixture of Tetramethylsilane. Temperature from 11,000 to 14,500 K and electron densities from 0.5 to $2.0 \cdot 10^{23}$ m^{-3}, are achieved in the radiation cooling region behind the reflected shock wave. Electron densities are determined with a two-wavelength (4880 and 6328 Å), two phase Michelson interferometer with photoelectric detection.

I. INTRODUCTION

The chemical composition of stars is known through their spectra, if the most important transition probabilities of all the atomic species present in their atmospheres are known. But under certain circumstances, the Stark effect plays a very important role in that determination [1]. It is when the Stark width is larger than the Doppler width and in the case of very strong lines where most of the absorbed energy lies in the wings of the Voigt profile, being determined by the Stark width only. For instance, for a B-type star standard atmosphere at a depth of $\tau_{5000} = 1$, the Stark width for the Si II lines at 4201 Å is comparable or even larger than the Doppler width. The calculated values of the equivalent widths become very sensitive to the adopted Stark broadening, as soon as the core begins to saturate. That is why an accurate determination of the Stark width is essential [2].

In previous publications [3,4] we have pointed out that there are still discrepancies concerning the Stark parameters of the five prominent visible Si II multiplets, especially for the multiplet (1), where both semiclassical predictions and a number of experimental determinations disagree by a factor of two [5-14]. Chiang and Griem's measurements [6] as well as Griem's predictions [9] and Lesage's experimental values [7] are by a factor of two larger than Sahal's calculations[7]. The early measurements done by Konjevic [10] as well as the later by Puric [11-13] and by Lesage [14] provided values which are all in closer agreement with Sahal's predictions than with Griem's. Semiempirical calculations [15] done using different approximations [5,8,10,14] and the fully quantum mechanical calculations done by Blaha (in [6]) agree with Sahal's predictions. The purpose of the present contribution is to show how a new method of deconvolution enable us to provide more reliable experimental data than those previously published and to unfold the weakest λ 3854 Å Si II (1) line, which has never been measured before.

II. RESULTS

Line profiles are observed with a REOSC Ha spectrograph and an intensified RETICON detector in the focal plane. They are corrected for self absorption. After subtracting the continuum, the observed experimental profile, which is affected by the apparatus function of the spectrograph, is unfolded by applying Biraud's method.

Present linewidths values are compared with previous experimental data. They are in good agreement with the measurements of Konjevic, Puric and Lesage. However, earlier values measured by Lesage and by Chiang are almost by a factor of two larger. The optical depth and inhomogeneity problems, already mentioned earlier could explain these large line width values.

Compared to present experimental results the semi-classical calculations done by Griem yield very high values, whereas Sahal's semi-classical calculations, the semi-empirical calculations done by Konjevic, Hey, Jones and Lesage are in good agreement. Blaha's distorted wave calculations provides a value in very close agreement to Sahal's results.

III. REFERENCES

1 T. Lanz, M. S. Dimitrijevic, M. C. Artru, Astron. Astrophys. **192**, 249 (1988).
2 T. Lanz, M. C. Artru, in: Elemental Abundance Analyses, S. J. Adelman, T. Lanz, Editors (Lausanne, 1988).
3 M. Sokoll, J. Mitsching, D. Meiners, A. Lesage, Proc. ICPIG XXI (1993).
4 A. Lesage, M. Depiesse, F. Wollschlaeger, D. Meiners, J. Richou, Proc. ICPIG XXII (1995).
5 J. D. Hey, JQRST **18**, 425 (1968).
6 T. Chiang, H. R. Griem, Phys. Rev. **A 18**, 1169 (1978).
7 A. Lesage, S. Sahal-Brechot, M. H. Miller, Phys. Rev. **A 16**, 1617 (1977).
8 W. Jones, Phys. Rev. **A 7**, 1826 (1973).
9 H. R. Griem, Spectral Line Broadening by Plasmas (Academic, New York, 1974).
10 N. Konjevic, J. Puric, L. J. Cirkovic, J. Labat, J. Phys. **B 3**, 999 (1970).
11 J. Puric, S. Djenize, J. Labat, L. J. Cirkovic, Z. Phys. **267**, 71 (1974).
12 J. Puric, S. Djenize, J. Labat, L. J. Cirkovic, I. Lakicevic, in: Proc. VIIIth Summer School on Phys. Ion. Gases, Dubrovnik (1976).
13 J. Puric, A. Lesage, V. Knezevic, in: Proc. IXth Summer School on Phys. Ion. Gases, Dubrovnik (1978).
14 A. Lesage, B. A. Rathore, I. S. Lakicevic, J. Puric, Phys. Rev. **A 28**, 2264 (1983).

Observation of the Giant Coulomb Broadening in the Gas Discharge Plasma

A.A.Apolonsky, S.A.Babin, S.I.Kablukov, S.V.Khorev,
E.V.Podivilov, A.I.Chernykh, D.A.Shapiro

*Institute of Automation & Electrometry, Russian Academy of Sciences,
1 University Ave, Novosibirsk, 630090, Russia*

Abstract. The Coulomb broadening is defined by the velocity change due to the diffusion during level lifetime. Laser upper levels 4p of ArII with lifetime about 5 nsec led to the broadening by factor 3–4. Special experiment with $3d'$ metastable ArII levels with lifetime 40 nsec gains the effect. The broadening by 100 times is observed in probe-field spectrum. The width of the Bennett hole is comparable with the Doppler one; its shape becomes cusp-like in agreement with the theory.

Earlier the maximal Coulomb broadening observed was by factor 3-4 [1] compared to the line homogeneous width. Theory predicted not only the strong broadening for level with long lifetime, but also the change in shape. The profile of nonlinear resonance should become cusp-like [2] proportional to $\exp(-2|\Omega|\ln 2/\Delta)$, where $\Omega = \omega - \omega_0$ is the frequency detuning from the line centre. When the lower level lifetime τ_n is much longer than that of upper level, the diffusion width Δ is defined by collisional frequency

$$\Delta = kv_T\sqrt{\frac{\nu_{ii}\tau_n}{2}}, \quad \nu_{ii} = \frac{16\sqrt{\pi}Ne^4\Lambda}{3m^2v_T^3}.$$

Here k is the wavenumber, v_T is the thermal velocity, N is the ion density, e is the elementary charge, Λ is the Coulomb logarithm, m is the ion mass.

In present paper we study the saturation resonance in the probe-field spectrum with long-lived metastable level ArII $3d'\ ^2G_{7/2}$ providing the giant broadening owing to the diffusion in the velocity space.

The strong field on transition 617.2 nm ($3d'\ ^2G_{7/2} - 4p'\ ^2F^o_{5/2}$) and counter-propagating probe wave were generated by single-frequency dye laser pumped

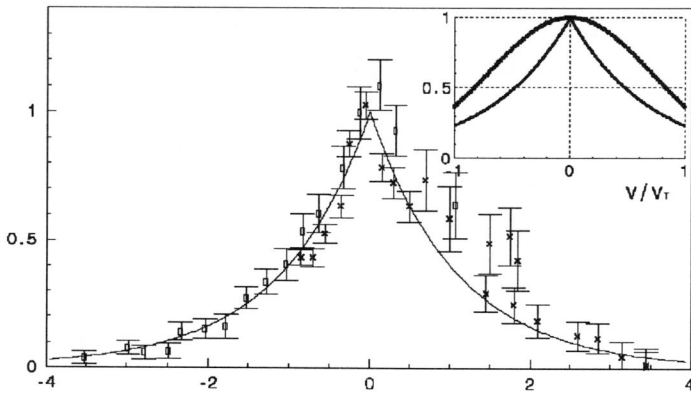

Figure 1: Profile of the resonance in the probe-field spectrum versus detuning $\Omega/2\pi$ (GHz) fitted into the experimental points by maximum-likelihood method. Crosses and boxes denote two different data sets. For comparison, the Doppler velocity distribution (in v_T units) is shown. The discharge length is $l = 100$ cm, electron density $N_e \simeq 1.7 \times 10^{-14}$cm^{-3}, electron and ion temperatures $T_e \simeq 4$ eV, $T_i \simeq 0.8$ eV, respectively.

by the argon laser. The observed profile of resonance is shown in fig.1. The homogeneous width was about 35 MHz, then the observed broadening is by 100 times. We measure independently the ion density 1.7×10^{14} cm^{-3} in plasma and found the diffusion coefficient to be 4.2×10^{17} cm^2/sec^3. The lifetime of the metastable determined by electron deactivation occurred to be 40 nsec. The probe-field resonance started from the similar argon ionic level $3d'$ $^2G_{9/2}$ had been observed [3], but without charged particle density diagnostics. The measured FWHM of the Bennett hole $\Delta = 3$ GHz is comparable with the Doppler width 5.3 GHz.

The theory for two-level system allows not only to describe the lineshape, but also to calculate the saturation dependence of absorbed power on the incident intensity. The prediction was that the velocity diffusion both enhance the absorption and changes the behaviour of the saturation curve from inhomogeneous to homogeneous shape [4]. To compare with the experiment we should do three improvements of the theory:

1. Include the degenerate magnetic sublevels with different dipole moments.

2. Consider the dynamic friction force significant for wide velocity distribution.

3. Take into account the optical thickness of plasma column.

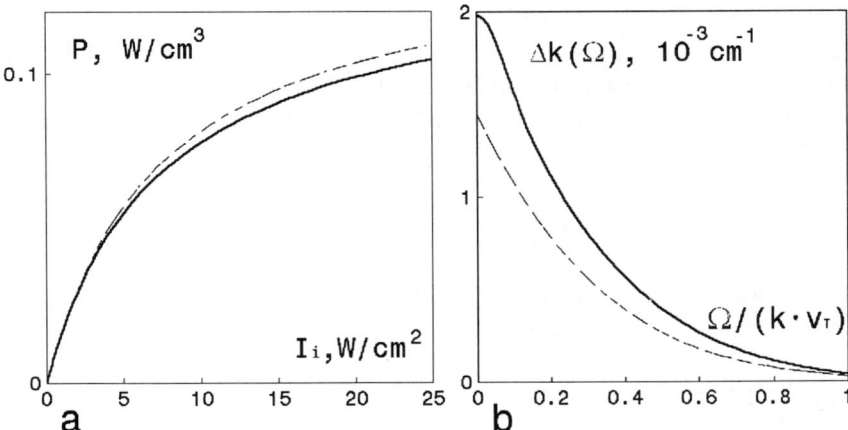

Figure 2: a — Dependence of the absorbed power P upon the incident wave intensity I_i in an optically thin medium for a two-level system (dashed) and a numerical simulation for magnetic sublevels with friction force included (solid). b – The variation of the absorption coefficient caused by the strong field at the intensity $I = 0.1 I_S$: solid curve plots the result of numerical simulation, dashed curve is the approximation without friction.

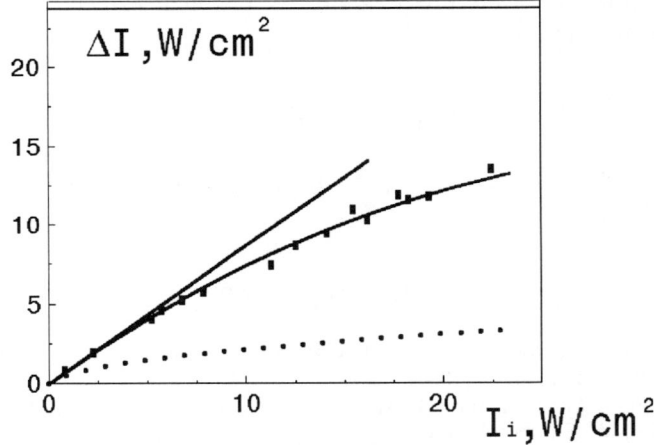

Figure 3: Dependence of the absorbed intensity $\Delta I = I_i - I_f$ on the transition $3d'\,^2G_{7/2} \to 4p'\,^2F^o_{5/2}$ upon the incident intensity I_i. Boxes correspond to experimental data. Theoretical dependence (solid curve) is fitted with the maximum-likelihood method including optical thickness. Dots correspond to inhomogeneous saturation while Coulomb diffusion neglected. The strait lines show to the derivative from the saturation curves at zero point, and to the asymptote of maximum possible absorption. Plasma parameters are the same as in fig. 1.

Some numerical results shown in fig.2 demonstrate that the magnetic sublevels and friction force do not almost change the absorption. Meanwhile the friction force enhance the gain coefficient while the Coulomb broadening is comparable with Doppler width. The optical thickness is accounted analytically. Obtained expression is

$$\ln \frac{I_i}{I_f} + \frac{I_i - I_f}{I_S} = \kappa_0 l,$$

where I_i, I_f are the incident and final intensities, l is the length of plasma column, κ_0 is the small-signal absorption coefficient,

$$\kappa_0 l \simeq \frac{\lambda^3 A_{mn}}{8\pi^{3/2} v_T} \left(N_n \frac{g_m}{g_n} - N_m \right), \quad I_S = \frac{8\pi \sqrt{2\nu_{ii}} k v_T \hbar c}{\lambda^3 A_{mn}} \left(\Gamma_m^{-1/2} + \Gamma_n^{-1/2} \right)^{-1}$$

is the saturation intensity, $\lambda = 2\pi/k$ is the radiation wavelength, A_{mn} is the Einstein coefficient, N_j, g_j, Γ_j is the population, statistical weight, and relaxation constant of level $j = m, n$, respectively, $c, \hbar$ are the speed of light and Plank constant.

Fig.3 illustrates the comparison of saturation curves with and without the diffusion. The fitting of the saturation curve to experiment allows us to estimate the lower level lifetime $\tau \sim 20$ nsec in rough agreement with the spectroscopic measurement.

ACKNOWLEDGMENTS

This work is supported in part by the International Science Foundation, grant RCN300.

REFERENCES

[1] Babin, S.A., Donin, V.I., Rodishevsky, A.V., and Shapiro D.A., *Sov. J. Quant. Electr.* **18**, 796–801 (1988).
[2] Babin, S.A. and Shapiro, D.A., *Phys. Rep.* **241**, 119–216 (1994).
[3] Elbel, M., Simon, M., and Welp, H., *Quantum Opt.* **2**, 351–364 (1990).
[4] Kurlayev, K.B., and Shapiro, D.A., *Quant. Electr.* **24**, 1003–1007 (1994).

Classical Descriptions of Rydberg Spectral Line Shapes

L.Bureyeva[*], V.Lisitsa[**], H.van Regemorter[***] and Ousname Motapon[***]

[*] Scientific Council on Spectroscopy of the RAS, 13/15, Vetoshnii per., Moscow, 103012, Russia
[**] RRC "Kurchatov Institute", Kurchatov Sq., 46, Moscow, 123182, Russia
[***] Department Atomes et Molecules en Astrophysique, Observatoire de Paris, Place Jules Janssen 5, 92195 Cedex, Meudon, France

Abstract. A general approaches to the line shape calculations in a frame of classical mechanics is based on Fourier analysis of Rydberg electron trajectory in electric fields of two types, namely atomic (in particular Coulomb) one and the fields which are responsible for broadening phenomena. Three cases to apply a classical approach for line shape calculations are considered, namely: 1) adiabatic approximation, 2) binary approach and 3) Model Microfield Method (MMM).

Rydberg spectral line shapes are of great interest both in astrophysic and laboratory plasmas. Their calculations on the basis of the standard quantum broadening in the impact approximation have been suggested by H. Van Regemorter [1]. Nevertheless the quantum theory faces with the problem of taking into account an interaction between a large number of atomic states. To avoid this difficulties it is natural to use a classical approach for calculations of such line shapes. We'll formulate lower general approaches to a construction of classical broadening theory.

A general approach to the line shape calculations in a frame of classical mechanics is based on Fourier analysis of Rydberg electron trajectory in electric fields of two types, namely atomic (in particular Coulomb) one and the fields which are responsible for broadening phenomena. The Fourier transforms of the electron trajectory may be separated , namely the Fourier harmonics of the motion in atomic potential determine the intensities of specific radiation transitions whereas the Fourier components in broadening electric fields are responsible for broadening of the specific components of the radiative transition. We arrive to a general picture of broadened spectra as a series of components with specific intensities and

© 1997 American Institute of Physics

shapes determined by corresponding Fourier harmonics of electron motion. The problems of specific calculations are connected with possibilities of determination of perturbed electron trajectory. These calculations may be performed with three following approximations.

1) **Adiabatic approximation** is based on slow nature of electron or ion perturbations which make it possible to use a static solution of the problem with parametric dependence of perturbations on time. The well known adiabatic solution for spectral line shapes takes a typical structure:

$$I(\omega) = \sum_{i,k} d_{i,k}^2 U_{i,k}(\Delta\omega) \qquad (1)$$

where $U_{i,k}$ is a Fourier transform of the evolution operator $U(t)$, $d_{i,k}$ are Fourier components of electron dipole momentum on the atomic trajectory. For the static ion field U is expressed in terms of the electric field distribution function. The calculations of eq. (1) in static approximation may be sharply simplified for Rydberg states with the help of classical mechanics using the results of Bureyeva [2] and Davydkin-Zon [3] in spherical coordinate system or Born [4] in parabolic ones. The distribution function in eq. (1) may be change in adiabatic theory by Fourier transform of adiabatic evolution operator as it follows from adiabatic broadening theory (see, for example, Sobelman [5]).

2) The possibility of line shape calculations (1) in the case of **binary collisions** follows from a possibility to reduce the evolution of the radiating atom to the evolution of the atom in crossed electrical and magnetic fields [6]. The solution of last problem is well known in classical mechanics [4].

3) At last the third possibility of solution of broadening problem is based on **MMM-theory** [7]. It is connected with the simple relations between MMM and static solutions. The corresponding line shape formula looks like (1) where instead of electric field distribution function the MMM propagator $\{U\}_{MMM}$ is used.

So at least in three pointed cased it is possible to apply a classical approach for line shape calculations. The only problems are performing of a summation over i,k array of specific quantum numbers (Fourier harmonics) which in the case of Rydberg states can be changed by an integration procedure .

REFERENCES

1. H. Van Regemorter, Ann.Rev.Astronom.Astrophys. **3**, 71 (1965)
2. L. Bureyeva, Sov.Astronom. **12**, N3, 962 (1968)
3. V. Davydkin and B. Zon, Opt.Spectr. USSR, **51**, 13 (1981)
4. M. Born: The mechanics of the Atom (Bell, London 1927)
5. I. Sobelman: Introduction to the Theory of Atomic Spectra (Pergamon, Oxford, 1979)
6. L. Bureyeva and V. Lisitsa, XIV ICAP, Book of Abstr., 1Q-9, Boulder, USA (1994)
7. A. Brissaud and U. Frish, JQSRT, **11**, 1767 (1971)

Laser Generated Plasma of Li, Zn and Li-Zn Mixture

S. Gogić and S. Milošević

Institute of Physics, P.O.Box 304, HR-10000 Zagreb, Croatia

Abstract. We have produced by 308 nm laser ablation lithium, zinc and lithium-zinc plasma and have studied their spectral, temporal and spatial characteristics with the aim to use its expansion in vacuum for the formation of the LiZn* and LiZn$^+$ excimers.

Recent progress in the study of intermetallic excimers of alkali-IIB group has pointed to the need of preparing excimers in the molecular beams (1). Since particular alloys are difficult to melt and vaporize the laser ablation and expansion of the generated plasma may be a way to overcome the problem. This method has been already used for many different molecular systems (2),(3),(4).

We have undertaken the present study in order to characterize the plume formed by 308 nm laser ablation of lithium, zinc and lithium-zinc targets.

The experimental setup is shown in Fig. 1. A XeCl excimer laser at 308 nm was used for ablation. The laser energy varied from 20 to 100 mJ per pulse and the pulse halfwidth was about 20 ns. The laser beam is focused by a quartz lens of 30 cm focal length. The target was placed inside a vacuum chamber which could be filled with different gases (Ar, He, H_2) up to 1 atm. The image of the plume was projected with the lens into the plain of opening of a bundle optical fiber (diameter 3 mm). The optical fiber was connected to a 0.6 m scanning monochromator. Dispersed fluorescence was detected by a photomultiplier (Hamamatsu R2949). The signal was sent to a boxcar averager where it was processed and the output was sent to an IBM compatible PC. Spatial, temporal and spectral analysis of the plume was performed.

We used as a targets pure lithium or zinc metal or their alloys. The alloys were prepared previously in the heat-pipe ovens. Targets (especially lithium) were rotated in order to get pulse to pulse stable plume.

© 1997 American Institute of Physics

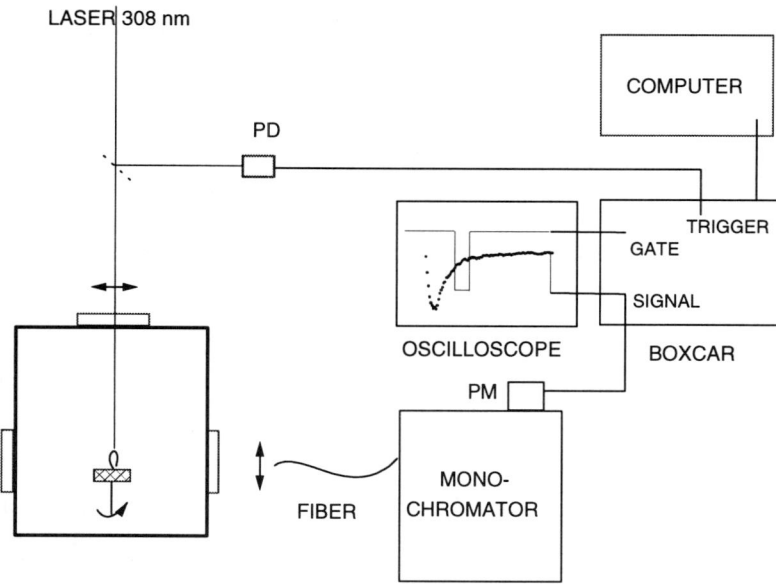

FIGURE 1: Experimental setup for laser ablation. PD is a fast photodiode.

When the laser pulse hits the surface of the target it vaporize and ionize the material. The light emitting plasma is getting different shapes in the space depending on the pressure and kind of a buffer gas. Under a vacuum conditions ($<10^{-3}$ mbar) bright white spot is observed at the place were the laser hits the target. Very tinny emission (blue, green, red for Zn, LiZn and Li respectively) extends all around filling almost a whole vacuum chamber. As a buffer gas is introduced into the chamber a region of this weak emission starts to shrink and forms a small bright ball of emitting light (10 to 15 mm at 2 - 5 mbar). This ball can have a shells of different colors. As a buffer gas pressure is increased more (100 mbar) the ball shrinks more into a small (3 mm) very bright white spot on the target surface. In the direction of incoming laser beam the reflection of the light from a small particles could be seen with naked eye.

In Fig. 2 we present spectra of lithium plasma in the region from 390 nm to 690 nm for different time delays after the beginning of the laser pulse. Basically only atomic spectral features could be seen. For a short times after the laser pulse (up to 100 ns) practically continuum emission is observed due to the spectral overlap of highly broadened atomic lines. At this conditions the most prominent effect is a linear Stark broadening on the $n\ ^2D - 2\ ^2P$ transitions, as well as, the appearance of

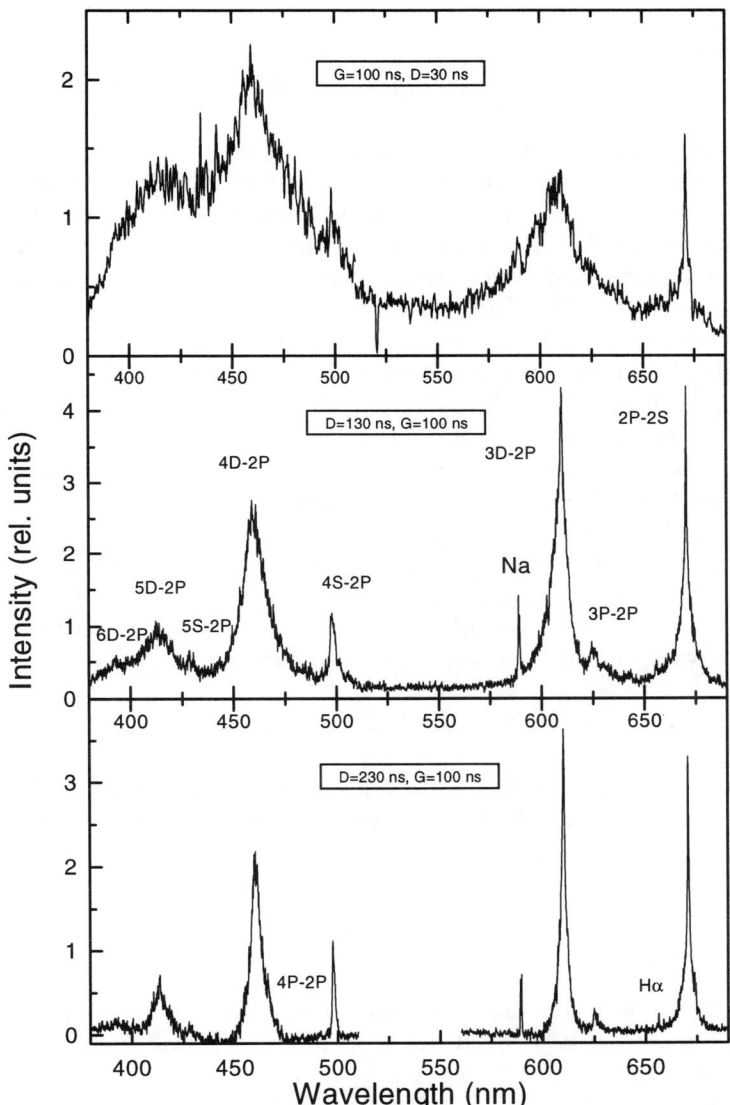

FIGURE 2: Spectra of lithium plasma for different time delays at Ar buffer gas pressure of 5 mbar. The light from complete plume of 1 cm diameter is fed to the monochromator. G is a gate and D is a delay setting of a box-car averager.

the forbidden 3,4 ^{2}P - 2 ^{2}P transitions, indicating a high electron density within the first few hundreds of nano seconds after the laser pulse. This points towards the similar plasma conditions as those obtained by Ya' akobi (3) by the method of an exploded lithium wire. Preliminary comparison shows that the electron densities of 10^{17}-10^{18} cm^{-3} are obtained in the present case. However we note that the self-reversal on the line shapes (4) is not observed for a low buffer gas pressures but only above 500 mbar. The linewidths were found to increase almost linearly with increasing laser pulse energy or a buffer gas pressure and decrease exponentially with increasing the time delay. The effective lifetimes have been measured within the line profile of the 4D -2P transition at 460.3 nm. Typical lifetimes it the wings of the lines are about 1 μsec whereas in the center of the line they amout 5-7 μsec.

In a pure zinc both atomic and ionic lines are observed within 300 ns after the laser pulse. The ionic transition 4 ^{2}F$_{7/2}$- 4 ^{2}D$_{5/2}$ at 492.3 nm also shows pronounced Stark broadening suitable for diagnostics. It appears that recombination processes responsible for the spectral emission from plasma are faster in zinc compared to the pure lithium. This is the main reason for the spatial regions of the plume with different colors in the case of laser ablation of the Li-Zn alloys (7).

The experiments are in progress in which the Li-Zn laser generated plasma is expanded with the flow of argon into vacuum in order to enhance possible formation and observation of the LiZn* or LiZn*$^+$ excimers and detect them by means of a cavity ringdown laser absorption spectroscopy (8).

ACKNOWLEDGMENTS

This work was financially supported by the Ministry of Science of the Republic of Croatia. We wish to thank Dr. Goran Pichler for valuable discussions and continuous support of this work.

REFERENCES

1. Milošević, S., "Diffuse bands of intermetallic excimers", in AIP Conference Proceedings 328, Spectral Line Shapes, Vol. 8 , 1995, pp. 391-405; Azinović, D., Li, X., Milošević, S., and Pichler, G., Phys. Rev. A **53**, 1323-1329 (1996).
2. Dietz, T. G., Duncan, M. A., Powers, D. E., and Smalley R.E., J. Chem. Phys. **74**, 6511 (1981).
3. Bondybey, V. E., and English, J. H., J. Chem. Phys. **74**, 6978-6979 (1981).
4. Kuo-mei Chen et al., Chem. Phys. Lett., **187**, 198- 202 (1991).
5. Ja' akobi, B., Phys. Rev. **176**, 227-230 (1968).
6. Ja' akobi, B., J. Quant. Spectrosc. Radiat. Transfer. **9**, 1097-1103 (1969).
7. Gogić, S. and Milošević, S., to be published
8. Paul, J. B., Scherer, J. J., Collier, C. P., and Saykally, R. J., J. Chem. Phys. **104**, 2782 (1996)

ATOMS AND MOLECULES
IN STRONG LASER FIELDS

Spectral Analysis of Ultrashort Pulses after Nonlinear Propagation

R. Lange, J.-F. Ripoche, M. Franco, G. Grillon, B. Prade and A. Mysyrowicz

LOA, ENSTA, Ecole Polytechnique, 91125 Palaiseau, France, mysy@ensta.enstay.fr

The spectrum of intense femtosecond optical pulses after propagation through transparent nonlinear medium contains precious information on the phase and amplitude of the optical pulse, provided the nonlinear response of the medium is known.
Alternatively, knowing the phase and amplitude of the optical pulse, it is possible to extract with high accuracy the nonlinear Kerr response of the medium. Examples of pulse characterisation and determination of Kerr coefficients of different solids, liquids, and gases will be presented.
A particular interesting phenomenon occurs in air, where spontaneous revivals of the nonlinearity are observed well after the passage of the pulse.

INTRODUCTION

Femtosecond optical pulses lend themselves to the study of ultrafast material dynamics. Indeed sub-picosecond pulses have been applied to a wide range of experiments in such diverse fields as ligand dynamics in Hemoglobin (1), coherent control of reaction dynamics in chemistry (2), electron relaxation in semiconductor physics (3) and X-ray generation in plasma physics (4). Sub-picosecond pulses inherently possess a large spectrum - it typically lies between 10 and 20 nm for a 100 fs pulse. The large spectrum undergoes modifications when the pulse propagates through a nonlinear, transparent medium (5). There is an obvious interest in performing line shape analysis of such picosecond pulses as these spectra contain a wealth of information. The analysis of lineshapes allows to characterise the optical pulse in its amplitude and phase. This provides a useful diagnostic tool in the development of CPA laser systems as sources of intense sub-picosecond pulses (6,7). The ENSTA phase-retrieval method (8) presented below is one amongst other methods of pulse characterisation (9-12). Its advantage resides in the experimental simplicity: the method requires only a spectrometer in order to record the spectrum before and after the nonlinearity and a standard desk computer to perform the spectral analysis. In this article we will first review the basic nonlinear effect of self-phase modulation which is at the core of our approach. In the second paragraph we present the ENSTA-phase retrieval method (8) which provides an efficient diagnostic tool for optical pulses. Once the pulse is

© 1997 American Institute of Physics

characterised the method allows to analyse unknown nonlinearities as shown in the third section. In the fourth section we will extend spectral analysis after propagation in order to decipher complex temporal pulse responses by cross-phase-modulation-techniques (XPM).

NONLINEAR LINESHAPING

A pulse is completely characterised by the spectrum and the phase of the various frequency components with respect to the central frequency ω_0. One commonly describes the initial phase in terms of the chirp of the pulse:

$$\phi(\omega) = \phi_2(\omega_0)\frac{(\omega-\omega_0)^2}{2} + \phi_3(\omega_0)\frac{(\omega-\omega_0)^3}{6} + \phi_4(\omega_0)\frac{(\omega-\omega_0)^4}{24} + ...$$

The spectral chirp of nth order corresponds to the $(n+1)^{th}$ coefficient in the expansion in the frequency domain. A similar expansion in the time domain leads to temporal chirp of various orders. The chirp represents the change of the frequency over the duration of the pulse, for instance blue frequency components of the spectrum arriving before the red spectral components.

Spectra corresponding to various initial phases of a 60 fs pulse are shown in Figure 1. Figure 1 (a) describes the case of a pulse with an initial flat phase (dotted line), corresponding to a bandwidth limited pulse. The modified spectrum obtained after propagation through a given nonlinear material is shown as a fat continuous line. The same initial spectrum is shown in graph (b) of Figure 1, but for a different initial phase, corresponding to a negative linear chirp. This negative linear chirp induces a contraction of the spectrum after propagation through the nonlinearity. The graph (c) shows again the same initial spectrum, but this time with a third order chirp leading to asymmetric spectral changes in the propagated spectrum.

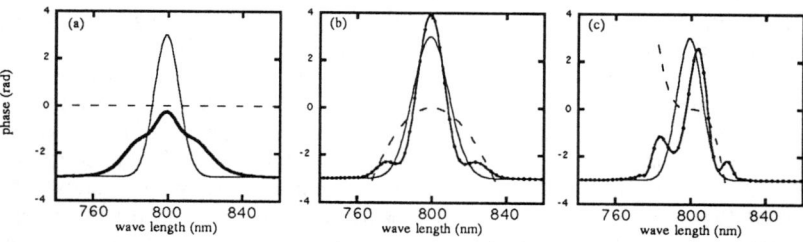

Figure 1 Input phase (dashed), reference (thin) and modulated (fat) spectrum as an illustration for various SPM lineshapes. Different chirps attributed to the same initial pulse lead to different line shapes: (a) flat phase, (b) first chirp and (c) quadratic chirp.

The physical origins of the spectral changes are well known (13) and are due to self phase modulation via the Kerr effect. The Kerr effect induces a nonlinear refractive index:

$$n_2 = \frac{12\pi}{n_0} \chi^{(3)}(-\omega,\omega,\omega,-\omega)$$

where $\chi^{(3)}$ is the third order susceptibility tensor for linearly polarised light in an isotropic medium and n_0 the refractive index of the medium.

Formally the propagation of a complex field $E(\tau, z=0)$ with its spectrum $|E(\omega, z=0)|^2$ and phase $\phi(\omega)$ over a distance Δz amounts to a multiplication of the complex field $E(t,z=0)$ with a complex phase factor in the time domain

$$E(t,z) = E(t,0) \cdot e^{i\phi_{NL}}$$

with the nonlinear phase $\phi_{NL} = \frac{2\pi}{\lambda_0} \cdot n_2 \cdot \Delta z \cdot |E(t,0)|^2$ and λ_0 as the centre wavelength of the pulse.

This self-phase-modulation (SPM) in the time domain induces spectral changes in the line shape. The time dependent phase $\phi_{NL}(t)$ generates new frequencies $\omega_0 + \Delta\omega$:

$$\Delta\omega = -\frac{d\Phi_{NL}}{dt} = -\frac{2\pi}{\lambda_0} \Delta z \cdot n_2 \frac{d}{dt} I(t)$$

The same values for $\Delta\omega$ may be generated during different times of the laser pulse envelope and lead to interference structures in the propagated spectrum. We define the B-integral in order to characterise the nonlinearity of a given medium for a known intensity I_{max} of the pulse:

$$B = \frac{2\pi}{\lambda_0} \cdot \int n_2(z) \cdot I_{max} \cdot dz$$

Different B-integrals induce different line shapes of the spectrum after propagation, depending also on the initial chirp of the input spectrum as shown in Figure 1.

It is thus possible to determine for a given nonlinearity the exact lineshape after propagation through a nonlinear medium once the spectrum and phase of the initial pulse are known. In the following we make use of the inverse situation: starting from the reference spectrum of the initial pulse and the modulated spectrum of the propagated pulse we intend to reconstruct the phase of the light pulse.

DETERMINING PULSES THROUGH PHASE RETRIEVAL

The numerical model of nonlinear propagation to be described can be used to reconstruct the entire pulse in its amplitude and phase from two given spectra. An optimisation algorithm adjusts the unknown spectral initial phase until the phase and the spectral amplitude leads to a complete recovery of the modulated spectrum.

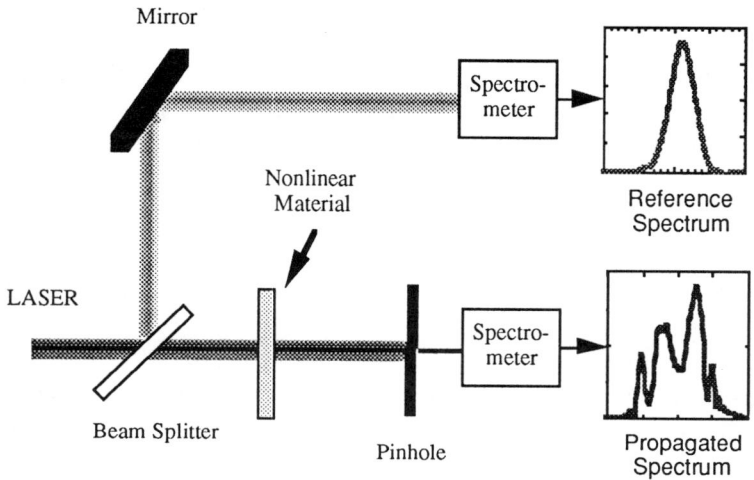

Figure 2   Experimental set up for the ENSTA phase retrieval method.

The experimental set up is shown in Figure 2 . The reference spectrum is taken from a reflection of the initial pulse, the propagated spectrum is recorded after transmission through the nonlinear material. With the knowledge of the nonlinearity n_2 we run the optimisation algorithm in order to find the spectral phase which permits the reconstruction of the modulated spectrum. The energy of the pulse is either known experimentally or obtained as the value corresponding to the best fit of the optimisation algorithm. The described set up allows a single shot operation which is important whenever strong fluctuations in the laser pulses are expected. Under stable conditions the method can be run in multiple shot operation.

The reference and modulated spectra provide the input data for an optimisation routine. In this case we use a substitution algorithm (14) sketched schematically in Figure 3. Starting with an assigned flat phase and the experimental reference spectrum at INPUT 1 in Figure 3, the field $E(\omega,0)$ is Fourier transformed in order to reconstruct the pulse in the time domain. The field is propagated to $E(t,z)$ and thereby acquires a nonlinear phase modulation. A Fourier transform yields the amplitude of the calculated spectrum after propagation. At INPUT 2 in Figure 3 the calculated phase is retained whereas the square root of the experimental spectrum replaces the calculated amplitude. This forces the algorithm towards the correct solution. The new field $E(\omega,z)$ is then transformed into the time domain. A reversed nonlinear propagation and subsequent Fourier transform gives the calculated initial field $E(\omega,0)$. Again the calculated phase is retained and the calculated amplitude is replaced by the experimental amplitude of the reference spectrum. This cycle is repeated over a number of iterations until the calculated spectra converge towards the experimental spectra.

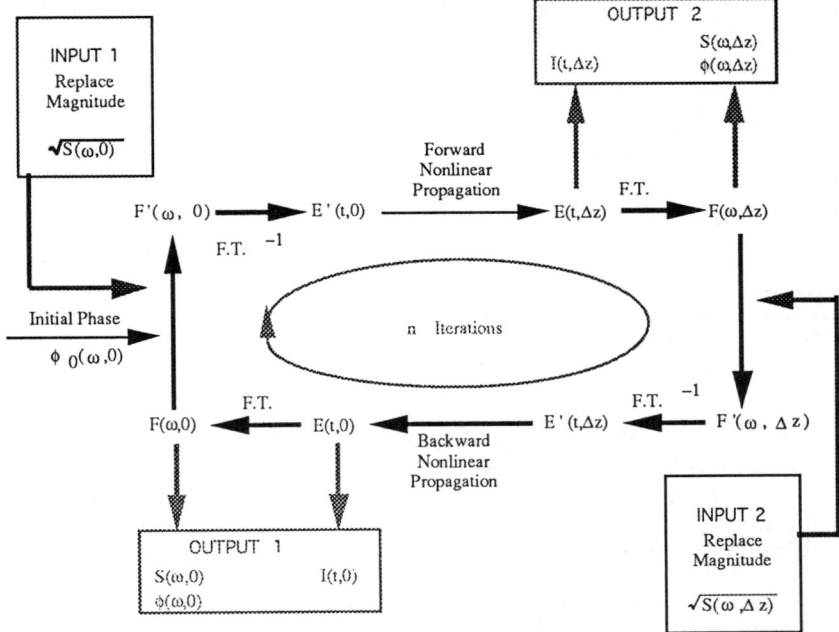

Figure 3   Schema of the ENSTA phase retrieval algorithm.

In order to monitor the quality of the retrieved solution we define the distance between the calculated $|E_{calc}|^2$ spectrum and the propagated spectrum $|E_{exp}|^2$ after the nth iteration as:

$$D(n) = \frac{\sqrt{\sum_{i=1}^{N}\left[|E_{exp,i}|^2 - |E_{calc,i}|^2\right]^2}}{\sum_{i=1}^{N}|E_{exp,i}|^2}$$

where the summation runs over all points N of the discretised fields. A solution is accepted if the measure of distance is inferior to $1.e^{-2}$.

The model of propagation assumes that the self-phase modulation is the dominating effect during propagation. That is phase shifts due to 1) diffraction, 2) Kerr lensing and 3) spatial variation in intensity remain negligible. The large beam diameter of a typical 1.5 cm fulfils condition 1). The nonlinear medium is chosen sufficiently thin in order to satisfy 2). We eliminate effects of spatial variation in intensity 3) by a pinhole. The pinhole is placed after the nonlinearity and it selects only a portion of the beam over which the intensity can be considered constant. This set-up can be used to measure the spatial variation of the initial phase of a laser beam by displacing the pinhole over the beam diameter.

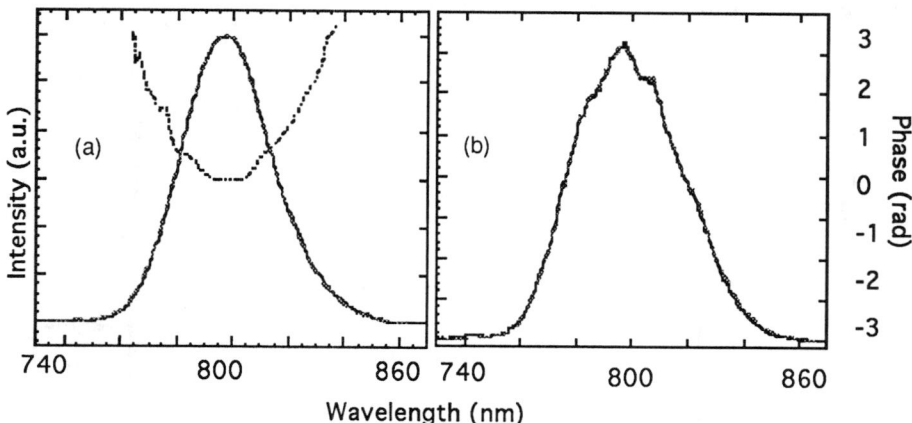

Figure 4   Pulse characterisation on a CPA Ti:Sapphire oscillator: (a) superposition of calculated and experimental reference spectrum and retrieved quadratic initial phase, (b) superposition of calculated and measured propagated spectrum. The pulse duration obtained with the ENSTA method yields 35 fs, the measured autocorrelation gives 40 fs.

An experimental case in Figure 4 of the measurement of a 35 fs pulse on a CPA laser illustrates the method. The pulse duration measured with an autocorrelator gives 40 fs in good agreement with the result of the ENSTA-method.

The validity of the method depends formally on the proof of unicity for a retrieved phase. One has to show that a given initial phase is the only phase which allows to reconstruct a given modulated spectrum from its reference spectrum. We have not found a proof of unicity. However all tested cases for phase excursions inferior to 0.2 rad at the FWHM of the spectrum converged with the ENSTA-algorithm - even for very complex initial phases. For very strong phase excursions and for reference spectra which show strong nonlinear modulation some numerical cases indicate a problem of ambiguity. In these cases the algorithm either diverges or produces a solution of a flat phase, i.e. the algorithm restitutes the bandwidth limited pulse.

The use of the method in applications discussed below has demonstrated the straightforward experimental implementation and the rapidity of the numerical analysis. The method has been used as well as a powermeter for the measurement of single pulses. Extended versions of the algorithm allow also to treat dispersive transparent media.

MEASUREMENT OF NONLINEARITIES WITH KNOWN PULSES

Once pulses from a stable laser source have been characterised with a known nonlinearity by the above mentioned method, one can use the knowledge of the pulse to determine the Kerr nonlinearity of an unknown material. For this purpose the phase retrieval algorithm calculates the phase for a given pair of experimental reference and propagated spectra generated with a known nonlinearity. Then the

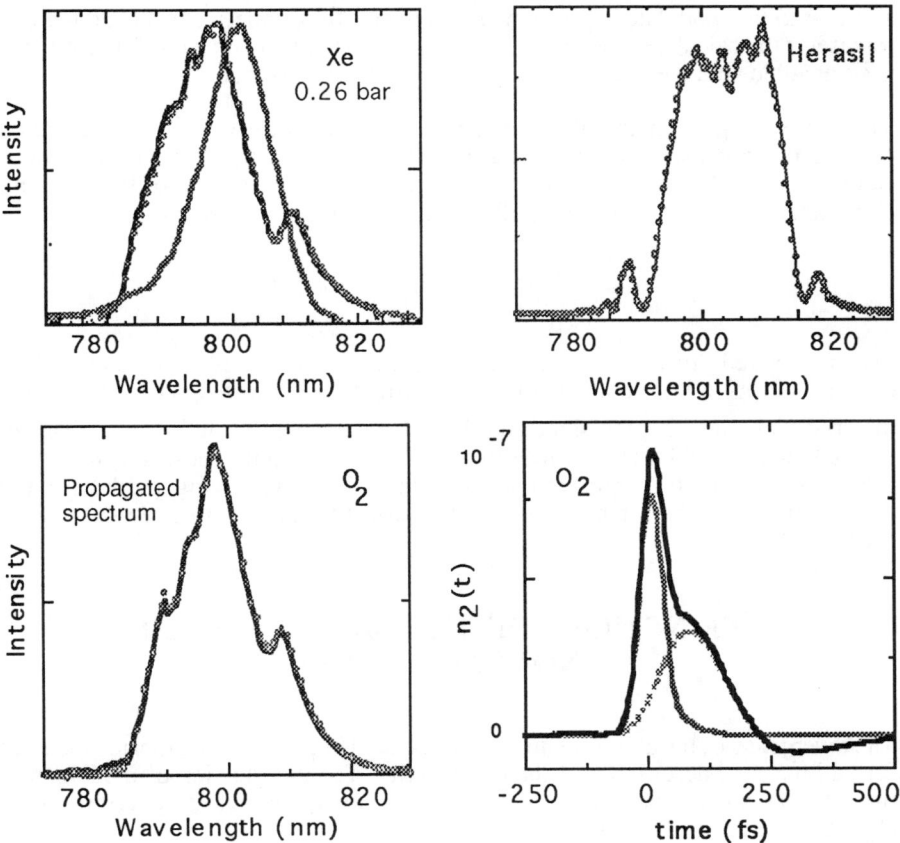

Figure 5 Measurements of n_2: Xenon and Herasil are fitted with instantaneous response and Oxygen with a quantum mechanical model for the molecular contribution to the nonlinearity. In the case of Xenon the reference spectrum is displayed as well. The experimentally obtained propagated spectrum is plotted as a solid line. The retrieved propagated spectrum is indicated by a dotted line. For Oxygen the electronic (high narrow peak) and molecular contributions (broad curve) are combined to give total nonlinear refractive index (fat).

known nonlinear material is replaced with the an unknown material and a third modulated spectrum after propagation through the unknown material is recorded. The nonlinear index n_2 is then fitted until the calculated spectrum converges to the third propagated spectrum.

The results of a study of various gaseous, liquid, and solid materials have permitted to determine coefficients of nonlinear materials for femtosecond pulses (15). Figure 5 shows the determination of the $n_2=(2.1\pm0.1)10^{-23}$ m^2/W for Xenon at 0.26 bar and of $n_2=(3.2\pm0.2)10^{-20}$ m^2/W for Herasil as examples for instantaneous responses. In the case of Xenon the laser output is displayed in addition to the modulated spectrum. The retrieval of the spectrum (the dotted line represents the calculated spectrum) is satisfactory and the nonlinear refractive index is compatible with the findings obtained with other methods (16).

The retrieval algorithm initially diverged in a number of cases. A closer inspection of the propagated spectra revealed a spectral red-shift of the laser. This red-shift can be understood in terms of a non instantaneous response $n_2(t)$ of the medium: The rising edge of the pulse E(t) shifts frequencies by

$$\Delta\omega = -\frac{d\Phi_{NL}}{dt} = -\frac{2\pi}{\lambda_0}\Delta z \frac{d}{dt}(n_2(t)\cdot I(t))$$

that is to the red side of the spectrum. On the other hand the response on the falling edge of the pulse is smoothened by the retarded nonlinear response $n_2(t)$ of the material, thus reducing the slope and therefore reducing the shift $\Delta\omega = -d\Phi_{NL}/dt$ on the blue side of the spectrum. If this shift is present in the measured spectrum, the algorithm in its simple version diverges. The underlying model for the nonlinearity has to be extended in order to account for the retarded response of the nonlinearity.

REFINEMENT OF THE MODEL OF THE NONLINEARITY

The divergence of the algorithm in case of red-shifted line shapes indicated the need for an improved description of the nonlinearity. The introduction of models for a time dependent response function leads to a convergence of the phase retrieval algorithm for red-shifted spectra.

The time dependent contribution to the nonlinear refractive index describes the non-instantaneous response of the material:

$$n_{2,non-inst}(t) = \int_0^\infty R(u)I(t-u)du$$

In a first step the non-instantaneous response function R(u) was approximated with a simple exponential decay for gaseous and liquid media:

$$R(u) = \frac{n_{2,nuc}}{T_{n_2}} \exp\left(-\frac{u}{T_{n_2}}\right)$$

As the exponential approximation for the finite response remains purely phenomenological, a more elaborated approach is necessary to establish the link between the macroscopic nonlinear refractive index $n_2(t)$ and the microscopic molecular response in gases. The physical effect responsible for the finite response time is in our case the coupling of the intense laser field to the rotational motion of the molecules. Molecules with an anisotropic polarisability are reoriented in the strong electrical field of the laser pulse. This orientation is slow compared to the fast polarisation of the electronic motion inducing the finite response time in the nonlinearity.

The distribution of rotational energy into the discrete rotational states requires a quantum mechanical description of the coupling process (17). The response function is then given by (18,19):

$$R(u) = n_{2,rot} \sum_{J=0}^{\infty} F_J \sin(-\omega_J u)$$

The factor $n_{2,rot}$ measures the macroscopic magnitude of the effect and depends on the density of the gas, the anisotropy of the molecular polarisability and the linear refractive index. The frequency ω_J is the transition energy between two rotational levels. F_J takes account of the statistical weights of each transition due to the thermal occupation of rotational states.

This model explains thus the redshift observed in the lineshapes of gases such as Nitrogen (N_2), Oxygen (O_2) and Carbondioxide (CO_2) molecules - Figure 5 (c) illustrates the case of Oxygen. The algorithm reconstructs well the propagated spectrum. The electronic and molecular contribution to the nonlinear refractive index are plotted as a function of time. But this model goes beyond the explanation of the red-shift as it predicts the revivals of the macroscopic nonlinear response long after the femtosecond pulse has passed.

In order to study the response dynamics on a picosecond time scale without loosing the femtosecond resolution we have modified our experimental method. Instead of line shape modulations of a single pulse (Self-Phase-Modulation: SPM) we monitor the modulations of line shape in an optical pulse triggered by a second intense pulse with a delay τ (Cross-Phase-Modulation: XPM).

REVIVALS IN ROTATIONAL POLARISABILITY

The nonlinear XPM phase factor Φ_{NL} due to the interaction of a pump and probe pulse can be written as:

$$E_{probe}(t,z) = E_{probe}(t,0) \cdot e^{i\phi_{NL}}$$

$$\phi_{NL} = \frac{2\pi}{\lambda_0} \cdot n_2(t) \cdot \Delta z \cdot |E_{pump}(t-\tau,0)|^2$$

The cross-phase-modulation induces a varying change in the line shape as a function of the delay τ. This change in line shape leads to a shift in the centre of gravity of the spectrum at each delay τ. A detailed analysis shows that the shift in the centre of gravity of a spectrum is proportional to the derivative of the autocorrelation. The dependence of the shift on the delay time τ provides therefore an independent method of pulse characterisation (20). Figure 6 displays the shift of the centre of gravity of the probing pulse spectrum as a function of the delay with air as the nonlinear medium. The line shape shifts after t = 6 ps attributed to Oxygen (O2) and t = 4 ps attributed to Nitrogen (N2) correspond to the first revival of the rotational contribution to the nonlinear refractive index, predicted by the model based on reference (18,19).

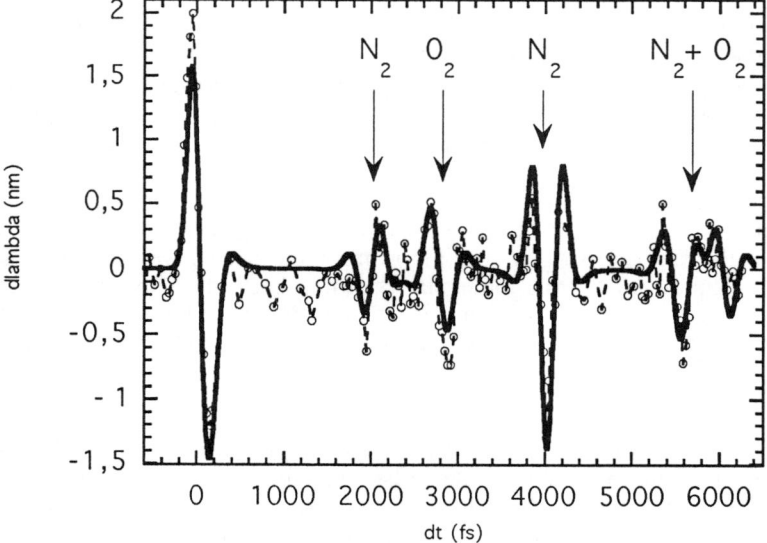

Figure 6   Calculated (solid line) and measured (open circles) revivals of the rotational contribution to the nonlinearity in air.

The physical picture for this effect is linked to the coherent rotation of the molecules with different rotational frequencies. The molecules are initially set into a coherent superposition of many rotational modes by the pump pulse. The absence of irreversible dephasing processes due to collisions on a picosecond time scale leads to their spontaneous revival.

PERSPECTIVE AND NEW APPROACHES

Current research in the area of spectral analysis after propagation concerns the resolution of the ambiguity problem in the ENSTA phase retrieval. Approaches to a solution follow two main lines: on the one hand the improvement of the optimisation routine itself, making use of simulated annealing and other optimisation routines. On the other hand the injection of additional experimental data into the phase retrieval algorithm, such as additional spectra or the autocorrelation, open ways to lift the ambiguity.

In the field of analysis of the nonlinear response of matter the pulses shorter than 20 fs and with broader lineshapes open new venues to study fast molecular dynamics. One example is the analysis of vibrational molecular motion in the time domain (21). Together with efficient phase control of the used pulses one can perform experiments with a chirped intense light source on a scale of less than 20 fs. In these experiments the imposed chirp of the pulse translates into a tuning of the optical excitation energy over the pulse duration. Another application of this technique is for instance the study of multi-photon ionisation (22).

ACKNOWLEDGMENT

R. Lange gratefully acknowledges the support by the Human Capability and Mobility Programme of the European Community (grant ERBFMBICT950065).

REFERENCES

1. J.-L. Martin and M. H. Vos,"Femtosecond Biology", Annu.Rev.Biomol.Struct., Vol. **21**, (1992) 199
2. P. Brumer and M. Shapiro,"Laser control of molecular processes", Annu. Rev. Phys. Chem. **43**, (1992) 257
3. A. Alexandrou, V. Berger, D. Hulin,"Direct observation of electron relaxation in intrinsic GaAs using femtosecond pump probe spectroscopy", Phys. Rev. B **52**, (1994) 4654
4. M.M. Murnane, H.C. Kapteyn, M.D. Rosen,and R.W. Falcone, "Ultrafast x-ray pulses from laser produced plasmas", Science **251**, (1991) 531
5. Q.Z. Wang, Q. D. Liu, D. Liu, P.P. Ho, and R.R. Alfano"High-resolution spectra of self-phase modulation in optical fibers", J.Opt.Soc.Am.B **11**, (1994) 1084
6. M.D. Perry, G. Mourou, "Terrawatt to Pettawatt subpicosecond lasers", Science 264, (1994) 917

7. C. Spielmann, P.F.Curley, T.Barbec, and F.Krausz,"Ultrabroadband femtosecond lasers",IEEE J. Quantum Electron. **30**, (1994) 1100
8. E.T.J. Nibbering, M.A.Franco,B.S.Prade, G.Grillon,, J.-P. Chambaret, and A. Mysyrowicz,"Spectral determination of the amplitude and the phase of intense ultrashort optical pulses", J. Opt.Soc.Am.B, Vol.13, No.2, (1996) 317
9. Kenneth W. DeLong, and Rick Trebino: "Improved ultrashort pulse-retrieval algorithm for frequency-resolved optical gating", JOSA A Vol.11, No. 9, (1994) 2429
10. J.L.A. Chilla and O.E.Martinez,"Direct determination of the amplitude and the phase of femtosecond light pulses", Opt.Lett. 16, (1991) 39
11. J. Paye, "How to measure the amplitude and phase of of an ultrashort light-pulse with an autocorrelator and a spectrometer", IEEE J. Quantum Electr. **30**, (1994) 2693
12. S.Diddams,S.Prein, and J.C. Diels,"Measuring femtosecond pulses with linear optics and nonlinear electronics", Ultrafast Phenomena, May 28 - June 01, San Diego 1996, Technical Digest, Vol **8**, 317
13. Y.R. Shen,"The principles of nonlinear optics", John Wiley & Sons, New York (1984), p.324
14. J.R. Fienup,"Phase retrieval algorithms: a comparison", Appl. Opt. 21, (1994) 2758
15. E.T.J. Nibbering, M.A.Franco,B.S.Prade, G.Grillon,, C.LeBlanc, and A. Mysyrowicz,"Measurement of the nonlinear refractive index of transparent materials by spectral analysis after nonlinear propagation", Optics Comm. 119 (1995) 479
16. D. McMorrow, W.T. Lotshaw and G.A. Kenney-Wallace,"Femtosecond optical Kerr studies on the origin of the nonlinear responses in simple liquids", IEEE Journal of Quant. Electron., Vol. 24, No. 2, (1988) 443
17. J.F.Ripoche, M. Franco,G. Grillon, E. Nibbering, B. Prade, and A. Mysyrowicz, "Dynamical studies of nonlinear refractive index by cross phase modulation", Ultrafast Phenomena, May 28 - June 01, San Diego 1996, Technical Digest, Vol **8**, 152
18. M. Morgen, W. Price, L. Hunziker, P. Ludowise, M. Blackwell and Y. Chen, "Femtosecond Raman-induced polarization spectroscopy studies of rotational coeherence in O_2, N_2 and CO_2", Chem.Phys.Lett. 209, (1993) 1
19. M. Morgen, W. Price, P. Ludowise and Y. Chen, "Tensor analysis of femtosecond Raman-induced polarization spectroscopy: application to the study of rotational coherence", J.Chem.Phys. 102 (1995) 8780
20. J.F.Ripoche, B. Prade, M. Franco, G. Grillon, R. Lange, and A. Mysyrowicz, "Determination of the duration of UV femtosecond pulse", submitted
21. C.J. Bardeen, Q.Wang, and C.V. Shank,"Selective excitation of vibrational wave packet motion using chirped pulses", Phys. Rev. Lett., Vol. **75**, No. 19 (1995) 3410
22. T. Baumert and G. Gerber, Adv. in At. Molec. and Opt. Phys. **35**, (1995) 163

Optical Collisions in an Atomic Beam Experiment

J. Grosser, O. Hoffmann, S. Klose

Institut für Atom- und Molekülphysik, Universität Hannover,
30167 Hannover, Germany

Abstract

The optical collision $Na(3s) + Kr + h\nu \longrightarrow Na(3p) + Kr$ has been investigated in an experiment with crossed atomic beams and with a differential detection of the excited collision products. Three groups of experimental results have been obtained, which all demonstrate unprecedented possibilities for the observation of atomic collisions. 1. The oscillatory structure of the differential cross sections allows a sensitive test of the molecular potentials. 2. Polarization experiments give direct information about the geometric characteristics of the collision complex. 3. State analysis of the collision products provides a tool for the observation of elementary nonadiabatic processes. A number of possible future experiments, which make use of the new ideas, is discussed, in addition.

1 INTRODUCTION

In an optical collision, $A + B + h\nu \longrightarrow A^* + B$, a pair of colliding atoms absorbs a photon during the collision. Optical collisions have been successfully used since more than twenty years as a tool for the study of molecular potentials and nonadiabatic interactions [3, 7, 16, 17, 26]. Normal collisions, $A + B$ or $A^* + B$ are governed by the same interactions, and the investigation of normal collisions offers another possibility for such studies. Optical collision experiments are fundamentally different from normal collision experiments, however, because they probe the collision itself and not only the initial and final states. For standard collision experiments, on the other hand, numerous powerful experimental techniques, like atomic beam and differential scattering techniques, are available. Optical collisions have never been studied with these techniques, only gas cell experiments have been reported so far. It appears promising to combine the advantages of the two techniques.

We report here the first [14, 15, 18, 22] atomic beam differential scattering study of an optical collision. The experiment faced two major difficulties. 1. Atomic beams have typical densities, which are significantly lower than those used in gas cell experiments. The experimental signal is correspondingly lower, and there is a correspondingly greater chance that the signal is obscured by competing processes. This difficulty was overcome in our experiment by the use of high density atomic beams and by a careful study of the various possible sources for competing signals [14, 22]. 2. The excited collision products have a short lifetime, and they do not travel macroscopic distances in this time. This seems to prohibit the use of a distant rotatable detector for the registration of angular distributions. The difficulty was overcome by a stabilization technique.

2 OPTICAL COLLISIONS

We studied the process

$$Na(3s) + Kr + h\nu \longrightarrow Na(3p) + Kr, \qquad (1)$$

Fig.1 shows the NaKr potential curves V(r). The optical transition occurs when the atoms are close to a distance r = r_c (the "Condon radius"), such that the resonance condition

$$h\nu = V^*(r_c) - V(r_c) \qquad (2)$$

is fulfilled. The condition is illustrated for a typical case by the arrow of length $h\nu$ in the figure. As usual, we characterize the photon energy by the detuning $\Delta\nu/c = \nu/c - \nu_{res}/c$, that is, by the difference between the actual wavenumber

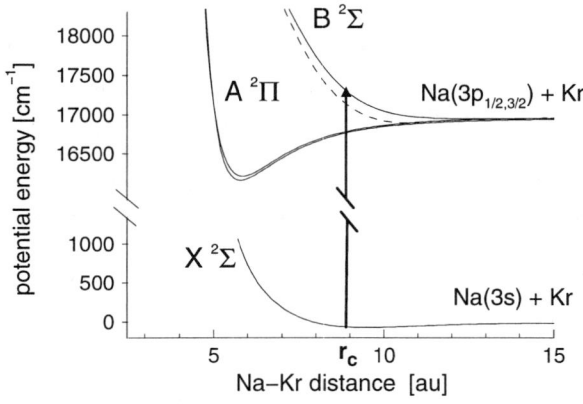

FIGURE 1. The NaKr potentials. The X and the A curves are spectroscopic results, the solid B curve is from the present work, the dashed B curve is a theoretical result. The arrow shows the photon energy.

and that one which belongs to the resonance transition Na(3s) - Na($3p_{1/2}$). The present experiments were all carried out with a positive detuning in the range between $100 cm^{-1}$ and $500 cm^{-1}$. Under these conditions, only the repulsive $B^2\Sigma$ state of the NaKr system can be excited, and the condition (2) is fulfilled only for one value r_c.

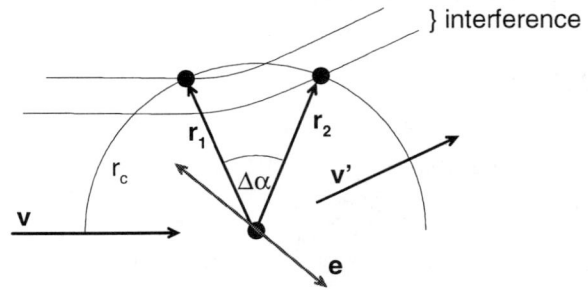

FIGURE 2. Geometrical representation of the optical collision.

Fig.2 shows a classical representation of the optical collision. The curves are classical trajectories for the vector **r** connecting the Na and Kr atoms, they were numerically computed from the potentials of fig.1. The half circle represents the Condon radius. There is one trajectory, for which the transition occurs at the first passage through r_c, and another one, for which the transition occurs at the second passage. The transition points are denoted as the Condon points, the corresponding vectors $\mathbf{r_1}$ and $\mathbf{r_2}$ as the Condon vectors. For the conditions of the present experiment, there are never more than two trajectories leading to a given scattering angle (note that the potential and hence the deflection inside r_c is different for the two trajectories). The signal contributions from the two trajectories interfere. This leads to an oscillatory pattern in the angular distribution, see sec. 4.

It is worth to note that fig.2 is not just a qualitative illustration, but can be used for a practically rigorous discussion of the process. Using well known semiclassical procedures [8], fig.2 can be translated into a simple expression for the differential cross section σ,

$$\sigma = |\Sigma f_i|^2 \quad \text{with} \quad f_i = \sqrt{b_i db_i/(\sin\theta d\theta)}\; p_i\; exp(i\Phi_i)\; (\mathbf{er_i}) \quad (3)$$

The summation is over the two trajectories, $\theta(b)$ is the classical deflection function with b the impact parameter, p_i measures the transition probability, Φ_i is a classical phase expression and **e** is the polarization vector of the excitation laser. It turns out that cross sections calculated with this formula are in general in a very good agreement with full quantum mechanical results. We conclude that the basic ideas in fig.2 - classical trajectories plus interference -

essentially yield a correct description of the optical collision.

For the comparison between experimental and theoretical results, we usually use a full quantum mechanical treatment of the collision dynamics [27]. It starts with given molecular potentials and nonadiabatic couplings. The photons and the motion of the atoms are treated quantum mechanically ("dressed collision pair"). The corresponding coupled channel equations are solved numerically, the cross sections are calculated in the form of partial wave sums.

3 EXPERIMENT

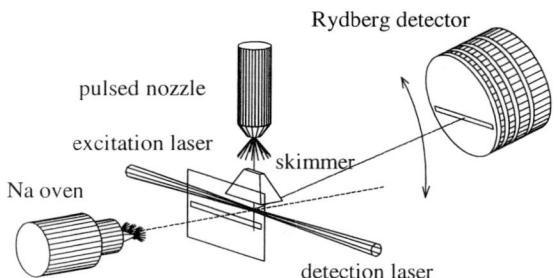

FIGURE 3. The principle of the experimental set-up.

The experimental set-up (fig.3) consist of an effusive Na source, a pulsed nozzle Kr source, two laser beams (excitation and detection laser), and a detector unit. The beams cross each other in a scattering volume of approximately 1mm x1mmx10mm, the detector is at 70mm from the scattering volume and can be rotated around it. The atoms and the light of the excitation laser interact to produce excited Na(3p) atoms. In order to allow a selective detection with a distant detector, the Na(3p) atoms are stabilized by further optical excitation to a Rydberg state,

$$Na(3p) + h\nu' \longrightarrow Na(nl). \tag{4}$$

nl is typically 30d, the photons are provided by the detection laser. Rydberg atoms arriving in the detector are field ionized and counted after amplification by a channeltron multiplier. Typical intensities are:

Na-atoms:	$3 \cdot 10^{17}/m^3$
Kr-atoms:	$10^{20}/m^3$
Excitation laser:	$3 \cdot 10^{-4}$ J
Detection laser:	10^{-4} J
Optical collision signal:	10 events/s

The characteristic times in the experiment are:

Duration of the collision: 10^{-12} s
Laser pulse length: 10^{-8} s
Na(3p) lifetime: 10^{-8} s
Na(nl) time of flight: 10^{-4} s

The pulse duration of the lasers is long compared to the collision time. The lasers are on (or off) during the entire collision. The detection transition occurs long after the collision, on the average. The use of pulsed lasers allows to determine the product velocity by measuring the time delay between the laser pulses and the detector signal. Together with the use of a Kr nozzle beam, which fixes the Kr velocity within narrow limits, the measurement of the Na product velocity vector is sufficient to determine the entire collision kinematics, that is, all velocity vectors before and after the collision.

The reliability of the Rydberg detection scheme is crucial for the experiment. On their way from the scattering volume to the detector, the Rydberg atoms interact with the target and residual gas (10^{-5} hPa), the black body radiation, and the laser radiation fields. The most likely result of these interactions is, however, not the destruction of the Rydberg atom, but rather a change of the quantum number l. Using literature data [30] for the probabilities of these processes and of the optical decay, we estimate that more than one out of ten Rydberg atoms, which start in the scattering region, is actually detected [22]. This result was confirmed by a careful comparison of the measured signal intensity with that expected theoretically. Similarly, there occurs no scattering, which might significantly change the velocity or velocity direction of the Rydberg atoms.

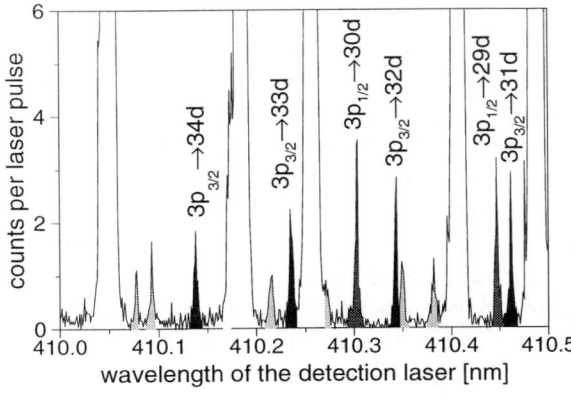

FIGURE 4. The signal at the Rydberg atom detector as a function of the detection laser wavelength. The small unlabeled peaks represent transitions 3p→ns, for the large peaks see text.

Fig.4 shows a spectrum of the detector signal versus the wavelength of the detection laser. The small peaks correspond to process (4), the large peaks represent two photon excitation of ground state Na atoms,

$$Na(3s) + h\nu + h\nu' \longrightarrow Na(nl), \qquad (5)$$

where one photon is from the excitation laser, and one from the detection laser. The most important competing processes are:
1. Process (5). It is supressed by an adequate choice of the laser wavelength.
2. Resonant excitation of ground state atoms, e. g. by amplified spontaneous emission (ASE). The excitation laser passes a Na cell in order to supress ASE; as the cell emits resonant light, there is a distance of 12m between the Na cell and the crossed beam apparatus.
3. Hyper-Raman excitation of ground state Na,

$$Na(3s) + 2h\nu \longrightarrow Na(3p) + h\nu', \qquad (6)$$

with two photons from the excitation laser. The process is identified by a quadratic variation of the signal with the laser intensity, and it is suppressed by working at low laser intensity.
4. Na + Na optical collisions. The process is identified by a quadratic variation of the signal with the Na intensity, and it is supressed by working at low Na intensity.
The competing signals are kept within the order of a few percent, such that the signal can easily be corrected.

4 DIFFERENTIAL CROSS SECTIONS

Fig.5 shows measured angular distributions. Throughout secs. 4 and 5, all results refer to a detection of the Na($3p_{1/2}$) state. The oscillatory structure is due to the interference between the two trajectories, which lead to the same scattering angle. The mechanism causing this structure is identical to the mechanism causing the so-called Stueckelberg oscillations in normal inelastic collisions, we refer to the present structures as Stueckelberg oscillations as well. Fig.6 collects the results of measurements as in fig.5, however with a variable velocity v'_{Na} and a variable detuning. In place of the entire $\sigma \sin\theta$ curves, only the positions of the oscillation maxima are shown. The data of fig.6 are very sensitive to the molecular potentials. We use spectroscopic results for the NaKr $X^2\Sigma$ and $A^2\Pi$ potentials [6]. For the $B^2\Sigma$ potential, no spectroscopic data are available, we used the results from a model potential calculation [10] for a first attempt, this potential is included as a dashed line in fig.1. The numerical results (full quantum mechanical calculations plus convolution over

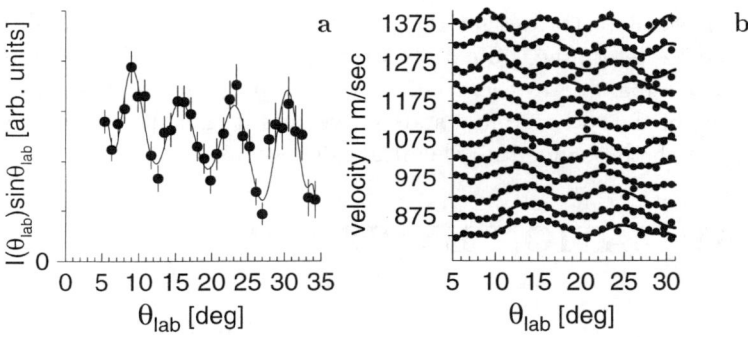

FIGURE 5. a. A typical differential cross section, multiplied by $\sin\theta$, b. $\sigma \sin\theta$ for different velocities of the Na(3p) atoms. The lines are polynomial fits.

the experimental conditions) are shown as dashed lines in fig.6. The agreement is poor. The solid line $B^2\Sigma$ potential in fig.1 is the result of a systematic trial-and-error search. It yields the solid lines in fig.6, which agree much better with the experimental data.

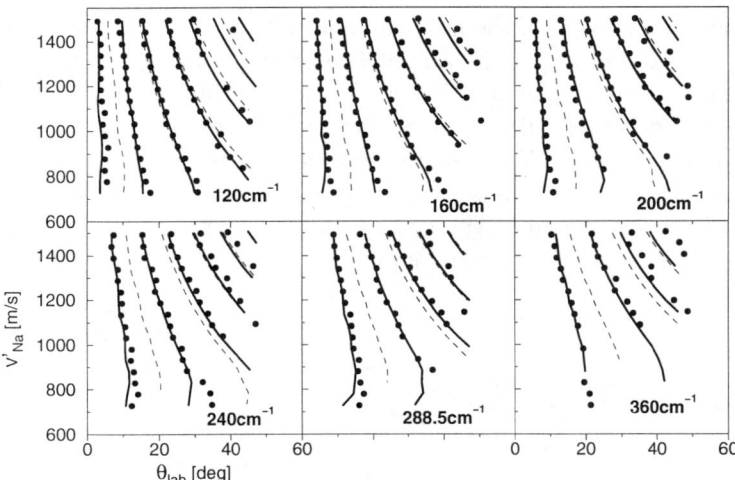

FIGURE 6. The position of the oscillation maxima. The excitation laser was linearly polarized at 66^0 relative to the Na beam direction. Dots experiment, dashed lines theory with the tentative, solid lines with the new $B^2\Sigma$ potential.

We found that the Stueckelberg structure of fig.6 probes essentially the potentials at r below 9a.u. (inside the Condon radii involved). The comparison between the two versions of the $B^2\Sigma$ potential and the corresponding theoretical curves in fig.6 demonstrates that the present method has the capability to test, and maybe determine, potentials with an accuracy of the order of 10cm^{-1}. This is comparable to the accuracy of the spectroscopic results for the repulsive part of the NaKr ground state curve, and it is of the same order as the accuracy of the best quantum chemical calculations.

5 POLARIZATION EXPERIMENTS

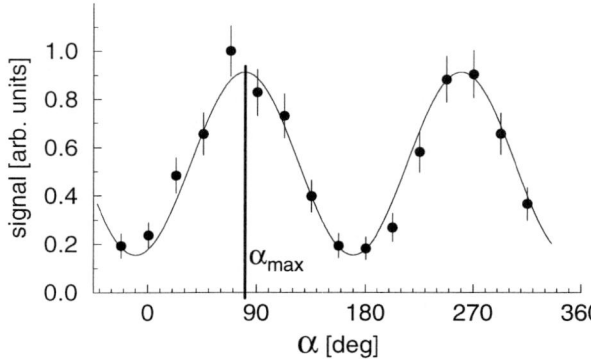

FIGURE 7. An example for the variation of the differential signal with the linear polarization of the excitation laser. $\alpha = 0$ is the Na beam direction. The solid line is the best fit to a sine curve.

Fig.7 shows how the intensity of the differential signal varies, when the linear polarization of the excitation laser is rotated. The polarization angle α is measured with respect to the Na beam direction. These experiments were carried out with a poor velocity resolution for the Na atoms, such that the Stueckelberg structure is supressed. We evaluated the polarization angle α_{max}, for which the intensity is largest. Fig.8 shows, how this angle varies with the scattering angle and the Na velocity. The symbols are the experimental results, the lines are the theory. The agreement is excellent.

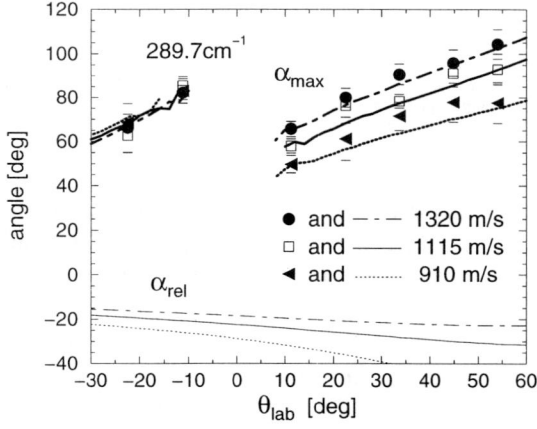

FIGURE 8. The angle α_{max} defined in fig.7, as a function of the scattering angle and for different velocities: symbols experiment, lines theory. $\alpha=0$ is the Na beam direction, α_{rel} indicates the direction of the relative velocity vector before the collision.

For a better understanding, consider fig.2 and eq.3. For a transition at a single Condon point $\mathbf{r}_i$, the transition probability is proportional to $(\mathbf{r}_i \mathbf{e})^2$, with $\mathbf{e}$ the polarization vector. This is a consequence of the dipole approximation and of the fact that the transition dipole is parallel to the internuclear axis for a $\Sigma - \Sigma$ transition. For a single Condon vector $\mathbf{r}_i$, the signal intensity is largest when the polarization is parallel to $\mathbf{r}_i$ and it is zero when the polarization is at right angles to $\mathbf{r}_i$. For the actual situation with two Condon vectors $\mathbf{r}_1$ and $\mathbf{r}_2$, maximum intensity is expected when $\mathbf{e}$ is between $\mathbf{r}_1$ and $\mathbf{r}_2$. Fig.9 shows computed classical trajectories with the Condon vectors, and, as additional lines, the experimental polarization direction of maximum intensity. The lines are between the Condon vectors, as expected. They are not in the middle, because one Condon point actually contributes about twice as much to the signal as the other one. Fig.9 demonstrates that the present technique allows the direct observation of geometric characteristics of the collision pair, here, of the weighted average of the Condon vectors.

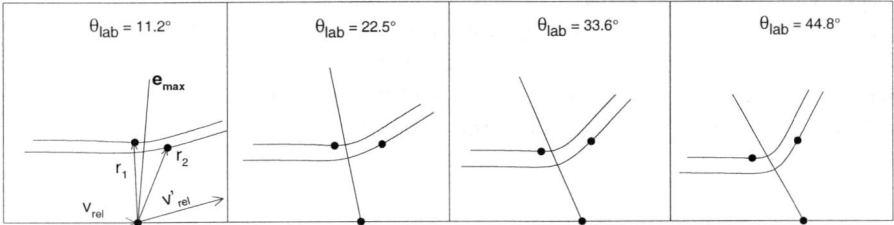

FIGURE 9. Computed classical trajectories and Condon points. The straight line shows the experimental result for the polarization direction, at which the largest intensity is found. Detuning 289.7cm^{-1}, v'$_{Na}$=1320m/s.

There is more geometric information, which can be obtained experimentally. Fig.10a shows the Stueckelberg oscillations for right and left circular polarization. They are shifted with respect to each other. Fig.10b summarizes the positions of the maxima for different velocities. Again, a good agreement with the theory is achieved. Within the framework of the semiclassical theory, the shift of the oscillation pattern arises, because the phase difference between the two interfering contributions is different for right and left circular polarization. The difference between the phase differences amounts to

$$\Delta\Phi_r - \Delta\Phi_l = 2\Delta\alpha, \tag{7}$$

where $\Delta\alpha$ is the angle between the Condon vectors (fig.2). We used eq.7 to derive this angle from the experimental results. Fig.11 compares experimental $\Delta\alpha$ values to those derived from classical trajectory calculations. The agreement is good.

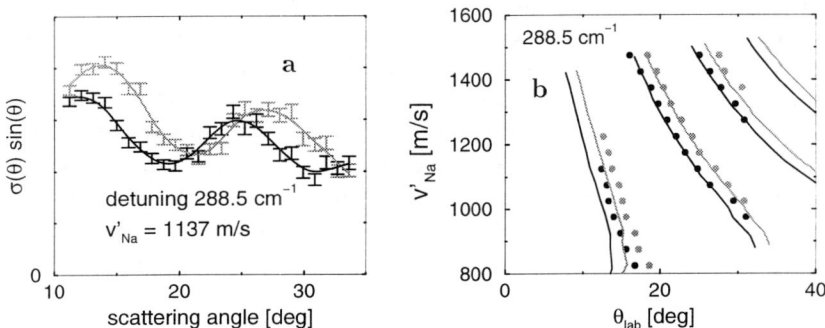

FIGURE 10. Stueckelberg oscillations with circular polarization. a. Cross sections multiplied by $\sin\theta$, b. the position of the oscillation maxima for different velocities, both for right (black) and left (shaded) circular polarization of the excitation laser. The symbols are the experimental results, the lines in fig.5a are polynomial fits, those in fig.5b show the theoretical results.

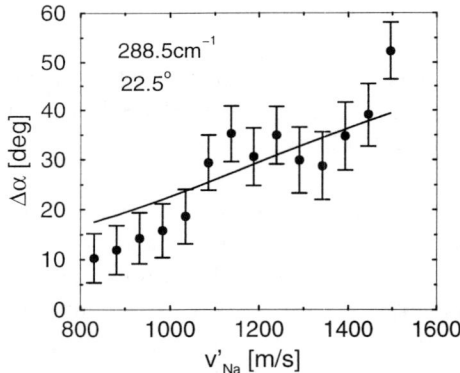

FIGURE 11. The angle between the Condon vectors. Symbols: experimental values derived from the shift of the Stueckelberg oscillations, line: classical trajectory result.

It is possible in principle to determine both Condon vectors separately. For instance, this might be done using the contrast between minimum and maximum in fig.7. In the present experiment, the depth of the minimum is determined by a lack of resolution (e. g., for the scattering angle). The data cannot be used to obtain further geometric information, at the moment.

6 FINAL STATE ANALYSIS

The detection process (4) allows the determination not only of the population, but also of the alignment and the orientation of the Na(3p) products. Separate determination of the population of the fine structure levels Na($3p_{1/2,3/2}$) is achieved by tuning the detection laser to a corresponding wavelength, see fig.4. Alignment and orientation can be obtained from measurements, in which the linear, respectively circular polarization of the detection laser is varied [4]. The experiments are rather difficult. In order to avoid systematic errors due to a beginning saturation of the detection transition, the detection laser has to be operated at extremely low intensity. So far, systematic measurements have been performed only for the population ratio of the fine structure levels (fig.12).

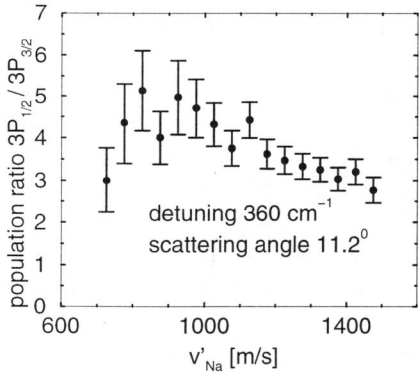

FIGURE 12. The population ratio of the Na fine structure levels (preliminary experimental results).

On the way from small to large internuclear distances, the collision pair passes from a region in which Hund's coupling case a applies to a case e region. The passage occurs between 10 and 12au, where the potential curves (fig. 1) approach each other. It is accompanied by a strong nonadiabatic coupling [12, 20, 24], which can cause transitions between the different excited molecular states. The most straightforward way to study nonadiabatic behaviour is to populate a single electronic state at small internuclear distance, and to measure the composition of the electronic state at large internuclear distance, after the passage through the coupling region. This is exactly what is done in the type of measurement discussed here. The present experiment provides an ideal tool for the study of nonadiabatic processes, which is superior to other methods. Photodissociation experiments, for instance, can be used for the same purpose. In the usual photodissociation experiment, however, an alignment or an orientation will not be observable, in contrast to the method discussed here.

7 DISCUSSION

The combination of two well established experimental techniques, the optical collision and the atomic beam techniques, has resulted in a powerful tool for the observation of atomic collisions.

Among the new experimental possibilities, we believe the possibility of geometric observations to be of the greatest importance for future developments. Of course, once the potentials of a collision system have been determined experimentally, a separate measurement of geometric properties is no more necessary, because they can be calculated. For diatomic systems, as in the present work, geometric observations do indeed not significantly extend our knowledge about the collision. Already for triatomic systems, however, the situation is entirely different, because there is no much hope to derive the potential surface from experimental observations. Geometric observations of the present type might therefore develop into an important tool for the observation of atom - molecule (e. g. chemically reactive) collisions.

Besides optical collisions, a number of related techniques have been developped, in which photons are used as probes for collision pairs and similar systems. Including optical collisions, these are

1. Optical collisions, where the detection of the excited product occurs long after the collision.
2. Fractional collision experiments [25]. These are optical collisions, where a second photon is absorbed during the same collision (pulse - probe collision experiments with CW light).
3. Pulse - probe collision experiments with ultrashort laser pulses.
4. Photodissociation experiments [28] with the product detection long after the photodissociation process (half collision experiments).
5. Photodissociation experiments, combined with the absorption of a second photon during the dissociation process [5] (Half collision pulse - probe experiments with CW light).
6. Pulse - probe experiments with ultrashort laser pulses, applied to photodissociation [19].
7. Photoassociation experiments with thermal [9, 29] or ultracold [1, 2, 11] atoms.
8. Time resolved photoassociation experiments [21, 23].

All these experiments, with very few exceptions, have been performed so far only under gas cell conditions (from the point of view of the present discussion, a magnetooptical trap provides gas cell conditions, as well). Differential cross sections have not been measured so far, except for the present experiments. The progress reported in the present paper is achieved, because the relative velocities **v** and **v'** before and after the collision are experimentally fixed. The possibility, for instance, to observe the collision geometry in a simple polarization experiment (fig.9) arises, because with **v** and **v'**, the Condon vectors are also fixed in space. Note that it would not be sufficient to fix only **v** or only **v'**! Similar ideas can be applied to the techniques described above. In cases 2 and 3, it would be necessary to fix the relative velocities before and after the collision, just as in the present paper. For the other cases, one would have to fix a velocity vector and the angular momentum vector of the stable molecule, which is the starting point (cases 4 - 6) or the result (cases 7, 8) of the process. One can expect further novel, direct insights into the mechanisms of these processes. This can go far beyond the present results. Results like in fig.9 have a crude similarity with a photographic picture of the collision (very fuzzy, admittedly). When time resolution is achieved in addition, one might end up with experimental results, which have the character of a movie showing the collision process [13].

ACKNOWLEDGEMENTS

The quantum mechanical calculations were carried out by F. Rebentrost (Max-Planck-Institut für Quantenoptik, Garching). We are grateful for a very good cooperation. Financial support from the Deutsche Forschungsgemeinschaft is gratefully acknowledged.

References

[1] Abraham, E. R. I., Ritchie, N. W. M., McAlexander, W. I., Hulet, R. G., *J. Chem. Phys.* **103** (1995) 7773-8

[2] Bagnato, V. S., Weiner, J., Julienne, P. S., Williams, C. J., *Laser Physics* **4** (1994) 1062-5

[3] Behmenburg, W., Makkonnen, A., Findeisen, M., *Z. Phys. D* **25** (1993) 315-21

[4] Blum, K.: *Density Matrix Theory and Applications.* New York: Plenum Press 1981.

[5] Bovensmann gen. Schroer, H., Tiemann, E., *Z. Phys. D* **35** (1995) 19-26

[6] Brühl, R., Kapetanakis, J., Zimmermann, D., *J. Chem. Phys.* **94** (1991) 5865-74

[7] Burnett, K., *Phys. Rep.* **118** (1985) 339-401

[8] Child, M. S.: *Molecular Collision Theory.* New York: Academic Press 1974

[9] Czajkowski, A., Kedzierski, W., Atkinson, J. B., Krause, L., *Chem. Phys. Letters* **238** (1995) 327-32

[10] Düren, R., *private communication*, 1982

[11] Gardner, J. R., Cline, R. A., Miller, J. D., Heinzen, D. J., Boesten, H. M. J. M., Verhaar, B. J., *Phys. Rev. Letters*, **74** (1995) 3764-7

[12] Grosser, J., *Z. Phys. D* **3** (1986) 39-58

[13] Grosser, J., *Comments At. Mol. Phys.* **21** (1988) 107-22

[14] Grosser, J., Gundelfinger, D., Maetzing, A., Behmenburg, W., *J. Phys. B* **27** (1994) L367-73

[15] Grosser, J., Hohmeier, D., Klose, S., *J. Phys. B* **29** (1996) 299-306

[16] Havey, M. D., *Phys. Rev. Lett.* **48** (1982) 1100-3

[17] Hedges, R. E. M., Drummond, D. L., Gallagher, A., *Phys. Rev. A* **6** (1972) 1519-44

[18] Klose, S., *Dissertation,* Hannover 1996

[19] Kundkhar, L. R., Zewail, A. H., *Ann. Rev. Phys. Chem.* **41** (1990) 15-60

[20] Lewis, E. L., *Phys. Rep.* **58** (1980), 1-71

[21] Machholm, M., Giusti-Suzor, A., Mies, F. H., *Phys. Rev. A* **50** (1994) 5025-36

[22] Maetzing, A., *Dissertation,*Hannover 1993

[23] Marvet, U., Dantus, M., *Chem. Phys. Letters* **245** (1995) 393-9

[24] Nikitin, E. E. in: *Atomic Physics,* zu Putlitz, G., Weber, E. W., Winnacker A., eds., Vol.4, p.529-57, New York: Plenum Press 1975

[25] Olsgaard, D. A., Lasell, R. A., Havey, M. D., Sieradzan, A., Phys. Rev. A. **48** (1993) 1987-96

[26] Petzold, H. C., Behmenburg, W., *Z. Naturf.* **33a** (1978) 1461-8

[27] Rebentrost, F. in: *Collision- and Interaction Induced Spectroscopy.* Tabisz, G. C., Neuman, M. N., eds., p.343, Dordrecht: Kluwer Academic Publishers 1995

[28] Schinke, R.: *Photodissociation dynamics.* Cambridge: University Press 1995

[29] Schloss, J. H., Jones, R. B., Eden, J. G., *J. Chem. Phys.* **99** (1993) 6483-94

[30] Stebbings, R. F., Dunnings, F. B., eds.: *Rydberg states of atoms and molecules,* Cambridge: University Press 1983

Interaction of Static and Polarization Channels in Spectra of Stimulated Emission in Ion-Electron Collisions in a Strong Laser Field

V.A.Astapenko*, A.B.Kukushkin** (+)

*Moscow Institute for Physics & Technology, Dolgoprudnyi,
Moscow region, 141700, Russia
**INF RRC "Kurchatov Institute", Moscow, 123182, Russia
(+) Presented by A.V.Demura**

Abstract. A universal analytic formula is derived which describes spectral line shape of the multiphoton stimulated Bremsstrahlung in electron-highly charged ion collisions, with allowing for both conventional ("static") and polarization channels. The effects of the interference of these two channels are investigated in dependence on laser frequency detuning from the closest atomic transition in the ion core.

The multiphoton stimulated emission/absorption in collisions of a charged particle with a highly charged ion (HCI) in the presence of a *strong* laser field is to be treated as an emission/absorption by a compound system {colliding particles + laser field} even for those non-resonant processes, which involve (virtual) excitation of colliding ion core with frequency detuning far outside the widths of atomic transitions in the ion core. In the case of radiative transitions in the compound system without excitation of the target atom/ion, this corresponds to the so called Polarization Bremsstrahlung [1]. The presence of a strong laser field stimulates "polarization" channels along with stimulating the "static" channel, i.e. conventional Bremsstrahlung.

In the case of the e-HCI collisions in a strong laser field -- but still not exceeding the strong Coulomb field of the HCI, it is possible to treat the problem -- within a broad range of parameters -- in the frame of the approximation of a fixed quantum/classical current for the incident particle motion in the field of the HCI. This enables us to derive universal analytic description for the multiphoton stimulated Bremsstrahlung - including both static and polarization channels - at arbitrary values of parameter $\xi \equiv Ze^2/\hbar v$ (Z is HCI charge; v, initial velocity of incident particle). Thus, for the probability W and cross-section σ of the emission/absorption of n photons, we have:

$$W_\Sigma(n) = J_n^2\left\{2\left(n_{k\lambda}^{Las} n_{k\lambda}^{stat}\right)^{1/2}|1-\delta|\right\}, \qquad \sigma(n) = \int d\sigma_{Coul}\, W_\Sigma(n), \qquad (1)$$

where J_n is Bessel function of n-th order; σ_{Coul} is elastic Coulomb scattering cross-section; n^{Las} and n^{stat} are photon occupation numbers for, respectively, laser field and spontaneous one-photon radiative transition of incident particle -- in the purely "static" channel -- between initial state $|E_i, \mathbf{n}_i>$ and the state $|E_i - \hbar\omega, \mathbf{n}_f>$, with $\mathbf{n}_f$ being the exact direction of incident particle momentum in final state; and factor $\delta = m_e \omega^2 \alpha(\omega, E_L)/Ze^2$ stands for the contribution of polarization channel (α is polarizability of the ion with allowing for the presence of a strong laser field E_L of frequency ω). The description of the polarization Bremsstrahlung obtained is valid in a broad near-resonance range of frequencies, wherein its contribution is of importance. For describing the Polarization Bremsstrahlung, the Fermi method of equivalent photons and the Kramers Electrodynamics method [2] are extended to the case of *multiphoton* radiative-collisional processes, with extending the Fermi method for spontaneous Polarization Bremsstrahlung /Recombination [3]. Equation (1) extends the results [4] for multiphoton Polarization Bremsstrahlung to the case of simultaneous allowing for static and polarization channels.

The effect of the interference of static and polarization channels is shown in FIG.1 for the conditions of interest to the problem of energy exchange between non-equilibrium recombining plasmas with the HCI and soft X-ray laser field. Here, the interference takes place in a broad spectral region, with laser frequency detuning Δ from the closest atomic transition in the ion core lying far outside the line width of this atomic transition (the contribution of the one of atomic levels of the near-resonance atomic transition is shown).

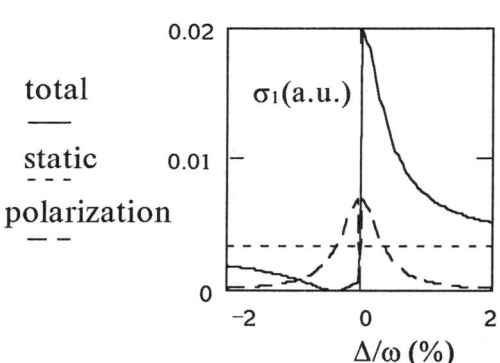

FIG.1. Electron velocity averaged cross-section of the one-photon stimulated emission/absorption in electron-HCl collision as a function of frequency relative detuning Δ/ω for laser frequency $\omega=3$, electron temperature $T_e=1$, $Z=10$, laser field strength $E_L=0.08$, oscillator strength $f_0=0.1$ (atomic units used here).

REFERENCES

1. Polarization Bremsstrahlung, Eds. V.N.Tsytovich & I.M.Ojringel (Plenum, N.Y. 1992).
2. Kogan V.I., Kukushkin A.B., Lisitsa V.S., Phys. Rep., 213 (1992) 1.
3. Kukushkin A.B., Lisitsa V.S., Chapter 11 in Ref.[1].
4. Astapenko V.A., Kukushkin A.B., Spectral Line Shapes, Vol.8 (AIP Conference Proc. #328, 12th ICSLS, Toronto, Canada, 1994), p.83.

HgIn Excimer and Bound-Bound Emission Following Laser Excitation of Atomic In in an Indium-Mercury Vapor Mixture

P. Bicchi[†], C. Marinelli[†], and R.A. Bernheim[*]

[†]INFM and Department of Physics, University of Siena,
Via Banchi di Sotto 55/57, 53100 Siena, Italy
[*]Department of Chemistry, Penn State University,
152 Davey laboratory, University park, PA 16802

Abstract. Laser excitation of the In($6^2S_{1/2}$ - $5^2P_{1/2}$) atomic transition in a Hg-In-Ne gaseous mixture results in HgIn excimer emission near the In atomic resonance lines at 410.3 nm and 451.1 nm as well as HgIn bound-bound transitions at 499 nm and 522 nm. The spectrum, including the fluorescence lifetimes of the various features, are interpreted in terms of analogous states extrapolated from high resolution spectroscopic observations of HgTl.

The mercury atom has received considerable attention as a partner atom in the formation of excimers with other metal species because of its $^1S_0$ closed subshell electronic ground state which results in mainly unbound electronic ground states in its interaction with many other atoms. Other advantages for its use in the generation of excimer radiation are its high vapor pressure and low-lying, easily accessible electronic states. The present contribution reports on a number of aspects of the HgIn excimer emission when generated by laser excitation of In atoms to the In($6^2S_{1/2}$) state in a Hg-In-Ne gaseous mixture.

Emission spectra of HgIn produced in electrical discharges[1-4] reveal excimer wings and satellites associated with the In atomic resonance lines at 410.3 nm and 451.1 nm as well as electronic band spectra at 499 nm and 522 nm. These were again recently observed in electrical discharges and discussed by Pichler, *et al.* in terms of a potential energy diagram based on HgTl *ab initio* calculations.[3,4] The most recent analysis of the bound-bound 522 nm band gives spectroscopic constants of v_e = 19 106, ω_e' = 198, and ω_e'' = 151 cm^{-1} with $\omega_e x_e$ having a value very close to zero implying deeply bound upper and lower states of this transition. In another study, the 522 nm band and another broad band centered at 505 nm were observed to be generated when InBr is photodissociated in the presence of Hg vapor if the In atoms are simultaneously excited to the In($6^2S_{1/2}$) state[5]. A conclusion of the experiment was that the 522 nm band originates from a HgIn state correlating with the $In(6^2S_{1/2})$ and $Hg(^1S_0)$ separated atomic states.

A different approach to the generation of HgIn has been used in the present experiments which were carried out in a quartz cell containing In metal (99.999% stated purity), a droplet of Hg, and 10 torr of Ne. At 1173K the number densities are 6.5×10^{13} for In, 8.2×10^{16} for Ne, and approximately 10^{19} cm^{-3} for Hg. The $In(6^2S_{1/2} - 5^2P_{1/2})$ transition was excited by 10 ns pulses of laser radiation at 410.3 nm and with 0.3 MW/cm^2 to 2.5 MW/cm^2 power densities. Emission spectra were obtained at 2 cm^{-1} resolution.

The emission spectra of the two In atomic resonance lines are shown in Fig.1. Of the features previously reported in electrical discharges, those that also appear with laser excitation are the red-wing satellites at 412.7 nm and 453.6 nm as well as the weak blue satellite at 447 nm. The other previously reported features close to the In resonance lines were not detected in these experiments.

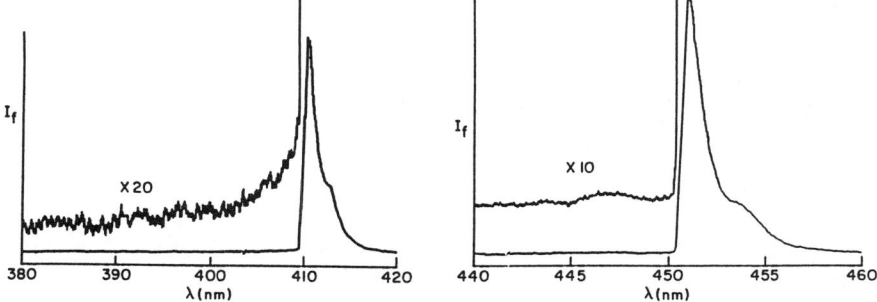

Figure 1. HgIn excimer emission, in arbitrary units, appearing as satellite wings on the $In(6^2S_{1/2} - 5^2P_{1/2})$ atomic line at 410.3 nm and the $In(6^2S_{1/2} - 5^2P_{3/2})$ atomic line at 451.1 nm following laser excitation at 410.3 nm in the presence of mercury vapor and neon at 1173K.

An excitation spectrum, obtained by observing the 451.1 nm fluorescence and scanning the exciting laser over the $In(6^2S_{1/2} - 5^2P_{1/2})$ transition at 410.3 nm, is shown in Fig. 2 together with an excitation spectrum of In vapor taken under identical conditions but in the absence of Hg. Careful examination of the region further to the red and beyond 412.7 nm revealed no distinct evidence of the satellite which is clearly observed in the fluorescence spectrum in Fig. 1. Excitation spectra taken by observing the low resolution emission at 522 nm and at 499 nm gave identical line shapes as that shown in Fig. 2 for the 451.1 nm emission.

The bound-bound transitions at 499 nm and 522 nm which were also produced by the laser excitation of $In(6^2S_{1/2})$ are shown in Fig. 3. The broad intense feature at 505 nm found in the InBr photolysis experiment was not present. A power dependence study of the 522 nm band revealed that its intensity was linearly dependent upon the laser power at 410.3 nm, despite the fact that the $In(6^2S_{1/2} - 5^2P_{1/2})$ transition is saturated. A temperature dependence study between 973K and 1173K gave a linear dependence on In atom density, which reinforces the conclusion that this band is due to emission of an electronic state of HgIn

Figure 2. Excitation spectrum taken by scanning the laser over the In($6^2S_{1/2}$ - $5^2P_{1/2}$) transition at 410.3 nm in a cell containing mercury vapor and neon at 1173K and detecting the In($6^2S_{1/2}$- $5^2P_{1/2}$) emission at 451.1 nm. For comparison, the narrow excitation spectrum of the same transition is shown which is obtained for In under the same conditions but in the absence of mercury vapor.

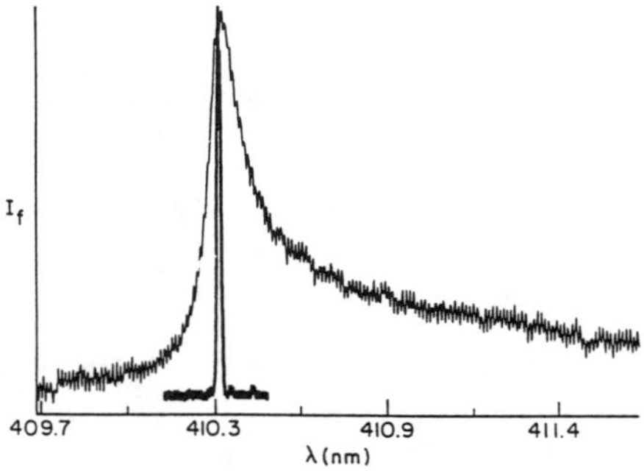

Figure 3. HgIn bound-bound fluorescence bands at 499 nm and 522 nm following laser excitation of the In($6^2S_{1/2}$ - $5^2P_{1/2}$) transition in the presence of mercury vapor.

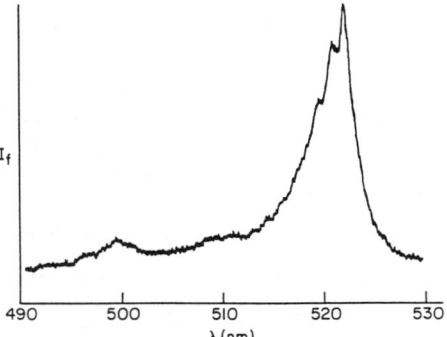

that is correlated with the In($6^2S_{1/2}$) + Hg($^1S_0$) separated atom states. However, the power dependence study reveals the possibility that the 522 nm band may, in fact, be a transition originating from a HgIn state above that which correlates with the In($6^2S_{1/2}$) + Hg($^1S_0$) separated atom states and which is populated by a multiple photon process. Further evidence for a multi-photon excitation is revealed by the presence of the In($5^2D_{3/2}$ - $5^2P_{1/2}$) and In($5^2D_{5/2}$ - $5^2P_{3/2}$) atomic emission transitions at 304 nm and 326 nm, respectively, which also exhibit *linear* power dependencies on the excitation of the In($6^2S_{1/2}$ - $5^2P_{1/2}$) transition at 410.3 nm under conditions where the latter is saturated.

A resolution of this conflicting evidence for the assignment of the 522 nm band can be obtained from the decay behavior of its fluorescence. At 1173K the time evolution of this fluorescence consists of a fast component with a time constant of 25 ns and a slow component having a time constant of 106 ns. Since the measured In($6^2S_{1/2}$) fluorescence lifetme (in the absence of Hg) at this temperature is 25 ± 2 ns, this observation would be consistent with emission from a HgIn state correlating with In($6^2S_{1/2}$) + Hg($^1S_0$) that is populated by two processes: (1) excitation at 410.3 nm of In colliding with Hg which would also radiate with a decay constatant close to 25 ns, and (2) a slower cascade from some reservoir state populated by a step-wise multiphoton process. The following elementary steps could be a contribution to the over-all process:

$$\text{In}(5^2P_{1/2}) + h\nu(410\,\text{nm}) \longrightarrow \text{In}(6^2S_{1/2}) \qquad (1)$$

$$\text{In}(6^2S_{1/2}) + \text{Hg}(^1S_0) \longrightarrow \text{HgIn}^* \qquad (2)$$

$$\text{HgIn}^* + h\nu(410\,\text{nm}) \longrightarrow \text{HgIn}^{**} \qquad (3)$$

$$\text{HgIn}^{**} \longrightarrow \text{HgIn}^* \qquad (4)$$

$$\text{HgIn}^{**} \longrightarrow \text{Hg}(^1S_0) + \text{In}(5^2D_{5/2,3/2,}) \qquad (5)$$

with (4) and (5) due to quenching or cascade processes.

A theoretical treatment of the potential energy curves for HgIn, with which one can interpret the spectral features, does not exist. Nevertheless, one can follow the example of Pichler, *et al.*[3,4] and extrapolate from the available information for the potential energy curves of HgTl. However, instead of using potential energy curves obtained from a theoretical treatment, a high resolution spectral analysis of HgTl is available[6] which reveals states having substantial bonding for HgTl correlating with the two lowest atomic states of Tl. Assuming that a corresponding scheme holds for HgIn, the 522 nm band would originate from the HgIn* state, HgIn($B\Sigma_{1/2}$), which correlates with In($6^2S_{1/2}$) + Hg($^1S_0$) and terminates on the HgIn($X_{3/2}$) state which correlates with In($5^2P_{3/2}$) + Hg($^1S_0$). The 499 nm band also originates from HgIn($B\Sigma_{1/2}$) but terminates on the HgIn($X_{1/2}$) state that correlates with the In($5^2P_{1/2}$) + Hg($^1S_0$) states. Furthermore, the red satellites at 412.7 nm and 453.6 nm would be expected to be due to exhibit bound-bound character as is the case for the corresponding features in HgTl.

In conclusion, the present work has produced the HgIn excimer emission by laser excitation of an In atomic resonance transition in a Hg-In-Ne gaseous mixture. The experiments confirm that the HgIn spectral features originate from a single HgIn state which correlates with the In($6^2S_{1/2}$) + Hg($^1S_0$) separated atom states and are consistent with HgIn potential energy curves having strong bonding in molecular states correlating with In($5^2P_{1/2,3/2}$) + Hg($^1S_0$) separated atom states.

Support from the National Science Foundation, NATO, and the Italian Consiglio Nazionale Ricerche is gratefully acknowledged.

REFERENCES

1. R. L. Purbrick, Phys. Rev. **81**, 89-94(1951).
2. S. Chilukuri, and J. G. Winans, J. Mol. Spectrosc. **16**, 309-324 (1965).
3. G. Pichler, D. Azinovic, and R. Beuc, Fiz. A **2**, 1-9 (1993).
4. V. Henc-Bartolic and G. Pichler, Fiz. A **5**, 1-10 (1996).
5. R. G. C. Thomson, I. Duncan, and T. Morrow, Chem. Phys. Lett, **187**, 396-400 (1991).
6. S. Chilukuri, W. J. Pearce, and J. G. Winans, J. Chem. Phys. **82**, 4405-4414 (1985).

Energy-Pooling Ionization Produced by The Collisions of In Atoms In Presence of a Resonant Laser Field

P. Bicchi, C. Marinelli, E. Mariotti, M. Meucci, L. Moi

INFM and Department of Physics, University of Siena, Via Banchi di Sotto 55/57 - 53100 Siena, Italy

Abstract. We report the production of electrons and ions in the collisions of In ($6^2S_{1/2}$) atoms resonantly excited by laser radiation. The cross section for the ions production is reported.

We have demonstrated, few years ago, that in the collision between indium atoms excited to the $6^2S_{1/2}$ level there is the possibility to produce atomic ions and population in Rydberg as well as in autoionizing levels[1]. This possibility arises from the level structure of indium atoms which have the first excited state at an energy which is almost half the ionization energy and a fine structure of the ground state $5s^25p5^2P_{1/2}$ which is large enough to allow two energy transfer processes following the collision of two In ($6^2S_{1/2}$) atoms:

$$In^* + In^* \longrightarrow In^{**} + In_{3/2} \pm \Delta E \tag{1}$$

$$In^* + In^* \longrightarrow In^+ + e^- + In_{1/2} + 2075.8 \text{ cm}^{-1} \tag{2}$$

where $In_{1/2, 3/2}$ denotes an In atom in the J = 1/2 or 3/2 ground state sublevel and In^+ denotes an InII ground state ion. Process (1) is the well known energy-pooling collision (EPC) which has been already studied in detail[2]. Process (2) is the energy-pooling ionization (EPI) process of which we got direct evidence in our experiment by monitoring directly the electrons produced by it.

Indium together with a few torr of buffer gas is kept in a steinless steel cross shaped heat-pipe oven (HPO) with quartz windows and is heated up to about 1000 °C. Along one of the HPO axis two electrodes have been inserted to create the electric field. The exciting laser beam crosses the HPO along the axis perpendicular to the electrodes at the center of their separation. It is produced by a pulsed dye-laser tuned to the indium resonance at 410.3 nm with a power density of typically 800 W/cm². Both the electron and the laser induced fluorescence signals can be detected and analyzed with an apparatus which is a standard one for this kind of experiment and which is described in detail elsewhere[2,3].

Figure 1 shows typical signals recorded while the laser is scanned through the resonance wavelenght. The continuous line is the fluorescence signal at 451.1 nm due to the transition $6^2S_{1/2}$ ---> $5^2P_{3/2}$ and the dashed-dotted line is the electron signal collected at the positive electrode. When the laser output is resonant with the atomic transition an electron signal appears. No electron is detected when the laser is out of resonance. At the low laser power density of the experiment the cross section for two- photon ionization, even on resonance, is totally negligeable[4], so this process is not responsible for the electron yeld. Also the production of electrons by stray effects such as the photoelectric one, has to be ruled out both because no electron is detected when the laser is out of resonance and because the electron signal shows a temperature threshold. When the latter is lowered below 850 °C the former disappears while the fluorescence signal remains. Besides when the laser power density is varied in the range 300 W/cm^2 - 1.5 KW/cm^2, the electron signal manifests a quadratic dependence on it. These characteristics are consistent with a signal caused by pooling collisions and the electrons are produced via the EPI process. From the electron signal, the density of electrons produced per unit time can be measured and the cross section for the EPI process derived[5]. We measured it to be σ_{EPI}= (6.0±1.8) x 10^{-17} cm^2.

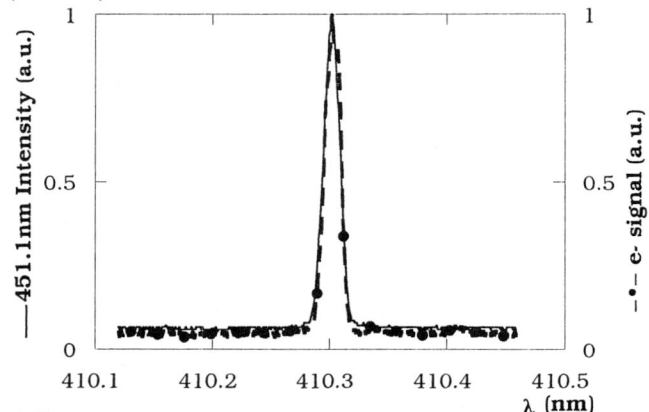

FIGURE 1. —) fluorescence signal at 451.1 nm and -•-) electron signal collected at the positive electrode while the laser is scanned through the resonance wavelenght. The signals are taken in a HPO containing In + 30 torr of He at 900 °C and a laser power density of 800 W/cm^2.

REFERENCES

1. P. Bicchi, A. Kopystynska, M. Meucci, L. Moi: Phys. Rev. A **41**, 5257 (1990)
2. P. Bicchi, C. Marinelli, E. Mariotti, M. Meucci, L. Moi: J. Phys.B:At. Mol. Opt. Phys. **26**, 2335 (1993)
3. P. Bicchi, C. Marinelli, E. Mariotti, M. Meucci, L. Moi: *Laser -Matter and Atomic and Molecular Interactions in Plasmas and their Applications*, Pisa: ETS press - in press (1996)
4. V.S. Letokhov: Optics and Laser Tech. **10**, 247 (1978)
5. P. Bicchi, C. Marinelli, E. Mariotti, M. Meucci, L. Moi: J. Phys.B:At. Mol. Opt. Phys. in press (1996)

Energy Pooling Collisions in Pure Cadmium Vapors

S. Barsotti, F. Fuso, A.F. Molisch* and M. Allegrini+

Unità dell'Istituto Nazionale per la Fisica della Materia,
Dipartimento di Fisica, Università di Pisa, Piazza Torricelli 2, I-56126 Pisa, Italy

Abstract. We have investigated energy pooling processes in pure Cadmium vapors excited by a pulsed laser beam resonant with the $Cd(5^1S_0) \rightarrow Cd(5^3P_1)$ intercombination transition. Starting from the measured fluorescence intensities, we have determined the cross sections for the processes populating the 5^3D_j levels using a rate equation system and accounting for the radiation trapping phenomenon.

Energy pooling (EP) is a very well known collisional process involving two excited atoms and yielding one highly excited atom and one in a low-lying state, typically the ground state. In particular, in our experiment we have excited the Cadmium intercombination transition $Cd(5^1S_0) \rightarrow Cd(5^3P_1)$ at 326.1 nm, and we have investigated the process:

$$Cd(5^3P_1) + Cd(5^3P_1) \rightarrow Cd(5^3D_j) + Cd(5^1S_0) + \Delta E \quad (1)$$

where j = 1–3 and ΔE, the energy balance factor, is $\approx$ 4–5 kT at the typical temperatures of our experiment (T $\approx$ 500 K).

In the past decades, EP has been widely investigated in alkali vapors, and, more recently, in elements belonging to the IIB column. For Cd, observations of EP as a function of the pressure of a buffer gas have been reported in [1]. Also in our laboratory, preliminary observations of EP in Cd have been performed using an heat-pipe oven filled with the metal [2]. However, at the best of our knowledge, no quantitative analysis of EP in Cd have been previously reported.

For the aim of the present work, i.e., the measurement of the cross sections for reactions (1), we have excited pure Cd vapors contained in a sealed quartz cell and detected the fluorescence emission. The cell temperature has been varied in the range T = 470–650 K, corresponding to a Cd density $n = 8 \times 10^{12} - 1 \times 10^{16}$ cm^{-3}, as deduced from Nesmeyanov vapor pressure tables. Excitation has been accomplished by exploiting a XeCl excimer pumped dye laser providing a maximum pulse energy $E_L \approx 100$ μJ (at 326.1 nm) in a 10 ns (FWHM) pulse. The fluorescence emission from the cell has been collected at right angles and analyzed by a 64 cm monochromator equipped with both single- and multi-

---

* Institut für Nachrichtentechnik und Hochfrequenztechnik, Technische Universität Wien, Gusshaustrasse 25/389, A-1040 Wien, Austria

+ Also at Dipartimento di Fisica della Materia e Tecnologie Fisiche Avanzate, Università di Messina, Salita Sperone 31, I-98166 Sant'Agata, Italy

channel detectors for time resolved and time integrated (in an adjustable gate width, τ_g) analysis, respectively.

Figure 1 shows fluorescence spectra in the range 280–400 nm at different cell temperatures acquired at $\tau_g = 500$ ns. The spectra exhibit the presence of transitions from $Cd(5^3D_j)$ levels (located around 340, 347 and 361 nm) at $T \geq 570$ K ($n \geq 1 \times 10^{15}$ cm^{-3}). At $T = 650$ K, other transitions are observed as a consequence of other population processes. The fluorescence intensity of different transitions has been measured as a function of the laser pulse energy and of the cell temperature in order to ascertain the collisional nature of the processes leading to the population of the $Cd(5^3D_j)$ levels [3]. Additional information on the dynamics of these processes has been found through time resolved analysis of the fluorescence signals, which indicates a relatively fast rise time for the fluorescence arising from $Cd(5^3D_j)$ levels, and a decay time approximately one half that of the $Cd(5^3P_1) \rightarrow Cd(5^1S_0)$ transition. These results give a further confirmation of the EP origin for the population of the $Cd(5^3D_j)$ levels.

The cross sections for the EP processes (1) have been deduced from the observed fluorescence intensities by solving a rate equation system involving the Cd levels relevant in the experiment. The effects of the radiation trapping phenomenon, which heavily affects the time behavior of the fluorescence signal for the intercombination transition and the density of the $Cd(5^3P_1)$ state, have been taken into account by numerical simulations of the radiation diffusion equations performed in the specific conditions of our experiment [3]. The measured EP cross sections at $T = 580$ K are 1.3 ± 0.7 Å^2, 1.1 ± 0.6 Å^2, and 0.7 ± 0.4 Å^2 for the $Cd(5^3D_j)$ levels with j = 1, 2, and 3, respectively.

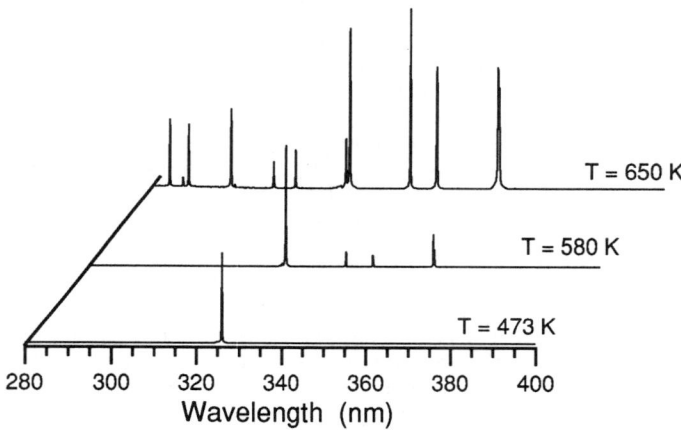

Fig. 1. Fluorescence spectra at different temperatures ($\tau_g = 500$ ns).

REFERENCES

1. H. Umemoto, J. Kikuma, A. Masaki and S. Sato, Chem. Phys. **127**, 227 (1988)
2. M. Musso, F. Fuso, G. De Filippo, M. Allegrini, D. Gruber and L. Windholz, submitted to J. Chem. Phys.
3. S. Barsotti, F. Fuso, A.F. Molisch and M. Allegrini, to be published

Multistep excitation of autoionizing Yb atom states and line shape of autoionizing resonance

I.M.Beterov, U.V.Brjzazovsky, V.L.Kurochkin, A.A.Chernenko, V.B.Alman

Institute of Semiconductor-Physics RAS, pr.Lavrentyev, 13, Novosibirsk, 630090, Russia

Abstract. The results of multiphoton ionization of the Yb atoms in long atomic beam near the autoionization (AI) level $4f^{13}6p^26s$ and determination of the Fano parametres of AI resonance are reported.

The great interest for the AI levels investigation is due to the possibility of the effectivity increasing a number of nonlinear processes as so as multiphoton ionization, the high order optical harmonic generation and so on. It's known from experimental date about existing AI resonance line width from 0,05 sm to hundred sm .First the AI states belonging electron configuration $4f^{14}7p7s$ of Yb atom were observed by V.S.Letokhov [1]. Lately authors [2] discovered and studied AI states belonging configuration $4f^{12}6p$nl, that were situated near the ionization limit.

Ionization of the Yb atoms in our experiments was produced by three photon excitation scheem under the optical fields of the colenear slightly collimated laser beams.On the first stage for the excitation of the resonant transition $6^1S_o - 6^3P_1$ (=555,6 nm) we used the continuously tuned Rhodamine 110 dye laser with the output power ~200 mW pumped by radiation of the argon ionic laser with pump power ~18 W. On the second and third stages we used the emissions of the pulse narrow line Rhodamine 6G dye laser pumped by the power Cu+ laser radiation (P 20 W, f~10 kz). On this way the second stage laser radiation frequency was tuned on the transition $6^3P_1 - (7/2,3/2)_2$ frequency of Yb atom (λ= 581,2 nm),and the third laser emission frequency was changed near the wavelengh ~582,8nm, this wavelengh corresponded to the excitation energy one of the AI levels. The Yb atomic beam was formed by using the effusion source with the split size ~ 1x10 mm2 and vacuum chamber with pressure P < 10(-4) torr, the max atom density was about ~10(12) at/sm3. The detecting system measured electron and ionic components of photo-current in the wide range of signal modification.The max registed ionic photocurrent was about 150 nA. The observed dependance of the ionic photocurrent from thirdstep laser radiation wavelengh is represented on

© 1997 American Institute of Physics

fig..(•-points). For comparison we show also the experimental date authors [2] obtained under the action of more lower exciting laser radiation intensity and more lower atomic beam density. One can see that dependances have resonance forme with typical for the AI levels assimetrical line form, but have the different line widths. The analysis of the exciting conditions according [3] had shown that AI resonance line form can be determined by the Fano's form [4] as:

$$\alpha(\Omega) = (q + \Omega)^2 / (1 + \Omega^2) \quad (1)$$

were Ω - the relative energy building up from the AI resonance line centre, q - Fano's parameter, that determing the attitude of the transition matrix elements of the discreet and continious electron energy spectra (interference of the transition channels).It's important the experimental determination of this parametra. From the computer processing of the experimental data according to the Fano's theory the next quantities of the line resonance width $\delta\lambda \approx 6,5$ Å and the Fano's parametre q= 10 - 11 were obtained. For the mentioned above experiment this quantities were $\delta\lambda$= 2,0 Å,q = 10-11. It's clear,that the same quantity of parameter q for two different experiments reflects the physics of AI process and the differences in the AI line width reflect the fact of dependence of this quantity from the such experimental conditions as the excited radiation power density,the neutral and charged partical density. More detailed determination the physical mechanisms of the AI resonance line width increasing observed will be studied else , because perfomed numerical estimations of influence on line width such physical factors as isotopical structure and optical field action arn't agree with observed line spread.

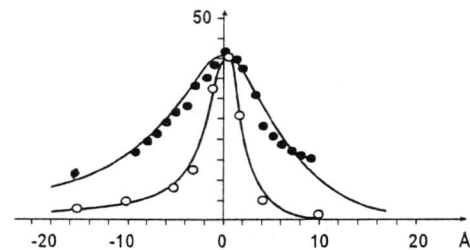

REFERENCES

1. Bekov G. I.,Vidolova-Angelova E.P.,Ivanov L.N.,Letochov V.S.,Mishin V.I.Opt.Comm.35,194,(1980).
2. Krinetsky B.B.,Mishin V.A.,Prochorov A.M.J.Appl.Spectr. 54,558,(1991).
3. Andrushin A.I., Kazakov A.E.,Fedorov M.V.JETF,88,1153, (1985).
4. Fano U.Phys.Rev.124,1866,(1961).

Excitation transfer and radiative decay in the n^2F and the $(n+3)^2S$ states of Rb

B. Bieniak, M. Głódź, J. Szonert

Institute of Physics, Polish Academy of Sciences, Al. Lotników 32/46, PL-02-668 Warsaw

Radiative decay and collisional processes for the n^2F (n=6,7,8) states and the closely lying $(n+3)\,^2S$ states of Rb has been investigated. The n^2F levels are populated by stepwise optical dipole- plus quadrupole- transitions or collisional transfer from the $(n+3)\,^2S$ states.

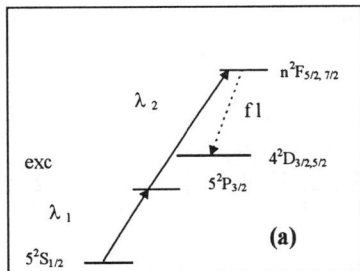 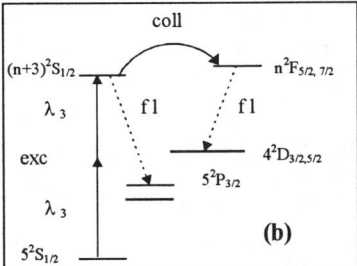

Fig. 1 Excitation/detection schemes for optical (a) and collisional (b) population of the n^2F states of Rb. λ_1 for cw diode laser excitation, λ_2, λ_3 - for pulsed excimer laser-pumped dye laser excitation.

Pure rubidium fluorescence cell is used, and operated in the low pressure regime (2×10^{-6} to 2×10^{-4} Torr). The time resolved fluorescence signals registered by means of photon counting enable to determine the radiative and quenching decay rates (1), (2). Cross section for the collisional transfer $Rb(5^2S)+Rb((n+3)^2S) \rightarrow Rb(5^2S)+Rb(n^2F)$ has been obtained as

$$\sigma_{(n+3)^2S \rightarrow n^2F} = (I_F / I_S)(A_{SP} / A_{FD})\Gamma_{nF}^{tot}/Nv,$$

© 1997 American Institute of Physics

where I_F and I_S are time integrals of the fluorescence signals from the n^2F and $(n+3)^2S$ states, A_{SP} and A_{FD} are Einstein coefficients for $(n+3)^2S_{1/2} \rightarrow 5^2P_{3/2}$ and $n^2F_{5/2,7/2} \rightarrow 4^2D_{3/2,5/2}$ transitions, respectively, Γ_{nF}^{tot} - total decay rate (radiative and collisional) of the n^2F state, N number density of the ground state Rb atoms, v mean relative velocity of collision partners.

The results has been corrected for the black-body radiation induced effects.

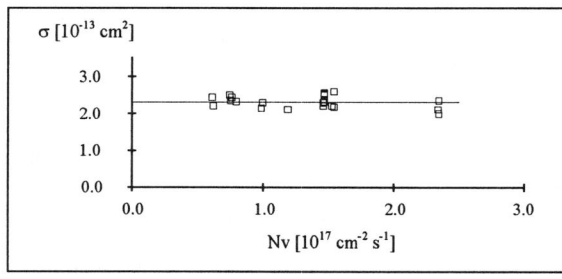

Fig. 2 Transfer cross section $\sigma_{Rb(10^2S \rightarrow 7^2F)}$ induced by collisions with Rb atoms, measured at different Nv (temperatures).

Table 1. Radiative lifetimes and cross sections for collisions in Rb vapour. The theoretical lifetimes of refs. (3) and (4) and geometrical cross sections are given for comparison. †Results published in ref. (1).

State	Radiative lifetimes (in ns, T = 0K)			Cross sections (in 10^{-13} cm^2)			
	Experiment	Theory		Total quenching		Transfer (n+3)S→nF	
	our results	Ref(3) J	Ref(4)	our results	Geometrical	our results n	
6^2F	171 (4)†	5/2 173.94 7/2 173.84	178.8	2(9)†	3	6	
7^2F	262 (15)†	5/2 267.11 7/2 266.91	275.0	17(24)†	5	7	2.3 (8)
8^2F	387 (12)†	5/2 389.72 7/2 389.40	401.6	15(11)†	9	8	
10^2S	424(5)	1/2 417.84	429.4	5.6(1.7)	5.5		

Cross sections $\sigma_{(n+3)^2S \rightarrow n^2F}$ for other n values are under study. This is a part of a broader project on collisional transfer and radiative lifetimes in alkali atoms.

REFERENCES

1. Szonert, J., Bieniak, B., Głódź, M., Piechota, M., Z. Phys. D 33, 177-180 (1995)
2. Bieniak, B., Fronc, K., Głódź, M., Szonert, J., Opt. Appl. 25, 15-24 (1995)
3. Theodosiou, C.E., Phys. Rev. A30, 2881-2909 (1984)
4. Hansen, W., J. Phys. B17, 4833-4850 (1984)

Time Behavior of the $7\,^3S_1 - 6\,^3P_2$ Transition Following Pulsed Laser Excitation of the $6\,^3P_1$ State of Mercury

K. Blagoev*, D. Gruber, A. Dreischuh**, A. Morozov***, U. Reiter-Domiaty, and L. Windholz

Institut für Experimentalphyik, Technische Universität Graz, Petersgasse 16, A-8010 Graz

Institute of Solid State Physics, Bulgarian Academy of Science, 72 Tzarigradsko Chaussee, BG-1784 Sofia

**Department of Quantum Electronics, Sofia University, J. Bourchier Blvd. 5, BG-1164 Sofia*

****Department of Optics and Spectroscopy, St. Petersburg State University, 198904 St. Petersburg*

Since mercury represents an often used additive in discharge lamps, investigations of the behavior of mercury transitions appear to be of practical interest. Furthermore a detailed knowledge of the interaction of mercury with laser fields can improve mercury pollution detection in the free atmosphere. Recently, the time behavior of the 546.0 nm spectral line was studied after two step pulse laser excitation of the $7\,^3S_1$ state via the $6\,^3P_1$ state [1].

We use a quartz cell to investigate the mercury vapor. The cell temperature is kept approximately 50 K higher than the temperature of a reservoir with a natural isotope mixture of mercury. Laser radiation is provided by a Nd:YAG laser pumped dye laser which output radiation is frequency doubled and used to populate the $6\,^3P_1$ state by a strong laser pulse.

The fluorescence spectra of the Hg vapor exhibit many lines from highly electronically excited states, which population partially can be ascribed to energy pooling (EP) processes [2,3].

© 1997 American Institute of Physics

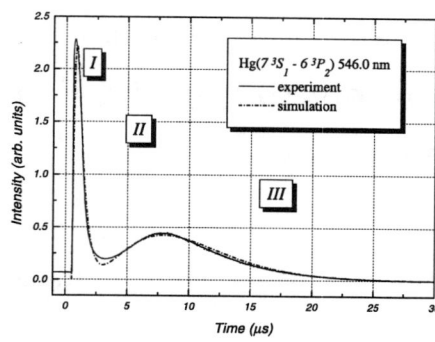
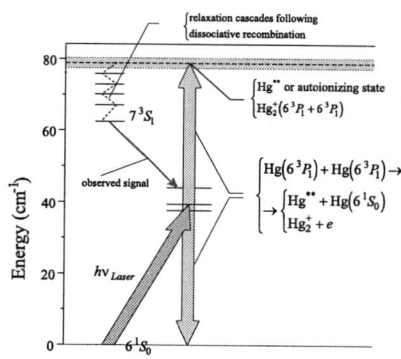

Fig. 1: Time behavior of the $Hg(7\,^3S_1 - 6\,^3P_2)$ line at 546.1 nm, following excitation of the $Hg(6\,^3P_1)$ state with 253.65 nm, at a temperature of 400 K

Fig. 2: Simplified energy level diagram of mercury representing the physical taken into account in the model

Figure 1 presents a typical time evolution of the fluorescence line at 546.0 nm, corresponding to the transition $7\,^3S_1 - 6\,^3P_1$, showing typical non-monotonic time behavior. The three regions indicated in the figure represent (*I*) the region, where the decay mainly is governed by radiation trapping and the spontaneous decay following EP, (*II*) a region, characterized by another increase in intensity peaking at a time delay of some 10 µs. Finally, region (*III*) shows a long living tail, that we ascribe to an associative ionization (AI) according to

$$Hg(6\,^3P_J) + Hg(6\,^3P_J) \to Hg_2^+ + e^- \qquad (1)$$

which then is followed by dissociative recombination (DR)

$$Hg_2^+ + e^- \to Hg^{**} + Hg(6\,^1S_0) + h\nu . \qquad (2)$$

A numeric model accounting for EP [4] and AI/DR (Fig. 2) is introduced. In our simulations we describe radiative cascading following (2) by use of 4 arbitrary intermediate states, which population and decay merely is caused by spontaneous processes. In order to reproduce the fast decaying part in region (*I*), we also take into account radiation trapping effects of all transitions considered. The dashed line in Fig. 1 represents the results of the calculation. The dependence of time evolution on laser detuning is discussed and will be published elsewhere.

[1] N. Omenetto, O. I. Matveev, W. Resto, R. Badini, B. W. Smith, and J. D. Winerfordner; Appl. Phys. B 58, 303 (1994)

[2] S. Majetich, E. M. Boczar, and J. R. Wiesenfeld; J. Appl. Phys. 66, 475 (1989)

[3] S. Majetich, C. A. Tomczyk, and J. R. Wiesenfeld; Phys. Rev. A 41, 6085 (1990)

[4] P. C. Johnson; J. Phys. B 11, 1877 (1978)

[5] This work was partially financed by the Fonds zur Förderung der Wissenschaftlichen Forschung, project no. P-9929-PHY, by the Jubiläumsfonds der Österreichischen Nationalbank, project no. 4873 and by the CEEPUS-program (network A-21).

The Effect of an Electromagnetically Induced Transparency on Laser-Assisted Collisional Ionization

Roberto Buffa

Dipartimento di Fisica, Università di Firenze, Largo E. Fermi 2, 50125 Firenze, Italy

Electromagnetically induced transparency (EIT) is the modification of the absorption and dispersion properties of an atomic transition when the upper level - usually decaying by autoionization - is coherently coupled to a third level by a relatively intense laser field [1]. It results from the combination of Autler-Townes splitting and quantum interferences of dressed states which decay into the same continuum of ionization. The aim of the present paper is to show how EIT can modify the dynamics of coherent ionization processes induced by laser radiation and atomic collisions [2].

The physical system considered in this work is shown in Fig.1. It is assumed that atom A can be described by two bound states $|\alpha_i\rangle$ and atom B by three bound states $|\beta_j\rangle$ and an autoionizing state $|\beta_4\rangle$. States $|\alpha_1\rangle$ and $|\alpha_2\rangle$, and $|\beta_1\rangle$ and $|\beta_2\rangle$ are coupled by dipole-allowed transitions, as well as states $|\beta_2\rangle$ and $|\beta_4\rangle$, and $|\beta_3\rangle$ and $|\beta_4\rangle$. State $|\beta_2\rangle$ is nearly resonant with state $|\alpha_2\rangle$ and the ionization potential of atom A is quite larger than the energy of state $|\beta_4\rangle$. Atom A, prepared in the excited state $|\alpha_2\rangle$, collides with atom B in the ground state, in the presence of a probe radiation field of Rabi frequency frequency χ_p and frequency ω_p, nearly resonant with the interatomic transition $|\alpha_2\rangle$ - $|\beta_4\rangle$. During the collision, the excitation energy of atom A can be transferred to atom B by the absorption of a photon, leading to a frequency dependent ionization of atom B. The peculiarity of the reaction, named laser-induced collisional autoionization (LICA) [3] is that three different kinds of interaction (collisional, radiative and Coulombic) occur simultaneously.

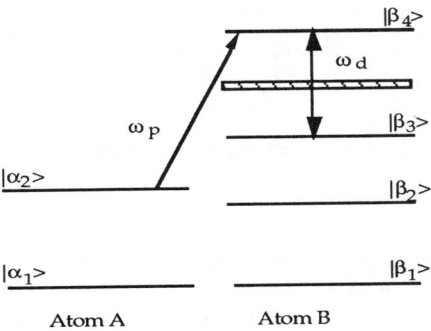

Fig.1 Schematic energy-level diagram of the model.

In the process considered here, an additional laser field, of Rabi frequency χ_d and frequency ω_d, nearly resonant with the atomic transition $|\beta_3\rangle - |\beta_4\rangle$, is used to create a pair of interfering dressed states which decay into the same continuum of ionization.

For a weak probe laser field, the cross section σ of the process can be written as a convolution integral [2]

$$\sigma = \chi_p^2 \int_{-\infty}^{+\infty} S(\omega)\, F(\omega_p - \omega, \omega_d)\, d\omega$$

where $S(\omega)$ reproduces the LICET (laser-induced collisional energy transfer) lineshape [4] and $F(\omega_p, \omega_d)$ the EIT absorption profile of the $|\beta_2\rangle - |\beta_4\rangle$ transition of atom B[1].

When no dressing laser field is applied, the cross section σ reproduces essentially the LICET line shape, broadened and slightly shifted by the finite linewidth of the final level $|\beta_4\rangle$ [2]. The effect of the EIT is visible in Fig.2 where the Rabi frequency of the dressing field - kept resonant with the $|\beta_3\rangle - |\beta_4\rangle$ transition - assumes the value $\chi_d = 2\gamma = 8\Delta_c$, where Δ_c is the half width at half maximum of the LICET line shape - corresponding to roughly the inverse of the average collisional time - and γ is the autoionization decay rate of level $|\beta_4\rangle$.

Ionization suppression of the order of 70 % is obtained at line center.

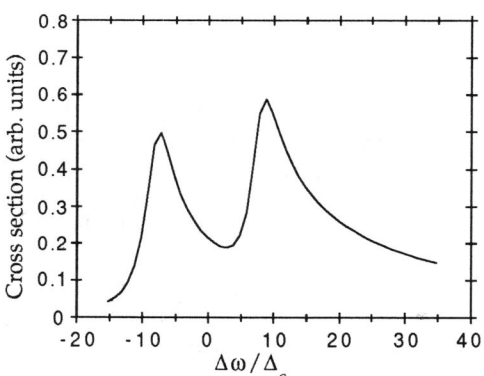

Fig.2. Cross section versus interatomic probe laser detuning $\Delta\omega$.

[1] K. J. Boller, A. Imamoglu, and S. E. Harris, *Phys. Rev. Lett.* **66**, 2593 (1991).
[2] R. Buffa, *Phys Rev. A* **53**, 607 (1996)
[3] R. Buffa, *Phys Rev. A* **46**, R1171 (1992); *Opt. Lett.* **20**, 204 (1995).
[4] A. Bambini, P. R. Berman, R. Buffa, E. J. Robinson and M. Matera, *Phys. Rep.* **238**, 245 (1994).

High Contrast Measurement of a Laser-Induced Structure in the Ionization Continuum of Sodium by Photoelectron Spectrometry

Roberto Eramo, Stefano Cavalieri, Lorenzo Fini, and Manlio Matera*

Dipartimento di Fisica and European Laboratory for Non-Linear Spectroscopy (LENS), Università di Firenze, Largo E.Fermi 2, I-50125 Firenze, Italy
*Istituto di Elettronica Quantistica, Consiglio Nazionale delle Ricerche, Via Panciatichi 56/30, I-50127 Firenze, Italy

Abstract. An experimental analysis of the structure induced by a strong laser field in the continuum of ionization of the sodium atom is reported. The experiment is performed on an atomic beam provided with a time-of-flight spectrometer, for an energy analysis of the photoelectrons. A substantial improvement in the evidence of the resonance has been obtained thanks to the selectivity of the detection system.

Quantum-mechanical interference between transition amplitudes involving the atomic ionization continuum has been a field of great interest in the last years. A discrete state embedded in the continuum provides a suitable scheme for the observation of such process, in a configuration where the direct decay of the atom into the continuum and its decay through the discrete state can interfere. Autoionization is the best known phenomenon of this kind in atomic physics but it is also known, theoretically since 1976, that the embedded state can be induced by a strong electromagnetic field (dressing field), coupling a bound state into the continuum. This process has been named Laser-Induced Continuum Structure (LICS) (1).

The LICS process is found to affect a number of non-linear interactions, such as induced birefringence and harmonic generation; nevertheless, the observation of the total ionization rate as a function of the frequency of a probe field provides the most direct evidence of this kind of quantum interference. Unfortunately, in photoionization experiments only weak modulations have been observed up to now, and the low signal to noise ratio has made the quantitative comparison with theory difficult (2).

This contribution reports an experimental analysis of a LICS structure observed in the one-photon ionization rate of sodium atom, following the excitation scheme of figure 1. The experiment has been performed on an atomic beam, by using a time-of-flight spectrometer to detect the ionization electrons. The spectrometer has

© 1997 American Institute of Physics

been designed to operate in the energy range 0.1÷ 10 eV. The flight tube consists of a 35 cm long, 5 cm diameter copper cylinder with gold-plated surfaces. The electrons are collected by three circular plates located at the entrance of the flight tube and are detected at the exit by a microchannel plate. The acquisition system is based on a time-to-digital converter, which is capable to detect up to eight stop pulses for each start pulse, with a resolution of 1 ns and a dead time of 10 ns.

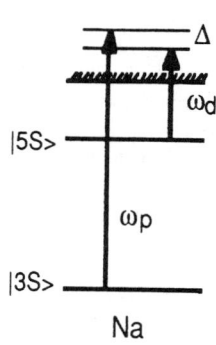

FIGURE 1. Scheme of the LICS process. ω_p: probe laser tunable around 234.7 nm; ω_d: strong laser at 1064 nm; Δ: detuning.

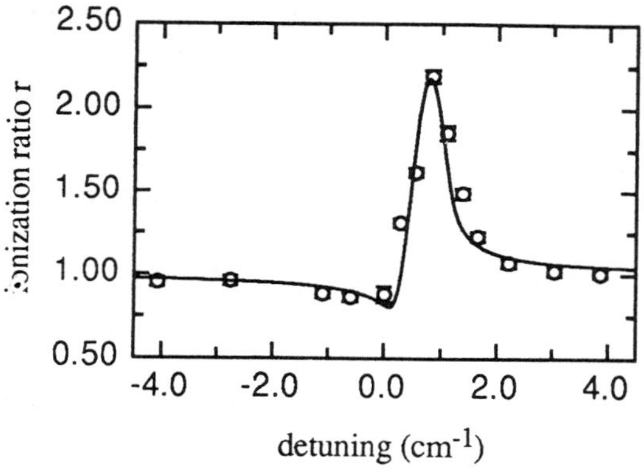

FIGURE 2. The ionization rate enhancement due to the LICS process. The continuous line is a theoretical LICS profile calculated taking into account a spatial and temporal distributions of the fields.

The normalized ionization yield measured in the presence of the dressing laser field is reported in figure 2. Background counts due to electron photoextraction induced by the probe field on the spectrometer electrodes and to multiphoton ionization of sodium dimers have been subtracted from the data.
The ionization profile, strongly asymmetric, shows an ionization enhancement of more than a factor two, and a clear effect of ionization suppression in the region of negative detuning. A theoretical LICS profile, taking into account temporal and spatial distributions modelling the laser fields, has been plotted with the data. A deeper analysis is now in progress with the aim of obtaining a more stringent test of the LICS model based on the known atomic parameters (3).

REFERENCES

1. Heller, Yu. I. and Popov, A. K., *J. Sov. Laser Res.* **6**, 1 (1985), Knight, P. L., Lauder, M. A., and Dalton, B. J., *Phys. Rep.* **190**, 1 (1990) and references therein.
2. Hutchinson, M. H. R., and Ness K. M. M., *Phys. Rev. Lett.* **60**, 105 (1990), Shao, Y. L., Charalambidis, D., Fotakis, C., Zhang, J., and Lambropoulos, P., *Phys. Rev. Lett.* **67**, 3669 (1991), Cavalieri S., Pavone, F. S., and Matera, M., *Phys. Rev. Lett.* **67**, 3673 (1991).
3. Cavalieri S., Eramo R., and Fini L., *J. Phys. B: At. Mol. Opt. Phys.* **28**, 1793-1801 (1995).

Time Resolved Spectra of Emission Signals from Strontium Vapour after Switching off Radiation Trapping with a Strong Laser Pulse

Lorenzo Fini*, Marina Mazzoni† and Ruggero Vaia†

†Istituto di Elettronica Quantistica del CNR, via Panciatichi 56/30, Firenze
*Dipartimento di Fisica, Largo E. Fermi,2, Firenze.

Abstract. We analyzed spectroscopically the fluorescent pulse detected in concomitance with a strong laser pulse applied in a radiation trapping setting with the use of two fast sampling oscilloscopes.

We used a strong laser to couple the upper level of an allowed atomic transition to the ground level, to a higher level, in a ladder type configuration. This was done in a radiation trapping setting, that resulted modified by the strong field application. We used strontium vapour in a heat-pipe made of two crossed tubes of 3 cm diameter, which was filled with 10 mbar of argon buffer gas. Atoms in the $Sr(5s5p)^1P_1$ state were prepared by means of near-resonant scattering of a pulsed dye laser (pump laser). The characteristic times of the dye laser pulse were measured with a fast photodetector connected with a 1.5 GHz analog bandwidth 9362 LeCroy digitizing oscilloscope. The duration between the points at 5% of the maximum pulse height resulted 12.5 ns. The pulse spectral bandwidth was 0.2 cm^{-1} (half-maximum-full-width). The coupling source is a second pulsed dye laser, with similar temporal and bandwidth characteristics, pumped by the third harmonic of the same Nd-YAG laser, and tuned near to the $Sr(5s5p)^1P_1$-$Sr(5p^2)^1D_2$ resonance transition frequency. The coupling beam was mildly focused inside the oven to a focal spot of 3 mm diameter, overlapping the radiation trapping region of about 250 µm. It was delayed 17 ns respect to the pump dye. In the centre of the heat-pipe we estimate a maximum Rabi frequency of 18 cm^{-1}. The main features of the process were observed by collecting fluorescence from the intermediate state emitted at 90° with respect to the laser beam axis. Radiation was filtered by a 0.85-m double monochromator and detected by a high gain photomultiplier with a typical rise time of 2.2 ns. In Figure 1 we show six traces recorded at $\lambda = 21700$ cm^{-1} in correspondence of six different oven temperatures, from 630 °C (1) to 680 °C (6), at increments of ten degrees. The time between two successive samplings is 0.2 ns. The first peak corresponds to the fluorescence maximum due to the presence of the pump laser and the second one to the presence of the coupling laser only. Radiation trapping is characterized by the exponential decay detected after the first pulse.

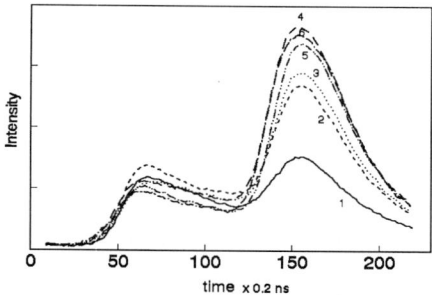

FIGURE 1. Fluorescence signals at $\lambda = 21700$ cm^{-1} obtained with a sampling oscilloscope (9362 LeCroy) at six different temperatures ranging from 630 °C (6) to 680 °C (1), at increments of ten degrees.

The temporal traces of the second pulse were recorded as a function of the double monochromator position with a 10 Gs/s digitizing oscilloscope (Tektronix TDS 680B). The high sampling rate of the oscilloscope allowed us to peak up signals collected at many different delays for each selected wave number, and to construct spectra at different selected instants. The spectra shown in Figure 2,

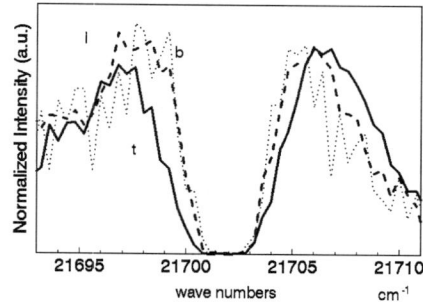

FIGURE 2. Normalised spectra obtained at the beginning (i) and at the top (t) of the fluorescent pulse by a computer program developed by us, processing the data collected with the Tektronix TDS 680B digitizing oscilloscope. Trace b is the reference signal collected without coupling laser.

obtained at a working temperature of 630 °C (strontium density $\delta = 1.25 \cdot 10^{15}$ atoms/cm^3), are relative to an initial instant (i) and to a peak instant (t). The signal detected in the absence of the strong laser doesn't vary and it shown (b) for comparison. The differences in the lineshapes were analysed with a best fit program to evaluate the medium modifications as a function of the laser intensity.

Experimental and Numerical Study of the Collision Induced Stimulated Effects in Alkali Vapors

Z. Konefał

Institute of Experimental Physics, University of Gdańsk,
Ul. Wita Stwosza 57, 80-952 Gdańsk, Poland.

Abstract. A computational and experimental study of the optical propagation of light in a collisionally broadened three-level Raman amplifier is presented. The Stokes light consists of two distinct spectral components: of resonance stimulated Raman scattering (SRS) and on-resonance stimulated fluorescence (ASE). The intensity dependence of these effects on the buffer gas and on the intensity and detuning of the exciting radiation are calculated. We have carried out the calculation over physically realistic propagation distance. The theoretical model for a description of ASE i SRS signals generation in such a system was developed. The numerical calculations based on the rate equations agree with the experimental results.

Let us consider three-level alkali atoms with energy levels E1<E2<E3 in the field of laser radiation that is quasiresonans with the 1 →3 transition ($3S_{1/2} \to 3^2P_{3/2}$ transition in sodium atoms). Transition 1→ 3 and 1 → 2 are allowed in the dipole approximation. The collisions with buffer atoms make the transition between levels 2 and 3 possible. When the laser frequency is detuned from the $3S_{1/2} \to 3^2P_{3/2}$ resonance, two different physical processes can occur:ASE and SRS. The ASE is generated on the $3^2S_{1/2} \to 3^2P_{3/2}$ transition, the SRS signal is due to the inversion of the population between the $3^2P_{1/2} \to 3^2P_{3/2}$ transition. In the experiments the excited state $3^2P_{3/2}$ is first populated by the laser pulse and the collisions with buffer gas bring the population to the $3P_{1/2}$ state, then stimulated ASE and SRS scattering starts from this excited state. These events are displaced against each other in time. For the case of resonant pumping these two contributions to fluorescence cannot be separately identified. For off-resonant pumping, however, the two processes are spectrally distinguishable because of a difference in the frequencies of scattered photons. The relationship which theoretically explains the population inversion between $3^2P_{1/2}$ and $3^2S_{1/2}$ levels follows from the detailed balance principle (1).

The theoretical model for a description of the ASE i SRS signals generation in such a system was developed. The numerical calculations based on the rate equations agree with the experimental results (2-5). Figure 1a show the SRS puls envelope for the differential propagation distance, the signal is generated by probe pulse. Figure 1b presents the time evolution of the SRS signal for two temperatures of the cell (422 °C and 395 °C).

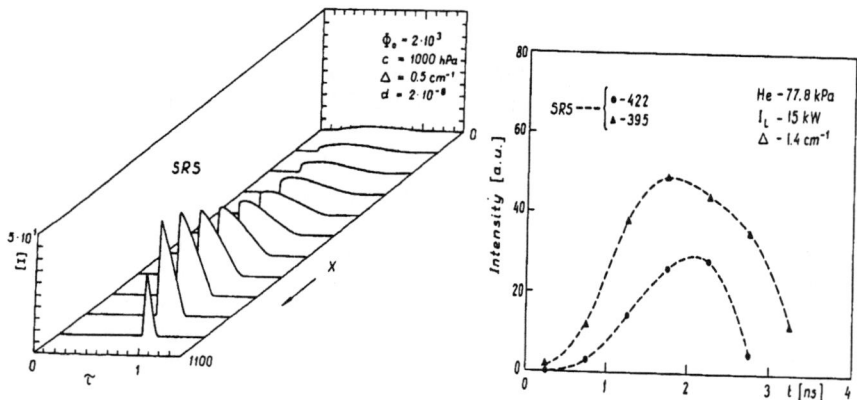

Figure 1. a) The SRS pulse envelope for different propagation distances.
b) Temporal behaviour of the SRS signal for two temperatures of the cell.

From figures 1a one can see that Raman signal envelope changes its width and intensity with the propagation distances. The pulse is shortened due to the propagation. The SRS signal increases with the increases of the propagation distances, but for the longer propagation distance, when the laser pulse is absorbed in the cell the intensity of the SRS signal decreases with the further increase of the propagation distances. This same behaviour of the SRS pulse was measured in the experiments (figure 1b).

ACKNOWLEDGMENTS

This work was partially supported by the University of Gdansk grant no.BW-5200 -5- 0073-6.

REFERENCES

1. J. Czub, J. Fiutak, and W. Miklaszewski, Z. Phys. D 27(1986).
2. Z. Konefal and M. Ignaciuk, Appl. Phys. B 51,285-291(1990).
3. Z. Konefal and M. Ignaciuk, Appl. Phys. B 61,101-110 (1990).
5. Z. Konefal, M. Ignaciuk, Z. Phys. D 27,49-54(1993).
6. Z. Konefal, M. Ignaciuk, Optical and Quantum Electron. 28, 169-180 (1993).

THEORETICAL INVESTIGATION OF THE CsHg MOLECULE

R. Polly[(*)], D. Gruber[(*)], L. Windholz[(*)], B. A. Heß[(**)]

(*) Institut für Experimentalphysik, Technische Universität Graz, Petersgasse 16, A-8010 Graz,
(**) Institut für Physikalische und Theoretische Chemie, Universität Bonn, D-53115 Bonn,

Abstract. We performed all-electron *ab initio* calculations on the CsHg potential energy curves on the CASSCF/MRDCI level, including kinematical relativistic effects as well as spin-orbit effects. Furthermore, we calculated the oscillator strenghts between ground and excited states of the CsHg molecule.

All electron *ab initio* calculations have been developed into a powerful tool of molecular physics recently. The data obtained by means of theoretical calculations facilitate the assignment and understanding of observed transitions in investigated molecules. Therefore, the calculation of the CsHg potential energy curves is of great experimental interest. The resulting fluorescence spectra of a Cs/Hg vapor mixture are expected to be very complicated and an assignment without knowledge of the potential energy curves would be very difficult. Moreover, on the theoretical side CsHg attracts a lot of interest since the molecule is composed by two heavy atoms and a careful consideration of relativistic effects is necessary.

We performed relativistic all-electron *ab initio* calculation of the CsHg molecule on the Complete Active Space Self-Consistent Field (CASSCF) / Multi Reference Configuration Interaction (MRCI) level. In these calculations we employed an [*24s17p12d2f*] atomic natural orbital (ANO) basis set. The kinematical relativistic effects as well as the spin-orbit interaction are included in the calculations. The Hamiltonian used is obtained by a series of controlled approximations from the Hamiltonian of Quantum Electrodynamics in the Schrödinger picture [1-4] and was used in former relativistic all-electron ab-initio calculations of the LiHg [5] and NaHg [6] molecule.

On the CASSCF level we optimized the orbitals for the $X^2\Sigma^+$ ground state of the CsHg molecule. The active space contains the 6s orbital of Cs and the 6s, $6p_z$, $6p_x$, $6p_y$ orbitals of Hg; three electrons are active.

© 1997 American Institute of Physics

The MRCI calculations for the ground state was performed with the averaged coupled pair functional (ACPF) method. For the calculation of the excited states we used the multireference single and double excitation method (MRD-CI). In both MRCI calculations we correlated 19 electrons, namely the $5d^{10}$, $6s^2$ electrons of Hg and the $5p^6$ and $6s^1$ electrons of Cs.

The results on the CsHg potential energy curves are encouraging and the calculated spectroscopic parameters are in good agreement with previous experimental investigations and thus together with the calculated oscillator strengths allow the prediction of spectral positions and relative intensities of chemluminiscence of the CsHg molecule. Further, the calculations stress the importance of the relativistic effects in the calculation of potential curves of molecules composed by heavy atoms.

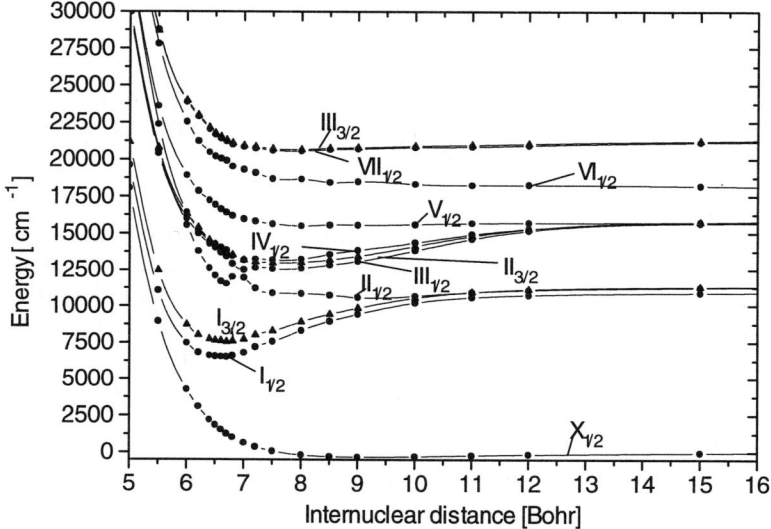

CsHg potential energy curves including spin-orbit interaction

[1] M. Douglas and N. M. Kroll, Ann. Phys. 82, 89 (1974).
[2] J. Sucher, Phys Rev. A 22, 348 (1980).
[3] W. Buchmüller and K. Dietz, Z. Phys. C 5, 45 (1980).
[4] G. Jansen and B. A. Heß, Phys. Rev. A 39, 6016 (1989).
[5] M. M. Gleichmann and B. A. Heß, J. Chem. Phys. 101, 9691 (1994).
[6] D. Gruber, U. Domiaty, X. Li, L. Windholz, M. Gleichmann and B.A. Heß, J. Chem. Phys. 102 (13), 5174, (1995).

COLLISION DISTRIBUTION OF PARTICLES EXCITED BY THE LASER LIGHT

Sylwia Zielinska-Kaniasty

Department of Physics II, Institut of Mathematics and Physics,
Technical and Agriculture University of Bydgoszcz, PL 85-791 Bydgoszcz, Poland

The velocity changing collisions (vcc) play an important role in the description of velocity redistribution of particles caused by nearly resonant interaction with the laser light and collisions (1). Collision relaxation rates are conveniently described by collision kernels which, in turn, depend on interatomic potential and scattering amplitudes
and is proportional to the collision cross section. The population kernel gives the probability density per unit time that an active particle in state | i > undergoes collisions with perturbers (usually a noble gas atom or impurity), which changes the active particle velocity from v' to v. Due to the fact that perturber-active particle interaction is state dependent the collisions cause different changes in the ground and excited states of active particle. Although the shape of the populations distributions resembles those from well-known Bennett hole theory, the distribution of the total population has two peaks and obviously is asymetrical (2). Vcc together with velocity selective laser excitation lead to the asymmetry of the ground and excited state population. The flux of excited particles has properties which are different from that of ground state because of interaction collision potentials for both states. The intensity of the laser light strongly influences the velocity distributions of the ground and excited state populations.

It should be stressed that mechanism described above can be applied as well for atoms or electrons as active particles. As it was emphasised a near perfect analogy exists between velocity distributions of the two-level-atom populations induced by the laser beam in presence of collisions and light induced electron drift effect (3).

It seems that correct description of the discussed phenomenon requires that the full set of equations for the density matrix has to be taken into account. This set of equations has to be full in the sense that it should not only include velocity dependent populations of the ground and excited states $\rho_{11}(v), \rho_{22}(v)$, but also coherences. A method based on the kinetic equation for the density matrix of the system, consists of the active particeles excited by the laser light and perturbers, is presented and the influence of relaxation processes and the pumping rate on the velocity distributions of the total population is discussed. Figures 1 and 2 shows the examples of asymmetry of total populations velocity redistribution for two-level-atoms and electrons which interact nearly resonantly with the laser.

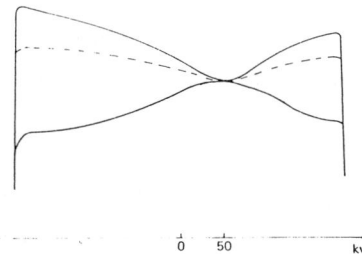

Fig. 1. Carriers in semiconductors. The lower (1) and upper (2) stste distributions. The broken line represents half of the total populations.

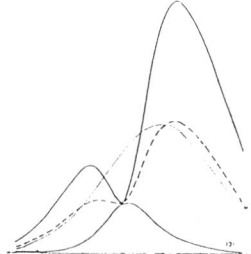

Fig. 2, see fig. 1, the atomic case.

REFERENCES

1. Rautian, S.G., Shalagin A.M., *Kinetics problems of Nonlinear Spectroscopy*, North Holland, Amsterdam ,1991
2. Zielinska-Kaniasty, S., *Physica B*, **168**, 219, (1991)
3. Woerdman, P., *Phys. Rev. Lett.* **59**, 1624, (1987)

The LiHg $X^2\Sigma^+ \to 2^2\Pi_{3/2}$ transition: Rotational analysis and potential curves

X. Li, D. Gruber, P. Pircher, and L. Windholz

Institut für Experimentalphysik, TU Graz, Petersgasse 16, A-8010 Graz, Austria

Abstract. We present the rotationally-resolved excitation spectrum of the LiHg $X^2\Sigma^+ \to 2^2\Pi_{3/2}$ transition. Different vibronic transitions are assigned. Rotational analysis is made for $\upsilon'=1 \leftarrow \upsilon''=0$, $\upsilon'=0 \leftarrow \upsilon''=0, 1, 2$ vibronic bands and molecular constants are obtained. Based on the Inverted Perturbation Approach (IPA), the potential curves of the excited and ground electronic states are calculated.

INTRODUCTION

Spectra of Ia (Li, Na, K, Rb, Cs)-IIb (Zn, Cd, Hg) group molecules have been observed long time ago [1], yet no rotationally-resolved spectrum has been reported previously [2]. This is mainly due to the peculiar properties of the Ia-IIb molecule, that is, the ground electronic states are weakly bound and bound-bound transitions are only important for the few lowest vibrational levels of the excited states [3]. In addition, the small concentration of the Ia-IIb molecules in the metal vapor mixture and the spectral overlaps with alkali dimer and alkali-hydride bands both contribute to the difficulties to obtain rotationally resolved spectra.

Previously we have reported the vibrationally resolved LiHg $2^2\Pi_{3/2} \to X^2\Sigma^+$ transition [4,5]. Because of the low resolution of the monochromator and the overlapping Li$_2$ B-X and LiH A-X transitions, rotational structure remains unresolved. In this letter we report the excitation spectrum of the $X^2\Sigma^+ \to 2^2\Pi_{3/2}$ transition obtained by detecting fluorescence in the 430-450 nm spectral region where the LiHg $2^2\Pi_{3/2} \to X^2\Sigma^+$ transition concentrates. The strong background fluorescence from Li$_2$ B-X and LiH A-X transitions is largely rejected due to unfavorable Franck-Condon factors of the Li$_2$ A-X transition in this spectral region and large vibrational spacing of the LiH ground state.

EXPERIMENTAL SETUP

The lithium-mercury vapor mixture was generated in a crossed heat-pipe oven, both of natural isotopic composition. A modified ring dye laser (CR699-21), tunable in a single mode operation from 425 to 460 nm, is used to excite the LiHg $X^2\Sigma^+ \rightarrow 2^2\Pi_{3/2}$ transition. The wavelength of the dye laser is monitored by a lambadameter and a temperature stabilized marker etalon. The absolute accuracy of the spectral lines are estimated to be better than 0.03 cm^{-1}.

The image of the fluorescence is rotated 90° by a set of mirrors onto the entrance slit of a monochromator. The entrance slit is set to 100 µm, whereas the exit slit is set to about 1.5 mm. Combined with a 150 groves/mm grating, this gives a band pass from 430 to 450 nm when the central wavelength is set to approximately 440 nm. The dispersed fluorescence is detected by a photo-multiplier and the signal is sent to an EG&G 5210 lockin amplifier. The analog intensity signal from the marker etalon is digitized by one of the A/D ports of the lockin amplifier, which in return is connected to a PC 486 through an IEEE-488 interface. A RS 232C interface is used to collect the signal from the lamdameter.

RESULTS AND ANALYSIS

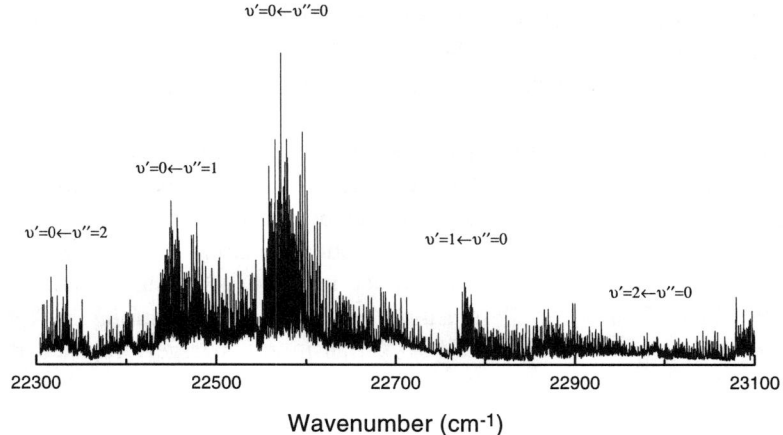

FIGURE 1. Excitation spectrum of the LiHg $X^2\Sigma^+ \rightarrow 2^2\Pi_{3/2}$ transition

In Figure 1 we present the excitation spectrum of the LiHg $X^2\Sigma^+ \to 2^2\Pi_{3/2}$ transition. Vibronic assignments are marked on the top of the figure based on previous studies [4,5,6]. Various vibronic bands are analyzed and rotational analysis is made. In Figure 2 we present the Fortrat diagram of the $\upsilon'=1 \leftarrow \upsilon''=0$ vibronic transition as an example.

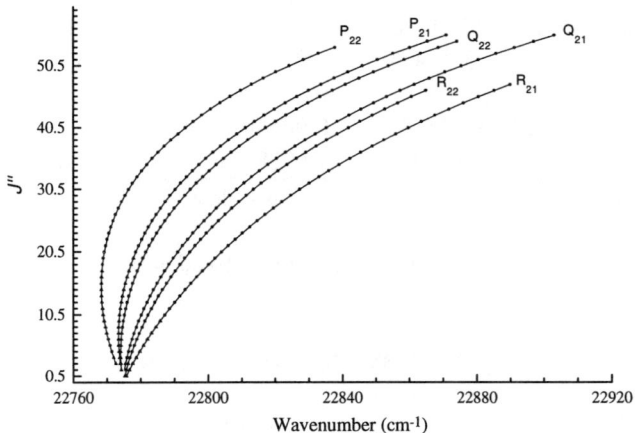

FIGURE 2. Fortrat diagram of the $\upsilon'=1 \leftarrow \upsilon''=0$ transition.

The identified rotational transitions are fitted to the known Hamiltonians of a $^2\Sigma^+ \to {}^2\Pi_{3/2}$ transition [7] using a general linear least square fitting program. Molecular constants are obtained. In Table 1 we present the molecular constants for the $2^2\Pi_{3/2}$ electronic state, in Table 2 we present those of the ground state.

TABLE 1. Molecular constants of the $2^2\Pi_{3/2}$ state[a]

υ'	$G'(\upsilon')-G''(0)$	$B_{\upsilon'}$	$10^6 \times D_{\upsilon'}$	$\delta_{\upsilon'}$
0	22558.520(8)	0.3079975(933)	2.4270(197)	0.0[b]
1	22774.735(14)	0.3038685(1180)	2.3638(326)	0.0[b]

[a] Units in cm^{-1}. Numbers in brackets are one standard deviation.
[b] Undetermined by the fit and consequently fixed at 0.

Using the identified rotational transitions, we have calculated the potential curves of the excited and the ground electronic states using the Inverted Perturbation Approach (IPA). They will be presented elsewhere together with detailed discussions of the characteristics of the excited and the ground electronic states.

Table 2: Molecular constants of the ground electronic state[c]

Quantity	$v''=0$	$v''=1$	$v''=2$
$G''(v'')-G(0)$	0.	122.12(1)	231.21(2)
B_v''	0.278555(103)	0.264550(151)	0.249681(197)
$10^6 \times D_v''$	4.5950(79)	5.8002(806)	6.5629(1340)
$10^{10} \times H_v''$	-8.5397(4890)	-4.4651(1330)	-5.4362(3340)
$10^{13} \times I_v''$	2.1813(1340)	0.0[d]	0.0[d]
$10^{17} \times K_v''$	-3.3871(1320)	0.0[d]	0.0[d]
$10^1 \times \gamma_0$	1.06831(518)	0.96540(500)	0.85818(639)
$10^4 \times u^2$	1.3651(439)	1.6928(456)	1.9680(866)

[c] Units in cm^{-1} except u^2 which is dimensionless. Numbers in brackets are one standard deviation.

[d] Undetermined by the fit and consequently fixed at 0.

ACKNOWLEDGEMENTS

This work was supported by Austrian Science Foundation Project No. P-9929-PHY, and by the Jubiläumsfonds der Österreichischen Nationalbank, Project 4873. X. L. acknowledges the receipt of a Lise Meitner postdoctoral stipend from the Austrian Science Foundation (No. M-108-PHY).

REFERENCES

[1] S. Barrat, Trans. Faraday Soc. 25, 758 (1929).

[2] S. Milosevic, in *Spectral Line Shapes*, edited by A. D. May, J. R. Drummond and E. Oks, AIP conf. Proc. No. 328 (AIP, New York, 1995), pp.391-405.

[3] X. Li, S. Milosevic, D. Azinovic, G. Pichler, R. Düren and M. C. van Hemert, Z. Phys. D 30, 39 (1994).

[4] D. Gruber and X. Li, Chem.Phys.Lett. 240, 42 (1995).

[5] D. Gruber, X. Li, L. Windholz, M. M. Gleichmann, B. A. Hess, I. Vezmar and G. Pichler, accepted for publication in J.Phys.Chem.

[6] M. M. Gleichmann and B. A. Hess, J.Chem.Phys. 101, 9691 (1994).

[7] G. Herzberg, *Spectra of Diatomic Molecules* (Van Nostrand, Princeton, 1950).

ON THE VALIDITY OF DIFFERENT INTERCOMBINATION RULES FOR THE PARAMETERS OF THE GROUND STATE (X^1O^+) POTENTIAL

G. D. ROSTON and M. S. HELMI

Alexandria University, Faculty of Science, Department of Physics, Alexandria Egypt, e-mail GDANIEL@Alex.eun.eg

The intermolecular potential parameters (α_{12}, ϵ_{12} and R_{12}) occurring in exp-six model,

$$U(R) = \frac{\epsilon}{1-(6/\alpha)} \left[\frac{6}{\alpha} e^{\alpha[1-(R/R_m)]} - (R_m/R)^6\right]$$

for a pair of unlike molecules from a knowledge of those for the corresponding like-pairs have been tested for the molecules consisting of the 2nd group metals (Hg, Cd and Zn) with inert gases (Xe, Kr, Ar and Ne) and also for the inert gas systems using six different groups [1-6] of itercombination rules. The input data for the like pairs Hg-Hg, Cd-Cd, Zn-Zn, Xe-Xe, Kr-Kr, Ar-Ar and Ne-Ne were taken respectively from [7-13].

Table I: The obtained results for the well depth (ϵ_{12} cm^{-1}) and its position R_{12} (Å) for the ground state potential.

Rules Ref.	Molec	Mason + Rice (1)		Sriva-stava (2)		Hoger-vorst (3)		Calvin + Reed (4)		Sikora (5)		Kong+Chakrabarty (6)		Exp	
		ϵ_{12}	R_{12}	ϵ_{12}	R_{12}	ϵ_{12}	R_{12}	ϵ_{12}	R_{12}	ϵ_{12}	R_1	ϵ_{12}	R_{12}	ϵ_{12}	R_{12}
Xe	Hg	294	4	285	4	261	4	285	4	278	—	282	4.2	240[14]	4.1
	Cd	207	4.1	267	3.9	252	3.9	267	3.8	223		229	4.2	192[15]	4.3
	Ze	225	4.5	225	4.5	221	4.5	225	4.5	220	—	227	4.5	162[16]	4.4
	Kr	206	4.2	160	4.2	159	4.2	160	4.2	156		158	4.2	137[11]	4.2
	Ar	140	4	135	4.1	130	4.1	136	4.1	125	—	130	4.1	132[17]	4.1
	Ne	86	3.6	74.	3.9	51	3.9	74	3.7	48	—	58	3.9	51[17]	3.9
Kr	Hg	248	3.9	247	3.8	211	3.9	247	3.8	241	—	246	4	200[18]	4
	Cd	197	3.9	231	3.7	204	3.8	231	3.6	206	—	196	3.9	130[19]	3.8
	Zn	196	4.3	195	4.3	184	4.3	195	4.3	179	—	191	4.3	115[20]	4.2
	Ar	118	3.9	117	3.9	116	3.9	117	3.9	116	—	117	3.9	116[21]	3.9
	Ne	70	3.5	64	3.5	48	3.7	64	3.5	50	—	56	3.6	51[11]	3.6
Ar	Hg	209	3.7	209	3.7	162	3.9	209	3.7	198	—	209	3.7	138[22]	3.9
	Cd	182	3.7	196	3.6	158	3.8	196	3.5	177	—	159	3.7	106[23]	4.3
	Zn	170	4.1	165	4.1	146	4.2	165	4.1	139	—	156	4.2	96[24]	4.2
	Ne	57	3.4	54	3.4	46	3.5	54	3.5	47	—	51	3.5	45[11]	3.5
Ne	Hg	120	3.3	114	3.4	56	3.8	114	3.4	81.5	—	102	3.4	51[25]	3.9
	Cd	134	3.2	107	3.3	55	3.7	107	3.2	85.5	—	65.6	3.6	38[26]	4.1
	Zn	104	3.7	90	3.8	54	4.1	90	3.7	49	—	68.7	3.9	—	—

© 1997 American Institute of Physics

The results of these calculations for the ground state parameters of the mentioned systems are given with their experimental data in table I.

From this table the following conclusions can be made:

1- The interacting parameters (ϵ and R_*) for the inert gas systems and also the positions of the well depths for the 2nd group metals with the inert gases can be obtained successfully using any one of these different rules. This is in agreement with Ref.[6,22].

2- Concerning the well depths for the systems consisting of 2nd group metals + inert gases there are some rules which can give reasonable results with the experimental data especially for heavy interacting atoms (e.g Hogervorst [3], Sikora [5] and Kong-Chakrabarty [6]).

REFERENCES

1- E.A.Mason and W.E.Rice,J.Chem.Phys.,22,522,1954.
2- B.N.Srivastava and K.P.Srivastava,J.Chem.Phys. 24 ,1275,1956.
3- W. Hogervorst, Physica,51,77,1971.
4- D.W.Calvin and T.M.Reed.III,J.Chem.Phys.,56,2484,1972.
5- P.T.Sikora,J.Phys.B:At.Mol.Phys.3,1475, 1970.
6- C.L.Kong and M.R.Chakrabarty,J.Phys.Chem.,77,2668,1973
7- E.W.Baylis,J.Phys.B:At.Mol.Phys.,10,L 583,1977.
8- M.S.Helmi,T.Grycuk and G.D.Roston,J.Spectrochemica Acta part B, 51, May, 1996
9- M.Czajkowski,R.Bobkowski and L.Krause,Physical Rev. A41,277,1990
10-G.C.Maitland,E.P.Smith,Chem,Phys.Lett.,22,443,1973.
11-D.W.Gough,E.B.Smith and G.C.Maitland,Mol.Phys.,27 ,867 ,1974
12-R.A.Aziz,H.H.Chen,J.Chem.Phys.,67,5719,1977.
13-B.Brunetti,F.Pirani,F.Vecchiocattivi and E.Luzzatti, Chem.Phys.Lett.,58,504,1978.
14-T.Grycuk and M.Findeisen,J.Phys.B:At.Mol.Phys.,16,975 ,1983.
15-M.S.Helmi,T.Grycuk and G.D.Roston,Accepted for Publication in J.Chem.Phys.1996
16-I.wallace, J.Koup and W.H.Breckenridge,J.Phys.Chem. 95,8060,1991
17-J.M. Parson, J.P.Schafer,P.E.Siska,F.P.Tully,Y.C. Wong and Y.T.lee,J.chem.Phys.53,2123,1970
18-T.Grycuk,E.Czerwosz,Physica,106 C,431,1981.
19-C.Bousquet,J.Phys.B:At.Mol.Phys.,19,3859,1986.
20-I.Wallace, J.Ryter and W.H.Breckenridge, J.Chem.Phys, 96,136,1992
21-R.A.Aziz,J.Preslay,U.Buck andJ. Schleuserner,J.Chem. Phys.,70,4737,1979
22-C.Bousquet,N.Bras and Y.Majdi,J. Phys.B:At. Mol.Phys., 17,1831,1984
23-M.Czajkowski,R.Bobkowski and L.Krause,J.Phys. Rev A45,6451,1991
24-I.Wallace,R.R.Bennett and W.H.Breckenridge,Phys.Chem .Lett.153,127,1988.
25-W.E.Baylis.J.Chem.Phys.51,2265,1969.
26-R.Bobkowski,M.Czajkowski and L.Krause,J.Phys.Rev A41 , 243 ,1990

Interatomic potentials for X $^1 0^+$ and $B^3 1$ states of the intercombination cadmium line 326.1 nm broadened by Ar pressure

G. D. Roston*, T. Grycuk** and M. S. Helmi*

* Alexandria University, Faculty of Science, Department of Physics, Alexandria, Egypt, e-mail:Gdaniel@Alex.eun.eg

** Warsaw University, Institute of experimental Physics, Warsaw , Poland, e-mail : Tgrycuk@fuw.edu.pl

Interatomic potentials for the Cd-rare gas van der Waals molecules in their ground and excited states have been the subject of several spectroscopic investigations [1-3]. In our previous paper [4] the interatomic potentials of the ground ($X^1 0^+_g$)and the excited ($^3 0_u^+$ and $B^3 1_u$) states were obtained from the analysis of the temperature dependence of the Cd 326.1 nm absorption line broadened by cadmium pressure.

This paper is mainly devoted to the determination of the interaction potentials of the ground ($X^1 0^+$)and the excited ($B^3 1$) states from the analysis of the cadmium 326.1 nm absorption line profile broadened by argon pressure. The effect of Cd-Cd interactions on the Cd-Ar mixture was reduced using the profiles obtained in [4]. The well known inversion procedure [5] based on the classical formulation of the quasimolecular theory of spectral line broadened by neutral perturber has been used.

The temperature dependence for the Cd-Ar blue wing profile was analysed as in our previous paper [4]. This is illustrated in Fig.(1). The experimental data showing the ground state potential V_g as a function of $\Delta\nu$ is illustrated in Fig.(2) This figure shows that the well depth of V_g is equal to about 109.7 cm^{-1} which lies at the frequency separation $\Delta\nu$ = 72 cm^{-1} . Using this value of $\Delta\nu$ with the corresponding internuclear separation (R_m = 4.3 Å) deduced by Bobkowski [3] from the molecular beam experiment, it was possible to obtain the difference potential $\Delta V(R)$ using the profile extrapolated to infinite temperature [see Fig.(1)]. As $\Delta V(R)$ and $V_g(R)$ are known, the excited state potential $V^1(R)$ is thus obtained. These potentials are illustrated in Fig.(3). The potentials $V_g(R)$ and $V^1(R)$ are fitted to Morse potentials with appropriate parameters.

The well depths (ϵ in cm^{-1}) and the minimum positions (R_m in Å) deduced in this work for the ground and excited states are compared with the results obtained

© 1997 American Institute of Physics

from other experimental methods. This is illustrated in table I.

Table I: Comparison of our data with other determinations.

$X^1 0^+$		$^3 1$		References
R_m (Å)	ϵ (cm^{-1})	R_m (Å)	ϵ (cm^{-1})	
4.3	109.7 ± 3	4.77	63.2 ± 2	This work
--	106	--	59.7	Kowalski [2]
4.3	106.5	5.03	56	Bobkowski [3]

It is seen from this table that our results are in close agreement with those obtained before [2,3].

Fig.(1): The temperature dependence for the blue wing.

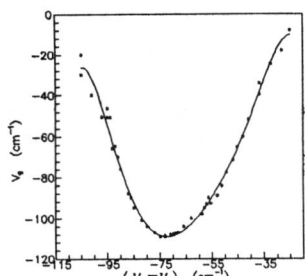

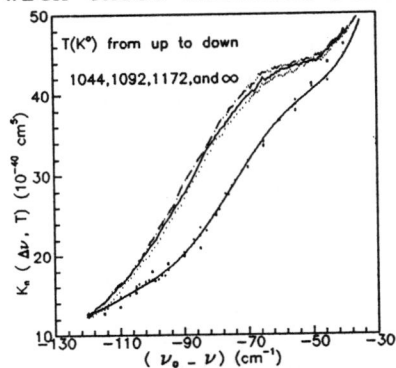

Fig.2: The ground state V_g as a function of $\Delta\nu$

Fig.3: The potentials V_g, ^1V and the Δ^1V for cd-Ar.

This work is supported by the Polish Committee for Scientific Research within the project no 2 0261 91 01.

REFERENCES

1- C. Bousquet, J. Phys.B : At. Mol. Phys. **19**, 3859, 1986
2- A. Kowalski, M. Czajkowski and W.H. Breckenridge, Chem. Phys. Lett., **121**, 217, 1985
3- R. Bobkowski, M. Czajkowski and L. Krause, Phys. Rev. **A41**, 234, 1990
4- M. S. Helmi, T. Grycuk and G. D. Roston, To be published in Spectrochemica Acta B, **51**, (6), (1996)
5- A. Gallagher, Atomic Physics, 4 (new York: Plenum)

Van-der-Waals-Interaction Constant

Dörte Neundorf

Lichttechnisches Institut, Universität Karlsruhe, D-76128 Karlsruhe, Germany
Doerte.Neundorf@etec.uni-karlsruhe.de

Abstract. Based on the interaction energy of two particles and calculations for hydrogen (1), the constant of the interaction law is calculated for a wide range of atoms. It is done by generalising an approach with quantum mechanical perturbation theory in its dipole approximation for neutral atoms. The atoms are modelled by a meta particle, which has the same dipole moment as the real atom. Two major cases are considered: the general case for symmetric atoms and the self resonance case. The constant is given numerically for mercury.

The **interaction energy** $U(R)$ of two particles at distance R is proportional to R^{-p}. In the following, the proportionality constant C_p is calculated for $p=6$ and $p=3$. By application of quantum mechanical perturbation theory, it is possible to get information on how two atoms interact. The two isolated atoms are taken as unperturbed system; thus the interaction energy is zero. The energy eigenvalues are zero, too. The interaction operator $\hat{W}$ is the perturbation; the eigenvalue correction is the interaction energy. The result depends on the energy level in view.

In quantum mechanics, the first order **eigenvalue correction** of a perturbation operator $\hat{W}$ is $\varepsilon_1 = \langle \varphi_n | \hat{W} | \varphi_n \rangle$, the average of $\hat{W}$. The second order correction is $\varepsilon_2 = \sum_{p \neq n} |\langle \varphi_p | \hat{W} | \varphi_n \rangle|^2 / [E_n - E_p]$. The operator $\hat{W} = [\delta_A \delta_B - 3(\delta_A \mathbf{n})(\delta_B \mathbf{n})]R^{-3}$ describes the interaction of two dipoles.

If the particle's electron distribution is spherically symmetric, its mean dipole moment and the first order correction vanishes ($\Rightarrow p=6$). Atoms with configurations ending on s^2, s and p^6 are symmetric, if all inner orbits are complete. Only one of the atoms must be in a symmetric state. A special effect occurs, if the atoms are identical but in different states. The state exchange between the particles adds a degree of **degeneration** ($\Rightarrow p=3$) because the first order correction does not vanish. The calculation of the interaction constant shall be as simple as possible. A hydrogen-like model can be developed out of the dipole moments. The dipole of an atom is $\delta_A = e \sum \mathbf{r}_j$. Now a **generalised position** $\mathbf{r}_A$ is introduced by $\mathbf{r}_A = \sum \mathbf{r}_j$, so that the dipole moment can be written as $\delta_A = e\mathbf{r}_A$. The generalised position $\mathbf{r}_A$ can be seen as the position vector of a replacing *meta electron* with charge $N \cdot e$. The **meta particle**, consisting of the meta electron and the nucleus, induces the same dipole moment as the atom in view. Its dipole moment can be used instead. Because the interaction depends only on the dipole moment, it is sufficient to use such a model. The **Bohr atom model** can be generalised for the meta particle. The meta electron ($q_1 = Ne$, $m_{e^*} = Nm_e$) circles around a nucleus ($q_2 = Ze, m_n$).

The resulting **generalised Bohr radius** is $a_0 = \hbar^2 / \mu q_1 q_2$.

For simplicity the calculations starts with the **case of symmetry**. As stated above, p=6. The energy of the ground state is assumed to be the negative ionisation energy. The state in view is the ground state of both atoms: $E_0 = E_{Ion} = E_{Ion,A} + E_{Ion,B}$.

$$C_6 = -\varepsilon_2 r^6 = e^4 \sum_{p \neq n} \frac{|\langle \Phi_p | X_A X_B + Y_A Y_B - Z_A Z_B | \Phi_0 \rangle|^2}{E_0 + E_{A,p} + E_{A,p}}$$

The energies of the higher levels are smaller than those of the ground state, so they can be neglected: $E_0 + E_{A,p} + E_{A,p} \approx E_0$. Laws of quantum mechanics show

$$C_6 = \frac{e^4}{E_n^0} \langle \Phi_0 | (X_A X_B + Y_A Y_B - Z_A Z_B)^2 | \Phi_0 \rangle.$$

Because of symmetry, the mean values of the cross terms vanish. In the mean value, each of the squared position operator's components is one third of the squared radius. With $\langle \Phi_0 | R^2/3 | \Phi_0 \rangle = (a_0)^2$, this finally results in $C_6 = 6e^4 E_0^{-1} (a_0^A)^2 (a_0^B)^2$.

Interaction of **Hg** and **Ar**, correction to ground state of Hg $\Rightarrow C_6 = 0.6660 \cdot 10^{-126}$. For a better approximation, we use $E_{n,p} = E_0 + E_{A,p}^0 + E_{A,p}^0$. E_{np} can be calculated. The mean value of these energies replaces E_0. In the example, C_6 does not change. It is possible to consider the **energy levels** in more detail. This results in

$$C_6^* = e^4 \sum_p \hat{W}_{np}^2 / E_{np} = C_6 - \sum_p Q_{np} \text{ with } Q_{np} = \hat{W}_{np}^2 (E_{np} - E_0)/E_0 E_{np}$$

$\sum \hat{W}_{np}^2 = a_0^4$, so approximately $\hat{W}_{np}^2 = a_0^4 / s$. The result is $C_6^* = 0.3756 \cdot 10^{-61}$.

As stated above, for the **self resonance case** first order correction applies (p=3). It can be expanded in terms like $\langle x_A \phi_n^A | \phi_{n'}^A \rangle \langle \phi_{n'}^B | x_B \phi_n^B \rangle$. Both atoms are identical, the wave functions are real (meta particle). This results in

$$\langle x_A \phi_n^A | \phi_{n'}^A \rangle \langle \phi_{n'}^B | x_B \phi_n^B \rangle = \langle x \phi_n | \phi_{n'} \rangle^2 = \hat{X}_{nn'}^2.$$

With $\langle \phi_n | \hat{X}^2 | \phi_n \rangle = \sum \hat{X}_{nj}^2$, a real number $S_x < 1$ exists, so that $\langle \phi_n | \hat{X}^2 | \phi_n \rangle S_x = \hat{X}_{nn'}^2$. The mean value $\hat{X}^2$ can be replaced by $\hat{R}^2/3$ and the scalar product a_0^2. Finally

$$\hat{X}_{nn'}^2 = S_x \langle \phi_n | \hat{R}^2/3 | \phi_n \rangle = S_x a_0^2.$$

This can be done for the other coordinates, too. Put together, the constant is

$$C_3 = e^2 [\hat{X}_{nn'}^2 + \hat{Y}_{nn'}^2 - 2\hat{Z}_{nn'}^2] = e^2 a_0^2 [S_x + S_y - 2S_z].$$

$[S_x + S_y - 2S_z] \approx 1 \Rightarrow C_3 = e^2 a_0^2$. (**Hg**: $C_3 = 0.4213 \cdot 10^{-73}$)

Further investigation on S_x, S_y, S_z show, that the upper value is actually too large. It has to be decreased the more, the higher the state difference in view is. If the atom is known in detail, quantum mechanical methods can be used to calculate the matrix elements of the operator.

REFERENCES

1. Cohen-Tannjoudi, C., Diu, B., Laloe, F., Quantum Mechanics, New York: John Wiley, 1977, Volume 2
2. Landau, L. D., Lifschitz, E. M., Quantenmechanik, Berlin: Akademie-Verlag, 1967

RADIATION SPECTROSCOPY OF HEAVY RARE-GAS EXCIMERS AND THEIR MIXTURES

N. A. Kryukov, P. A. Saveliev, M. A. Tchaplyguine[1]

The radiation spectra density dependence of $Xe(^3P_{1,2}) - Xe$ and $Xe(^3P_{1,2}) - Kr$ excimers has been measured. The excimers were created in the low energy gas discharge in pure Xe and in $Xe - He$ -, $Xe - Kr$ - mixtures. A high resolution vacuum monochromator with $0.83 nm/mm$ linear dispersion has been used. In the experiment with pure Xe the current was $5mA$, the Xe pressure ranged from 1 to 66 Torr (Fig. 1). In the experiment with Xe-He mixture the current was $3mA$, the atomic temperature — about 300K, the pressure — 4.5 Torr, with He pressure ranging from 0 to 143 Torr (Fig. 2). In the experiment with Xe-Kr mixture the Xe pressure was 1.2 Torr, the Kr pressure ranged from 0 to 65 Torr (Fig. 2). The current and the atomic temperature was the same as for the Xe-He mixture. The Xe_2^* and Xe^*Kr excimer spectra have been also calculated using the quasistatic approach and the high density approximation, i.e. the approximation of equilibrium vibrational distribution of the excimers and equilibrium distribution between two radiating electronic states (0_u^+ and 1_u for Xe_2^*; 0^+ and 1 for Xe^*Kr). The calculation has been done using the excited state potentials, taken from [1, 2] (for Xe_2^*) and [3] (for Xe^*Kr), the ground state potentials taken from [4](Xe_2^*) and [5](Xe^*Kr) and radiation widths Γ from [6](Xe_2^*) and [7](Xe^*Kr). The calculation of Xe_2^* spectra based upon the potential from [1] is shown in Fig. 1 and 2 by dashed line, the one based upon [2] — by dotted line. The calculation of Xe^*Kr spectrum is shown in Fig.2 by a solid line. The experimental Xe_2^* spectra shape becomes closer to the theoretically calculated one with pressure increasing. The calculated Xe^*Kr spectra shows a bad correspondence with the experimental one because the Xe^*Kr excited state potential has been not enough studied yet.

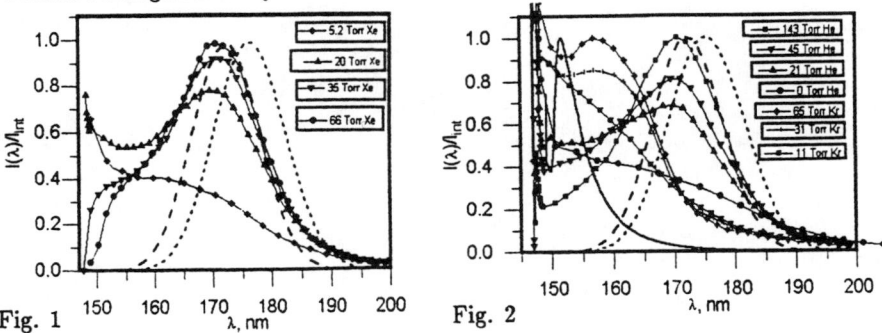

Fig. 1   Fig. 2

References:
[1] I.Messing, D.J.Eckstrom and D.C.Lorents // J. Chem. Phys., 1990, **93**(1), p.34
[2] R.H.Lipson, P.E.LaRocque and B.P.Stoicheff // J. Chem. Phys. 1985, **82**, p.4470
[3] G.Nowak and J.Fricke //J. Phys. B: At. Mol. Phys., 1985, **18**(7), p. 1355
[4] Aziz R.A., Slaman M.J. // Mol. Phys., 1986, **57**, p. 825
[5] Aziz R.A., Van Dalen R. // J. Chem. Phys., 1983, **78**(5), p. 2402
[6] A.Z.Devdariani, A.L.Zagrebin // Opt. and Spectr., 1992, **72**(3), p. 571(in Russian)
[7] A.L.Zagrebin, N.A.Pavlovskaya // Opt. and Spectr., 1990, **69**(5), p. 976(in Russian)

---

[1] Department of Optics and Spectroscopy, St.Petersburg State University, 198904 St.Petersburg, Russia. Fax: (812)4287240, e-mail: PSavel@niif.spb.su, apver@onti.phys.lgu.spb.su

DOPPLER-FREE
AND ULTRA-FINE SPECTROSCOPY

High-Resolution Measurements and Multichannel Quantum Defect Analysis of Spectral Line Shapes of Autoionizing Rydberg Series

Kiyoshi Ueda

Research Institute for Scientific Measurements, Tohoku University, Sendai 980-77, Japan

Abstract. Spectral line shapes for autoionizing Rydberg series are briefly reviewed within the framework of multichannel quantum defect theory (MQDT). Recent high-resolution measurements and MQDT analysis for the spectral line shapes are reviewed for the $mp^5(^2P_{1/2})ns'$ and nd' $J=1$ odd spectra of the Ar, Kr, and Xe atoms ($m=3$, 4, 5 for Ar, Kr, and Xe) and the $3p^5(^2P_{1/2})nd'$ $J=2$ and 3 odd spectra of Ar*$3p^54p$ excited atoms. Some results are also discussed for the Ca $4p(^2P_{1/2,3/2})ns$ and nd $J=1$ odd spectrum and the Ba $5d(^2P_{5/2})nd$ $J=1$ odd spectrum.

INTRODUCTION

Since the pioneering work by Beutler [1] and Fano [2], autoionization spectra have been investigated by a number of workers/groups for over half a century and still attract many atomic and molecular spectroscopists, perhaps partly because of their beautiful spectral line shapes. It is well known that the photoabsorption cross section for an isolated autoionization line can be decribed by [3]

$$\sigma(\epsilon) = \sigma_a \frac{(q+\epsilon)^2}{1+\epsilon^2} + \sigma_b, \tag{1}$$

where

$$\epsilon = (E - E_r)/\frac{1}{2}\Gamma \tag{2}$$

indicates the departure of the incident photon energy E from the resonance energy E_r, in a scale normalized against the resonance half width $\frac{1}{2}\Gamma$; σ_a and σ_b correspond to the resonant and non-resonant portions of the cross sections, respectively, and q is a profile index which characterizes the spectral line shape. In spite of the limited validity of the assumption in Eq. (1) (i.e., the autoionization line concerned should lie in effect infinitely far from other

© 1997 American Institute of Physics

lines), it has been widely used to parametrize the observed spectral line shapes of autoionization lines and to extract physical quantities such as the resonance width Γ and the resonance energy E_r.

Breakdown of Eq. (1) can be seen in cases where the resonance width Γ becomes comparable to the energy separation between adjacent Rydberg levels. The autoionization spectrum in such a situation is well described by multichannel quantum defect theory (MQDT) developed by Seaton and coworkers [4] and reformulated by Fano and others [5-8]. In the present article, elements of MQDT are reviewed, focusing on the description of autoionization; then some recent experimental results are discussed. The first example is $J = 1$ odd spectra of the rare gas atoms (Ar, Kr, and Xe). Special attention will be focused on extrapolation of the energy-normalized oscillator strength of the autoionizing Rydberg series to the region above the threshold. The second example is Ar $J = 2$ and 3 odd spectra observed by two-step laser excitation from the Ar* $3p^5$ $4s$ $^3P_{0,2}$ metastable levels. I discuss how to resolve overlapping resonances. The third example is a Ca $J = 1$ odd spectrum, illustrating how the spectrum becomes complex if more than two thresholds are involved. Finally an absolute photoabsorption cross section measurement of Ba in the autoionization region is briefly described.

THEORETICAL BACKGROUND

Consider a Rydberg series converging to an ionization threshold. Then the term value E_n is given by the well-known Rydberg formula:

$$E_n = I - \frac{Ry}{(n-\mu)^2} \, , \qquad (3)$$

where I is the ionization energy, Ry is the Rydberg constant, and μ is the *quantum defect*. The quantum defect is a measure of the deviation of the relevant Rydberg level from the hydrogenic level. This term is caused by the short range non-Coulomb interaction between the ion core and a Rydberg electron that happens to come very close to the ion core. The quantum defect changes very slowly as a function of energy. The key point of quantum defect theory (QDT) is that the extrapolation of the quantum defect to the threshold gives the phase shift (relative to the Coulomb phase) of the continuum wave function at the threshold [4,8,9]. Because of this continuity, the dipole amplitude also goes through the threshold without any abrupt change. Consequently, we can estimate the photoionization cross section at the threshold by plotting the oscillator strengths for the transitions to the discrete levels normalized to unit energy and extrapolating the curve thus obtained to the threshold [8]. In QDT, a Rydberg series and its associated continuum are called a *channel* and are treated in a unified way.

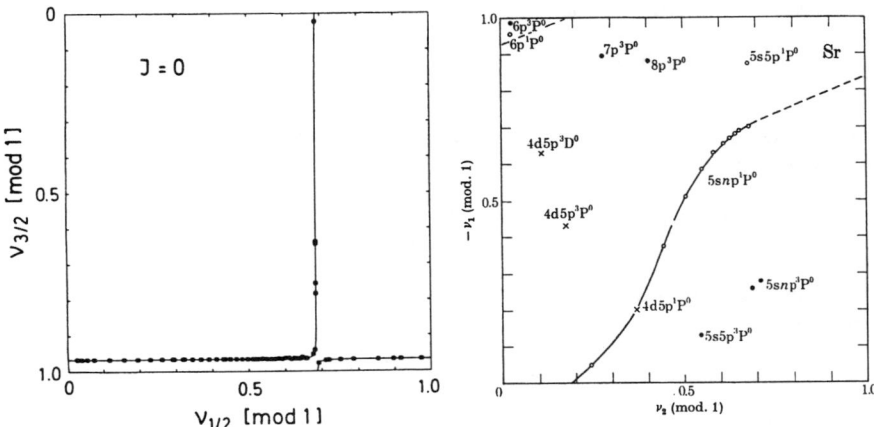

Figure 1: Lu-Fano plots for Ne $J = 0$ odd spectrum [11] and Sr $J = 1$ odd spectrum [12].

Consider a two-channel case in which two Rydberg series converge to two different thresholds. Below the first threshold, the Rydberg series interact with each other and rearrangement of the energy levels occurs. Between the two thresholds, the Rydberg levels converging to the second threshold autoionize to the continuum associated with the first threshold. The same channel interaction governs these two different phenomena. As in the one-channel case, the basic physical parameters that describe the two channels and the interaction between them should be continuous through these two thresholds. Multichannel quantum defect theory (MQDT) is a method to describe discrete, autoionization, and continuum spectra in a unified way using the same set of parameters [4-8].

In order to visualize the two channels and their interaction, we often draw a Lu-Fano plot [10]. If we have two thresholds I_1 and I_2, we can calculate two effective principal quantum numbers ν_1 and ν_2 for each discrete level of term energy E:

$$E = I_1 - \frac{Ry}{\nu_1^2} = I_2 - \frac{Ry}{\nu_2^2} \,. \tag{4}$$

The Lu-Fano plot is a plot of the fractional part of ν_1 versus ν_2. Figure 1 shows Lu-Fano plots for the Ne $J = 0$ odd spectrum [11], which represents the case of weak interaction, and for the Sr $J = 1$ odd spectrum [12], which represents the case of strong interaction. The data points are fitted to the curve for each case. This curve includes all the information about the two channels and their interaction.

There are several ways to describe this curve. We can express it using the

2×2 reactance matrix R_{ij} as [4]

$$\begin{vmatrix} \tan(\pi\nu_1) + R_{11} & R_{12} \\ R_{12} & \tan(\pi\nu_2) + R_{22} \end{vmatrix} = 0. \quad (5)$$

Alternatively, we can use the eigenchannel representation developed by Fano and others [5,6]:

$$\begin{vmatrix} U_{11}\sin[\pi(\nu_1 + \mu_{\alpha 1})] & U_{21}\sin[\pi(\nu_2 + \mu_{\alpha 1})] \\ U_{12}\sin[\pi(\nu_1 + \mu_{\alpha 2})] & U_{22}\sin[\pi(\nu_2 + \mu_{\alpha 2})] \end{vmatrix} = 0, \quad (6)$$

where the eigen quantum defects μ_α and a transformation matrix $U_{i\alpha}$ are defined by

$$R_{ij} = \sum_\alpha U_{i\alpha} \tan(\pi\mu_\alpha) U^\dagger_{\alpha j}, \quad (7)$$

where $\sum_i U^\dagger_{\alpha i} U_{i\beta} = \delta_{\alpha\beta}$.

Another representation to be discussed here is the *phase-shifted* representation discussed by Giusti-Suzor and others [7,13,14]. By shifting the phases of the Coulomb base functions in a way in which the diagonal elements R'_{11} and R'_{22} of the reactance matrix become null, we can obtain the following expression:

$$\begin{vmatrix} \tan[\pi(\nu_1 + \mu'_1)] & \xi \\ \xi & \tan[\pi(\nu_2 + \mu'_2)] \end{vmatrix} = 0, \quad (8)$$

where the phase shifts $\pi\mu'_i$ adopted here in fact give the asymptotic values of the two quantum defects (i.e., the quantum defects of the *unperturbed* case) and $\xi = R'_{12}$ describes the channel coupling.

This phase-shifted representation has some advantages in describing the autoionization spectrum. In this representation the phase shift δ_1 of the energy eigenstate in the autoionization region can be described by [7,14]

$$\begin{vmatrix} -\tan(\delta_1 - \pi\mu'_1) & \xi \\ \xi & \tan[\pi(\nu_2 + \mu'_2)] \end{vmatrix} = 0. \quad (9)$$

Analyzing this equation, we find that the resonance position is given by

$$\tan[\pi(\nu_2 + \mu'_2)] = 0 \quad (10)$$

and the resonance width is related to ξ by the relation [7]

$$\xi^2 = \frac{\pi}{4}\left(\frac{\Gamma}{Ry}\right)\nu_2^3. \quad (11)$$

Introducing the dipole amplitudes D'_1 and D'_2 for the relevant phase-shifted channels, we can obtain a line shape formula which is analogous to the Fano formula Eq. (1) for autoionization but takes account of the periodic structure of the Rydberg series using MQDT [7]:

$$\sigma = \frac{4\pi^2\alpha\omega}{3} D'^2_1 \frac{(q+\epsilon)^2}{1+\epsilon^2}, \quad (12)$$

where
$$\epsilon = \tan[\pi(\nu_2 + \mu'_2)]/\xi^2 \tag{13}$$
is the normalized periodic energy scale,
$$q = -D'_2/D'_1\xi \tag{14}$$
is the Fano profile index, and ω is the photon energy.

Extension of this 2-channel treatment to the case of one closed channel (the n-th channel) with many $(n-1)$ open channels is straightforward. With appropriate orthogonal transformation and phase shifts we can transform the $n \times n$ reactance matrix to the form [13,15]

$$R' = \begin{pmatrix} 0 & \cdots & 0 & R'_{1n} \\ \vdots & & \vdots & \vdots \\ 0 & \cdots & 0 & R'_{(n-1)n} \\ R'_{1n} & \cdots & R'_{(n-1)n} & 0 \end{pmatrix}. \tag{15}$$

Then the line shape formula is given by [15]
$$\sigma_n = \sigma_{an}\frac{(q_n + \epsilon_n)^2}{1 + \epsilon_n^2} + \sigma_b, \tag{16}$$
where
$$\epsilon_n = \tan[\pi(\nu_n + \mu'_n)]/W_n, \tag{17}$$
$$W_n = \sum_{i=1}^{n-1} R'^2_{in}, \tag{18}$$
$$q_n = -D'_n/\sum_{i=1}^{n-1} R'_{in}D'_i, \tag{19}$$
$$\sigma_{an} = \frac{4\pi^2\alpha\omega}{3}(\sum_{i=1}^{n-1} R'_{in}D'_i)^2/W_n, \tag{20}$$
and
$$\sigma_b = \frac{4\pi^2\alpha\omega}{3}\sum_{i=1}^{n-1} |D'_i|^2 - \sigma_{an}. \tag{21}$$

(Notice the misprint in Eq. (19) in Ref. [15].)

Another convenient way to describe autoionization is the use of a complex quantum defect [4,16]. In the case in which $n-1$ channels are open and the channel n is closed, the matrix element $\chi_{nn} = \exp[2\pi i(\alpha_n + i\beta_n)]$ of the $n \times n$ global scattering matrix χ corresponds to the part that describes the scattering from the closed channel to the closed channel. The real part of its phase α_n gives the resonance position ($\alpha_n = \mu'_n$), and the imaginary part β_n gives the

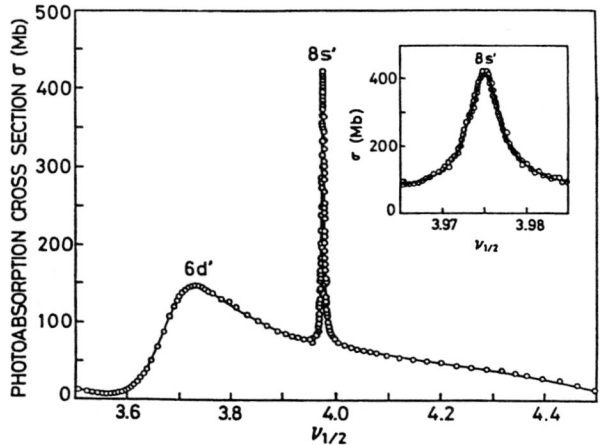

Figure 2: Photoabsorption cross section of Xe across the $6d'$ and $8s'$ plotted as a function of $\nu_{1/2} = [Ry/(I_{1/2} - E)]^{1/2}$ [17].

resonance width ($\tanh \pi\beta_n = W_n$). Extension of the use of complex quantum defects to many degenerate closed channels (i.e., associated with the same new threshold) is also straightforward. In such cases, complex quantum defects are defined as the phase terms of the eigen values of the submatrix corresponding to the closed-closed part χ_{CC} of the global schattring matrix χ [4]:

$$\sum_{j,k} X^\dagger_{ij} \chi_{CCjk} X_{kl} = \exp[2\pi i(\alpha_i + i\beta_i)]\delta_{il} , \qquad (22)$$

where $\sum_j X^\dagger_{ij} X_{jk} = \delta_{ik}$.

RARE GAS $J = 1$ ODD SPECTRA

Photoabsorption of the rare gas atoms in the $mp^6\ {}^1S_0$ ground state ($m = 3$, 4, and 5 for Ar, Kr, and Xe, respectively) leads to $J = 1$ odd levels consisting of five Rydberg series which can be expressed, using jK coupling notation, as $mp^5(^2P_{3/2})ns[3/2]_1$, $(^2P_{3/2})nd[1/2]_1$, $(^2P_{3/2})nd[3/2]_1$, $mp^5(^2P_{1/2})ns'[1/2]_1$, and $(^2P_{1/2})nd'[3/2]_1$. The first three of these converge to the first threshold $I_{3/2}$ and the last two series ns' and nd' converge to the second threshold $I_{1/2}$. Between $I_{3/2}$ and $I_{1/2}$, ns' and nd' are subject to autoionization. Photoabsorption spectra for these autoionization resonances of the rare gases were first observed by Beutler [1] and theoretically analyzed by Fano [2].

High-resolution measurements for the absolute photoabsorption cross sections of Ar, Kr, and Xe in the autoionization region were carried out by Maeda et al. [17] at the Photon Factory, a synchrotron radiation facility in Japan,

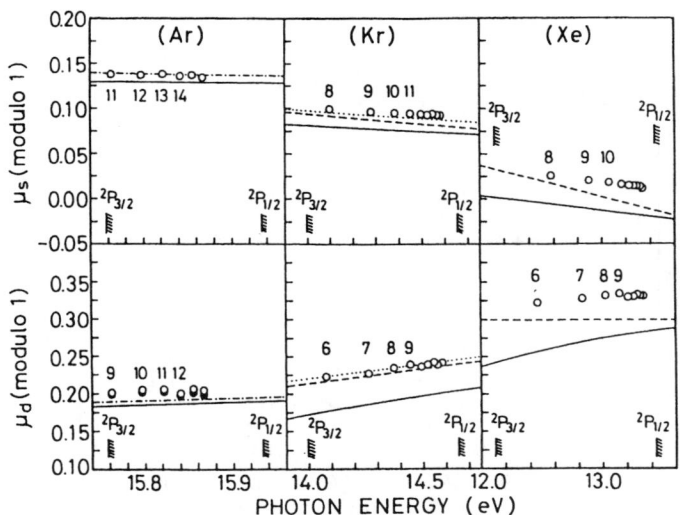

Figure 3: Quantum defects for the ns' and nd' $J = 1$ odd resonances of Ar, Kr, and Xe [17]. The chain curves for Ar are based on the semiempirical MQDT parameters from Lee and Lu [6], the dashed curves for Kr and Xe are those from Geiger [18], the dotted curves for Kr are those from Aymar et al. [19], and the solid curves are based on the relativistic ab initio MQDT parameters from Johnson et al. [20].

using a 6.65-m high-resolution spectrometer 6VOPE. Figure 2 gives a portion of the measured cross section for Xe as a function of $\nu_{1/2} = [Ry/(I_{1/2} - E)]^{1/2}$, where E is the excitation energy. The spectral line shape in Fig. 2 can be described by the following expression

$$\sigma = \sigma_{as}\frac{(q_s + \epsilon_s)^2}{1 + \epsilon_s^2} + \sigma_{ad}\frac{(q_d + \epsilon_d)^2}{1 + \epsilon_d^2} + \sigma_b, \qquad (23)$$

where

$$\epsilon_l = \tan[\pi(\nu_{1/2} + \mu_l')]/W_l \qquad (24)$$

for $l = s$ and d. The result of the fitting to the experimental data points is shown by the solid curve in Fig. 2. The fitted curve passes through the experimental data points almost completely, illustrating that use of Eqs. (23) and (24) is adequate from the viewpoint of parametrization, although the influence of $s - d$ coupling is neglected in the above treatment.

μ_l and W_l thus obtained are plotted in Figs. 3 and 4, where the open and closed circles correspond to the results before and after the deconvolution of the instrumental functions. For comparison, μ_l and W_l were calculated [17] from the semiempirical and ab initio MQDT parameters available in the literature, using complex quantum defects. In the figures the chain curves for Ar are based on the semiempirical MQDT parameters from Lee and Lu [6], the dashed

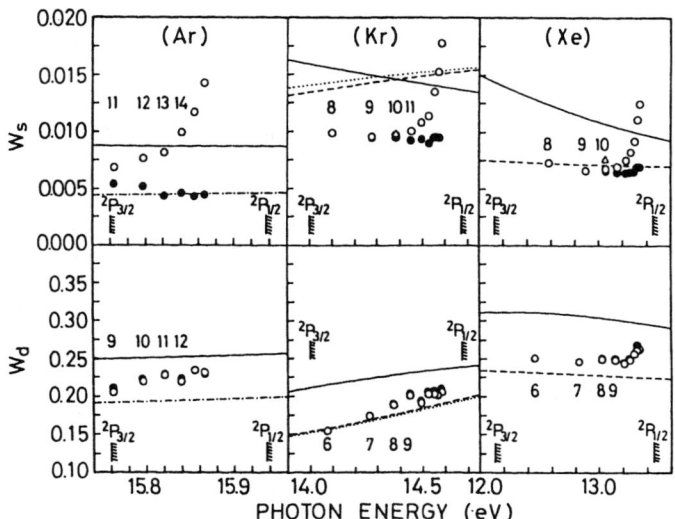

Figure 4: The width paramters for the ns' and nd' $J = 1$ odd resonances of Ar, Kr, and Xe [17]. The open and closed circles corresponds to the data points before and after deconvolution of the instrumental function. See the caption of Fig. 3 for further detail.

curves for Kr and Xe are those from Geiger [18], the dotted curves for Kr are those from Aymar et al. [19], and the solid curves are based on the relativistic ab initio MQDT parameters from Johnson et al. [20]. The agreement between the measurements and the theoretical predictions is in general reasonable.

Note that the quantities $\sigma_{as}W_s q_s^2$ and $\sigma_{ad}W_d q_d^2$ correspond to the *partial* cross sections $\sigma_{l'}$ for the phase shifted s' and d' channels associated with the second threshold $I_{1/2}$:

$$\sigma_{l'} = \sigma_{al}W_l q_l^2 = \frac{4\pi^2 \alpha \omega}{3} \mid D_l' \mid^2 , \qquad (25)$$

and σ_{as} and $\sigma_b + \sigma_{ad}$ correspond to the *partial* cross sections for the phase shifted s and d channels associated with the first threshold $I_{3/2}$. In Fig. 5, $\sigma_{as}W_s q_s^2$ and $\sigma_b + \sigma_{ad}$ are plotted as a function of photon energy. These quantities are not sensitive to the instrumental function and can be easily extrapolated smoothly in the region above the threshold. These quantities are not exactly equivalent but close to the quantities that can be measured by means of spin resolved photoelectron spectroscopy. Indeed the extrapolation of $\sigma_{l'}$ results in reasonable agreement with the s and d photoelectron partial wave cross sections obtained by Heinzmann [21] by means of spin resolved photoelectron spectroscopy, illustrating a close relation between the two different experimental approaches. (An exact comparison becomes possible after some algebra if a complete set of the MQDT parameters μ_α and $U_{i\alpha}$, or a complete set of the reactance matrix elements R_{ij}, is known.)

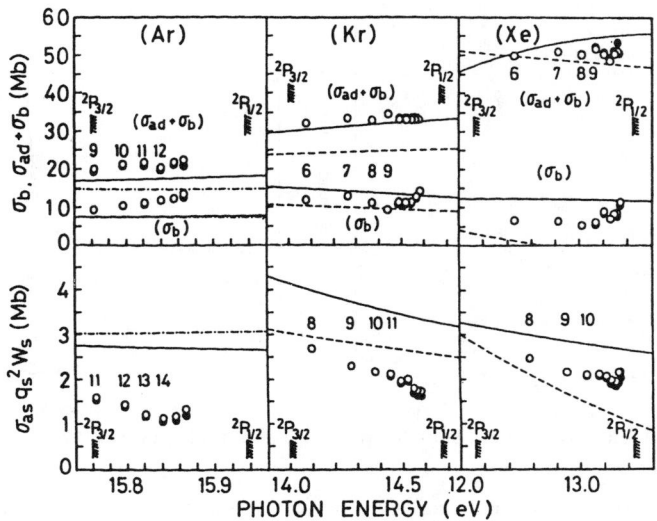

Figure 5: $\sigma_{as}W_s q_s^2$, σ_b, and $\sigma_b + \sigma_{ad}$. See the caption of Fig. 3 and text for the detail.

Ar $J = 2$ AND 3 ODD SPECTRA

Consider two-step laser excitation from the Ar* $3p^54s$ $^3P_{0,2}$ metastable levels. With an appropriate choice of the intermediate level $3p^54p$, we can reach final states $3p^5ns$ and nd with J from 0 to 4. Here we focus on the $J = 2$ and 3 Rydberg series. The $J = 2$ levels consist of five Rydberg series which can be expressed as $3p^5(^2P_{3/2})ns[3/2]_2$, $(^2P_{3/2})nd[5/2]_2$, $(^2P_{3/2})nd[3/2]_2$, $3p^5(^2P_{1/2})nd'[5/2]_2$, and $(^2P_{1/2})nd'[3/2]_2$. The latter two nd' series converge to the second threshold $I_{1/2}$ and thus are subject to autoionization between $I_{3/2}$ and $I_{1/2}$. The $J = 3$ levels consist of three Rydberg series which can be expressed as $3p^5(^2P_{3/2})nd[7/2]_3$, $(^2P_{3/2})nd[5/2]_3$, and $3p^5(^2P_{1/2})nd'[5/2]_3$. The last nd' series converges to the second threshold $I_{1/2}$ and thus is subject to autoionization between $I_{3/2}$ and $I_{1/2}$. In this section we discuss how we can observe and analyze these three autoionization resonances $3p^5(^2P_{1/2})nd'[5/2]_2$, $(^2P_{1/2})nd'[3/2]_2$, and $(^2P_{1/2})nd'[5/2]_3$.

The $J = 2$ and 3 odd discrete spectra were observed recently by Weber et al. [22] at Kaiserslautern University in Germany, by means of two-step excitation from the $3p^54s$ $^3P_2$ metastable level via the $3p^54p$ $^3D_3$ intermediate level using two single-mode cw lasers. Figure 6 shows the Lu-Fano plots for the $J = 2$ and 3 odd levels thus measured. These spectra were analyzed using a semiempirical MQDT procedure described by Lee and Lu [6] and the eigenchannel MQDT parameters were determined [22]. From the MQDT parameters we can estimate the quantum defects μ' and the width parameters

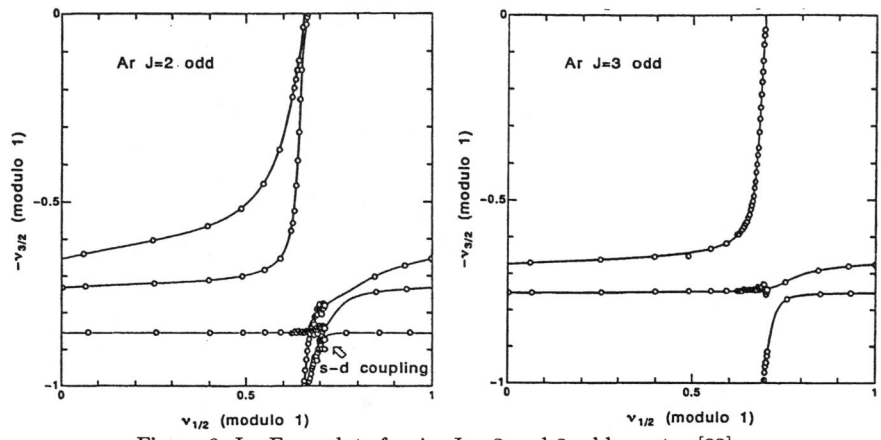

Figure 6: Lu-Fano plots for Ar $J = 2$ and 3 odd spectra [22].

Table 1: μ' and W for the $nd'[3/2]_2$, $nd'[5/2]_2$, and $nd'[5/2]_3$ resonances.

	$nd'[3/2]_2$	$nd'[5/2]_2$	$nd'[5/2]_3$
μ'_{MQDT}	0.357	0.353	0.307
μ'_{exp}	0.355	0.352	0.316
W_{MQDT}	0.191	0.042	0.058
W_{exp}	0.186	0.036	0.051

W of the three autoionization resonances using the complex quantum defect analysis. The results are summarized in Table 1. From these values we can immediately see that the $[3/2]_2$, $[5/2]_2$, and $[5/2]_3$ resonances overlap strongly.

In order to measure the resonance energies (quantum defects) and the resonance widths of these three resonances separately, an experiment was carried out by Klar *et al.* [23,24] at Kaiserslautern University in Germany, in which selectivity was introduced into the excitation process through a proper choice of the intermediate state (see *e.g.* Ref. [25]). Using the jK coupling scheme the transition matrix element from the intermediate $(j_i l_i = 1)[K_i]J_i$ states to the $(j = \frac{1}{2} \, l = 2)[K]J$ resonance states is given by [25,26]

$$\langle (\frac{1}{2}2)K\frac{1}{2}J \mid D^{(1)} \mid (j_i l_i = 1)K_i \frac{1}{2} J_i \rangle = \hat{K}_i \hat{K} \hat{J}_i \hat{J} (-1)^{J_i} \left\{ \begin{array}{ccc} K & J & \frac{1}{2} \\ J_i & K_i & 1 \end{array} \right\}$$

$$\times \left\{ \begin{array}{ccc} K & 2 & \frac{1}{2} \\ 1 & K_i & 1 \end{array} \right\} \langle d \| D^{(1)} \| p \rangle , \quad (26)$$

where $\langle d \| D^{(1)} \| p \rangle$ is the relevant reduced dipole matrix element and $\hat{J} \equiv (2J+1)^{1/2}$. Calculating the geometric factors (besides $\langle d \| D^{(1)} \| p \rangle$) we find that $4p[1/2]_1 \to nd'[3/2]_2$, $4p[3/2]_1 \to nd'[5/2]_2$, and $4p[3/2]_2 \to nd'[5/2]_3$

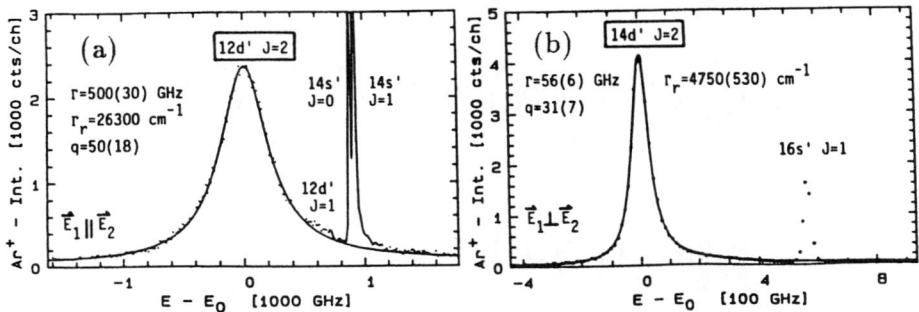

Figure 7: Two-photon ionization signals [23,24]. (a) Ar $3p^54s\ ^3P_2 \to 3p^5(^2P_{1/2})4p[1/2]_1 \to 12d', 14s'$. (b) Ar $3p^54s\ ^3P_0 \to 3p^5(^2P_{1/2})4p[3/2]_2 \to 14d', 16s'$.

transitions are prominent as pointed out by Wang and Knight [27]. Figure 7 (a) and (b) are the spectra which predominantly show the $4p[1/2]_1 \to nd'[3/2]_2$ and $4p[3/2]_1 \to nd'[5/2]_2$ resonances, respectively. We can clearly see the difference in the reduced widths $\Gamma_r \equiv \nu_{1/2}^3 \Gamma$ for the two $J = 2$ resonances. The quantum defects μ' and the width parameters $W = \frac{\pi}{4}(\Gamma/Ry)\nu_{1/2}^3$ thus measured for the three resonances are compared in Table 1 with the theoretical predictions based on the semiempirical MQDT parameters. The agreement is remarkably good, illustrating the consistency of our analysis.

Ca AND Ba $J = 1$ ODD SPECTRA

In the case of the rare gases we have just two thresholds to be considered. In the case of the alkaline earth atoms there are many thresholds to be considered and the spectrum becomes much more complex. Consider the Ca Rydberg series converging to the $4p\ ^2P_{1/2,3/2}$ excited levels of the Ca$^+$ ion. Below the $^2P_{1/2}$ threshold, five Rydberg series converging to these two thresholds interact with each other and simultaneously autoionize to the continua associated with the Ca$^+$ $3d_{3/2,5/2}$ and $4s_{1/2}$ levels.

Figure 8 shows a portion of the absorption spectrum just below the $^2P_{1/2}$ threshold. The spectrum was measured photographically using the high resolution spectrometer 6VOPE at the Photon Factory in Japan [28]. Although the spectrum is very complex, it is still possible to analyze it. The lower spectrum is the one calculated using the ab initio MQDT parameters given by Greene and Kim [29]. We can clearly see that all the observed peaks are well reproduced by the calculation.

In order to discuss spectral line shapes in detail, however, it is necessary to carry out a photoelectric measurement instead of a photographic measurement. Furthermore, in order to extract quantitative information as we have done for

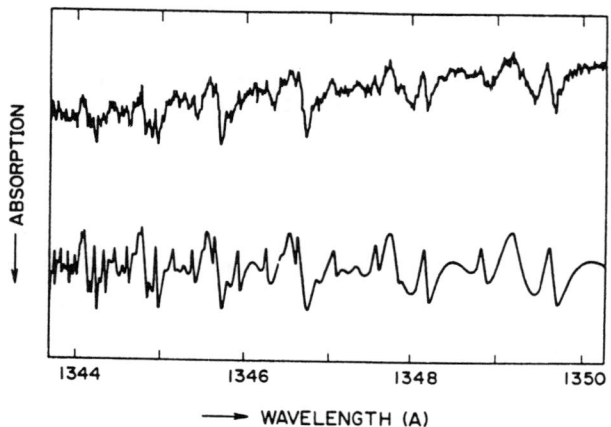

Figure 8: Absorption spectrum of Ca just below the Ca$^+$ $4p$ $^2P_{1/2}$ threshold [28]. The upper spectrum, a densitometer trace of the observed spectrogram. The lower spectrum, calculated from the ab initio MQDT parameters given by Greene and Kim [29].

the rare gas atoms, it is highly desirable to carry out an absolute cross section measurement. We have just started such measurements on Ba atoms [30].

In the case of Ba, the first threshold corresponding to Ba$^+$ $6s$ $^2S_{1/2}$ is at 238 nm and the first autoionization line $6s^2$ $^1S_0 \to 5d8p$ $^1P_1$ sits on this threshold. We have observed this autoionization line using a second spectrometer (II) just before and after the high-resolution photoelectric absorption measurement using the high-resolution spectrometer 6VOPE. This autoionization line can be used as a monitor of the column density Nl (i.e., the number density integrated over the line of sight) of the barium vapor. To do so, we have carried out a calibration measurement at the laboratory. We measured the absorption of this line using the same spectrometer (II) as used on the beamline and simultaneously measured the barium column density Nl very precisely using an improved hook method [31].

We have measured the spectral line shapes of the prominent $5dnp$ series converging to the upper $5d$ $^2D_{5/2}$ threshold. The width parameter W drops sharply at $n \sim 13$ illustrating that this series is strongly perturbed. This perturbation is mainly due to the interloper $7s6p$. Abutaleb et al. [32] pointed out also the effect of the $7s6p$ interloper on a weak $5dnp$ series converging to the lower $5d$ $^2D_{3/2}$ threshold. Further analysis and measurements are currently in progress.

ACKNOWLEDGMENTS

The work on $J = 1$ rare gas and alkaline earth spectra was carried out with the approval of the Photon Factory advisory committee, in collaboration with K. Maeda, K. Ito, T. Matsui, H. Chiba, M. Okunishi, K. Ohmori, Y. Sato, T. Namioka, and J. B. West. The work on Ar $J = 2$ and 3 was carried out in collaboration with D. Klar, J. M. Weber, S. Baier, M.-W. Ruf, and H. Hotop, supported by the Deutsche Forschungsgemeinshaft through Sonderforschungsbereich (SFB91). The author is grateful to his coworkers for fruitful collaboration and to J. B. West for critical reading of this manuscript.

REFERENCES

1. Beutler, H., *Z. Phys.* **93**, 177 (1935).
2. Fano, U., *Nuovo Cimento* **12**, 156 (1935).
3. Fano, U., and Cooper, J. W., *Phys. Rev.* **137**, A 1364 (1965).
4. Seaton, M. J., *Rep. Prog. Phys.* **46**, 167 (1983).
5. Fano, U., *Phys. Rev.* A **2**, 353 (1970).
6. Lee, C.-M., and Lu, K. T., *Phys. Rev.* A **8**, 1241 (1973).
7. Giusti-Suzor, A., and Fano, U., *J. Phys.* B **17**, 215 (1984).
8. Fano, U., and Rau, A. R. P., *Atomic Collisions and Spectra*, Orlando: Academic, 1986.
9. Seaton, M. J., *Man. Not. R. Astron. Soc.* **118**, 504 (1958).
10. Lu, K. T., and Fano, U., *Phys. Rev.* A **2**, 81 (1970).
11. Harth, K., Raab, M., and Hotop, H., *Z. Phys.* D **7**, 213 (1987).
12. Lu, K. T., *Proc. R. Soc. Lond.* A **353**, 431 (1977).
13. Cooke, W. E., and Cromer, C. L., *Phys. Rev.* A **32**, 2725 (1985).
14. Wintgen D., and Friedrich, H., *Phys. Rev.* A **35**, 1628 (1987).
15. Ueda, K., *Phys. Rev.* A **35**, 2484 (1987).
16. Ueda, K., *J. Opt. Soc. Am.* B **4**, 424 (1987).
17. Ueda, K., Maeda, K., Ito, K., and Namioka, T., *J. Phys.* B **22**, L481 (1989); Maeda, K., Ueda, K., Ito, K., and Namioka, T., *Phys. Scripta* **41**, 464 (1990); Maeda, K., Ueda, K., Namioka, T., and Ito, K., *Phys. Rev.* A **45**, 527 (1992); Maeda, K., Ueda, K., and Ito, K., *J. Phys.* B **26**, 1541 (1993).
18. Geiger, J., *Z. Phys.* A **282**, 129 (1977).
19. Aymar, M., Robaux, O., and Thomas, C., *J. Phys.* B **14**, 4255 (1981).
20. Johnson, W. R., Chen, K. T., Huang, K.-N., and Le Dourneuf, M., *Phys. Rev.* A **22**, 989 (1980).
21. Heinzmann, U., *J. Phys.* B **13**, 4367 (1980).
22. Weber, J. M., Ueda, K., Kreil, J., Ruf, M. W., and Hotop, H., *Abstracts of Contributed Papers of 19th ICPEAC*, Whisler, Canada (Michell, J. B. A., McConkey, J. W., and Brion, C. E., eds.) 1995, pp. 648; full note in preparation for publication.
23. Klar, D., Harth, K., Ganz, J., Kraft, T., Ruf, M.-W., and Hotop, H., *Proceedings of UK/USSR Seminar*, Leningrad (Amusia, M. Ya., and West, J. B., eds.) DL/SCI/R29 SERC Daresbury, 1990, pp. 78.
24. Hotop, H., Klar, D., and Schohl, S., *Inst. Phys. Conf. Ser.* No. 128, Section 2, Yew York: IOP Publishing, 1992, pp. 5.
25. Klar, D. Ueda, K., Ganz, J., Harth, K., Bußert, W., Baier, S., Weber, J. M., Ruf, M.-W., and Hotop, H., *J. Phys.* B **27**, 4897 (1994).
26. Cowan R. D., and Andrew, K. L., *J. Opt. Soc. Am.* **55**, 502 (1965).
27. Wang, L.-G., and Knight, R. D., *Phys. Rev.* A **34**, 3902 (1986).

28. Ueda, K., Ito, K., Sato, Y., and Namioka, T., *Phys. Scripta* **41**, 75 (1990).
29. Greene, C. H., and Kim, L., *Phys. Rev.* A **36**, 2706 (1987);
    Kim, L., and Greene, C. H., *Phys. Rev.* A **36**, 4372 (1987).
30. Maeda *et al.*, in preparation for publication.
31. Ueda, K., Sonobe, O., Chiba, H., Sato, Y., Namioka, T., and Ito, K., *Rev. Sci. Instrum.* **63**, 1690 (1992).
32. Abutaleb, M., de Graaff, R. J., Ubachs, W., Hogervorst, W., and Aymar, M., *J. Phys.* B **24**, 3565 (1991).

THEORY OF QUANTUM OSCILLATIONS IN SELF-BROADENING OF 4^2S-n^2S RYDBERG TRANSITIONS IN POTASSIUM

R. M. Herman and M. E. Henry

Department of Physics, Penn State University
104 Davey Laboratory, University Park, PA 16802 USA
(Fax: 814-865-3604; e-mail: rmh@phys.psu.edu)

Abstract

A pseudohamiltonian impact theory for describing Rydberg transition line shapes is proposed. Preliminary results give lineshapes in reasonable agreement with experiment, including the quantum oscillations in alkali self broadening and shifting, and rare gas broadening.

An outstanding problem of fundamental importance in the theory of spectral lines is that of developing a systematic theory allowing the pressure width and shift data for Rydberg transitions[1] to be systematically described through an impact theory of lineshapes. A related problem is that of extending free-electron scattering theory to include a description of scattering between weakly bound electron states. To date, the best model[2] simply utilizes the ^3P free electron-neutral K scattering resonance as if it were a stable state whose energy dynamically crosses that of the Rydberg level in question, leading to coherence destruction. Notwithstanding its success in predicting widths, there seems to be no systematic theory emerging from this model as is illustrated in part by the fact that there is no obvious way to extend the theory to include the line shifts.

In our approach we extend the free-electron scattering theory for equal incoming and outgoing waves (which therefore lead to real stationary states) to the derivation of a perturbational pseudohamiltonian of form

$$H'_{eff} = -\frac{8\pi^2\hbar^2}{mk} \sum_{\ell,m;s=0}^{1} sin\delta_{\ell,s}(k) cos\delta_{\ell,s}(k) T^*_{\ell m} \delta(\vec{x} - \vec{X}(t)) T_{\ell m} \delta_{s,S} \quad (1)$$

where $\vec{x}$, $\vec{X}(t)$ and $\vec{k}$ specify electron and perturber positions, and local electron propagation vectors within the Rydberg orbital at position $\vec{x}(t)$; $\delta_{\ell,s}(k)$ is the partial wave scattering phase shift associated with the ℓ partial wave for singlet and triplet scattering, and $T_{\ell m}$ are operators which select the incoming and outgoing ℓ, m partial wave amplitudes from a wavefunction consisting of any Fourier superposition whatever. They are defined through the relation

$$T_{\ell m} \mid e^{i\vec{k}\cdot\vec{x}} > = Y_{\ell m}(\hat{k}) \mid e^{i\vec{k}\cdot\vec{x}} > \qquad (2)$$

with $(ik)^\ell T_{\ell m}$ being a combination of Cartesian components of ∇ which is identical with the combinations of (x,y,z) which make up $r^\ell Y_{\ell m}(\theta,\varphi)$. The factor $\delta(\vec{x}-\vec{X}(t))$, with $\vec{X}(t)$ describing the perturber classical path, provides a time dependence that allows H_{eff} to enter line shape calculations in standard fashion. The quantum oscillations arise by virtue of the ^{3}P resonance occurring at positions which coincide alternately with large and small derivatives (which govern the P-state scattering amplitudes) of the Rydberg state wavefunctions as one goes from one to another upper $n^2 S$ state.

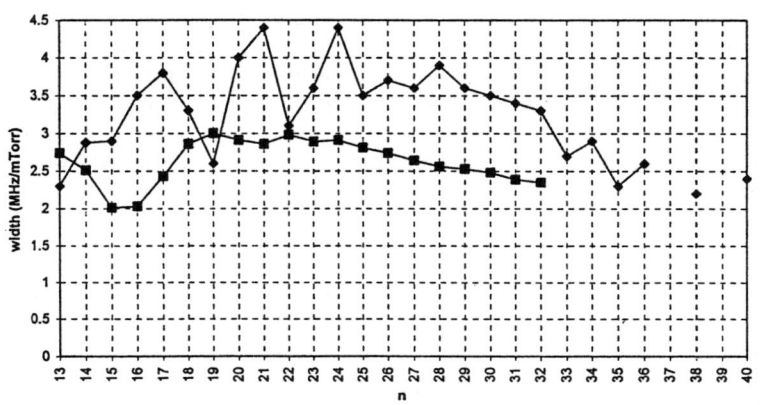

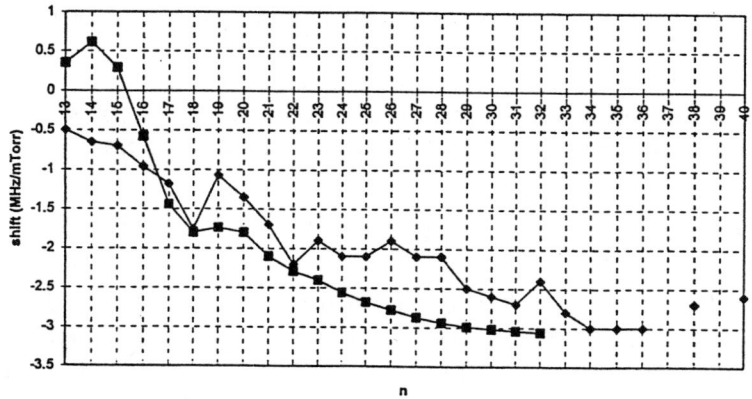

Diamonds, experimental results of Stoicheff et al; squares, theoretical results of Herman and Henry.

In our calculations we have employed standard impact theory with Murphy-Boggs type cutoff conditions[3]. Preliminary results, using the phase shifts of Fabrikant[4], are shown in Figs. 1 and 2 for the widths and shifts of the self-broadened $4^2S - n^2S$ transitions. While there are several points of agreement between these (preliminary) calculations and experiment, we have observed that if one neglects the interference between incoming and outgoing scattered waves so that the factor $cos\delta_{\ell,s}(k)$ would be omitted from eq. (1) the results, as shown in figs. 3,4, are more encouraging.

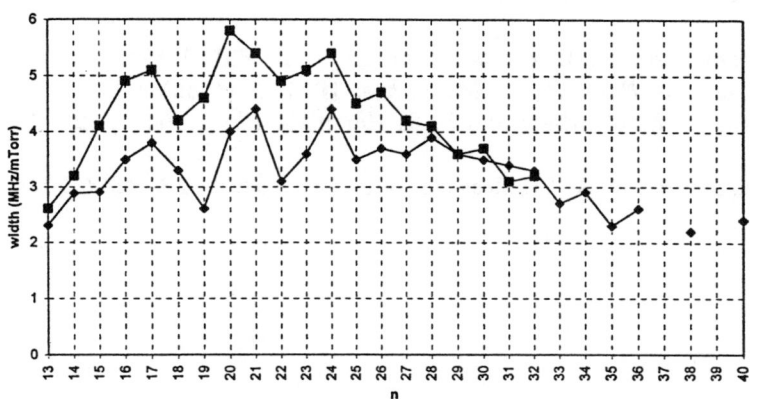

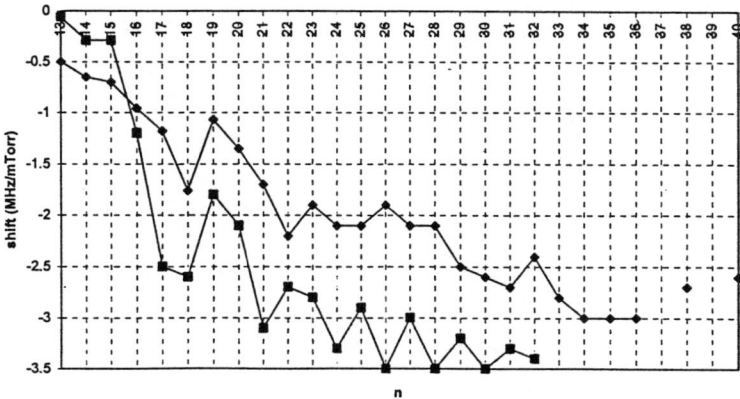

Diamonds, experimental results of Stoichoff et al; squares, theoretical results of Herman and Henry

Finally, we have applied the same theory with H'_{eff} again given by eq. (1) to rare gas (specifically Ar) broadening of Na 3^2S-n^2S transitions with results that are highly encouraging, the widths having the same qualitative dependence upon n as seen experimentally[5], with overall magnitudes of the widths lying within a factor two of those observed, while differing by a factor of the order 10^2 from the self broadening cases. The shifts have a qualitative dependence previously[6] reported for Cs-Ar single photon transitions.

1. D. C. Thompson, E. Weinberger, G-X. Xu and B. P. Stoicheff, *Phys. Rev.* *A***35**, 690 (1987).
2. V. M. Borodin and A. K. Kazansky, *J. Phys. B: At. Mol. Phys.* **25**, 971 (1992).
3. J. S. Murphy and J. E. Boggs, *J. Chem. Phys.* **47**, 691 (1967).
4. I. I. Fabrikant, *J. Phys. B: At. Mol. Phys.* **19**, 1527 (1986).
5. R. Kachru, T. W. Mossberg and S. R. Hartman, *Phys. Rev. A* **21**, 1124 (1980).
6. M. A. Mazing and N. A. Vrubleskaya, *Sov. Phys. JETP* **23**, 228 (1966).

Lineshapes for a pure three level system: quantum coherence effects on absorption and birefringence of helium

F.S. Pavone, F. Bassani*, G. Bianchini**, P. Cancio***, F.S. Cataliotti****, T.W. Hänsch****, M. Inguscio****

European Laboratory for Non-linear Spectroscopy, Università di Firenze (LENS)
L.go E. Fermi 2, I-50125, Firenze
Italy

*   Scuola Normale Superiore
Pisa, Italy
**   Istituto Nazionale di Fisica della Materia
sezione di Firenze, Italy
***   Permanent address: Instituto de Estructura de la Materia CSIC
Madrid, Spain
****   Dipartimento di Fisica
Università di Firenze, Italy

Abstract. We have investigated atomic coherence effects in the ladder scheme of the triplet transitions 2S-2P-3S of helium. A powerful dressing laser at 706nm in resonance with the 2P-3S transition induces a transparency on the 2S-2P transition at 1083nm probed by a weak laser. We have characterised this induced transparency also known as Electromagnetically Induced Transparency (EIT), demonstrating the possibility of reaching 100% of bleaching. Also, we have evidenced the presence of the two photon effect on the EIT process by studying its dependence on the mutual probe-dress polarisation angle. With such a simple level scheme a quantitative comparison with the theory has been made possible and many of the predicted effects have been investigated. Furthermore, polarisation of the dressing laser field has allowed for the first time the study of the electromagnetically induced birefringence which causes a change in the probe laser polarization. The experiments were performed not only in the clean environment of a metastable atomic beam but also in a weak radiofrequency discharge; this configuration could be suggestive for non inverted gain phenomena through incoherent population of excited levels.

Quantum coherence effects in atomic three level systems interacting with two monochromatic laser fields have long provided a rich source of interesting phenomena in laser spectroscopy. Their theory has been developed extensively in

© 1997 American Institute of Physics

early works related to Doppler-free laser spectroscopy[1]. They are responsible for the non-absorbing resonances demonstrated by Alzetta et al.[2] in a λ scheme. More recently, the phenomenon of electromagnetically induced transparency (EIT) has attracted considerable attention[3]. EIT permits laser without population inversion[4] and it can lead to a large refraction index without absorption[5,6]. In essence, a coherence between two atomic levels is created by a strong coupling field. This affects the absorption of a weak probe beam connected with a third level. Experimentally, this phenomenon has been observed in all three possible level configurations (λ[7], V[8], ladder[9]).

In our experiment, we study a ladder level scheme in atomic helium. Metastable helium atoms in a RF discharge are "dressed" with a strong laser field near 706nm, coupling the levels 2^3P and 3^3S, and the absorption is probed on the 2^3S-2^3P transition near 1083nm. Unlike previous studies with alkali atoms, our helium experiment is not encumbered by hyperfine splitting. Moreover the discharge environment suggests the possibility of a laser without inversion pumped by collisions.

By using a theoretical derivation[10] of the non-linear susceptibility, we have studied the lineshape of the induced transparencies. We have also characterized the phenomenon by studying its dependence on laser power, magnetic fields, gas pressure, pump detuning, etc. Furthermore, we have evidenced the contribution of the two-photon transition amplitude which, interfering with the two-step process, is at the base of EIT. This was done by changing the mutual probe-dress polarization angle which affects only the two-photon process amplitude.

Here, we demonstrate also optical birefringence due to quantum coherence in an atomic medium driven to transparency by a strong coupling field. Although light-induced birefringence is well known for polarization spectroscopy, its role in EIT seems to have gone unexplored so far, despite the potential for the development of novel optical sensors, light modulation or control system.

In the experiment, the powerful dressing radiation is provided by a Ti:S laser and powers up to 3W (400μm waist) are available also thanks to the use of a build-up ring cavity.

Weak probe counter propagating radiation is provided by a semiconductor diode laser which is scanned to observe the transmitted signal. The pump laser, instead, is kept fixed on resonance. Reduced absorption is observed as predicted by the theory and even complete transparency can be observed at high dressing powers. A typical probe transmission showing an induced transparency is shown in figure 1. Here a good overlap with a theoretical shape has been obtained by using the non-linear susceptibility[10]:

$$\chi(\omega;\omega_2,\varepsilon_2) = 1 + \frac{N|\langle 0|\mu|1\rangle|^2}{E_{10} - \hbar\omega - i\Gamma_1 - \frac{\hbar^2\alpha_2^2}{E_{20} - \hbar\omega_2 - \hbar\omega - i\Gamma_2}} \qquad (1)$$

where μ is the dipole moment, the levels are called respectively 0, 1 and 2, Γ is the level width, ω is the probe frequency, ω_2 is the pump frequency. In (1) the antiresonant term has been neglected. The Rabi frequency has been expressed as

$$\alpha_2 = \frac{\langle 0|\mu|1\rangle \varepsilon_2}{2\hbar} \qquad (2)$$

where ε is the field amplitude.

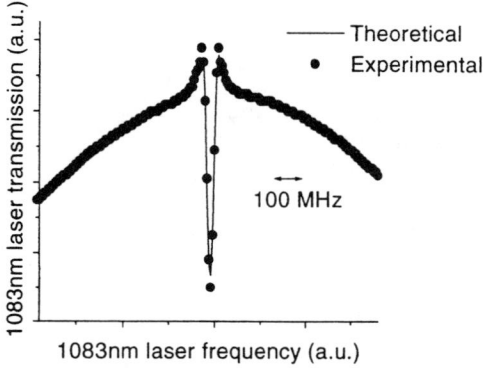

Figure 1. A typical absorption lineshape showing induced transparency at 1083nm fitted with a theoretical curve on the (1-0) component

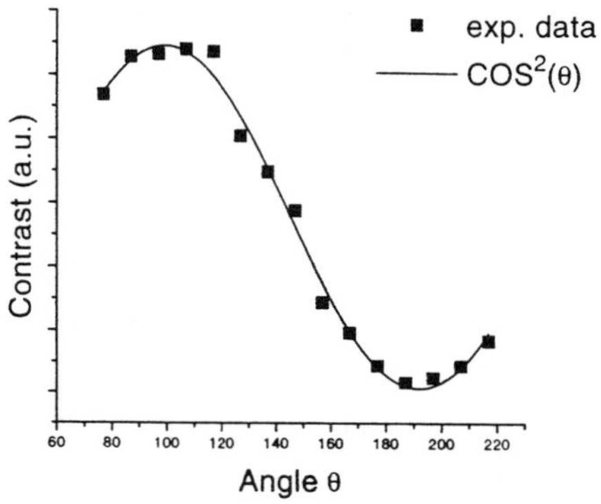

Figure 2. Evidence of the two photon contribution on EIT: effect of the change of the mutual pump-probe polarization angle. Transparency of the 1-2 component at 1083nm

We have separated the real and imaginary part of susceptibility and integrated it over the velocity Doppler profile of the transition.
As discussed, in order to evidence the two-photon contribution, we have investigated the amplitude of EIT signal as a function of the mutual pump-probe polarization angle.
In figure 2 is plotted the transparency versus the angle on the 1-2 component and the behaviour is well overlapped with the predicted $\cos^2(\theta)$, typical of the nature of two-photon process related to the conservation of angular momentum.
Also induced birefringence has been pointed out on the probe transmission by using a polarization scheme detection (two crossed polarizers at the entrance and at the exit of the absorption cell). In this way it was possible to record a signal analysing the transmitted light with a crossed polarizator.
In figure 3 is shown a typical recording on the 1-1 component compared with a theoretical result obtained using the same non-linear susceptibility. We explain the peaks as the Autler-Townes splitting: the main ones are due to the resonant contribution while the other peaks are due to the off-resonant contribution on the other component.
It is of interest also to note that the discharge process, utilized to produce the metastable atoms do not perturb the effect investigated. So, we think that the

observed phenomenon opens the possibility of using the atomic medium as optical tool for laser beam manipulation.

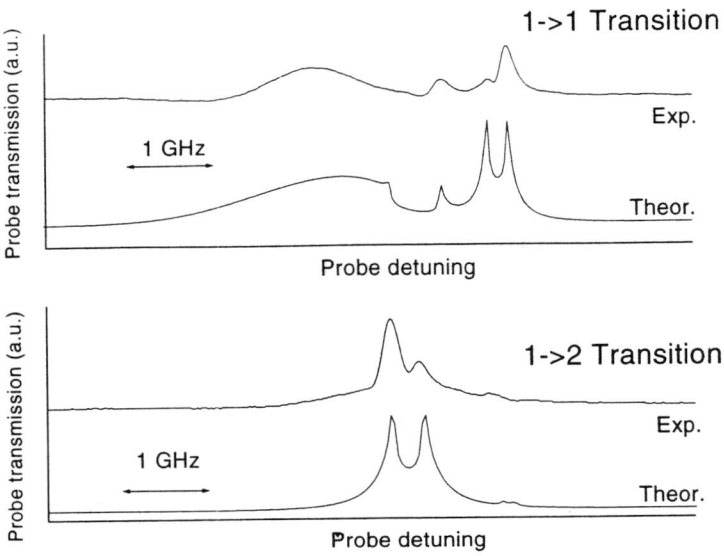

Figure 3. Crossed polarizator transmissions on the 1-1 (a) and 2-2 (b) component (pump 15mW)

Finally, the presence of a weak radiofrequency discharge could be suggestive for non inverted gain phenomena through incoherent population of excited levels.

REFERENCES

1. M. S. Feld and A. Javan, Phys. Rev. 177, 540 (1968); T. Hänsch et al., Z. Phys. 226, 293 (1969); I.M. Beterov and V.P. Chebotaev, Sov. Phys. JEPT lett. 9, 127 (1969).
2. G. Alzetta, A. Gozzini, L. Moi, G. Orriolis, Nuovo Cimento 36B, 5   (1976).
3. A. Imamoglu and S.E. harris, Opt. Lett. 14, 1344 (1989).
4. A. S. Zibrov, M.D. Lukin, D.E. Nikonov, L. Hollberg, M.O. Scully, V. Velichansky, and H.G. Robinson, Phys. Rev. Lett. 75, 1499 (1995).
5. M. Xiao , Yong-qing Li, Shan-zheng Jin, and Julio Gea-Banacloche,   Phys. Rev. Lett. 74, 66 (1995).
6. O. Schmidt, R. Wynands, Z. Hussein, and D. Meschede, Phys. Rev.   A53, R27 (1996). A.S. Zibrov, M.D. Lukin, L. Hollberg, D.E. Nikonov, M.O. Scully, H.G. Robinson, and V.L. Velichansky, Phys. Rev. Lett. 76, 3935 (1996).
7. K.J. Boller, A. Imamoglu, and S.E. Harris, Phys. Rev. Lett. 66, 2593   (1991).
8. A. Weiss, F. Sander, and S.I. Karnosky, in IEEE Technical Digest, 5th European Quantm Electronics Conference '94 (Optical Society of   America, Washington, DC, 1994), pg. 252.
9. J. E. Field, K.H. Hahn, and S.E. Harris, Phys. Rev. Lett. 67, 3062 (1991).
10. F. Bassani and S. Scandolo, Phys. Stat. Sol. (b) 173, 263 (1992) and reference therein.

Nonlinear Interference Effect in Ionic Zeeman Laser

S.A.Babin, S.I.Kablukov, M.I.Kondratenko, D.A.Shapiro

Institute of Automation & Electrometry, Russian Academy of Sciences,
1 University Ave, Novosibirsk, 630090, Russia

Abstract. The output of single-frequency argon laser is studied as a function of the axial magnetic field. The narrow dip (100 MHz) is detected at linear polarization against the Doppler background of the width 7 GHz. The effect is interpreted as the nonlinear interference.

Coulomb broadening of saturation resonance had been studied in details [1], while the broadening of two-photon peak was analyzed only in theory and indirect experiment [2]. In present paper we directly study it in the Zeeman laser for the wave of linear polarization on ionic transitions.

Due to the Zeeman effect and selection rules a pair of three-level systems of V-type arise (see figure). Then the narrow upper level makes the two-photon process dominant. The profile of two-photon resonance under Coulomb scattering is calculated in terms of function

$$\mathrm{Re}[Dk^2(\Gamma_m - i\Delta)]^{-\frac{1}{2}} \propto \sqrt{\frac{\Gamma_m + \sqrt{\Gamma_m^2 + \Delta^2}}{\Gamma_m^2 + \Delta^2}}, \qquad (1)$$

where $\Delta = 2\mu_B g_m B/\hbar$ is the magnetic detuning between waves of orthogonal circular polarization, Γ_m and g_m are the upper level relaxation rate and the Lande factor, respectively, μ_B is the Bohr magneton, $\hbar$ is the Plank constant, k is the wavenumber, D is the diffusion coefficient in the velocity space.

The experiments have been performed with a single-frequency Ar$^+$-laser placed in a longitudinal magnetic field. The discharge tube of 5 mm bore is explored at current 100 A and pressure close to optimum for blue-green lines. The measured output power W vs magnetic field B (gauss) shows quite different behaviour for transition $J_m = 1/2 \to J_n = 1/2$ (458 nm) and $3/2 \to 1/2$ (529 nm). In the latter case a narrow resonance (B<50 gauss) appears due to the nonlinear interference effect owing to the interaction of orthogonal circular polarizations $\Delta M = \pm 1$ (see figure). The profile of the resonance corresponds to predictions (1), that include Coulomb scattering. At higher field B>200 gauss a well known magneto-plasma and saturation effect determined by homogeneous width $\Gamma \gg \Gamma_m$ are observed for both transitions [3]. Note that the narrow resonance remains at the multimode regime, but it is not observed for unpolarized light.

The investigated line 529 nm is convenient because it has a negligibly small Einstein coefficient, $A_{mn} \ll \Gamma_m$, that simplifies the theory. Nevertheless, the narrow resonance is also observed for other Ar$^+$-laser lines with $A_{mn} \sim \Gamma_m$ and $J_m = 3/2, 5/2 \to J_n = 1/2, 3/2$.

This work is supported by the Russian Foundation of Basic Research, grant No 96-02-19052, and R&D Programme "Optics and Lasers Physics" of the Russian Ministry of Science.

REFERENCES

[1] Babin, S.A., Shapiro, D.A., *Phys. Rep.* **241**, 119–216 (1994).
[2] Lebedeva, V.V. et al., *Zh. Prikl. Spektr.* **41**, 385-388 (1984).
[3] *Hanle effect and level crossing spectroscopy*, eds. G.Moruzzi and F.Strumia. New York: Plenum Publ. Co, 1990, ch. IV, pp.123–235.

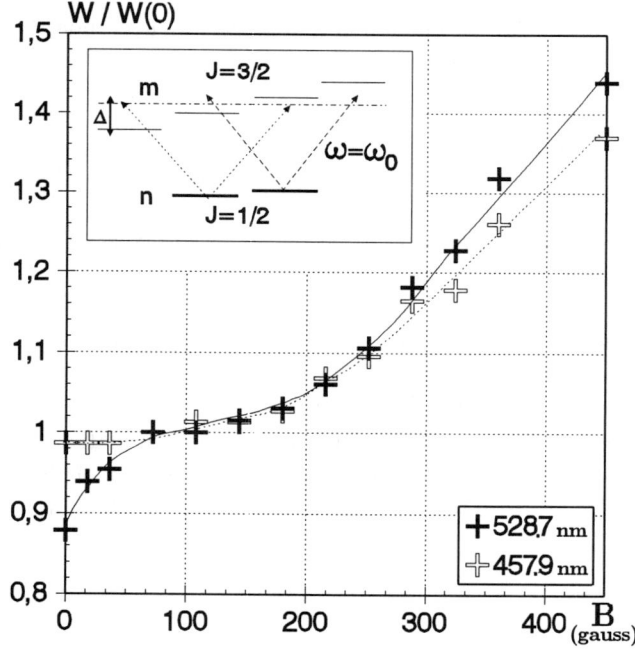

NON INVERTED GAIN LINESHAPES OF THE CESIUM RESONANCE TRANSITION AT 894 nm

F. S. Cataliotti(*)(**), C. Fort(**), M. Prevedelli(*)(**),
T.W. Hänsch(*)(**), M. Inguscio(*)(**)

(*) Department of Physics, University of Florence, L.go E. Fermi 2, 50125 Firenze, Italy
(Fax: ++39-55-224072; E-mail: laserspec@fi.infn.it)
(**) European Laboratory for Non-Linear Spectroscopy LENS, L.go E. Fermi 2, 50125 Firenze, Italy

Abstract: We report on elettromagnetically induced transparency (EIT) in a V-type system in cesium. We investigated the induced EIT as a function of the pump-laser power for different hyperfine components on the D1-D2 lines. Adding repumping light on the D1 transition we observed 2% single pass gain.

Nonlinear effects in three levels systems interacting with two light fields have been the object of many theoretical and experimental studies [1] since 1969 [2].
Recently [3] Hollberg and co-workers have reported laser oscillation without population inversion (LWI) in rubidium atoms starting from the observation of the electromagnetically induced transparency (EIT) in a V-scheme.
With respect to rubidium, cesium offers the advantage of a fully resolved hyperfine structure in the probe transitions also in the Doppler limited recordings. This indeed allowed a systematic investigation of the effect as a function of the hyperfine sublevels involved.
In our experimental setup a strong (20 mW, $\Omega_{rabi}=17\Gamma$) coupling diode laser at 852 nm is locked on a hyperfine component of the D2 line. A weak (less than 80 μW, $\Omega_{rabi}<1.3\Gamma$) probe beam from a second diode laser at 894 nm [4] is superimposed on polarizing cube to the pump beam. In this way the pump and probe beams are copropagating (this allows to observe almost Doppler free nonlinear structures) with orthogonal polarizations. The probe absorption in a cesium cell is then recorded while the laser is scanned through the D1 line. A third laser diode (up to 5 mW) at 894 can be used as a repumper to increase transparency and eventually achieve gain. The schematic of the experiment and the Cs energy levels involved are shown in Figure 1.

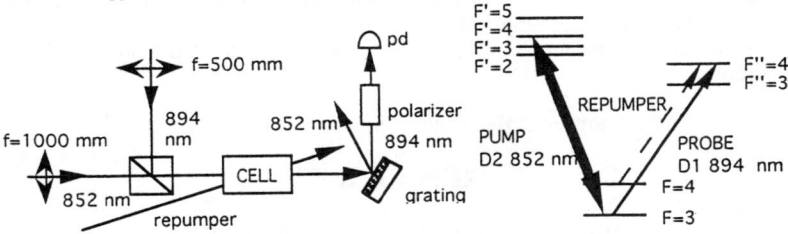

FIGURE 1. Experimental setup and level scheme of the involved transitions.

In presence of the pump laser the gaussian absorption of the probe shows a subDoppler transparency dip whose position depends on the pump detuning. Transparency increases with the pump power: the maximum value depends on the involved hyperfine transitions. The best results (96%) are obtained for the pump laser locked on the F=3 -> F'=4 hyperfine component of the D2 line and the probe laser scanning the F=3 -> F"=3,4 component of the D1 line. Such a result indicates that, to fully understand the phenomenon, it is necessary to take into account the Zeeman structure of the levels.

To observe gain we added repumping light coupling the upper level of the probe transition to the uncoupled hyperfine level of the ground state to alter the levels populations [5]. This also make the transparency feature narrow and increases probe absorption in the rest of the gaussian profile. In order to further reduce optical pumping we added a small magnetic field perpendicular to the repumper polarization.

Figure 2 shows a typical data set for such optimized conditions.

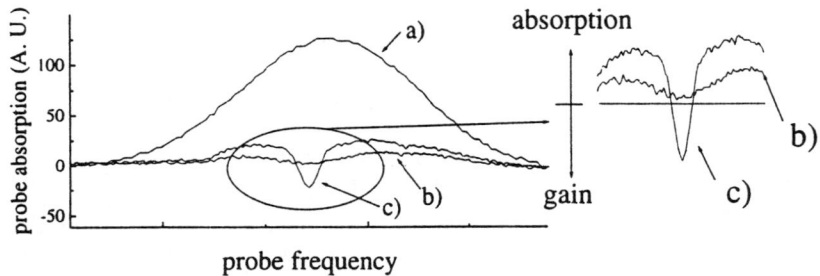

FIGURE 2. Typical absorption spectra on the D1 line (F=3 -> F"=4,3): (a) simple absorption, (b) EIT effect without the repumper, (c) gain when the repumper is present.

We performed a systematic investigation of the transparency (gain) as a function of pump probe and repumper power. We also checked the effect of cesium pressure, of the pump/probe polarization angle and of the different. hyperfine components. Gain was not sensitive to polarization and angle of incidence of the repumping beam.

A maximum gain of 2% was observed on the F=3-> F"=3 D1 component when the pump is locked on the F=3 -> F'=4 D2 line in a 4 cm long cell kept at a temperature of 40° C. Probe and pump laser polarizations were perpendicular. Gain saturated above a pump power of 15 mW.

Work is in progress for the construction of a cavity that could allow CW laser operation.

We gratefully acknowledge the loan of a 894nm laser diode from A. Weiss.

REFERENCES

[1] D. J. Futon et al. Phys Rev A **52**, 2302 (1995) and references therein.
[2] T.W. Hänsch et al. Z. Phys. **226**, 293 (1969)
[3] A.S. Zibrov et al. Phys. Rev. Lett. **75**, 1499 (1995)
[4] S. B. Ross et al. Opt. Comm. **120**, 156 (1995)
[5] M. Fleischhauer et al. Phys. Rev. A **46**, 1468 (1992).

Quantitative Analysis of Second Derivative Absorption Lineshapes in Difficult Environments: Detection of Molecular Oxygen using a DFB–Laser at 761 nm

C. Corsi,* M. Gabrysch**, M. Prevedelli***, and M. Rosa-Clot***

European Laboratory of Non-Linear Spectroscopy (LENS), Largo E. Fermi 2,
50125 Florence, Italy

* Dipartimento Scienze Neurologiche, Università di Firenze, Viale Pierracini, 50139 Florence, Italy
** Universität Heidelberg, Physikalisches Institut, Philosophenweg 12, 69120 Heidelberg, Germany
*** Dipartimento di Fisica dell'Università di Firenze, Largo E. Fermi 2, 50125 Florence, Italy

Abstract

The optical detection of molecules using semi–conductor diode lasers offers many advantages with respect to other techniques. In particular, the intrinsic amplitude stability of these lasers offers the easy performance of high sensitivity detection methods like for example wavelength modulation spectroscopy. A careful analysis of the detected second-derivative lineshapes allows to extract parameters like the concentration or the temperature of the examined gas, which is of great importance to industrial and environmental applications.

Low-wavelength modulation-(LWM), frequency-modulation (FM)- and two-tone (TTFM) spectroscopy applied to semi-conductor diode lasers provide a powerful tool to detect small concentrations of molecules [1]. In these methods

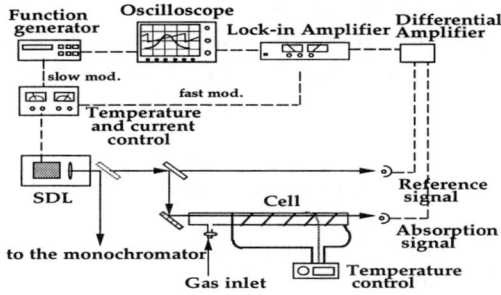

Figure 1: Experimental setup for LWM.

the laser current is modulated at a frequency ν_{mod} and the detected laser light is demodulated at a multiple n of the modulation frequency (see Fig.1). As the laser wavelength is scanned over the absorption profile one detects for $n=2$ the second derivative of the in general Doppler- and collision-broadened line [2] resulting in significant gain of the signal to noise ratio [3].

To extract useful information like for example the amplitude and the linewidth

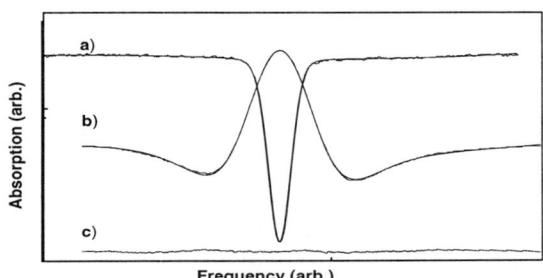

Figure 2: Typical absorption signals of molecular oxygen at 761nm: a) Direct absorption profile with fitted Voigt function. b) Second derivative with fitted function. c) Difference between experimental data and fit-function.

from the recorded data a good understanding of the 'second derivative' lineshapes is necessary. For this reason several fitting-procedures including Fast-Fourier-Transform (FFT) were developed, which allow to calculate the Lorentzian width γ_L and the Gaussian width σ_G of the second derivative signal.

To test the fitting procedures laboratory and in situ measurements in a combustion chamber using a Distributed-Feedback-Laser at 761nm were performed. Typical absorption lines of the electric dipole forbidden $b^1\Sigma_g^+(\nu' = 0) \leftarrow X^3\Sigma_g^-(\nu'' = 0)$–band of O_2 are shown in Fig. 2. The temperature of the gas could be changed in a range of 20ºC to 1000ºC. The data analysis showed a different behavior of the linewidths for different temperature regions. For example, in the region of between 20ºC and 300ºC the temperature of the gas sample could be measured with a accuracy of 25ºC. In the region above 350ºC the data analysis is more difficult since the theoretical assumed absorption profile, which is based on a Voigt function, seems to be no more valid and has to be replaced with an extended theoretical model which will be tested in further measurements.

REFERENCES

[1] see for instance, M. Inguscio in: Frontiers in Laser Spectroscopy (p.41), Elsevier Amsterdam, North–Holland (1994)
[2] C. Corsi, M. Gabrysch and M. Inguscio, Opt. Comm., in press
[3] F.S. Pavone and M. Inguscio, Appl.Phys.B **56**, 118 (1993)

DIODE LASER SPECTROSCOPY IN A MAGNETRON DISCHARGE

C. Csambal and V. Helbig

Universität Kiel, Institut für Experimentalphysik, D 24098 Kiel, Germany

Diode laser spectroscopy is a powerful and versatile tool for plasma diagnostics. The availability of tunable, narrow-bandwidth laser diodes that can be completely controlled by a computer enables one to use them in classical absorption experiments as well as in set-ups for Doppler-free spectroscopy.

Free-runnning laser diodes are characterized by two features that limit their use for spectroscopic purposes. Without further measures their spectral bandwidth is of the order of 100 MHz and the tuning range of the wavelength is interrupted by mode-hops caused by the difference of the temperature coefficient of the gain profile and the index of refraction of the active medium. There has been a number of suggestions to improve the behaviour of the diodes in this respect. Most successful have been the experiments using an external Fabry-Perot cavity or a grating in Littrow-mount for stabilization. Band-widths of the order of 100 kHz and continuous tuning ranges of up to 70 nm are reported. We used for our stabilization the Littrow set-up with a holographic grating with 1800 l/mm. We operated the device in the external cavity mode, e.g. we did not antireflex-coat the front face of the diode. This has the advantage of simpler operation, however, it will not suppress the mode-hops. For our purpose the obtained band-width of 4.5 MHz and the tuning range of approximately 25 GHz was completely sufficient.

Absorption spectroscopy is a sensitive method to obtain number densities and temperatures in gas discharge research. This technique especially allows access to the first four excited ns levels of the rare gases that are important for the discharge mechanism. Two of those are connected to the ground state and give rise to the resonance transitions in the VUV-region that are difficult to observe in emission. The other two levels are metastable and the only way to measure their population is the use of absorption techniques. Because of the large population density of the metastable states the absorption lines usually are influenced by optical thickness. This requires special care when reducing the data to obtain the desired number densities. In the measurements reported in this paper the plasma of a cylindrical magnetron discharge was investigated. In this case an additional difficulty in the interpretation of the spectra was caused by the Zeeman splitting of the lines originating from the magnetic field applied to the discharge.

As we observe the plasma along the magnetic field lines we deal with the longitudinal Zeeman effect. It causes a splitting of the line in σ^+- and σ^-- components. Depending on the J-values of the levels involved each σ-line consists of a number of components. Due to Doppler-broadening the different components overlap and the resulting sum profile exhibits only two peaks. A least-squares fitting routine was used to fit the expected pattern to the recorded profile. The splitting and the relative intensities of the components were taken from the theory. The magnetic field strength was taken from the current of the coils that has been calibrated by means of a Hall probe. Fitting parameters were the Doppler width, the intensity of one component and the dispersion. The latter could also be deduced from the transmission peaks of a Fabry-Perot interferometer which were recorded simultaneously in order to check the results from the fitting procedure.

Due to the Zeeman splitting the description of the absorption has to be modified compared to the case without magnetic field. We probe the discharge with linearly polarized light that can be described as consisting of left-hand and right-hand circularly polarized components. Therefore the absorption for the σ^+- and σ^-- components has to be treated separately. We assume both circularly polarized beams do have the same intensity. Then each σ- component will at most absorb half of the intensity of the incoming beam. Only in the very center where the two Zeeman subgroups overlap partially one will observe an absorption larger than $I_0/2$. This leads in the case of moderate magnetic fields to a line shape that does not exhibit the expected dip but that is peaked instead. To verify the given interpretation of the experimental line profiles and to show that there are no central π- components which too could have caused the observed central peak we recorded Doppler-free spectra of the lines that indead resolved the different σ- components and revealed no π- components. The observed line profiles for different lines of the 1s - 2p transition array of argon will be discussed in the poster.

SPECTROSCOPIC INVESTIGATION OF THE RUBIDIUM RESONANCE LINE

T. Rieper, T. Rose and V. Helbig

Universität Kiel, Institut für Experimentalphysik, D 24098 Kiel, Germany

D. Veza

University of Zagreb, Institute of Physics, POB 304, 4100 Zagreb, Croatia

Diode laser spectroscopy has been used to investigate the pressure broadening and the saturation behavior of the rubidium resonance line at 780 nm. A grating stabilized diode laser in Littrow mount was used to obtain absorption profiles as well as Doppler-free spectra where the hyperfine structure of the rubidium D_2 line could be resolved.

We used an open cell of 15 cm length connected to an ultra-high vacuum system. The purity of the rubidium was better than 99.5%. The rubidium vapor pressure at room temperature was sufficient to obtain a good signal to noise ratio for the absorption profiles as well as for the saturation spectra. For the investigation of the foreign gas broadening by argon the pressure was adjusted by means of a high precision needle valve.

For the absorption measurements the pressure and the laser power were varied by seven and five orders of magnitude, respectively. The absorption spectra exhibit four lines that are separated completely at low pressure. Always two of them result from the 85 and 87 rubidium isotope, respectively. Each of the four lines is composed of three hyperfinestructure lines resulting from the splitting of the upper level. Using Lambert Beer's law we converted the absorption profiles to the spectrum of the absorption coefficient. For the data reduction we used a least-squares computer routine that fitted to two lines at a time a model function that contained six Voigt profiles. The wavelength separation and the relative intensities of the hyperfinestructure components as well as the Gaussian widths were taken from theory. From plots of the absorption coefficient versus laser power and argon pressure one obtains saturation intensity [1] and saturation pressure. Plotting the shift and the Lorentzian width of the lines versus pressure one obtains the broadening and the shift parameters. Since no other experimental data for the resolved lines seem to exist in literature we took results for the unresolved D_2 line for comparison [2, 3]. The absolute values for the broadening and the shift as well as the sign of the shift compare well with our results.

For extrem values of the pressure and/or the incident laser intensity the absorption profiles drastically change their shape. They become asymmetric and show a dip in the line center. We set up a simple model to explain the strange behavior qualitatively. The basic idea is that since the oscillator strength is distributed differently across the line profile complete saturation occurs first in the line center. In the model this is taken into account by the introduction of a weighting function of Gaussian shape.

The Doppler-free spectra obtained using saturation spectroscopy exhibit in addition to the resolved hyperfine structure lines cross-over resonances. This yields six components for each of the four lines instead of three. It therefore is difficult to measure shift and width caused by the foreign gas atoms. By means of a least-squares fitting routine the desired line parameters were determined from observed profiles.

The fit only gives accurate results as long as the rubidium number density is not too high to guarantee that the lines are opically thin and the argon pressure is low enough so that the six components are reasonably well resolved. However, pressure broadening only becomes the dominating broadening mechanism beyond the mentioned argon pressure. This is the reason that we have obtained only preliminary results for the pressure broadening and shift of the single hyperfinestructure lines so far.

[1] K. Shimoda, "High Resolution Laser Spectroscopy", Springer, Berlin, Heidelberg, New York (1976)
[2] Ch. Ottinger, R. Scheps, G.W. York and A. Gallagher, Phys. Rev. A **11**(6), 1815 (1975)
[3] E.L. Lewis, Phys. Reports **58**, 3 (1980)

Shape of D line of sodium by single-mode laser excitation

J.H. Xu- - Scuola Normale Superiore - Pisa
R.M. Celli- Dipartimento di Fisica- Pisa

Experiments on optical pumping and coherent trapping of sodium atoms in selected hyperfine ground states have been performed in cells that have been coated with a polydimethylsiloxane polymer (PDMS). The polymer coated cell, when illuminated by light, shows a strong photodesorption of sodium atoms from the surface producing a gas phase density of 10^{11}-10^{12} atoms/cc at room temperature. Moreover due to residual vapour pressure of the PDMS polymer and the excellent non disorienting properties of the coated walls, very long spin relaxation times can be achieved. When examining in such type of coated cell the sodium atomic fluorescence by sweeping a single mode dye laser, with linear polarization, instead of the two Doppler broadened hyperfine lines with F=1 and F=2, (fig. 1), a single narrow fluorescence line appears as soon as the laser power is greater than a few milliwatts, (fig. 2).

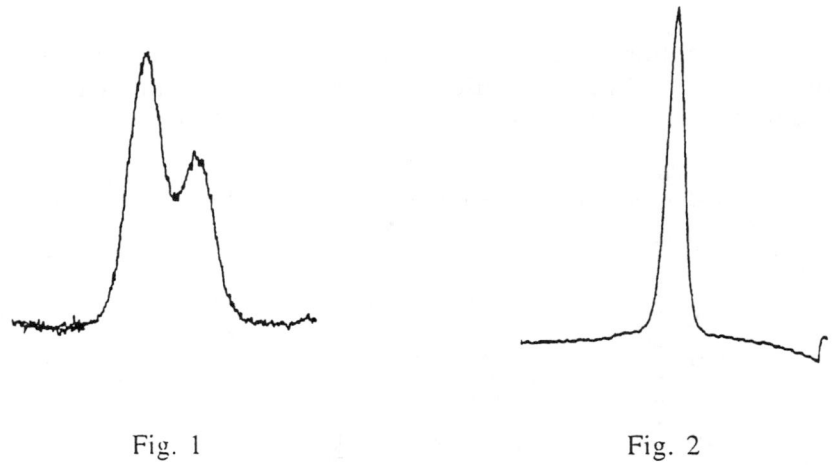

Fig. 1   Fig. 2

Experimental observation of D_1-line fluorescence at 3 mW/cm^2 (fig. 1) and 300 mW/cm^2 (fig. 2)

If N_1 is the population of level F=1 and P_1 a gaussian-type velocity distribution along the direction of the laser beam, in the hypothesis that a small fraction of the laser intensity is absorbed inside the cell, the rate equations for the populations N_1 and N_2, can be written:

$$\frac{dN_1}{dt} = -a N_1 P_1 I + b N_1 P_1 I + c N_2 P_2 I + d (N_{10}-N_1)$$

for the level F=1 and, by interchanging 1 with 2, the analogous equation for the level F=2. The first term, with coefficient a, accounts for absorption probability to P state. Stimulated emission and decay back to the same level are included in the second term; c accounts for the decay from P state to the F=1 level due to excitation of the F=2 level, N_{10} being the population of F=1 state at thermal equilibrium and d for thermal relaxation between the two ground levels. If the relaxation term can be disregarded due to the long relaxation times in PDMS coated cells, at equilibrium we should have: $N_1 P_1 = N_2 P_2$. When this is verified, the fluorescence F, will be proportional to: $I (N_1 P_1 + N_2 P_2)$.
As $N = N_1 + N_2$, N being the total population of S ground state, it follows:

$$F \propto I N \frac{P_1 P_2}{P_1+P_2}$$

The plot of this function together with the plot of the absorption profile of D1 line of sodium, when the laser intensity is weak enough to prevent hyperfine optical pumping, is shown in fig. 3.

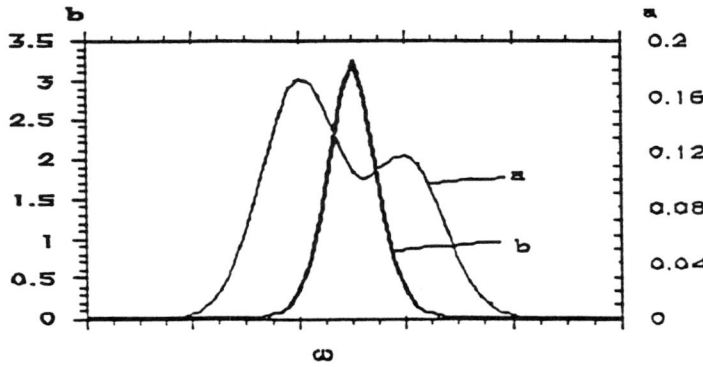

Fig. 3

Theoretical curves for low (a) and high (b) laser power

Incomplete Optical Shielding In Cold Sodium Atom Traps

Vladimir Yurovsky and Abraham Ben-Reuven

School of Chemistry, Tel Aviv University, 69978 Tel Aviv, Israel

Abstract. A simple two-channel model, based on the semiclassical Landau-Zener (LZ) approximation, with averaging over angle-dependent exponents, is proposed as a fast means for accounting for the incomplete optical shielding of collisions, as observed in recent experiments conducted by Weiner and co-workers on ultracold sodium-atom traps, and its dependence on the laser polarization. The model yields a reasonably good agreement with the recent quantum close-coupling calculations of Julienne and co-workers. The remaining discrepancy between both theories and the data is qualitatively attributed to a partial overlap of the collision ranges at which loss processes and optical shielding occur.

Collisions in cold atom traps are a major source of instability of the trap. Cooling of the atoms is usually attained with a laser shifted slightly to the red of the atomic cooling transition ($S_{1/2} \to P_{3/2}$ in the alkali-metal atoms). However, the excited atoms undergo during collisions processes (such as inelastic fine-structure transitions, resonance fluorescence, or associative ionization) that end up in an excess of translational energy leading to trap escape. These processes usually occur at atomic separations smaller than the Franck-Condon transition point associated with the red-shifted laser. As the ground-level collision potential is almost flat at this range, the level crossing occurs to one of the binding potentials in the singly-excited level of the collision pair.

A possible cure to this problem is the application of another laser, detuned to the blue, causing level crossing to one of the repulsive potentials. A proper choice of the laser frequency detuning would lead to level crossing at atomic separations outside the range at which trap loss is initiated. As the blue-shifted laser intensity is increased, avoided crossing will then cause a repulsion of the colliding atoms, preventing penetration into the range of trap loss.

The measure of shielding can be obtained by monitoring the associative ionization current (as an indicator of penetration effectiveness), as a function of the shielding-laser intensity. The penetration effectiveness should gradually decrease on increasing the shielding-laser intensity, eventually leading to complete shielding.

Recent measurements carried out on Na atom traps (1), and others, reveal a more complex situation, in which the most prominent observation is the lack of complete shielding at large laser intensities. We shall limit the discussion here to the Na results. The collision dynamics in Na is far from being simple. The multiplicity of molecular terms in the singly-excited electronic state, the presence of hyperfine structure, partial-wave centrifugal forces, and effects of polarization in real (3-dimensional) space, all contribute their share. A polarization effect is clearly observed in the experiments, producing different results when one uses a laser beam with circular polarization (CP), rather than one with linear polarization (LP).

An attempt to account for some of these complications was made by Julienne and coworkers (1,2). They use an optical collision model based on the quantum close-coupling method, taking both polarization and partial-wave effects into account (including the phenomenon of forbidden partial waves in the ground electronic state, and the associated angular-momentum selection rules). However, they consider only one electronic transition, and no hyperfine structure. The results of their calculations clearly show a tendency towards a plateau in plotting the penetration effectiveness as a function of the shielding laser intensity, especially with LP. However, the resulting plots lie appreciably below the experimental points showing a much poorer shielding. Furthermore, the experimental data indicate a secondary "revival" of shielding at higher laser intensities, most prominent with CP.

In an attempt to account for the discrepancy between close-coupling theory and experiment, more complex and costly calculations, involving many more collision channels, would be necessary, perhaps without enough prior knowledge of the relative importance of the various added features. It would therefore be very helpful to have a simple and fast way of estimating the outcome of adding various effects into the collision picture, before embarking on more complex calculations. It is suggested here to use the old semiclassical LZ method, extended in a simple way to multichannel problems in 3-dimensional space, as a fast estimation method.

Objection to the application of the LZ method to cold-atom collisions can be raised on three grounds:

(a) This method is valid only when the kinetic energy at the crossing point is "large" enough, compared to the submillikelvin energies of ultracold collisions.

(b) It can be possibly extended to the case of multichannel multiple crossings only if the crossing points lie sufficiently "far" apart.

(c) The extension to three dimensions is complicated by the nuclear rotational motion, so that a 1-dimensional LZ amplitude cannot be assigned at each point.

The answer to all these objections lies in the long-range nature of the processes involved (loss and shielding), usually occuring beyond 100 bohr separations, where the ground-level potential is practically *flat*. It follows from the recent studies of Nakamura and Zhu (3), extending the applicability of semiclassical method, that (a) the range of "forbidden" energies shrinks, and the traditional (unmodified) LZ exponents are valid, when one of the potentials is flat. Also, (b) the multichannel S matrix can be expressed as a product of matrices of single-crossing amplitudes, even for small separations. Furthermore, (c) the rotation of the nuclear axis at the long range concerned is sufficiently slow; hence the angle θ between the nuclear axis and the polarization direction can be considered as fixed, although its value varies arbitrarily from one collision to another. One can then use a LZ calculation, in which the radiative coupling strength depends on θ, ultimately averaging over an isotropic angular distribution.

This approach can be best tested by comparing it with extant quantum close-coupling models. We therefore applied it to the same case studied in Refs. 1 and 2. Best agreement is obtained by treating the translational motion as purely classical (i.e., ignoring partial-wave structure). A 2-channel model was thus used, in which polarization effects were incorporated by expressing the shielding ineffectiveness as a LZ exponential, with an angle-dependent exponent,

$$P_{AV} = \langle \exp[-2\pi\Lambda_0 F(\theta)] \rangle,$$

with $F(\theta) = \sin^2\theta$ for LP, and $F(\theta) = (1+\cos^2\theta)/2$ for CP.

Here Λ_0 is the ordinary (angle-independent) LZ exponent for the relevant molecular term (twice the atomic value in the present case of a $\Sigma \to \Pi$ transition), and the angular brackets denote the averaging over an isotropic distribution of the angle θ. The sine function is used with LP since the transition dipole to the Π state is perpendicular to the nuclear axis. All relevant parameters are as in Ref. 2.

The results for both polarizations are shown in **Fig. 1** (dash-dot curves). These are compared with the experimental results (squares), and with the close-coupling calculations (dashed curves). Also shown are the angle-independent (one-dimensional) LZ exponentials (dotted lines).

The qualitative agreement obtained between this simple model and the close-coupling calculations, especially at lower laser intensities, demonstrates the potential of using semiclassical methods as a means for obtaining fast estimations even for cold collisions. Clearly an explanation of the discrepacy between the data and both theoretical approaches should be sought elsewhere. We further suggest here that, owing to the many channels involved in the real situation, both processes (shielding and loss) occur over finite ranges, that may partially overlap. This may be qualitatively expressed by replacing the expression for P_{AV} above with

$$P = (1-X)P_{AV} + X$$

for the shielding ineffectiveness, with X representing the fraction of loss pathways lying outside the shielding range. The best fit to the data, obtained with $X = 0.4$, is also shown in **Fig. 1** (solid curves).

The physical origin of a major difference, obtained with both theories, between the effects of the two polarizations – namely, the relatively more efficient shielding of CP, compared to that of LP – is clearly understood by the present model. Only with LP there is an angle ($\theta = 0$) at which shielding is completely absent; no such angle exists with CP. The "revival" of shielding, observed in the experimental data for CP at higher intensities, remains unaccounted for with both theories.

ACKNOWLEDGMENTS

The authors are grateful to P. S. Julienne, H. Nakamura, J. Weiner, and C. Zhu, for helpful discussions and prepublication results.

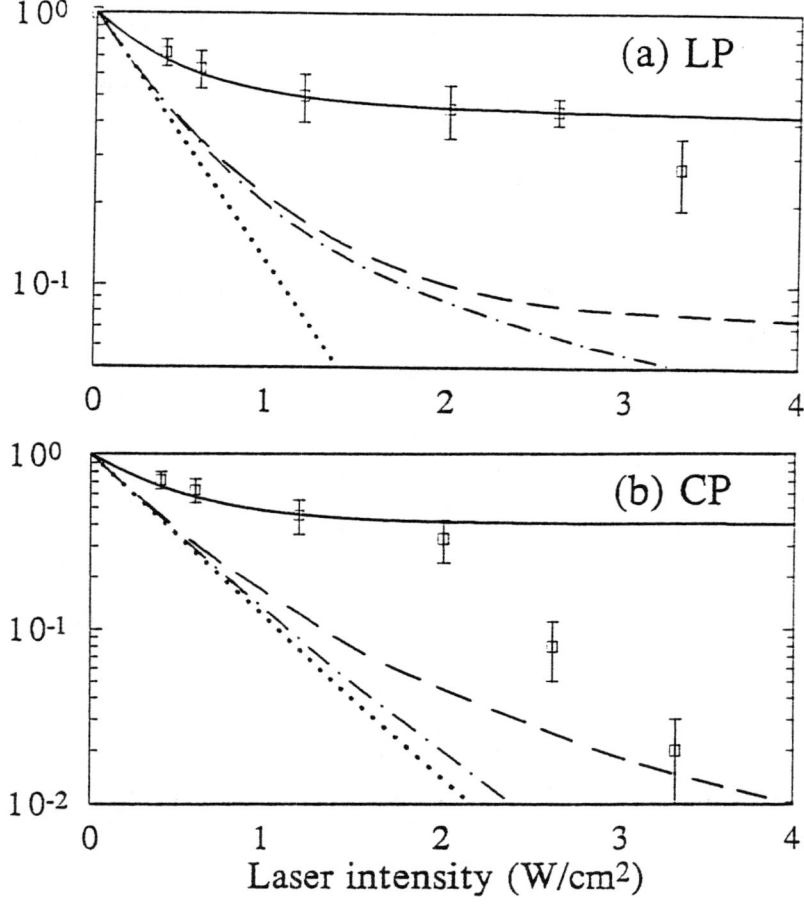

FIGURE 1. Shielding ineffectiveness in cold Na, as a function of shielding-laser intensity, with (a) linear, and (b) circular polarization. See text for further explanations.

REFERENCES

1. S. C. Zilio, L. Marcassa, S. Muniz, R. Horowicz, V. Bagnato, R. Napolitano, J. Weiner, and P. S. Julienne. *Phys. Rev. Lett.* **76**, 2033 (1996).

2. R. Napolitano, J. Weiner, and P. S. Julienne, *Phys. Rev. A* (in press).

3. C. Zhu. and H. Nakamura, *J. Chem. Phys.* **101**, 10630 (1994); H. Nakamura and C. Zhu, *Comments At. Mol. Phys.* **32**, 249 (1996).

Structured Continua of the Intermediate Long-range Cs₂ Molecules

Damir Veža, Robert Beuc, Slobodan Milošević and Goran Pichler

Institute of Physics, P.O.Box 304, HR-10000 Zagreb, Croatia

Abstract: New development in cold collisions has revived the interest in the process of photoassociation, because the free-bound transitions are very selective at extremely low temperatures. Some molecular systems where photoassociation has been observed in the past may now be studied in much greater detail than ever before. The photoassociation from the weakly bound Cs_2 molecules at elevated and ultra cold temperatures will form peculiar molecules in their excited states in the intermediate internuclear distances.

Alkali dimers with large spin-orbit interaction and their potential curves at intermediate long-range distances present a very interesting case where the influence of the resonance interaction at long range region is still large, the short-range forces are not predominant and the spin-orbit interaction is certainly non negligible. At these ranges the resonance interaction becomes smaller than the short-range interaction forces. Some of the potential curves undergo an avoided crossing which may form bound potential curves in the region of about 15 a_0. Thus, the 0_g^+(np $^2P_{3/2}$ + ns $^2S_{1/2}$) electronic state appears to become a double minimum potential in the case of K_2 and Rb_2. In the case of K_2 both minima lie above the np $^2P_{3/2}$ + ns $^2S_{1/2}$ asymptote. The outer minimum in the case of Rb_2 may have many bound states, which were never observed before. These double minimum potentials retain double minimum shape even when making the difference potential with the ground electronic states stemming from the ns $^2S_{1/2}$ + ns $^2S_{1/2}$ asymptote. This means that to each extrema in the difference potential curve one satellite band may be observed in the very far quasi-static wings of the self-broadened first resonance lines of K, Rb and Cs.

All above mentioned electronic states of Cs_2 molecules may serve as intermediate states for the two-photon ionisation and/or further excitation into some Rydberg states of alkali dimers and intermetallic excimers. Cold atom collisions open up completely new field of spectroscopic investigations in which details of the molecular structure could be seen with high accuracy.

Here we present an absorption study of the structured continua appearing in the far wings of the caesium resonance lines. Figure 1 presents a typical result of our measurements, together with the assignment of these structured continua (1). Using the available results for the Cs_2 potential energy curves (2), we calculated the contribution to the total absorption profile from the most peculiar potential curve, $^3\Pi(0_g^+)$. This contribution explains very well some fine details in the measured absorption profile (3).

REFERENCES:

1. Pichler, G., Movre, M., Veza, D., and Niemax, K., Proceedings of the 33rd Symp. on Molec. Spectroscopy, p. TF11, (Columbus, Ohio, USA) 1978.
2. Spiess, N., PhD thesis, Universität Kaiserslautern (Fachbereich Chemie), 1989; S. Magnier, These, Universite Paris-Sud, Centre d'Orsay, Paris, 1993.
3. Veza, D., Beuc, R., Milosevic, S., and Pichler, G., prepared for publication.

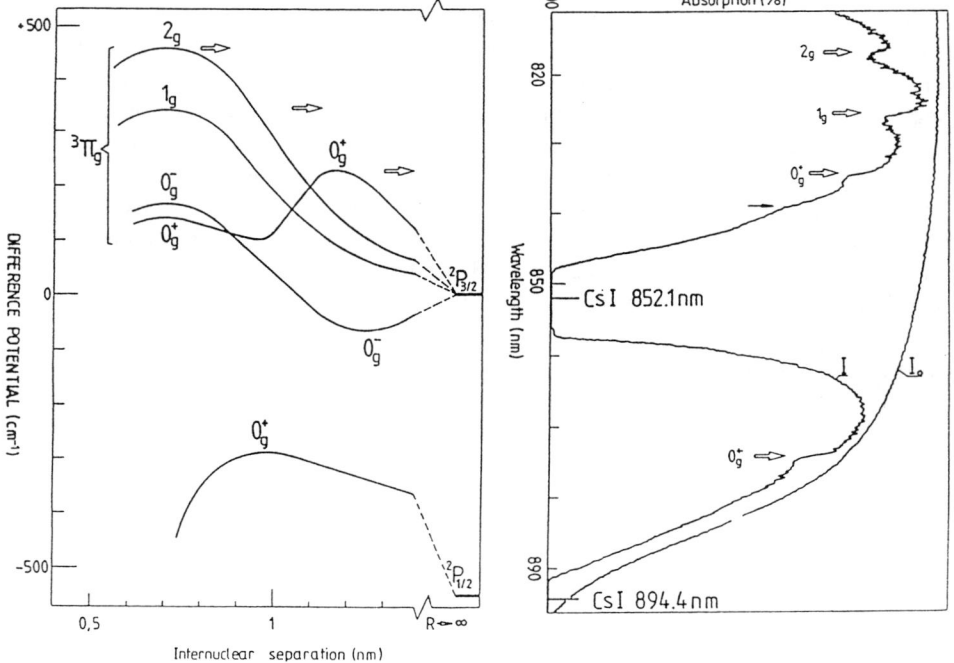

Two step resonances in a Rb magneto-optical trap

C.Gabbanini, A.Evangelista, S.Gozzini, A.Lucchesini

Istituto di Fisica Atomica e Molecolare-C.N.R.,Via del Giardino 7, 56127 Pisa

Abstract. We show two color spectroscopic measurements that we have done on cold rubidium atoms confined in a magneto-optic trap. The measurements have been done on both Rb isotopes by exciting the $8s_{1/2}$, $5d_{3/2}$ and $5d_{5/2}$ states from the trapping $5p_{3/2}$ level. The obtained lineshapes show the presence of the Autler-Townes splitting induced by the strong trapping laser. The hyperfine constants of the 5d levels have been derived.

KEYWORDS: atomic trap, spectroscopy, hyperfine structure

The development of vapor cell traps[1] for alkali atoms has allowed to collect dense atomic clouds near or even below the Doppler limit temperature with semplified experimental setups. The spectroscopic resolution attainable in these cold atomic samples is higher than in thermal atoms because first and second order Doppler effects are eliminated. So a double optical resonance, that can investigate the structure of high energy bound states having the same parity of the ground state, is favored by the use of cold atoms.

We have performed spectroscopic measurements on two step resonances in a Rb magneto-optical trap. The rubidium atoms are trapped in a σ^+-σ^- magneto-optic trap realized in a pyrex cell with quartz windows pumped by an ion pump. The trapping laser is a diode laser injected by an extended cavity diode laser; it is frequency locked to the red side of the F=3-F'=4 D_2 transition for ^{85}Rb (F=2-F'=3 for ^{87}Rb). The laser beam, having a circular shape with a diameter of 8 mm, is divided in three arms and retroreflected. On two arms a free running diode laser is superposed to the trapping laser to avoid optical pumping by exciting the other hyperfine level of the ground state. Two coils in the anti-Helmoltz configuration give the magnetic field gradient. A cloud of more than 10^7 atoms is trapped at the center of the beam crossing region in a volume of a fraction of mm^3.

The excitation pathway of the cold atoms is composed by the $5s_{1/2}$-$5p_{3/2}$ transition at 780 nm, induced by the trapping laser, followed by the excitation to the $8s_{1/2}$ state by a c.w. ring dye laser at 616 nm or to one of the $5d_{3/2}$ and $5d_{5/2}$ states by an extended cavity diode laser operating at 776 nm. In both cases we detect the $6p_{3/2}$-$5s_{1/2}$ cascade fluorescence at 420 nm using a photomultiplier faced by an interference filter. Lasers giving the second excitation step photons are swept on the resonance; the frequency sweep is monitored by a Fabry-Perot interferometer and recorded together with the fluorescence.

In atomic traps the AC electric field created by the strong trapping laser splits the transition lines in two components; this is known as the Autler-Townes splitting[2]. The fluorescence line shapes in fact show this effect; in our experiment the splittings correspond to about 16 MHz as shown in Figure 1. The hyperfine structure of the 5d levels is well resolved, so the values of the magnetic dipole constant a and that of the electric quadrupole constant b have been derived; they are shown in Table I. The a values are derived with more than 90 percent accuracy while for b just the order of magnitude can be extracted; these results are consistent with the more precise results obtained by Nez and al[3] by two photon spectroscopy.

The results have been compared with those obtained exciting thermal atoms in a separate cell where the two excitation step lasers counterpropagate, minimizing the Doppler effect. The resolution is not considerably different when the 5d levels are excited because the two step wavelengths are very close, resulting in a small residual Doppler width when thermal atoms are used. On the contrary, in the case of $8s_{1/2}$ excitation, the two color wavelengths are far and the advantage in spectroscopic resolution of cold atoms over thermal atoms is evident.

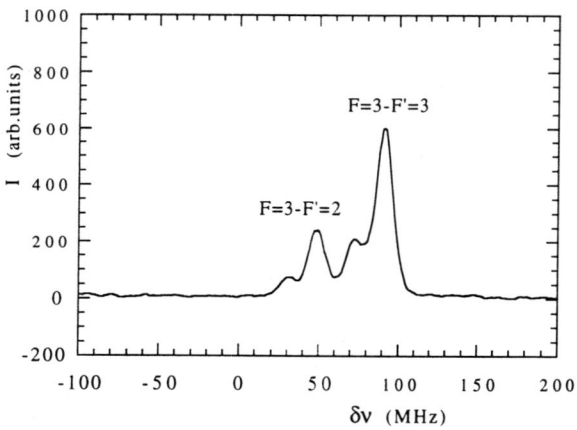

Figure 1: line shape of the $5p_{3/2}$-$5d_{3/2}$ transition of ^{87}Rb

TABLE 1: a and b values derived from this experiment

Level	Rb isot.	a (MHz)	b (MHz)
$5d_{3/2}$	85	4.43±0.28	1.7± 2.4
$5d_{3/2}$	87	14.65±0.30	0.8± 0.8
$5d_{5/2}$	85	-2.31±0.23	2.7± 2.7
$5d_{5/2}$	87	-7.27±0.28	0.9± 2.4

REFERENCES

1) C.Monroe, W.Swann, H.Robinson and C.Wieman, Phys.Rev.Lett. **65**, 1571 (1990)
2) S.H.Autler and C.H.Townes, Phys.Rev. **100**, 703 (1955)
3) F.Nez, F.Biraben, R.Felder and Y.Millerioux, Opt.Commun. **102**, 432 (1993); erratum Opt.Commun. **110**, 731 (1994)

Investigation of Radiation Trapping in a Cs Magneto-Optical Trap

A. Fioretti, J.H. Müller, M. Colla, A. Molisch* and M. Allegrini+

Unità dell'Istituto Nazionale per la Fisica della Materia,
Dipartimento di Fisica, Università di Pisa, Piazza Torricelli 2, I-56126 Pisa, Italy

Abstract. We have investigated the radiation trapping phenomenon in a standard Magneto-Optical Trap (MOT) for Cesium atoms. These measurements, obtained by monitoring the radiative decay of the excited trapped atoms, allowed an independent check for the density and the number of cold atoms in the MOT.

In a MOT, it is possible to achieve densities and total number of cold atoms such that the optical thickness of the medium is relevant. Cold and trapped atoms, interacting through radiation trapping, experience repulsive and heating forces that strongly determine the behavior of the MOT [1-4].

In the present experiment we monitor directly the effective radiative lifetime of an optically thick Cs MOT. The lengthening of the natural radiative lifetime of the $6^2P_{3/2}$ level due to radiation trapping has been measured for different MOT operating conditions.

The experiment starts by loading a standard MOT with some 10^8 Cs atoms with a peak density in the range 1-3 10^{11} cm^{-3} [5]. After the trapping laser is switched off, we excite the trapped atoms with a short pulse of a weak probe laser, frequency locked to the (F=4, F'=5) hyperfine component of the Cs D_2 line.

The probe operation starts after a short delay (~ 80 µs) from the switch-off of the cooling laser and has a duration short enough to neglect modifications of the MOT density distribution due to temperature and gravity. This short pulse (500 ns time duration and 30 ns extinction time) is generated from the cw probe diode laser by a fast electro-optic modulator. The resonant photons emitted by the trap are collected and detected by a fast fotomultiplier. The effective lifetime is obtained by an exponential fit of the temporal decay data of the fluorescence light.

The experimental result is an effective lengthening of the D_2 radiative lifetime (up to 3 times the natural one) that strongly depends on the trap parameters. A MonteCarlo (MC) simulation of the process as well as an analytical approach [6] have been used to interpret the observed dependence on the trap

---

* Institut für Nachrichtentechnik und Hochfrequenztechnik, Technische Universität Wien, Gusshausstrasse 25/389, A-1040 Wien, Austria
+ Also at Dipartimento di Fisica della Materia e Tecnologie Fisiche Avanzate, Università di Messina, Salita Sperone 31, I-98166 Sant'Agata, Italy

parameters. The MC approach simulates photons diffusion in a cloud of ground state atoms at rest, with partial frequency redistribution of emitted photons. The analitical approach gives a solution of the Holstein equation under some simplifying assumptions, i.e. Doppler width much less than natural one, not too large opacity, ellipsoydal geometry and uniform distribution of absorbers.

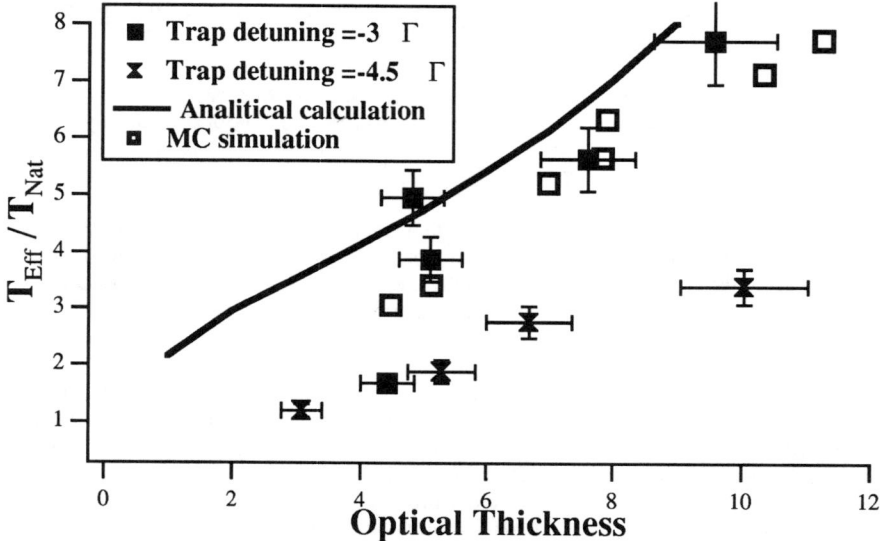

In the Figure experimental and theoretical results are shown. One set of data is well reproduced by both the theoretical results, while another set of data is off by a factor two. This can be explained by taking into account shape modifications of the trap while changing experimental parameters, i.e. detuning of the trapping lasers. Also our hypothesis of uniform excitation of the atoms is a very strong one, which is not always fulfilled in the experiment. For further interesting insights on the phenomenon, these preliminary results will be extended in the future by performing the experiment as a function of the detuning from resonance of the probe laser.

We acknowledge very useful suggestions and discussions by Philippe Verkerk and Alan Gallagher

REFERENCES

1. J. Dalibard, Opt. Commun. **68**, 203 (1988).
2. T. Walker, D. Sesko and C. Wieman, Phys. Rev. Lett. **64**, 408 (1990).
3. K. Ellinger, J. Cooper and P. Zoller, Phys. Rev. A **49**, 3909 (1994)
4. G. Hillebrand, C.J. Foot and K. Burnett, Phys. Rev. A **50**, 1479 (1994).
5. S. Grego, M. Colla, A. Fioretti, J.H. Müller, P. Verkerk and E. Arimondo, Opt. Commun. (1996), in press
6. A.F. Molisch, B.P. Oehry and G. Magerl, J. Quant. Spectrosc. Radiat. Transfer, **48**, 377 (1992).

NEUTRALS: ATOMS AND MOLECULES

Purely Quasi-Molecular Radiation in the Hg-Inert Gas Atom Collisions

A.Z.Devdariani*, M.G.Lednev[†], A.L.Zagrebin*, M.A.Tchaplyguine*

*Department of Optics, Institute of Physics, St.Petersburg State University,
198904 St.Petersburg, Russia
[†]Department of Physics, Baltic State Technical University, 198005 St.Petersburg, Russia

Abstract. The work deals with the experimental and theoretical investigation of spectra produced by collisions of metastable $Hg(^3P_2)$ atoms with rare gas atoms. Potential energy curves as well as radiation widths are calculated on the basis of semiempirical approach, spectra are estimated in the frame of quasistatical approximation. The results are in a reasonable agreement with experimental data obtained on collisions with Kr, Xe atoms.

1. Introduction

The present work deals with the experimental and theoretical investigation of the emission spectrum temperature dependence when radiation is initiated by the atomic interaction during thermal collisions

$$Hg(6s6p^3P_2)+RG \rightarrow \begin{cases} Hg(6s^2\ ^1S_0) + RG + \hbar\omega \\ HgRG(X^1\Sigma) + \hbar\omega \end{cases} \quad (1)$$

where RG is an atom of either Ne, Ar, Kr or Xe.

The optical transition $^3P_2 \rightarrow {}^1S_0$ is forbidden in a separated atom, that is why it is possible to consider the radiation in reaction (1) as a transition existing only for the Hg-RG quasimolecule formed during the collision, that is a purely quasimolecular transition. Depending on the details of quasimolecular potential curves and collision energy, the optical transition leads to the formation of the free atoms in the ground state and/or molecule in the ground electronic state.

The quasimolecular radiation has been intensively investigated in the past years in connection with a vacancy creation in inner shells, e.g.[1]. The investigation of the outer shell quasimolecular radiation is difficult because of relatively weak intensities to registrate and multiple radiation mechanisms. The outer shell quasimolecular radiation is sensitive to fine details of interaction and term structure. Therefore theoretical analysis of such radiation necessarily includes a quantum mechanics part, that is the calculation of the ground and excited terms involved, and the transition dipole momentum calculation. The latter strongly depends on interatomic distance. That makes purely quasimolecular transitions in question qualitatively different from those allowed in the limit of separated atoms. The peculiarity of the quantum mechanics part for the problem lies in the fact that the information about the terms and dipole momenta is necessary for the medium ($6a_0$) and large interatomic distances as optical transitions form the greater part of the spectrum exactly in these regions. The necessity to use a large amount of configurations for term and dipole momentum description for this region of distances in the framework of traditional quantum chemistry computational methods explains the absence of *ab initio* calculations in literature for the *Hg-RG* quasimolecule in spite of mercury being a popular object for experimental spectroscopic investigations.

As it was shown for the first time in [2] the radiation in reaction (1) can be described in terms of interaction between three quasimolecular states with full electronic angular momentum projection $\Omega=1$ that corresponds to atomic states $^1P_1$, $^3P_2$ and $^3P_1$ of one excited $6s6p$ configuration at large interatomic distances. The states $|^1P_1\Omega=1>$ and $|^3P_1\Omega=1>$ are connected by spin-orbital interaction which weakly depends on interatomic distance and is practically equal to the interaction in an isolated atom. The states $|^3P_1\Omega=1>$ and $|^3P_2\Omega=1>$ are connected by the atomic interaction strongly depending on the interatomic distance.

The quantitative consideration in [2] was based on asymphotic methods of the atomic interaction theory. The approach leads to reasonable results for the total cross section and the rate coefficient of quenching, but can not give a detailed description of spectrum temperature dependence. This was proved by a consequent experimental investigation of the process (1) for the *Hg-Xe* system [3-5]. The experimental data stimulated the more detailed investigation of the interaction potentials of *Hg-Xe*. The suggested semiempirical potentials made it possible to improve the agreement between calculated [6] and experimental [5] spectral distribution.

In the present work an approach that can be called semiempirical was used (see paragraph 2) in order to determine the terms of $Hg(6^3P_2)$ - *Ne, Ar, Kr, Xe* and the dipole moments of the radiative transition from the excited to the ground state. Briefly, the main idea of it is to use effective Hamiltonian method that makes it reasonable to select the long and short distance interaction parts in the matrix form taking into account the spin orbital interaction. The short distance part, being the most complicated one for calculation, is determined from the existing experimental data on the allowed $^3P_1$-$^1S_0$ spectral transitions. Then the matrix elements determined are used for the reconstruction of all terms formed by $6s6p$ configuration and also for the determination of dipole moment dependence versus interatomic distance. The calculations of spectra averaged over the atom energy distribution at different temperatures are presented in paragraph 3. In paragraph 4 an experimental set-up for purely quasimolecular radiation registration in the case of Hg-Kr, Xe is described.

2. Quasimolecular Potential Curves and Radiation Probabilities.

The information about radiating and ground state terms ($U^*(R)$ and $U_0(R)$ respectively) and optical transition probability $\Gamma(R)$ is necessary for the spectrum calculation. While interacting with an inert gas atom in the ground $^1S_0$ state the atomic metastable state $^3P_2$ generates three quasimolecular states 0^-, 1, 2 ($^3P_2$). The radiative dipole transition to the ground $0^+(^1S_0)$ state is possible from the $1(^3P_2)$ state.

Let us determine the energy of the radiating state and transition probabilities in the first approximation in the framework of only one $6s6p$ excited configuration. The point is that the admixing of the upper lying Hg^*-RG configuration wave functions to $Hg(6^3P_j)$-RG under the influence of atomic interaction in the region of large R is small. The most important mixing takes place between the wave function inside one $Hg(6s6p)$ -RG configuration. It is also significant that the resonance $Hg(6^{1,3}P_j)$ states with the largest probabilities of radiational transitions to the ground state belong exactly to this configuration.

In order to determine interaction potentials of excited $Hg(6s6p^3P_j)$ atoms with inert gas atoms RG ($^1S_0$) in all 6 possible states 0^-, 1 , $2(^3P_2)$, 0^+, $1(^3P_1)$, $0^-(^3P_0)$, let us use experimental potentials for two states - 0^+, $1(^3P_1)$ and semiempirical method for the quasimolecular term analysis [7]. The method allows to determine interaction potentials for all states of this configuration on the basis of interaction potentials for several quasimolecular states belonging to $Hg(6s6p)$-$RG(^1S_0)$ configuration.

The basis of semiempirical procedure for $Hg(6s6p^3P_j)$-$RG(^1S_0)$ interaction potential reconstruction using the experimental potentials for a few states is the method of effective Hamiltonian [8,9]. For $Hg(6s6p^3P_j)$-$RG(^1S_0)$ quasimolecule the effective Hamiltonian $\hat{H}$ is expressed as

$$\hat{H} = \hat{H}_{Hg^*} + \hat{H}_{RG} + \hat{V}, \qquad (2)$$

where $\hat{H}_{Hg^*}$ and $\hat{H}_{RG}$ are the Hamiltonians of free atoms, $\hat{V}$ is the effective operator for atomic interaction.

It is not necessary to present in details the Hamiltonian $\hat{H}_{RG}$ as the inert gas atom remains in the same $^1S_0$ state. In $\hat{H}_{Hg^*}$ we separate the terms responsible for the splitting of $6s6p$ atomic configuration levels.

$$\hat{H}_{Hg^*} = \hat{H}_{Hg^+} + \hat{V}^{Coul}_{eHg+} + \hat{V}^{exch}_{eHg+} + \hat{V}^{SO}_{Hg^*}, \qquad (3)$$

where $\hat{H}_{Hg^+}$ is the ion Hg^+ Hamiltonian that also does not require further refinement; $\hat{V}^{Qoul}_{eHg+}$ and $\hat{V}^{exch}_{eHg+}$ are effective operators for direct Coulomb and exchange interactions between an excited electron and Hg^+ ion; $\hat{V}^{SO}_{Hg^*}$ is an operator for spin-orbit interaction in Hg^* atom. Matrix elements of $\hat{V}^{Coul}_{eHg+}$ and $\hat{V}^{exch}_{eHg+}$ operators are expressed through Slater's integrals. Matrix elements of $\hat{V}^{SO}_{Hg^*}$ operator are expressed through spin-orbit interaction constants. Further it is considered that effective Hamiltonian $\hat{H}_{Hg^*}$ eigenvalues coincide with the energies $E(^{1,3}P_j)$ of $Hg(6s6p)$ configuration atomic levels.

The operator $\hat{V}$ includes effective operators of the exchange and dispersion atomic interaction [8]

$$\hat{V} = \hat{V}^{exch} + \hat{V}^{disp}. \qquad (4)$$

For $Hg(6s6p)$-$RG(^1S_0)$ quasimolecule matrix elements of $\hat{V}$ operator are expressed through $^{1,3}\hat{H}_\sigma(R)$ and $^{1,3}\hat{H}_\pi(R)$ potentials of $Hg(6s6p^3P_j)$-$RG(^1S_0)$ system in the states $^{1,3}\Sigma$ and $^{1,3}\Pi$ without taking into account spin-orbit splitting:

$$^{1,3}\hat{H}_\sigma(R) = <^{1,3}\Sigma^d(6s6p^{1,3}P)|\hat{V}|^{1,3}\Sigma^d(6s6p^{1,3}P)>;$$

$$^{1,3}\hat{H}_\pi(R) = <^{1,3}\Pi^d(6s6p^{1,3}P)|\hat{V}|^{1,3}\Pi^d(6s6p^{1,3}P)>,$$

where $|^{1,3}\Sigma^d(6s6p^{1,3}P)>$ and $|^{1,3}\Pi^d(6s6p^{1,3}P)>$ are diabatic coordinate wave functions of either singlet or triplet states with projections of the orbital momentum L on the interatomic axis equal to zero and one. In a single configuration approximation these functions are the products of coordinate atomic wave functions. It is necessary to mention that semiempirical dependencies $^{1,3}\hat{H}_{\sigma,\pi}(R)$ determined below on the basis of experimental data partially take into account the configuration interaction effect.

For $Hg(6s6p)-RG(^1S_0)$ quasimolecule effective Hamiltonian $\hat{H}$ matrix element calculation let us use $|^{1,3}P_j\Omega_{IC}>$ wave function basis of coupling type c which are the eigenfunctions for the Hamiltonian of noninteracting atoms $\hat{H}_{Hg^*} + \hat{H}_{RG}$. The quasimolecular wave functions $|^{1,3}P_j\Omega_{IC}>$ are the products of atomic wave functions: $|^{1,3}P_j\Omega_{IC}> = |Hg(^{1,3}P_j\Omega)_{IC}>|RG(^1S_0)>$. Atomic wave functions of intermediate coupling type $|Hg(^{1,3}P_j\Omega)_{IC}>$ are linear combinations of the LS-type wavefunctions $|Hg(^{1,3}P_j\Omega)_{LS}>$:

$$|Hg(^1P_1\Omega)_{IC}> = a|Hg(^1P_1\Omega)_{LS}> + b|Hg(^3P_1\Omega)_{LS}> \quad (5)$$
$$|Hg(^3P_1\Omega)_{IC}> = -b|Hg(^1P_1\Omega)_{LS}> + a|Hg(^3P_1\Omega)_{LS}> \quad (6)$$
$$|Hg(^3P_0\Omega)_{IC}> = |Hg(^3P_0\Omega)_{LS}> \quad (7)$$
$$|Hg(^3P_0\Omega)_{IC}> = |Hg(^3P_0\Omega)_{LS}>, \quad (8)$$

with the amplitudes $a = 0.979$ and $b = -0.203$ [6]. It should be noted that the antisymmetrization omitted while determining $|^{1,3}P_j\Omega_{IC}>$ wave functions is compensated by introducing the exchange interaction operator $\hat{V}^{exch}$ into the effective quasimolecular Hamiltonian (4).

Matrix elements of the effective Hamiltonian $\hat{H}$ for $Hg(6s6p)-RG(^1S_0)$ are expressed through $^{1,3}H_{\sigma,\pi}(R)$ potentials, atomic level energies $E(^{1,3}P_j)$ and amplitudes a and b. Interaction potentials (effective Hamiltonian matrix eigenvalues) for eight quasimolecular states $0^+(^1P_1)$, $1(^1P_1)$, $0^-(^3P_2)$, $1(^3P_2)$, $2(^3P_2)$, $0^+(^3P_1)$, $1(^3P_1)$ and $0^-(^3P_0)$ of $Hg(6s6p)-RG(^1S_0)$ configuration are expressed through four independent potentials $^{1,3}\hat{H}_\sigma(R)$ and $^{1,3}\hat{H}_\pi(R)$. That is why experimental interaction potentials for four states make it possible in principle to determine potentials for all eight adiabatic states. If we limit ourselves only to consideration of the triplet terms - $Hg(^3P_j)-RG(^1S_0)$ then it is possible to take $^1H_\sigma = ^3H_\sigma$ and $^1H_\pi = ^3H_\pi$ as the first approximation. It is justified by a relatively small contribution of $^1H_\sigma$, $^1H_\pi$ to the interaction potentials for the triplet states ($|a|^2 >> |b|^2$) and by approximately the same dependencies $^1H_\sigma(R)$ and $^3H_\sigma(R)$, $^1H_\pi(R)$ and $^3H_\pi(R)$.

Thus if we limit ourselves to the analysis of only $Hg(^3P_j)-RG(^1S_0)$ triplet terms, interaction potentials for all six states can be expressed through the experimental potentials for two states. The interaction potentials for $1(^3P_1)$ and $0^+(^3P_1)$ states were chosen as a basis for that simplified semiempirical procedure for quasimolecular term reconstruction as the most detailed spectroscopic data exist exactly for these radiating states. Experimental interaction potentials for $1(^3P_1)$ and $0^+(^3P_1)$ were used as corresponding eigenvalues of the effective Hamiltonian matrix and $^3H_\sigma(R)$,

$^3H_\pi(R)$ dependencies were determined. Then, through the diagonalization of the matrix, the semiempirical interaction potentials for the states 0^-, 1, $2(^3P_2)$, and $0^-(^3P_0)$ were determined.

Adiabatic quasimolecular wave functions $|\Omega\ (^{1,3}P_j)>$ being eigenfunctions of Hamiltonian (adiabatic correlation with a corresponding state in the limit of separated atoms is shown in brackets) are linear combinations of the diabatic functions $|(^{1,3}P_j\Omega)_{IC}>$. In particular,

$$|1\ (^3P_2)> = c_1(R)|^1P_1\ 1_{IC}> + c_2(R)|^3P_2\ 1_{IC}> + c_3(R)|^3P_1\ 1_{IC}> \quad (9)$$

$$|1\ (^3P_1)> = d_1(R)|^1P_1\ 1_{IC}> + d_2(R)|^3P_2\ 1_{IC}> + d_3(R)|^3P_1\ 1_{IC}> \quad (10)$$

$$|0^+\ (^3P_1)> = f_1(R)|^1P_1\ 0^+_{IC}> + f_2(R)|^3P_1\ 0^+_{IC}>, \quad (11)$$

where $c_i(R)$, $d_i(R)$ and $f_i(R)$ coefficients are determined by the effective Hamiltonian matrix diagonalization. Quasimolecular radiation transitions $1(^3P_2)$, $1(^3P_1)$, $0^+(^3P_1)$ - $0^+(^1S_0)$ take place as the result of the radiating diabatic $|^1P_1\ \Omega_{IC}>$ and $|^3P_1\ \Omega_{IC}>$ state contribution to the corresponding adiabatic states.

For $\Gamma(\Omega(^3P_j),R)$ probability calculation in (9)-(11) it is convenient to return back to **LS**-coupling diabatic basis according to (5)-(8), as among the states $|^{1,3}P_j\Omega_{LS}> = |Hg(^{1,3}P_j\Omega)_{LS}>|RG(^1S_0)>$ only $|^1P_1\Omega_{LS}>$ state is connected with the $|0^+(^1S_0)>$ state by a dipole transition. The dipole moment of the $^1P_1\Omega_{LS}$ - $0^+(^1S_0)$ transition is connected with the observed magnitude - the dipole moment of the atomic transitions: $<^1P_1\Omega_{LS}|\hat{d}|0^+(^1S_0)> = - b^{-1}<Hg(^3P_1\Omega_{IC})|\hat{d}|Hg(^1S_0\ 0^+)>$. That is why in particular $<1\ (^3P_2)|\hat{d}|0^+(^1S_0)> = - (ac_1+bc_3)b^{-1}<Hg(^3P_1\Omega_{IC})|\hat{d}|Hg(^1S_0\ 0^+)>$. It is taken into account that in the interatomic distance region under consideration the ground state adiabatic wave function practically coincides with the diabatic $|Hg(^1S_0)>|RG(^1S_0)>$ function, as the ground state is far away from other Hg - RG states. Introducing $\Gamma(^3P_1)$ probability for the atomic transition $^3P_1 \to {}^1S_0$ we finally obtain:

$$\Gamma(1(^3P_2),R) = \Gamma(^3P_1)\left(\frac{\omega(1(^3P_2),R)}{\omega(^3P_1)}\right)^3 \left(\frac{a}{b}c_1-c_3\right)^2 \quad (12)$$

where $\omega(^3P_1)$ is the $^3P_1 \to {}^1S_0$ atomic transition frequency and $\omega(1(^3P_1),R)$ is the frequency of the $1(^3P_1) \to 0^+(^1S_0)$ quasimolecular transition. The experimental value of $\Gamma(^3P_1)=8.5\ 10^6\ c^{-1}$ was used for calculations.

It is also convenient to introduce the relative radiational widths according to the expression

$$\gamma(\Omega(^3P_j),R) = \frac{\Gamma(\Omega(^3P_j),R)}{\Gamma(^3P_1)}\left(\frac{\omega(^3P_1)}{\omega(\Omega(^3P_j),R)}\right)^3 \quad (13)$$

These magnitudes do not depend on the quasimolecule ground state (that influences the $\omega(\Omega(^3P_j),R)$ frequency) and characterise the ratio of the quasimolecular transition dipole momentum square to atomic $^3P_1 \to {}^1S_0$ transition dipole momentum square. The $\gamma(\Omega(^3P_j),R)$ relative widths coincide with the $\Gamma(\Omega(^3P_j),R)/\Gamma(^3P_1)$ ratio with the accuracy of a frequency multiplier close to unit.

From the considerations listed above it is clear that the quality of the potential reconstruction procedure is determined by the accuracy of initial 0^+, $1(^3P_1)$ potentials received from experimental data. For this purpose the results of [10] for *Hg-Ne, Ar* pairs and [11] for *Hg-Kr, Xe* pairs were used as they were obtained on the basis of the uniform experimental approach - laser induced fluorescence - for investigation of molecular absorption band vibration-rotation structure.

Radiating state potentials in the Morse approximation valid for the distances close to the potential minimum are presented in [10, 11]. That is why the potentials reconstructed according to the procedure described above are reliable in the limited area.

Semiempirical terms and relative radiation widths based on [10, 11] data can be applied in the *R* region from $7a_0$ to $11a_0$ for *HgKr* and *HgXe*. For *HgAr* and *HgNe* the *R* regions where Morse approximation is valid for 0^+, $1(^3P_1)$ interaction potentials practically do not overlap. That is why semiempirical potentials and radiation widths for the systems and also corresponding values for Hg^*Kr, Hg^*Xe in the regions of $R<7a_0$ can be used only for reasonable estimations. For the spectral distribution calculation the ground state atom interaction potentials $U_0(R)$ from [10] for *HgNe, HgAr*, from [12] for *HgKr* and from [13] for *HgXe* were used. Difference interaction potentials $\Delta U = U - U_0$ and relative radiation widths are presented in Fig. 1 - 2.

3. Quasimolecular Spectra

.In the framework of the well-known quasistatical approximation, the rate coefficient $K(T)$ for the process (1) and photon spectral distribution $I(T,\Delta\omega)$ is determined by the following expression:

$$K(T) = \frac{2}{5} 4\pi \int_0^\infty \Gamma(R) \exp\left(-\frac{U^*(R)}{kT}\right) R^2 dR \qquad (14)$$

$$I(T,\Delta\omega) = \frac{2}{5} \frac{4\pi R_c^2 \Gamma(R_c) \exp\left(-U^*(R_c)/kT\right)}{K(T) d|\Delta U(R)/\hbar dR|_{R=R_c}}, \qquad (15)$$

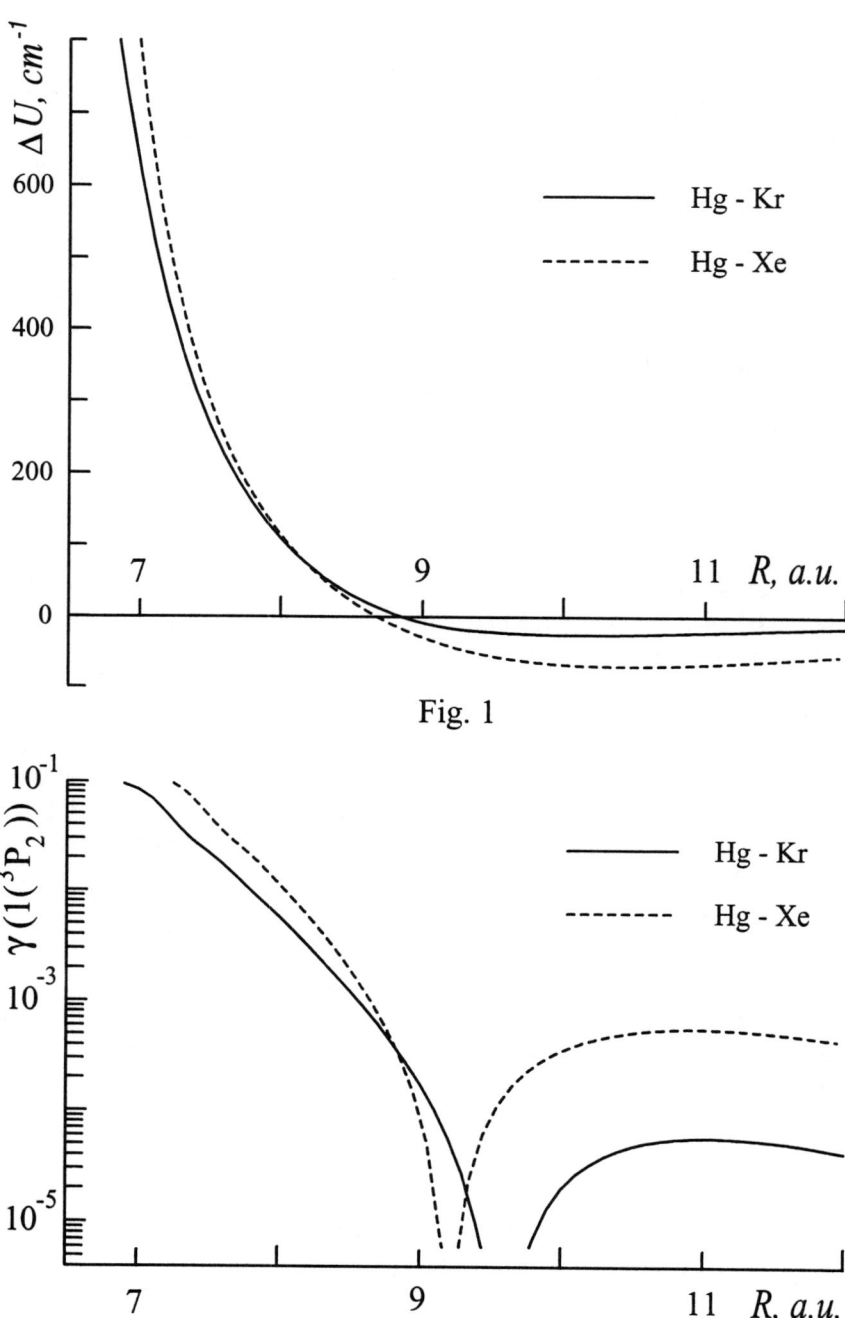

Fig. 1

Fig. 2

where the coefficient 2/5 defines the probability of the $Hg(6^3P_2)+Xe$ quasimolecule to be found in $1(^3P_2)$ state; $\Delta\omega=\omega-\omega_0$ - the quasimolecular transition frequency ω detuning from the frequency ω_0 of the forbidden atomic transition $^3P_2 \to {}^1S_0$; R_c - the Condon point determined by the condition $\Delta U(R_c) = \hbar\Delta\omega$ (this equation has only one root in the cases under consideration). Spectral distribution (15) is normalised according to the condition

$$\int I(T,\Delta\omega)d\hbar\Delta\omega = 1 . \qquad (16)$$

The calculated spectral distributions $I(T,\Delta\omega)$ at $T \approx 300K$ are presented on Fig.3-4, and the rate coefficient $K(T)$ - in Table 1.

TABLE 1

T, K	$K(T)$, 10^{-17} $cm^{-3}c^{-1}$				
	He [a]	Ne	Ar	Kr	Xe
300	0.17	0.15	0.44	0.88	2.13
400	0.28	0.23	0.57	1.07	2.25
500	0.40	0.31	0.72	1.26	2.45
600	0.52	0.40	0.86	1.47	2.69
700	0.65	0.49	1.03	1.68	2.95
800	0.78	0.59	1.19	1.89	3.22
900	0.92	0.81	1.35	2.11	3.49
1000	1.06	0.79	1.51	2.32	3.78

[a] - calculation [14]

As can be seen from the results of the calculation, the quasimolecular radiation formed as the result of the process (1) is a short-wave satellite of the forbidden atomic line $^3P_2 \to {}^1S_0$. As temperature grows, the maximum of the spectral distribution shifts to the short wave region; the spectral distribution becomes wider and its integral intensity ($K(T)$ magnitude) increases.

Expression (15) for spectral distribution is valid when thermodynamical equilibrium between the bound and quasibound states in the $1(^3P_1)$ state potential well and also between energy continuous spectrum states (for relative motion of $Hg(6^3P_2)$ and RG) is established. For the cases under consideration when the atomic state has large life-time and potential well depth is small ($D_e < kT$) it is this equilibrium situation that is realised in the experiments in a gas cell with inert gas concentration $N_{RG} > 10^{17}$ cm^{-3}. That is why, strictly speaking, the rate coefficient $K(T)$ defined by (14) and the corresponding effective cross-section $<\sigma(T)> = K(T)/v$ are not

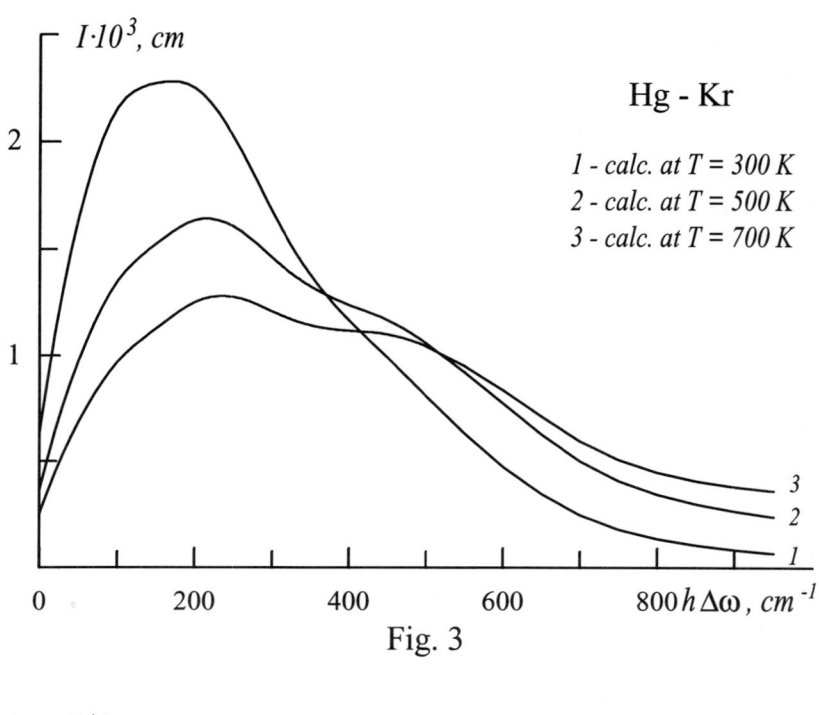

Fig. 3

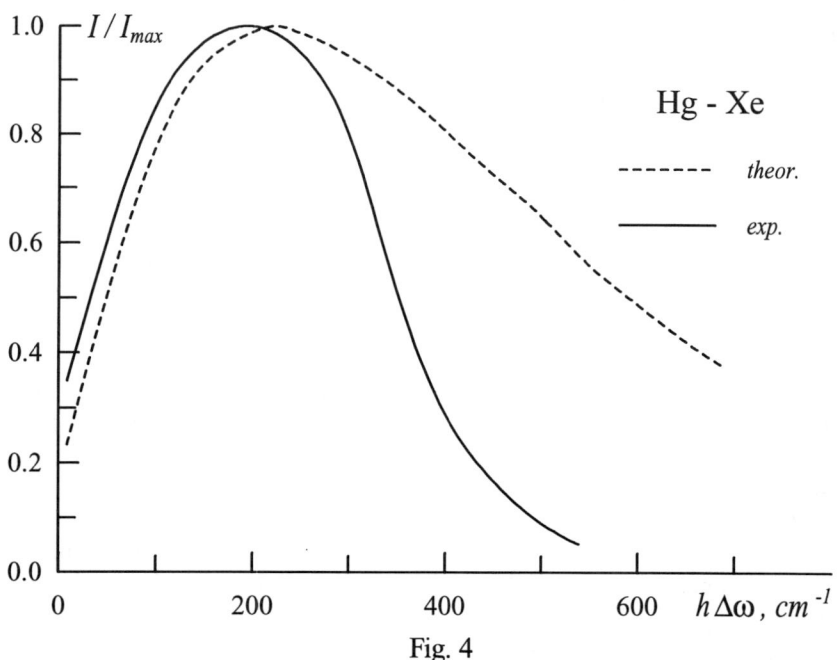

Fig. 4

characteristic of a binary collision processes as they take into account the contribution of radiation from the bound states.

For the systems considered above at $T=300K$ the bound state radiation contribution to the spectral distribution $I(\Delta\omega)$ is small for two reasons: i) the relative population of the bound states is not significant ($D_e < kT$) ; ii)The $\Gamma(R)$ dependence has such a character that quasimolecular radiation is formed mostly as the result of transitions in the region of the closest approach where $U^*(R)>0$ (in the R-region over the potential well the $\Gamma(R)$ magnitude is small, Fig. (1-2). This statement is confirmed by the calculation [7] where for the Hg-Xe system insignificant difference of the spectral distribution (15) from the spectral distribution formed only as the result of binary collisions is shown. The latter situation when the bound and quasibound states of the $Hg(6^3P_2)$-RG system are not populated is realised only in the case of low density ($N_{RG} \ll 10^{17} cm^{-3}$).

4. Optical Cell Experiments.
Radiation-collision Quenching of Metastable Mercury Atoms by Xenon and Krypton[5,15].

The investigated quasimolecules have been created in the low temperature plasma of the diffusion gas discharge in the mixture of mercury vapour and inert gas. A cylindrical quarz tube with two electrodes located perpendicular to the tube axis was used as the experimental volume. The tube had a side knee used as a reservoir for mercury. The reservoir had its own heating system providing necessary mercury concentration. UV-radiation from the tube was focused by an optical system on the entrance slit of a monochromator with a 1200s/mm grating. Monochromator linear dispersion was 20 A/mm. Dispersed light was registered by a photomultiplier and amplified in continuous regime. Parameters of the constant current discharge were chosen in order to avoid heating and contraction of plasma on the one hand and to provide maximum signal on the other hand. The range of the discharge current was in the interval 1-30mA. The 2280A atomic line of a cadmium spectral line served as a reference. It was possible to change the temperature of the gas mixture and to measure the concentration of atomic parent states by spectral line absorption method on the transition 6^3P_2 - 7^3S_1 (λ = 5460.7A). The temperature of the tube wall was measured by a thermocouple. The reference source for absorption measurements - a high frequency electrodeless lamp- was located in front of the tube back window. The lamp spectral contours were preliminarily investigated on a Fabri-Perot interferometer.

Absolute sensitivity of the registration system was preliminarily determined with the help of a hydrogen-deuterium lamp calibrated in the ultraviolet region using a synchrotron. Hydrogen UV continuum was used to recalculate the amplifier output voltage values into the radiation power density within the experimental volume.

The experiment was carried out in the following way. The vacuum system contained a vessel with an inert gas (xenon or krypton) that could be injected by small portions of a few Torrs into the tube. At a fixed value of temperature the inert gas pressure was scanned up to the values when contraction was becoming observable - about 50 Torrs. At each point of pressure the $Hg(^3P_2)$ state atom concentration was measured. Several sets of measurements were made for different temperatures in the range from 300 to 375K.

For the pressures of tens of Torrs the radiation spectrum was observed in the
interval from 2270 to 2240-2230A with a maximum at about 2260A for both gases,
Fig.3-4 . The peak was noticeable asymmetric with specta width of 15-20A at the half maximum. For xenon, a shift of the spectral maximum to the shorter wavelength side was registered. For both gases the width at half maximum was observed to increase with temperature.This spectrum was interpreted as quasimolecular radiation due to $1(^3P_2) - 0^+(^1S_0)$ transition. The integrated intensity of the spectrum was observed to increase with pressure and to decrease by increasing temperature. The $Hg(^3P_2)$ concentration also decreased by increasing the temperature. The comparison between the experimentally recorded spectra and the calculated ones is rather satisfactory, as it appears in Fig. 3-4.

ACKNOWLEDGMENTS

The research described in this publication was made possible in part by Grant N 96-03-33679a from the Russian Foundation for Basic Reserches.

REFERENCES

1. P.Vincent en: Quantum Electrodynamics of Strong Fields. ed. W.Greiner, NATO Adv.Study Inst., 1981.
2. A.Z.Devdariani, A.L.Zagrebin, Opt. Spectry **58** (1985) 752.

3. A.B.Callear and K.Du, Chem. Phys. Lett., **128** (1986) 141.
4. A.B.Callear and K.Du, Chem. Phys. **113,** (1987) 73.
5. N.A.Kryukov, N.P.Penkin, T.P.Redko, Opt. Spectr. **66** (1989) 564.
6. A.L.Zagrebin, M.G.Lednev, Sov. Tech. Phys. Lett., **18** (1992) 243.
7. A.Z.Devdariani, A.L.Zagrebin, K.B.Blagoev. Ann. Phys. Fr. **14** (1989) 467.
8. S.Ja.Umanskij, E.E.Nikitin. Theoret.Chim.Acta. (Berl.) **13** (1989) 91.
9. A.Z.Devdariani, A.L.Zagrebin, Sov. J. Chem. Phys (UK) **5** (1986) 237.
10. Yamanouchi K., Isogai S., Okunishi M., Tsuchiya S., J. Chem. Phys. **88** (1988) 205.
11. Okunishi M., Nakazawa H., Yamanouchi K., Tsuchiya S., J. Chem. Phys., **93** (1990) 7526.
12. M.Findeisen, T.Grycuk, J. Phys. B. At. Mol. Phys. **22** (1989) 1583.
13. T.Grycuk, M.Findeisen, J. Phys. B. At. Mol. Phys. **16** (1983) 975.
14. A.L.Zagrebin, M.G.Lednev, Opt. Spectry. **79** (1995) 912.
15. N.A.Kryukov, P.A.Saveliev, M.A.Tchapliquine, 5th EPS Conf. on At.Mol.Phys. Contributed Papers. Part 1. Edinburgh. 1995. P.160.

Action Spectra in Collisions Between Neutral Atoms and Molecules

Kenneth M. Sando and Yu-Ming Hung

Department of Chemistry, Iowa Advanced Technology Laboratory
University of Iowa, Iowa City, IA 52242

Abstract. Line shapes from the neutral broadening of atomic emission and absorption lines have been used successfully to gain information about atomic and molecular interactions. However, there is not enough information in the traditional atomic line shape to derive all of the details of the interactions, especially when potential energy surface crossings occur. In the action spectra, referred to here, the principles of atomic line broadening are used, but the incident light has a precise frequency and polarization, and the products may be measured state specifically. A large amount of information is obtained, but interpretation in terms of atomic interactions is a challenge to theory.

In our theoretical treatment we use a classical path/quantum close coupling method. The nuclei move in classical trajectories, but the coupling between atomic states is handled quantum mechanically. As compared to fully quantum mechanical methods, our approach has the advantages of being easily extended to new systems (even those involving molecular perturbers) and of producing physical insight into the process. The method will be outlined and applications to alkaline earth-rare gas state changing collisions and to alkali and alkaline earth-diatomic molecule reactive and non-reactive quenching collisions will be discussed.

INTRODUCTION

We are concerned with the absorption or emission of light by a neutral atom in collision with a neutral atom or molecule. The collision results in a transient exciplex, an excited complex between the active atom and the perturber with a lifetime on the order of the duration of a collision (about one picosecond). When it is first formed, the excitation energy of the exciplex is assumed to be primarily associated with the active atom, however, in its short life time, the exciplex may evolve in a variety of ways. It may dissociate to an excited atom plus a perturber, in which case the excitation energy remains associated with the active atom. It may radiate at a frequency near to an atomic line and dissociate to a ground (or intermediate) state atom plus a perturber. It may dissociate to an excited perturber plus a ground state atom, in which case the excitation of the active atom is quenched through a non-radiative energy transfer to the perturber. When the perturber is a molecule, the exciplex may dissociate into a reactive product and a residual fragment, in which case the excitation of the active atom is quenched by chemical reaction. Other processes are possible, such as radiative association.

In each of the various processes mentioned above, the exciplex evolves to products that are stable for a time long compared to the duration of the collision. Each product will be in a particular internal state. If we can measure concentrations of each product in a particular quantum state before those concentrations are altered by subsequent collisions or radiative decay, we can determine the nascent state-specific populations of the products of the evolution of the exciplex. Those populations as a function of the frequency and polarization of the incident light that created the exciplex are action spectra.

It is our goal to relate the action spectra to the detailed interactions that occur during the evolution of the exciplex. Often more than one potential energy surface is involved in the evolution of the exciplex. We seek to learn about those surfaces and about the interactions among them. We are particularly interested in situations in which surfaces cross. In those cases, transitions from one surface to another can be highly probable.

The focus of this paper is on a theoretical approach toward learning about fundamental interactions from action spectra. We seek an approach that facilitates a conceptual understanding of the processes involved, is readily applied to a variety of systems, and is at least semi-quantitatively accurate. A classical path/quantum close-coupling method will be outlined that partially fulfills these criteria.

Application to a few specific examples of state-changing atom-atom collisions and non-reactive and reactive atom-molecule quenching will be described.

AUTOCORRELATION FORMALISM

The dipole autocorrelation function formulation of the atomic line shape is important for this discussion because it illustrates most clearly the effects of the processes under discussion on the line shape and because it is a time-dependent formulation that matches well with the time-dependent classical path method to be described later.

The dipole autocorrelation function $C(\tau)$ is the inverse Fourier transform of the line shape function $I(\nu)$

$$C(\tau) = \int_{-\infty}^{\infty} I(\nu) \, e^{2\pi i \nu \tau} d\nu \quad . \tag{1}$$

It is defined by the average

$$C(\tau) = \langle\langle \mu(0) \cdot \mu(\tau) \rangle\rangle \quad , \tag{2}$$

where $\mu(\tau)$ is the transition dipole moment vector. Decay of the autocorrelation function with time leads to atomic line broadening. In the present context, three processes leading to the decay of the autocorrelation function can be identified: 1) Perturbation of the frequency of the atomic transition during the collision results in a dephasing of the dipole moment function; 2) Change in the orientation of the dipole moment vector due to the collision leads to a depolarization; 3) Quenching due to the collision leads to a decrease in the magnitude of the dipole moment.

In a measurement of the atomic line shape function, the effects of these processes become intermingled and difficult to distinguish. An example from

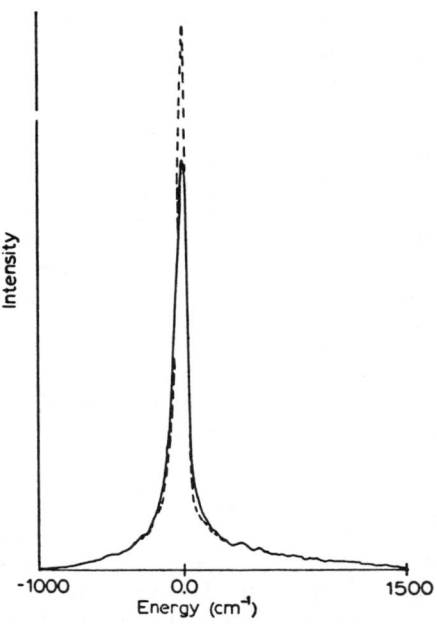

Figure 1. Simulations of the high-density line shape of sodium perturbed by argon with (solid line) relaxation of polarization and without (dashed line).

earlier work (1) is shown in Fig. 1. Here, the emission line shape of sodium perturbed by argon was simulated theoretically with and without consideration of the depolarization due to collisions. There is a clear broadening due to depolarization, but no distinctive features are added. In a comparison between an experimental observation and a theoretical calculation based upon imperfect potential surfaces, it would be very difficult to separate the broadening due to depolarization from that due to perturbation of the frequency.

It is important to measure the effects of each possible process separately. That is to determine and interpret the action spectra for the various processes.

CLASSICAL PATH FORMALISM

In order to relate observed action spectra to fundamental interactions in the exciplex, it is necessary to have a theoretical model capable of predicting the observations from the potential energy surfaces that control the evolution of the exciplex. In order to describe the formation and dissociation of the exciplex, we must follow the dynamics over a broad range of molecular geometries. There is usually no single representation in molecular electronic states that diagonalizes the Born-Oppenheimer Hamiltonian over the entire range of geometries. Therefore, we must deal with several closely lying and interacting potential energy surfaces. This represents a challenge to the dynamical model.

A fully quantum mechanical model, in which all degrees of freedom are treated in a quantum representation, would be the most accurate, but would also have disadvantages. It is very difficult to develop a computationally tractable quantum mechanical method for systems that have both multiple potential surfaces and three or more atoms. It is also difficult to interpret the results in terms of events that occur within the exciplex. Given initial conditions and potential energy surfaces, a quantum mechanical method can produce cross-sections, but does not directly reveal what happens during the evolution from the formation of the exciplex to its dissociation. Also, potential surfaces are often not known to sufficient accuracy to justify the use of a highly accurate dynamical theory.

We choose to use a classical path/quantum mechanical close-coupling model. In most cases this means that the nuclear motion is treated with Newtonian dynamics, but electronic states are treated quantum mechanically. In some cases, when vibrational energy levels are widely separated, the vibrational motion is also quantized.

When multiple electronic states are involved, there is a fundamental assumption underlying our method that prohibits it from producing exact results. We assume that classical dynamics on a single potential energy surface adequately describes the time evolution of the nuclear geometry. In fact, it can be shown that no such single potential energy surface exists. Various choices can be made for this non-existent potential, including a zero potential that leads to the straight-line trajectory model. We use a potential that is the average of the potentials for all of the electronic states involved. It has been argued that the average potential gives the most reliable results. We have tested our model by comparison with the results of fully quantum mechanical calculations. We have found surprisingly good agreement. One example will be presented later.

The classical path method is simple to describe and simple to use. If **R** represents the set of nuclear coordinates, we require **R** as a function of time **R**(t), which we find by allowing the nuclei to move according to Newtonian dynamics on the average potential. Given **R**(t), we can find the potential energy matrix **V**(t) as a function of time. The potential energy matrix has values of the potential energy surfaces along the diagonal and coupling between surfaces as off-diagonal matrix elements. We represent the electronic state of the exciplex as a time-dependent superposition of the electronic states used to represent the Hamiltonian,

$$\Psi(t) = \sum_j c_j(t) |j\rangle \quad . \tag{3}$$

We then use standard numerical methods to solve the time-dependent Schrödinger equation for the time-dependent coefficients $c_j(t)$,

$$i\hbar \frac{dc_j}{dt} = \sum_k V_{jk}(t) c_k(t) \quad . \tag{4}$$

The matrix **V** is the representation of the Hamiltonian in the basis $|j\rangle$. Any convenient representation can be used to solve Eq. (4) as long as the proper transformation to asymptotic states is made to impose initial conditions and calculate observables. We usually use a diabatic molecular representation, in which case we assume there is only potential coupling between states and no velocity cou-

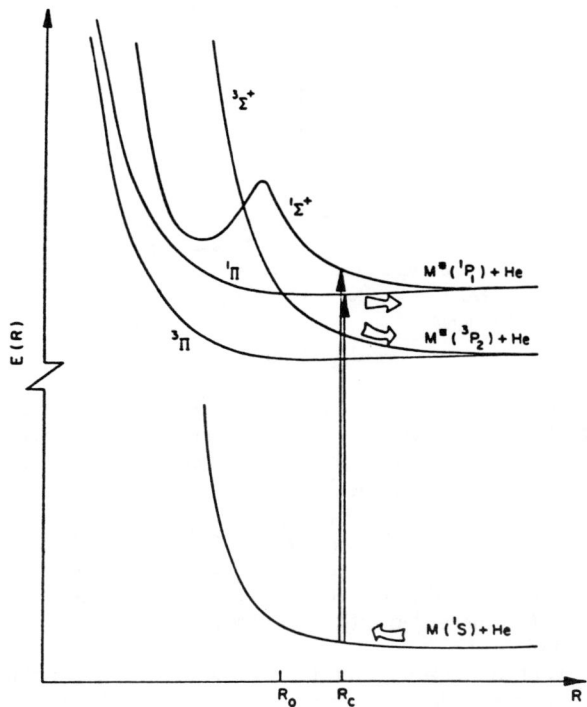

Figure 2. Relevant interaction potentials for the Ca-He system with vertical arrows indicating blue wing excitation and red wing excitation.

pling. The square magnitudes of the coefficients $c_j(t)$ can be calculated as a function of time and represent instantaneous electronic state populations. This information can help to interpret significant events that occur during the evolution of the complex.

SPECIFIC APPLICATIONS

Here, we present a variety of examples to illustrate the breadth of this line of study. For some, the theoretical interpretation is complete. For others, it is in varying degrees of progress.

Calcium Helium State-Changing Collisions

In this experiment (2), a transient ground state dimer is formed between a Ca and a He atom. The dimer is excited by a tunable laser to an excimer, which then evolves to produce an excited Ca atom and a ground state He atom. To insure that the light is absorbed by the dimer, and not by free Ca atoms, the laser is tuned to

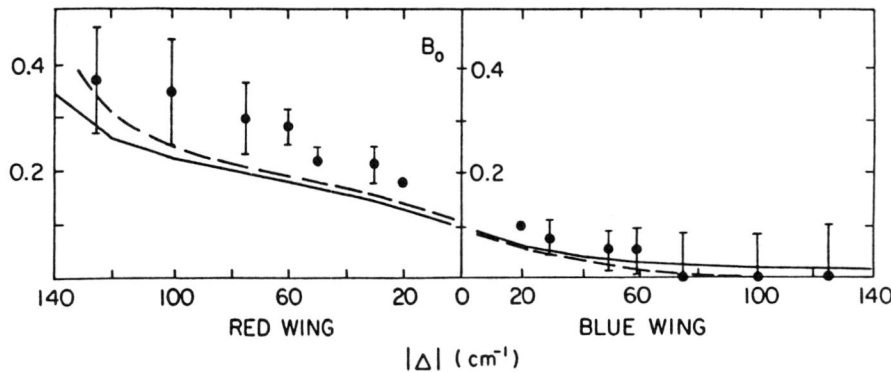

Figure 3. Nascent branching fraction $B_o = N(^3P)/(N(^1P)+N(^3P))$ as a function of detuning (Δ) in the CaHe system. Solid circles represent measurements. Lines are theoretical model predictions: fully quantum mechanical (solid lines); and classical path (dashed lines).

wave lengths near to, but detuned from the Ca ($4s^2\ ^1S - 5p\ ^1P$) line. An experiment of this nature is often referred to as a "half-collision experiment." Absorption is into the singlet manifold of molecular states, but dissociation can lead to ^1P or ^3P atomic states. The measured action spectra consist of ^1P and ^3P populations as a function of detuning.

From the relevant interaction potentials (shown in Fig. 2), it can be seen that detuning to the red of the atomic line excites the $^1\Pi$ diatomic state, whereas blue detuning excites the $^1\Sigma$ state. Thus, the alignment of the Ca p-orbital with respect to the molecular axis is determined by the detuning. We also note that the $^1\Pi$ state is crossed by the $^3\Sigma$ state. We can infer that formation of ^3P atoms will be more likely from the $^1\Pi$ state, and thus from red detunings. In fact, this is what is seen in the measurements (Fig. 3). The points in Fig. 3 represent measurements of the branching fraction (the fraction of the total excited state population found in triplet states) as a function of detuning.

Because this is a diatomic system, fully quantum calculations are tractable and have been done by Pouilly (3). The results of those calculations are shown as the solid line in Fig. 3.

In order to gain more insight into the state-changing process and in order to test the validity of our theoretical approach, we applied the classical path/quantum close-coupling method to this system. The calculation was treated as a scattering problem with 13 channels, corresponding to the 12 excited molecular states and a "dressed" ground state. We used the same interaction potentials chosen by Pouilly. The results are shown as the dashed line in Fig. 3.

We note that the branching fractions predicted by the classical path method are in excellent agreement with those predicted by the fully quantum mechanical method from the same interaction potentials. This gives us confidence to apply the classical path method to more complicated systems for which the fully quantum mechanical method is impractical. We also note that both calculations are in semiquantitative agreement with the measurements. What discrepancies there are can be attributed to the use of imperfect potential curves.

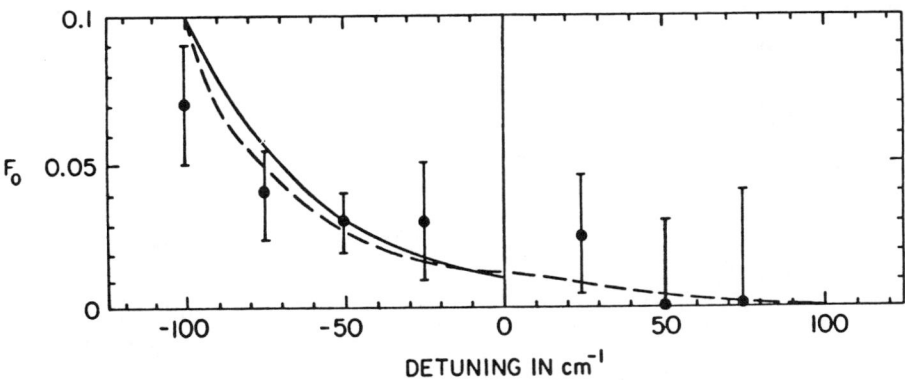

Figure 4. Nascent branching ratio $F_0 = N(^3P_0 + {}^3P_1)/N(^1P_1)$ as a function of detuning (Δ) in the SrHe system. Solid circles represent measurements. Lines are classical path model predictions: with angular velocity coupling (solid lines); and without (dashed lines).

Strontium Helium State Changing Collisions

The Strontium-Helium system is similar to the Calcium-Helium system, except that it is more complex. Many more final atomic states are accessible and, due to the larger spin-orbit splitting, the measurements are more detailed because individual fine-structure state populations can be measured. Observable final excited Sr states include $5s6p\,^1P_1$, $5s6p\,^3P_{2,1,0}$, $5p4d\,^3F_{4,3,2}$, and $5p4d\,^1D_2$. Also, in these experiments, polarization of the final state emission was measured relative to the direction of the polarization of the exciting laser.

No fully quantum mechanical calculation has been performed for this system. We have done classical path calculations within the same 13-state model described above for Ca-He at two levels of approximation: with and without angular velocity coupling. As no interaction potentials have been determined for the relevant states of Sr-He, we modeled them with curves chosen to reasonably reproduce the total absorption spectrum.

An example of a comparison between theoretical and experimental results is shown in Fig. 4. Shown is the branching ratio $N(^3P_0 + {}^3P_1)/N(^1P_1)$ as a function of the detuning Δ. The points are the measured values, the lines are the results of the model calculation. The dashed line represents results without angular velocity coupling, the solid line with. There is satisfactory agreement between the experimental and theoretical results. Again, we see a substantially greater population in the triplet-P manifold with red wing excitation. As in the case of Ca-He, this is due to a curve crossing. The small effect of rotational coupling is supportive of a conceptual model called "orbital locking and following." In this model, the p-orbital of Sr locks onto the molecular axis at a large internuclear distance in either the Π or Σ orientation. While the orbital orientation is locked to the molecular axis, the effects of rotational coupling are negligible. Rotational coupling must be

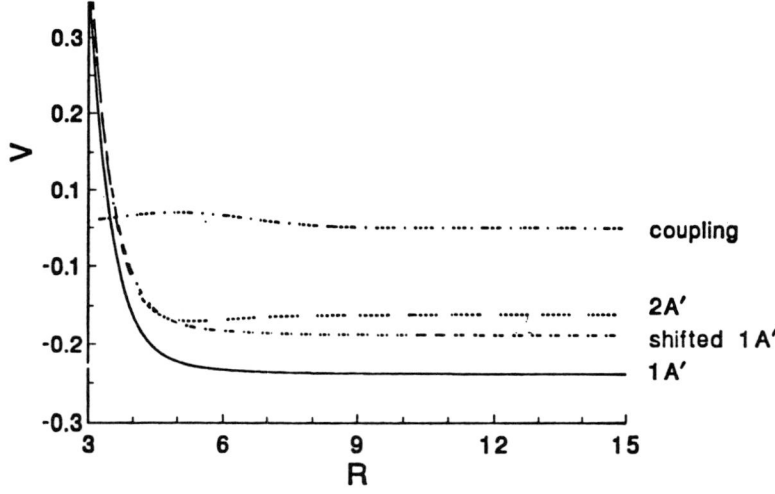

Figure 5. A cut through the potential energy surfaces of Na + N_2 for an approach angle of 45°. The shifted 1A' surface is 2 vibrational quanta above the bare 1A' surface. The coupling function couples the 2A' surface to the shifted 1A' surface to allow for quenching.

included, however, in order to predict the results of the polarization experiments. Those results will not be discussed here.

Broadening of the Sodium Resonance Emission Line by Nitrogen

When an atomic transition is perturbed by a molecule, rather than by another atom, new processes become possible, including quenching via electronic to vibrational energy transfer, and quenching via chemical reaction. We have not yet developed our theoretical methods to handle all possible outcomes in one calculation, but we are taking steps toward that goal. The first step is to treat an example in which chemical reaction does not occur.

Line wing intensities in the broadening of the Sodium resonance emission line by nitrogen molecule have been measured by Kamke, et al. (4). In this system, chemical reaction cannot occur, but non-reactive quenching plays a role in the line-broadening. An excited sodium atom colliding with a nitrogen molecule can lose its excitation energy by emission of a photon at a frequency perturbed by the nitrogen or it can lose its energy non-radiatively by energy transfer to the vibrational motion of the nitrogen molecule.

The vibrational energy level spacings in the N_2 molecule are large compared to rotational energy level spacings and to the thermal collision energy. We, therefore, choose to consider the vibrational motion of the nitrogen molecule to be quantized. We include the vibrational states by "shifting" the ground state potential energy surface by the quanta of vibrational energies. As shown in Figure 5 for an approach angle of the Na to the N_2 of 45°, if one vibrational state is includ-

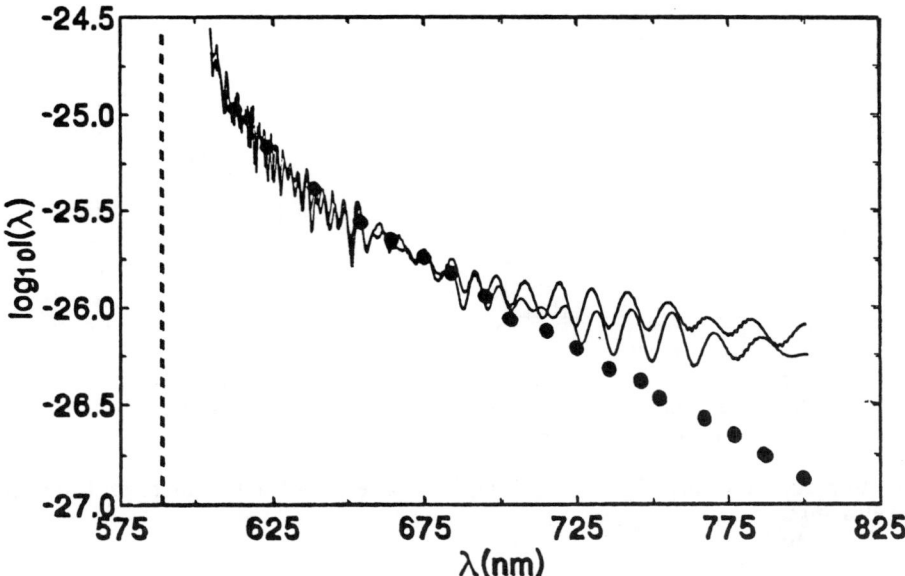

Figure 6. Red wing of the emission spectrum of the sodium resonance line perturbed by molecular nitrogen. The points are the experimental measurements. The lines result from classical path calculations. The upper line includes no quenching effects. The lower line includes quenching to the second vibrational state of N_2.

ed (in this case v=2), our classical path method couples 3 channels: the excited state; the bare ground state; and the shifted ground state.

The results of a sample calculation are shown in Fig. 6. The effect of quenching is seen clearly as a decrease in the intensity far in the line wing. The decrease in intensity results from a decrease in the population of the excited 2A' state as the exciplex evolves due to electronic to vibrational energy transfer. There is clearly still a discrepancy between the calculated intensity in the far wing and the measured intensity. In other calculations (results not shown here) we have included up to 7 vibrational channels (a 9 channel calculation) and have varied the strength of the coupling function with little effect. The disagreement with experimental results may be due to imperfect potential energy surfaces, but we also cannot rule out the possibility that our dynamical model may be inadequate in this case.

Reactive Quenching of Magnesium by Hydrogen Molecule

The most interesting and theoretically challenging example we will present here is that of magnesium atom plus hydrogen molecule (5). In the experiment, magnesium forms a transient ground state complex with hydrogen. The complex is excited by a tunable laser to an exciplex, which then evolves to produce a variety of products. To insure that the light is absorbed by the complex and not by free Mg atoms, the laser is tuned to wave lengths near to, but detuned from the magne-

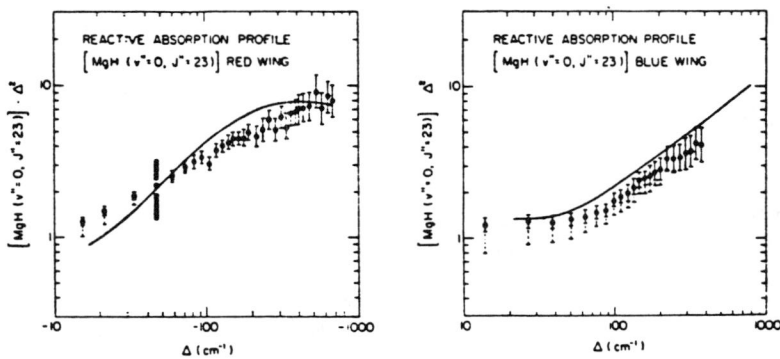

Figure 7. Action spectra for the formation of MgH(v=0,J=23). Here, the points are the measurements and the lines are the results of calculations based upon a quasistatic model.

sium resonance line Mg($3s^2\ ^1S - 3s3p\ ^1P$). The possible products include the following:
1) Mg(^1P) + H_2
2) Mg(^1S) + H_2(v,J)
3) MgH(v,J) + H

Of these, the population of Mg(^1P) is measured and the populations of MgH(v,J) are measured state-specifically. Representative examples of action spectra are shown in Figure 7.

In Figure 7, the lines represent the results of a quasistatic model that assumes a very high reaction probability for Mg + H_2. The profiles of the action spectra are thus determined primarily by the entrance channel. The excited potential surface for Mg(^1P) + H_2 crosses the ground state surface leading to MgH + H on some hypersurface. Whenever that region of space is reached by the exciplex, there is nearly unit probability of the formation of reactive products.

For this system, the classical path model must be adapted so that rotational state populations of the MgH product can be predicted. Calculations are in progress, but the results are too preliminary to be reported.

ACKNOWLEDGEMENTS

The work described here resulted from collaborations with Paul Kleiber, Sharath Ananthamurthy, Marjatta Lyyra, William Stwalley, and others.

REFERENCES

1. Erickson, G. J. and Sando, K. M., Phys. Rev. A **22**, 1500 (1980).
2. Ananthamurthy, S., Sando, K. M., and Kleiber, P. D., J. Chem. Phys. **101**, 10485 (1994).
3. Pouilly, B., J. Chem. Phys. **95**, 5861 (1991).
4. Kamke, W., Kamke, B., Hertel, I., and Gallagher, A., J. Chem. Phys. **80**, 4879 (1984).
5. Kleiber, P. D., Lyyra, A. M., Sando, K. M., Zafiropulos, V., and Stwalley, W. C., J. Chem. Phys. **85**, 5493 (1986).

Stark Effect in Some Lines of Br I

A. Bacławski, A. Goly, I. Książek and T.Wujec

Institute of Physics, Opole University, 45052 Opole, Oleska 48, Poland

The aim of this work is to add some experimental results of profile parameters for prominent lines of neutral bromine to the existing scarce data set, for which serious discrepancies between experimental values of different authors as well as between experimental and calculated values were observed. In some cases the ratio of measured to calculated Stark parameters attained 7.4. In the first paper concerning calculations of profile parameters for BrI lines [5] the ratio of measured to calculated full half-width was even higher (7.89) but in a later paper (treating apparently the same semi-classical theoretical approach) this ratio was two times smaller (see table 1- the value with the asterisk). The observed discrepancies between experiment and theory are too large, when for the experimental values [2] and [1] - both obtained in wall stabilized arcs - acceptable agreement may be constated. It is obvious, that the experimental value of Bengtson [3] coming from a shoc tube experiment is out of range of acceptable inaccuracy in measurements and should be remeasured. The theoretical results obtained by different assumptions differ strongly, too. For intercombination lines (4472 Å f. ex.) it may be expected that results of calculations are doubtful, but for the other lines the factor 2 shows the need to find better theoretical approaches for profile parameter calculations.

Figure 1 shows the used plasma source. For generating the bromine spectrum a

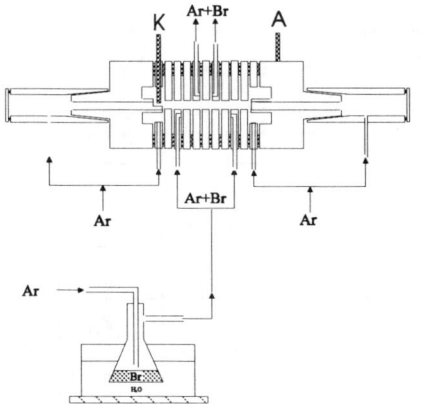

Fig.1. Plasma source for studies of the bromine spectrum.

© 1997 American Institute of Physics

mixture of argon and bromine vapours was introduced in the central part of a wall-stabilized arc. The arc was operated at currents in the range of 25-60 A. The diameter of the arc channel was D=4mm. The length of the Ar+Br plasma was about 55 mm. The amount of bromine vapours (about 10%) was changed by adjusting the temperature of the thermostat. The plasma column was imaged onto the entrance slit of the spectrometer applying a concave mirror (f=73cm). The spectra were recorded using a spectrograph PGS2, an optical multichannel analyzer OMA4 and a PC-386 computer. The plasma diagnostics was based on emission coefficients of argon and bromine lines and the half width of the Ar I 4300 Å line (LTE was assumed).

Table 1. Measured and calculated halfwidths and shifts of 3 BrI lines.
This Table contains the measured profile parameters (all available experimental values) compared to some theoretical results.

λ (Å)	Ref.	T (K)	$Ne*10^{-16}$ (cm^{-3})	w_m (Å)	d_m (Å)	w_m/w_{th}	d_m/d_{th}
4441.74	1	9800	3.25	1.02	0.39	0.59	0.61
	2,4	10750	4.5	1.26	0.52	0.44	0.48
		11530	7.0	1.70	1.19	0.37	0.70
	3,5	11000	10.0	10.40	-	4.53*	-
4477.72	1	9800	3.25	1.04	0.43	0.60	0.66
	2,1	10750	4.5	1.49	0.78	0.52	0.72
		11530	7.0	2.32	1.50	0.50	0.89
4472.61	1,4	9800	3.25	1.06	0.42	2.92	5.28
	2,1,4	10570	4.5	1.31	0.65	3.1	5.9
		11530	7.0	2.00	1.33	2.9	7.4

REFERENCES

1. S. Djurović, R. Konjević, M.Platiša and N. Konjević, *J. Phys. B: A.M. O.P.* **21**, 739 (1988)
2. *This work*
3. R.D.Bengtson, *Uni. of Maryland, Technical Note* BN-559, (1968)
4. S. Djurović, N. Konjević, M. S. Dimitrijević, *Z. Phys. D - At., Mol. and Cl.* **16**, 255 (1990)
5. M. S. Dimitrijević and N. Konjević, *JQSRT* **30,** 45(1983)

Acknowledgements: This work was supported in the frame of the Joint Polish-American M. Sklodowska-Curie Fund II (project MEN/NIST-94-166)

Light Absorption During a Resonant or Near-Resonant Collision: Study of the Cross Section in the Far-Wing

Stefano Cavalieri* and Milva Celli

Dipartimento di Fisica, Università di Firenze, Largo E. Fermi 2, I-50125 Firenze, Italy
*also with European Laboratory for Non-linear Spectroscopy (LENS), Università di Firenze, Largo E. Fermi 2, I-50125 Firenze, Italy

Abstract. We have studied the absorption of light induced by a resonant or near-resonant collision between two atoms. The calculations have been performed by taking into account also the magnetic sublevels, which makes our theoretical predictions more applicable to realistic cases. Analytical expressions for the far wing absorption cross-section have been obtained.

Absorption of light in the presence of collisions between similar or dissimilar species has been studied for many decades [1-3]. An important aspect of such a problem is the role of the finite duration of the collision which implies the breakdown of the impact approximation in the far wing absorption region. There have been experimental and theoretical studies about absorption induced by collisions of an alkali or alkali-earth atom with a noble gas atom [4,5], [6 and reference therein]. Over the past few years, collision has been reconsidered not only as a cause of decoherence but also as possible cause of extra resonance in non linear spectroscopy [7, 8]. In a previous work [9] the modification of the far-wing absorption profile due to coherence in a near-resonant collision has been analysed [10]. The result was obtained that a marked asymmetry in the absorption line shape might be present. Very recently such model was extended by Szudy and Baylis by the inclusion of the collisional shifts of atomic levels [11]. In these works, however, the magnetic degeneracy of the atomic states was fully ignored.

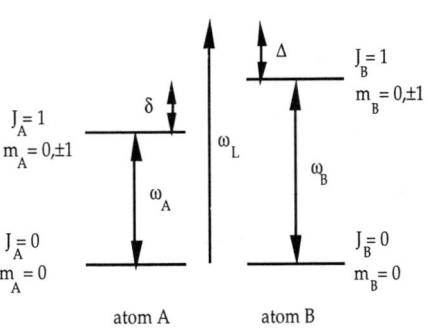

The possible influence of the magnetic degeneracy has been an important point in several treatments of collision as pointed out by Light and Szöke [12] for non resonant collisions and by Bambini and Geltmann for radiative collisions [13].

In this work we have extended the study of the process by including also the magnetic degeneracy. The results we obtain are more realistic and thus more suitable to be tested by experiments. We consider two colliding dissimilar atoms in the presence of a linearly polarised monochromatic laser field of frequency ω_L. (see figure).

We perform our analytical calculation by using the approximation of: binary collisions, first-order (i.e. dipole-dipole) inter atomic interaction only, classical rectilinear paths of atoms and adiabatic solution of collisional interaction. We solve the weak field radiative absorption by using the stationary phase approximation ($\Delta \gg 1/\tau_c$ for the blue side or $\Delta' = -(\Delta+\delta) \gg 1/\tau_c$ for the red side of the wing) and averaging over the impact parameter and orientation of the laser field polarization respect to the collision planes.

For $\delta \neq 0$ we have:

$$\sigma_{blue} \propto \frac{I_L}{\overline{v}} \frac{\left(2.5 d_A^2 \Delta + 2.5 d_B^2 (\Delta+\delta) + d_A d_B \Delta^{0.5}(\Delta+\delta)^{0.5}\right)}{\Delta^{1.5}(\Delta+\delta)^{1.5}} \quad (1)$$

$$\sigma_{red} \propto \frac{I_L}{\overline{v}} \frac{\left(2.5 d_B^2 \Delta' + 2.5 d_A^2 (\Delta'+\delta) - d_A d_B \Delta'^{0.5}(\Delta'+\delta)^{0.5}\right)}{\Delta'^{1.5}(\Delta'+\delta)^{1.5}} \quad (2)$$

and for $\delta=0$: $\sigma_{blue} \propto \frac{I_L}{\overline{v}} \frac{6 d^2}{\Delta^2}$; $\sigma_{red} \propto \frac{I_L}{\overline{v}} \frac{4 d^2}{\Delta'^2}$ (3)

The line shape (1) is in agreement with an experiment performed by Niemax [14]. The results obtained show asymmetry in the two far wings which is reduced in comparison to the works of references 9 and 11. This difference is due to the inclusion in the present calculation of the magnetic degeneracy allowing the influence of the angular dependence of dipole-dipole potential. The system studied is indeed closer to the actual experimental conditions. An experiment is planned for the verification of the present theoretical results.

REFERENCES

[1] P.R. Berman and W. Lamb, Phys. Rev. **187**, 221 (1969)
[2] J. Huennekens and A. Gallagher, Phys. Rev. **27**, 1851 (1983)
[3] K. Niemax and G. Pichler, J. Phys. B : At. Mol. Phys. **8**, 2718 (1975)
[4] A. Szöke and M.G. Raymer, Phys. Rev. A **15**, 1029 (1977)
[5] W.J. Alford, N. Anderson, K. Burnett and J. Cooper, Phys. Rev. A **30**, 2336 (1984)
[6] K. Burnett, Phys. Rep. **118**, 339 (1985)
[7] G. Grynberg and P.R. Berman, Phys. Rev. A **41**, 2677 (1990)
[8] P.R. Berman, Phys. Rev. A **51**, 592 (1995)
[9] S. Cavalieri, E. Arimondo and M. Matera, Phys. Rev. A **45**, 8005 (1992)
[10] By the term resonant or near-resonant collision we mean respectively collision where $\delta=0$ (it is the case of two identical atoms) or $0 < \delta \ll \omega_A, \omega_B$.
[11] J. Szudy and W. E. Baylis, Phys. Rev. A **53**, 2539 (1996)
[12] J. Light and A. Szöke, Phys. Rev. A **18**, 1363 (1978)
[13] A. Bambini and S. Geltman, Phys. Rev. A **50**, 5081 (1994)
[14] K. Niemax, Phys. Rev. Lett. **55**, 56 (1985)

Speed–dependent Galatry Profile

R.Ciuryło and J.Szudy

*Institute of Physics, Nicholas Copernicus University,
Grudziądzka 5/7, 87-100 Toruń, Poland.*

At low perturbing gas pressures the spectral line shapes are usually interpreted in terms of the Lorentzian profile which is derived as the Fourier transform *s–integral* of the impact–limit correlation function $\exp\{(i\Delta - \Gamma)s\}$. The Lorentzian parameters Γ and Δ which represent the HWHM and shift of the line peak with respect to the unperturbed frequency ω_0 are dependent on the emitter velocity v_E. The thermal motion of emitter gives rise to the phase shift $\vec{k}\vec{r}$ of the radiating dipole, where $\vec{k}$ is the wave vector of the emitted radiation and $\vec{r}$ is the time–dependent position of the emiter. Let $W_\beta(\vec{r}, s; \vec{v}_E)$ be the probability to find the radiating atom with the initial velocity $\vec{v}_E$ at the position $\vec{r}$ at time s. In the impact limit the correlation $\Phi(s)$ averaged over the Maxwellian distribution $f_{m_E}(\vec{v}_E)$ and the distribution $W_\beta(\vec{r}, s; \vec{v}_E)$ can be written as

$$\Phi(s) = \int d^3\vec{v}_E \, f_{m_E}(\vec{v}_E) \int d^3\vec{r} \, W_\beta(\vec{r}, s; \vec{v}_E) \, e^{i\omega_0 s + i\vec{k}\vec{r} + [i\Delta(v_E) - \Gamma(v_E)]s}$$

Here β denotes effective velocity–changing collision rate and m_E is emitter mass. If we neglect the velocity changes during collisions, i.e. for $\beta = 0$ when $W_\beta(\vec{r}, s; \vec{v}_E) = \delta(\vec{r} - \vec{v}_E s)$, the Fourier transform of this correlation function yields the speed–dependent Voigt profile [1, 2]. In order to take into account the velocity changing collisions and correlation between the collisional and Doppler broadening we adopt a diffusion model for description of the motion of emitting atom and assume $W_\beta(\vec{r}, s; \vec{v}_E)$ in the form used by Galatry [3]. The line shape obtained from this correlation function in this way is refered to as the speed–dependent Galatry profile (SDGP). This profile is shown in Fig. 1 where it is compared with the Voigt profile, speed–dependent Voigt profile (SDVP) [1, 2] and original Galatry profile [3]. Following Herbert [4] we used the reduced variables $u = (\omega - \omega_0)/\omega_D$, $g = \Gamma/\omega_D$, $d = \Delta/\omega_D$ and $z = \beta/\omega_D$ where $\Gamma = \int d^3\vec{v}_E \, f_{m_E}(\vec{v}_E) \, \Gamma(v_E)$, $\Delta = \int d^3\vec{v}_E \, f_{m_E}(\vec{v}_E) \, \Delta(v_E)$, $\omega_D = \omega_0\sqrt{2k_B T/(m_E c^2)}$, k_B is the Boltzmann constant, c is the speed of light, m_E is the emiter mass and T is the gas temperature. The calculations were performed with assumptions that the perturber mass $m_P \gg m_E$ and the interatomic interaction is described by the van der Walls potential. We have shown that our SDGP formula yields results close to those obtained from a model due to Duggan et al [5] based on the theory of Robert et al

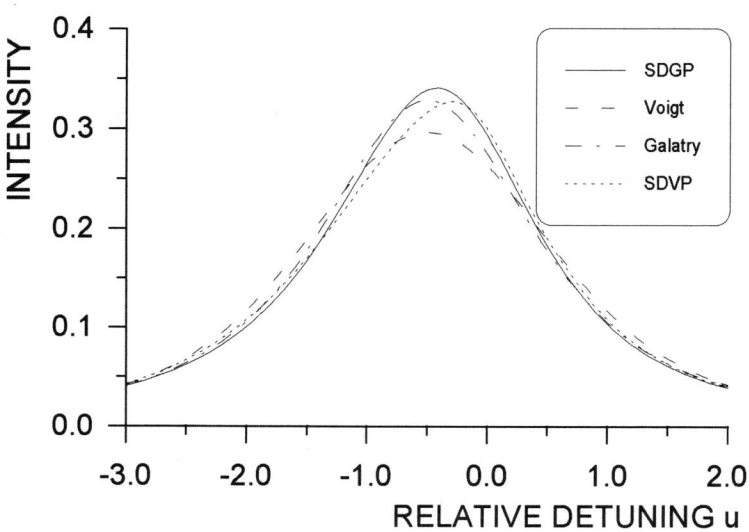

FIGURE 1: The comparison of the SDGP (full line) with the usual Voigt profile (dashed line), Galatry profile (dash–and–dot line) and SDVP (dotted line). Calculations were done for: $g = 0.7$, $d = -0.5$, $z = 0.8$.

[6], folding in the effect of Dicke narrowing. We should note that the formula derived in Ref. [5] yields good agrement with experimental data for the CO lines perturbed by heavy perturbers.

ACKNOWLEDGMENTS

When this work was done one of authors (R.C.) was the scholarship–holder of the Fundation for Polish Science. This work was partially supported by a grant No 673/P03/96/10 (2 P03B 005 10) from the State Committee for Scientific Research.

REFERENCES

1. P. R. Berman, *JQSRT* **12**, 1331 (1972).
2. J. Ward, J. Cooper and E. W. Smith, *JQSRT* **14**, 555 (1974).
3. L. Galatry, *Phys. Rev.* **132**, 1218 (1961).
4. F. Herbert, *JQSRT* **14**, 943 (1974).
5. P. Duggan, P. M. Sinclair, A. D. May and J. R. Drummond, *Phys. Rev. A* **51**, 218 (1995).
6. D. Robert, J. M. Thuet, J. Bonamy and S. Temkin, *Phys. Rev. A* **47**, R771 (1993).

Speed–dependent Correlations between Pressure and Doppler Broadening of the 748.8 nm Neon Line

R.Ciuryło, A.Bielski, J.Szudy, R.S.Trawiński

*Institute of Physics, Nicholas Copernicus University,
Grudziądzka 5/7, 87-100 Toruń, Poland.*

The Lorentzian and Doppler widths of the neon 748.8 nm ($2p^53d_3 - 2p^52p_{10}$) spectral line perturbed by neon and helium in the low–current 1.5 mA glow discharge have been investigated using a pressure–scanned Fabry–Peròt interferometer. The experimental set–up and the methods of line profile analysis were identical to those used in a previous work [1]. Measurements were performed in the pressure range between 1.6 and 40.0 Torr in the case of broadening by helium (Ne*–He), and between 0.8 and 100.0 for the selfbroadening (Ne*–Ne). Using a formula due to Ballik [2], the Lorentzian and Doppler widths were determined by a least–squares procedure and the Marquardt algorithm. We have found that in the case of selfbroadening, i.e. when the perturbers are the ground–state neon atoms, the Doppler width at higher pressures decreases with the increase of the perturber density. Such an effect was not observed in the case of the broadening by helium. A detailed analysis performed during the course of this study indicates that our observations can be interpreted as a manifestation of the correlation between the pressure broadening rate and the emitter velocity as discussed by Berman [3] and Ward et al [4] and experimentally studied by the Lewis group [5]. The correlation effects are negligible if $\alpha = m_P/m_E$, the ratio of perturber and emitter masses, is much less than one. This happens in the case of the broadening of the neon line by helium when $\alpha = 0.2$. In such a case the Doppler width is independent of the Lorentzian width as shown in Fig. 1. Contrary to that, for the broadening of the Ne line by the ground–state neon atoms, i.e. when $\alpha = 1.0$ the experimentally determined Doppler widths of the 748.8 nm line depend on the Lorentzian widths (see Fig. 1). In order to get more insight into the problem of the role of correlation effects for the Ne*–Ne system we have performed simulation of the dependence of the Doppler width upon the Lorentzian one using the speed–dependent Ballik profile formula [6] for $\alpha = 1$ with the assumption that the interaction potential is of the van der Waals form. Results of this simulation are presented in Figure 1 and plotted as a solid line. Experimental values are close to the results obtained from the simulation. This effect has not been observed earlier for rare gases probably because experiments were performed for too low pressures of perturbing

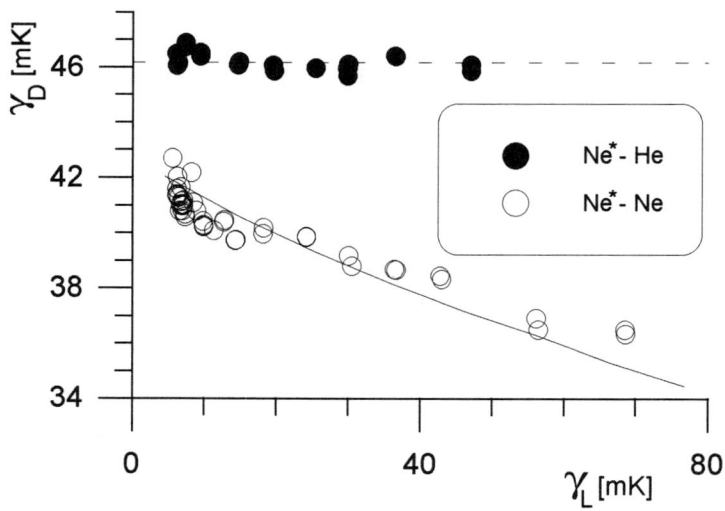

FIGURE 1: Dependence of Doppler width of 748.8 nm Ne – line upon Lorentzian widths. Experimental points: neon (o) and helium (•) as a perturber gas.

gases or the broadening of the lines under investigation had the resonance character. For the resonance broadening this efect does not appear [7, 8].

ACKNOWLEDGMENTS

When this work was done one of authors (R.C.) was the scholarship–holder of the Fundation for Polish Science. This work was partially supported by a grant No 673/P03/96/10 (2 P03B 005 10) from the State Committee for Scientific Research.

REFERENCES

1. R. S. Trawiński, A. Bielski, R. Ciuryło, J. Szudy, *Ann. Physik* **2**, 1 (1993).
2. E. A. Ballik, *App. Opt.* **5**, 170 (1966).
3. P. R. Berman, *JQSRT* **12**, 1331 (1972).
4. J. Ward, J. Cooper, E. W. Smith, *JQSRT* **14**, 555 (1974).
5. M. Harris, E. L. Lewis, D. McHugh, I. Shannon, *J. Phys.* B **17**, L661 (1984); I. Shannon, M. Harris, D. McHugh, E. L. Lewis, *J. Phys.* B **19**, 1409 (1986); M. Harris, E. L. Lewis, D. McHugh, I. Shannon, *J. Phys.* B **19**, 3207 (1986).
6. R. Ciuryło, A. Bielski, S. Brym, J. Jurkowski, *JQSRT* **53**, 493 (1995).
7. J. Cooper, D. N. Stacey, *Phys. Rev.* A **12**, 2438 (1975).
8. D. N. Stacey, R. C. Thompson, *J. Phys.* B **16**, 537 (1983).

On the Temperature Dependence of the Stark Broadening Parameters of Ar I 425.9 nm Line

S. Djurović[*], Z. Mijatović[*], R. Kobilarov[*], and N. Konjević[**]

[*]Institute of Physics, Trg Dositeja Obradovića 4, 21000 Novi Sad and [**]Institute of Physics, P.O. Box 68, 11080 belgrade, Yugoslavia

EXPERIMENT

We report results of an experimental study of the Stark widths and shifts of Ar I 425.9 nm line.

For the plasma source an atmospheric pressure stabilized electric arc is used. For diagnostic purposes 4% H_2 is mixed with pure argon. The current of 30 A was supplied to the arc by a current-stabilized power supply with stability of 0.3%. The plasma observations were performed side-on with a 1m monochromator and photomultiplier tube. The signals from the photomultiplier were led to the digitizing oscilloscope working in the averaging mode. For the shift measurements, the low pressure argon lamp with microvawe excitation is used as a source of unshifted spectral line.

An electron density in the range $(0.47 - 3.50) \times 10^{22}$ m^{-3} is determined from the width of Balmer H_β line (1). The electron tempera ture in the range (8900 - 11000) K is obtained from the plasma composition data (2).

RESULTS

The spectral line profiles from the stabilized arc were Abel inverted (3). Than Ar I 425.9 nm line profiles were treated by computer program developed for deconvolution of Gaussian and $j_{A,R}(x)$ profile (4) to obtain measured full halfwidth data. The spectral line profiles from the reference source, needed for shift measurements, were fitted by least square method to Gaussian profiles.

The measured widths and shifts data are compared with theoretical calculations. The values of width data are (20 - 30)% smaller and shift data are (15 - 25)% smaller than those theoretically predicted (see Table 1) and show gradual

Table 1 The measured and theoretical halfwidth w and shifts d

N_e $(10^{22}\,m^{-3})$	T_e (K)	w_m (nm)	w_t (nm)	d_m (nm)	d_t (nm)
3.50	11070	0.071	0.103	0.038	0.051
3.40	11040	0.070	0.100	0.037	0.049
3.30	10980	0.066	0.096	0.036	0.048
3.10	10890	0.062	0.090	0.034	0.045
2.70	10690	0.057	0.078	0.030	0.039
2.40	10540	0.049	0.069	0.025	0.035
2.00	10310	0.042	0.057	0.023	0.029
1.70	10120	0.035	0.048	0.019	0.025
1.40	9930	0.030	0.040	0.016	0.020
1.05	9680	0.024	0.029	0.012	0.015
0.90	9470	0.019	0.025	0.009	0.013
0.70	9260	0.014	0.019	0.008	0.010
0.60	9130	0.012	0.016	0.007	0.009
0.47	8900	0.010	0.013	0.006	0.007

increase of the discrepancy with temperature. Furthermore, comparisons with other experimental results (5 - 12) which are not numerous and which are not in the our electron density range, show similar discrepancy with theory. These comparisons indicate that further improvements of the theory are required.

REFERENCES

1. Vidal, C. R., Cooper, J., and Smith, E. W., Astrophys. J. Suppl. Ser. **25**, 37 (1973).
2. White, W. B., Jonson, S. M., and Dantzig, G. B., J. Chem. Phys. **28**, 751 (1958).
3. Djurović, S., Kelleher, D. E., and Roberts, J. R., to be published.
4. Mijatović, Z., Kobilarov, R., Vujičić, B. T., Nikolić, D., and Konjević, N., J. Quant. Spectrosc. Radiat. Transfer **50**, 339 (1993).
5. Gericke, W. E., Z. Astrophys. **53**, 68 (1961).
6. Powel, W. R., Ph. D. Thesis, The Jons Hopkins University (1966).
7. Schulz, P., and Wende, B., Z. Phys. **208**, 116 (1968).
8. Bues, I., Haag, T., and Richter, J., J. Astron. Astrophys. **2**, 249 (1969), Bues, I., Haag, T., and Richter, J., Lab. aus dem Inst. fur Experimental-physik der Univ. Kiel, Kiel (1969).
9. Morris, J. C., and Morris, R. U., Aerospace Research Laboratories Rport No ARL 70-0038 (1970).
10. Musielok, B., Musielok, J., and Wujec, T., Zesz. Nauk. Wyzsz. Szk. Pedagog. Opolu, Fiz. No17, 63 (1976).
11. Klein, P. and Meiners, D., J. Quant. Spectrosc. Radiat. Transfer **17**, 197 (1997).
12. Jones, D. W., Wiese, W. L., and Woltz, L. A., Phys. Rev. A **34**, 450 (1986).

Observation of collision-time asymmetry of the ^{114}Cd $\lambda = 326.1$ nm line perturbed by krypton

J. Domysławska and R.S. Trawiński

Institute of Physics, Nicholas Copernicus University,
Grudziądzka 5/7, 87-100 Toruń, Poland.

Experiment was performed using the ^{114}Cd isotope to avoid hyperfine and isotope structures of the studied $\lambda = 326.1$ nm ($5^1S_0 - 5^3P_1$) absorption line. The quartz cell containing the cadmium vapour and krypton was situated in an oven at temperature 750K. The side arm enabled us to controll cadmium density in the cell. The number density of the cadmium atoms was changed between 5×10^{13} and 5×10^{15} atom cm^{-3}. The number density of krypton was between 6.5×10^{18} and 20×10^{18} atom cm^{-3}.

If the pressure and Doppler broadening are independent the line shape can be written as a convolution of Doppler and pressure broadened profiles. In the core of pressure-broadened spectral line the line shape can be described in the framework of the impact approximation. The absorption coefficient is then written in the form $\mu(\tilde{\nu}) = \mu_C(\tilde{\nu}) \otimes \mu_D(\tilde{\nu})$, where $\mu_C(\tilde{\nu})$ and $\mu_D(\tilde{\nu})$ are the collisional and Doppler profiles, respectively, and symbol $\otimes$ denotes a convolution.

Collisional part of absorption coefficient has a form (see e.g. [1])

$$\mu_C(\tilde{\nu}) = \mu_0 \left(\frac{\gamma}{2}\right)^2 \frac{1 + B(\tilde{\nu} - \tilde{\nu}_0)}{(\tilde{\nu} - \tilde{\nu}_0)^2 + (\gamma/2)^2}$$

where $\tilde{\nu}_0$ – wave number for the maximum of the absorption line,
γ – collisional line width (FWHM), B – asymmetry factor,
$\mu_0 = \mu_C(\tilde{\nu}_0) = (f N_{Cd}/\gamma)(2e^2/mc^2)$ where f – oscillator strength, N_{Cd} – number density of Cd atoms.

In the framework of the Unified Franck-Condon treatment [1] the asymmetry parameter is given by

$$B = <\frac{4\pi N\hbar}{\mu k_i} \sum_{l=0}^{\infty}(2l+1)a_l \sin \eta(\infty) >$$

where symbol $< \ldots >$ denotes average over the initial wave numbers k_i or equivalently over initial perturber energies, μ is the reduced mass of the perturber-emitter system, N denotes perturber number density, and

$$a_l = \frac{1}{\hbar}\int_0^\infty du\, \Delta V(u) \sin \eta(u) u \quad \text{where} \quad \eta(u) = \frac{1}{\hbar}\int_0^u du'\, \Delta V(u')$$

© 1997 American Institute of Physics

Observed intensity distribution of the pressure and Doppler broadened spectral line can be written in the following form

$$O(\tilde{\nu}) = I(\tilde{\nu}) \otimes A(\tilde{\nu}) = I_0[\exp{(-l\,\mu_C \otimes \mu_D)}] \otimes A(\tilde{\nu})$$

where $A(\tilde{\nu})$ is an instrumental profile, l is the thickness of an absorber.

The $O(\tilde{\nu})$ profile was fitted to the measured profile using a least-squares algorithm given by Marquardt [2]. Fitting parameters were asymmetry factor B, Cd number density N_{Cd} and $\tilde{\nu}_0$. We assumed Doppler width to be $\gamma_D = 0.055 \text{cm}^{-1}$ and we calculated the collisional width (γ) using data from previous work [3].

We recorded about 40 profiles for every perturber pressure. The asymmetry factor B for each pressure of krypton is the weighted mean value of B factors determined for each of recorded profiles.

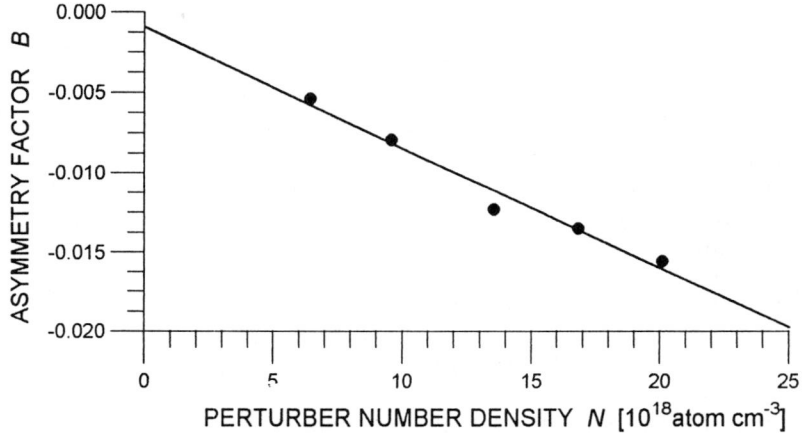

FIGURE 1: Dependence of asymmetry factor on the perturber number density N for ^{114}Cd $\lambda = 326.1$ nm absorption line perturbed by krypton.

From the slope of the straight line on figure 1 we obtained asymmetry coefficient $B/N = (-0.75 \pm 0.12) \times 10^{-4}$ (atom cm^{-3})$^{-1}$.

ACKNOWLEDGMENTS

This work was partially supported by a grant No 673/P03/96/10 (2 P03B 005 10) from the State Committee for Scientific Research.

REFERENCES

1. J. Szudy, W. E. Baylis, *J. Quant. Spectr. Rad. Transf.* **15**, 641 (1975).
2. D. W. Marquardt, *J. Soc. Industr. Appl. Math.* **11**, 431 (1963).
3. S. Brym and J. Domysławska, *Physica Scripta* **52**, 511 (1995).

Quasistatic Non-Binary Self-Broadening and Radiationless Transfer of Excitation in a System of Classical Oscillators

Ya.A.Korennoy[*], A.B.Kukushkin[*], and A.G.Zhidkov[**][(+)]

[*]*INF RRC "Kurchatov Institute", Moscow, 123182, Russia*
[**]*Institute for General Physics, Vavilova 38, Moscow, Russia*
[(+)] *Presented by V.S.Lisitsa[*]*

Abstract. Numerical modelling of the line shape in a system of classical oscillators under conditions of quasistatic non-binary self-broadening is carried out, and good agreement is found with the corresponding quantum results. An analytic solution and the scaling laws are derived for the non-steady-state problem of the excitation transfer in a 1D regular chain of classical oscillators, and the results show good agreement with the results of numerical modelling carried out.

The problem of radiationless transfer of excitation is analyzed, both analytically and numerically, for a system of classical oscillators under conditions of quasistatic non-binary self-broadening. In the frame of such a "classical" approach the line shape is formed by a set of eigenmodes of (collective) oscillations in the system of *like* oscillators, which interact each with other by the self-consistently produced dipole-dipole potential [1]. The quasistatic self-broadening implies steady-state spatial distribution of oscillator center-of-mass positions, either regular or random. The line shape is given by statistical averaging over this

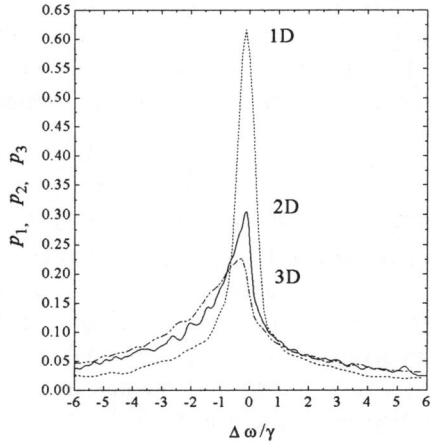

FIG.1. Line shapes for the non-binary self-broadening in 1D, 2D, and 3D systems of classical oscillators.

spatial distribution and thus takes a form of the probability $p(\omega)d\omega$ for the frequency eigenvalue to lie in the interval $\omega \div \omega + d\omega$. The most representative results of calculations are shown in FIG.1. The line shapes, obtained in classical oscillators approach, show good agreement with the corresponding quantum results [2] for 'isotropic' two-level atoms. This demonstrates classical nature of the line self-broadening mechanism, including effects of line width asymmetry and "red" shift of the line center. The far line wings have the following form in 1D, 2D, and 3D cases, respectively:

$$p_1 \approx 0.26\gamma^{1/3}\Delta\omega^{-4/3}, \quad p_2 \approx 0.53\gamma^{2/3}\Delta\omega^{-5/3}, \quad p_3 \approx 1.14\gamma\Delta\omega^{-2},$$

where δ and $\gamma = e^2/m\omega_0\delta^3$ are average distance and characteristic time of radiationless energy exchange between two neighbouring oscillators. The results show strong dependence of line shift and line width on the dimensionality of the system. This takes place because of essentialy different statistical weight of those oscillators whose perturbing field is not screened by the closest neighbors. The latter appears to be of special importance for excitation transfer.

An analytical investigation of the excitation transfer problem in a meduim composed of self-broadened oscillators is available in the frame of the "continualized" representation of meduim excitation in the form of an integro-differential equation. Such an equation can be easily derived in the case of regular static positions of oscillators when there is no contribution of space-time fluctuations of oscillator density. Here, in the case of infinite meduim one can solve this equation by the Fourier method for instant point source of the dipole momentum D_0. In the 1D regular chain, analytic solution for the excitation propagation gives scaling laws for the evolution of excitation energy in the following cases: namely, at excitation front, $x \sim \delta\gamma t$; behind the front, $x \ll \delta\gamma t$; and, far from the front, $x \gg \delta\gamma t$, respectively ($\gamma t \gg 1$):

$$\varepsilon(x_{fr}(t),t) \approx \mu \frac{m\omega_0^2}{4e^2}\frac{D_0^2}{(2\gamma t)^{2/3}}, \quad \varepsilon \approx \mu \frac{m\omega_0^2}{2e^2}\frac{D_0^2}{\pi\gamma t}, \quad \varepsilon \approx \mu \frac{m\omega_0^2}{2e^2}\frac{(t\gamma D_0)^2}{|x|^2},$$

where $x_{fr} = \delta\gamma t$ is front position; $\mu = 1/\delta$ is oscillator (linear) density, $F(x,t) \sim 1$ is a positive oscillating function of its arguments.

The 1D excitation non-steady state transfer in the case of small random deviations from regular positions of oscillators was analysed numerically. The results show substantial slowing down of energy transfer -- up to its suppression in the case of complete randomization of oscillators positions. However, in 2D and 3D systems numerical modelling shows that such a suppression doesn't occur.

REFERENCES

1. Breene R.G., The Shift and Shape of Spectral Lines, Pergamon (1961).
2. Leegwater J.A., Mukamel S., Phys. Rev. A, 49 (1994) 146.

Quantum Width and Shift of Spectral Lines in the Non-Impact Regime

W. C. Kreye* and J. F. Kielkopf[†]

*Research and Instructional Computer Center, Division of Computing and Telecommunications Services, Wright State University, Dayton, OH 45435 USA

[†]Department of Physics, University of Louisville, Louisvillle, KY 40292 USA

Abstract. A general quantum-mechanical theory of the perturbation shift and broadening of spectral lines in the non-impact regime has been extended and programmed and applied to two simple potentials--the square-well potential and a Morse potential. At low pressure, the non-impact and impact line shapes coincide, whereas at high pressures the non-impact line has acquired an asymmetry, a slight relative shift, and a width about twice that of the impact line.

Most fully quantum-mechanical treatments of perturbation shift and width of spectral line shapes in the line core are in the impact regime [e.g., Kreye and Kielkopf's (1) study of the Ar-perturbed K line]. The present non-impact-region study was deemed necessary because, for example, if an experimental line width is interpreted using impact theory alone, the predicted width of, say, a square-well potential (SWP) would be erroneously small. A semi-classical treatment of the non-impact regime was published by Kielkopf (2). The rigorous, elegant paper by Baranger (3) is the theoretical basis for the present study. Starting with the standard expression for dipole radiation, he develops a general expression for g(s) which appears in the Fourier transform of the line shape exp[-n · g(s)]. s has units of time and may be considered the approximate time between collision of perturber and radiator. His Eq. (29) is simplified here for the case where the final potential is zero, yielding a plane-wave function $\vec{k}$, and becomes

$$g(s) = is\langle \vec{k} | V_i | \psi_{ik}\rangle +$$

$$\text{const} \cdot \int d\,\Phi' \int \sin\theta'\, d\theta' \int d\varepsilon'\, \varepsilon'^{1/2} |\langle \vec{k'} | V_i | \psi_{ik}\rangle|^2 \cdot$$

$$\left\{ \frac{2\sin^2[(\varepsilon - \varepsilon')s/2]}{(\varepsilon - \varepsilon')^2} + i\, \frac{\sin[(\varepsilon - \varepsilon')s]}{(\varepsilon - \varepsilon')^2} \right\} \quad (1)$$

where he has introduced a complete set of intermediate states $\vec{k'}$ which are eigenstates of H_f. The line shape then becomes

$$F(\omega)2\pi = 4\int_0^{s_{max}} ds\, \exp\{n \cdot R[g(s)]\}\cos\{-n \cdot I[g(s)] + \omega s\} \quad (2)$$

© 1997 American Institute of Physics

Note that two singular points exist at $\varepsilon = \varepsilon'$. In Eq. (2) s_{max} is the largest value of the upper limit beyond which no further change in the integral occurs. In Eq. (1) the first term, $g_1(s)$, corresponds to the impact regime and the second, $g_2(s)$, to the non-impact contribution. The method of partial waves is used, where

$$\vec{k} = \sum_{\ell'} (2\ell' + 1) \, j_{\ell'}(kr') \, i^{\ell'} P_{\ell'}(\cos \theta) \tag{3}$$

$$\psi_{ik} = \sum_{\ell} j_{\ell}(\alpha r') \left[(2\ell + 1) i^{\ell} e^{-i\delta_{\ell}}\right] P_{\ell}(\cos \theta) \tag{4}$$

For this study, we used a narrow SWP (0.4 Å) which facilitates the numerical development and debugging; more importantly, its line shape exhibits all the characteristics of the non-impact regime. A Morse potential $V(r')$ now being studied makes the numerical methods considerably more demanding since the relatively simple function $j_{\ell}(\alpha r')$ in Eq. (4) is replaced by $R_{\ell}(r')$--the solution of the radial Schroedinger equation that includes the variable potential $V(r')$.

Figure 1a,b shows the effect of pressure on line shape; solid (——) and dashed (- - -) lines represent the non-impact and impact-regime line shapes, respectively.

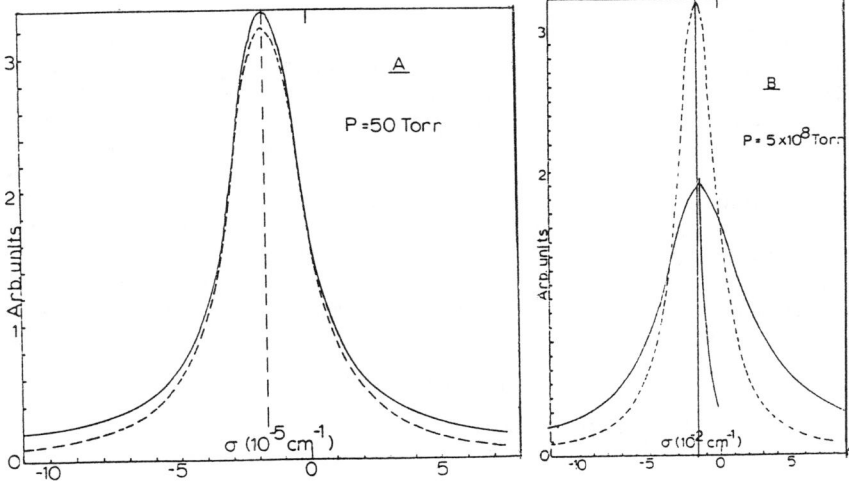

FIGURE 1. Effect of pressure. T = 400 K, K/Ar system.

At the lowest pressure of 50 Torr, the non-impact and impact line shapes coincide; the corresponding value of s_{max} is large. This large value corresponds to the Baranger (3) condition, namely, that the impact approximation holds for large values of s. At intermediate pressures, the width of the non-impact line increases to about twice that of the impact line (not shown). At the high pressure of 5×10^8 Torr, the non-impact line has acquired a large width and shift relative to the impact line and an asymmetry. At these high pressures s_{max} is very small and approaches the value of the "collision time" [Baranger (3)].

1. Kreye, W. C., and Kielkopf, J. F., "Non-adiabatic effects in the broadening and shift of the K 7s-4p transition by argon," submitted to *J. Phys. B: At. Molec. Phys.*
2. Kielkopf, J. F., *J. Phys. B: At. Mol. Phys.* **16**, 3149 (1983).
3. Baranger, M., *Phys. Rev.* **111**, 481 (1958).

Stark Effect in Some Lines of Neutral Argon Emitted from a Ferroelectric Plasma Source

Józef Kusz and Dariusz Mazur

Institute of Physics, Opole University, Oleska 48, 45-052 Opole, Poland

Abstract. The experimental results of profile parameters for selected lines of neutral argon emitted from a ferroelectric plasma source are presented. The obtained results show good agreement with the literature data confirming the applicability of the ferroelectric plasma source for investigations of the interatomic Stark effect.

Ferroelectric plasma source has been found useful for investigations in the field of optical spectroscopy as a stable and efficient source of low-temperature nonequilibrium plasma, the parameters of which can be easily regulated [1]. The easiness of introduction of substances into the ferroelectric plasma source and the broad range of its working parameters makes the source very useful for excitation of atoms and molecules of substances investigated by optical spectroscopy. It has been found to be especially useful for exciting the atoms and ions of rare earth elements. The aim of this work is to test whether the ferroelectric plasma source is also suitable for investigations of the Stark widths and shifts of spectral lines emitted from it. At first we have to confront the results obtained by using ferroelectric plasma source with the existing literature data.

In this work we present experimental results of profile parameters for selected lines of neutral argon emitted from a ferroelectric plasma source. Argon was choiced because an extensive number of experimental Stark broadening and shift data exist in the literature. Most of the results were obtained in wall-stabilised arcs and plasma jets, but other plasma sources were also used.

The plasma was generated in argon under atmospheric pressure with traces (below 1%) of hydrogen in a gap (0.16 mm width) between two ferroelectric plates with the diameter 32 mm and thickness 2 mm. The plates were made of barium titanate ceramics covered with silver electrodes on one side. The electrodes were supplied with alternating (sinusoidal) voltage of the frequency of 10 kHz and amplitude of 1250 V. The gas gap was projected by a quartz lens onto

© 1997 American Institute of Physics

the entrance slit of the spectrometer. The spectra were recorded by using a grating spectrograph (PGS2), an optical multi-channel analyser OMA4 and a PC-386 computer. The spectrum of the plasma was registered in the range 300 to 1500 nm. The value of the electron concentration, estimated from the measurement of the half-widths of the ArI 430.01 nm and of the H_β lines was about 4×10^{15} cm^{-3}. The ArI 415.86 nm, ArI 430.01 nm and ArII 434.81 nm, as well as ArI 667.73 nm and ArII 663.82 nm lines, were applied for determination of the ionisation equilibrium temperature of the plasma. The temperature was about 11100 K.

TABLE 1. Stark widths and shifts for selected ArI lines

No.	Transition array	Wavelength (nm)	w_m - (nm) this work	w_m - (nm) Ref. 5	d_m - (nm) this work	d_m - (nm) Ref. 5
1	4s-5p	416.418	0.0369	0.0181-0.181	0.004	0.010-0.079
2	4s-5p	419.832	0.0328	0.022-0.12	0.0096	0.018-0.095
3	4s-5p	420.067	0.0278	0.019-0.155	0.013	0.012-0.088
4	4s'-5p'	425.936	0.0285	0.0187-0.197	0.077	0.005-0.056
5	4s-5p	427.217	0.0293	0.02-0.169	0.013	0.012-0.089
6	4s-4p'	667.728	0.0381	0.04-0.082	0.0091	0.031-0.054
7	4p-4d	675.283	0.0704	0.076-0.214	0.038	0.025-0.086
8	4p-4d	687.129	0.0684	0.081-0.311	0.031	0.048-0.104
9	4s-4p'	696.543	0.0284	0.008-0.063	0.016	0.005-0.023
10	4s-4p'	714.704	0.0344	0.016-0.07	0.017	-
11	4s-4p'	727.293	0.0439	0.018-0.08	0.015	-
12	4s'-4p'	750.387	0.1424	0.028-0.060	0.0048	-
13	4s-4p	763.511	0.0380	0.020-0.050	0.016	0.017-0.0852
14	4s'-4p'	794.818	0.0391	0.020-0.053	0.022	-
15	4s'-4p	1047.005	0.0458	-	-0.012	-0.05
16	3d'-4f'	1463.411	0.0631	0.68	-	-0.3

We have obtained Stark widths and shifts for more than fifty ArI lines. The results for selected lines are collected in TABLE 1. Comparison data [2] were selected for nearly the same temperature and electron density as in the present work. The obtained results show in most cases good agreement with the literature data thus confirming the usability of the ferroelectric plasma source for investigations of the Stark effect. We have also obtained the Stark widths and shifts for more than twenty ArI lines not presented in paper [2].

REFERENCES

1. Kusz, J., Ferroelectric plasma sources and some of their applications, in *Advances in Low-Temperature Plasma Chemistry, Technology and Applications*, Vol. 4, H. V. Boenig ed. Technomic, Lancaster, PA, 1992, p. 237 - 314.
2. Konjevic, N., and Roberts, J. R., J. Phys. Chem. Ref. Data, Vol. 5, No. 2, 1976.

Pressure Broadening of $2p^5\,3s - 2p^5\,3p$ Neon Lines

P. J. Leo[‡], D. F. T. Mullamphy[†], G. Peach[‡], V. Venturi[†] and
I. B. Whittingham[†]

[‡]*Department of Physics and Astronomy, University College London,
London WC1E 6BT, UK.*
[†]*Department of Physics, James Cook University, Townsville, Australia, 4811.*

Abstract. We present a detailed theoretical calculation of collisional self-broadening in neon where the lower level is either a metastable or intercombination level, $3s'[\frac{1}{2}]_0$ or $3s[\frac{3}{2}]_{1,2}$, and the upper level is $3p'[\frac{1}{2}]_1$, $3p'[\frac{3}{2}]_{1,2}$, $3p[\frac{1}{2}]_{0,1}$, $3p[\frac{3}{2}]_{1,2}$ or $3p[\frac{5}{2}]_2$. The collisionally broadened width and shift are obtained for the temperature range $70\,K$ to $370\,K$ under the impact approximation using realistic interaction potentials and a fully quantum mechanical model of the scattering event. Parametric fits to the temperature dependence of the width and shift are reported which can be used to reproduce the theoretical results to within 0.5%.

RESULTS

For the details of the interaction potentials and the quantum scattering model we refer the reader to the papers (1) and (2). Comparison of existing experimental data to the theoretical results can also be found in these references and here we provide only the parametric fits to the temperature dependence of the Lorentzian half-halfwidth and shift.

The temperature dependence of the width and shift is particularly useful when making comparisons between theory and experiment as experimental data is generally collected at temperatures determined only subsequently from either the temperature of the bath containing the gas discharge, corrected for heating effects of the discharge current, or from the Gaussian component of the line profile, extracted from the measured Voigt profile. The fits in tables 1 and 2 reproduce the theoretical width and shift to within 0.5% for the temperature range $70\,K$ to $370\,K$ (except for transitions from the $3p[\frac{1}{2}]_1$ level where for temperatures less than $80\,K$ the fits are within 1.5%).

1. Leo P J, Mullamphy D F T, Peach G and Whittingham I B 1995 *J. Phys. B: At. Mol. Opt. Phys.* **28** 4449-4458
2. Leo P J, Mullamphy D F T, Peach G, Venturi V and Whittingham I B 1996 Submitted to: *J. Phys. B: At. Mol. Opt. Phys.*

Table 1: Fits to line widths (in units of $\times 10^{-21}$ MHz m^3/atom).

Transition	Width coefficients[a]			
	$c_1 \times 10^{-9}$	$c_2 \times 10^{-6}$	$c_3 \times 10^{-4}$	$c_4 \times 10^{-1}$
$3s'[\frac{1}{2}]_0-3p'[\frac{1}{2}]_1$	+1.128	−1.404	+9.272	+1.347
$3s[\frac{3}{2}]_1-3p'[\frac{1}{2}]_1$	+1.074	−1.353	+9.094	+1.366
$3s[\frac{3}{2}]_2-3p'[\frac{1}{2}]_1$	+1.074	−1.353	+9.093	+1.366
$3s[\frac{3}{2}]_1-3p'[\frac{3}{2}]_2$	+1.262	−1.530	+9.418	+1.391
$3s[\frac{3}{2}]_2-3p'[\frac{3}{2}]_2$	+1.275	−1.538	+9.436	+1.388
$3s'[\frac{1}{2}]_0-3p'[\frac{3}{2}]_1$	+0.720	−1.002	+7.147	+1.365
$3s[\frac{3}{2}]_1- 3p'[\frac{3}{2}]_1$	+0.764	−1.026	+7.116	+1.389
$3s[\frac{3}{2}]_1-3p[\frac{1}{2}]_0$	+1.037	−1.349	+9.036	+1.317
$3s'[\frac{1}{2}]_0-3p[\frac{1}{2}]_1$	+0.248	−0.197	+4.987	+1.561
$3s[\frac{3}{2}]_1-3p[\frac{1}{2}]_1$	+1.097	−0.823	+5.995	+1.600
$3s[\frac{3}{2}]_2-3p[\frac{1}{2}]_1$	+0.926	−0.681	+5.600	+1.640
$3s[\frac{3}{2}]_1-3p[\frac{3}{2}]_2$	+0.787	−1.084	+7.766	+1.308
$3s[\frac{3}{2}]_2-3p[\frac{3}{2}]_2$	+0.790	−1.087	+7.774	+1.306
$3s'[\frac{1}{2}]_0-3p[\frac{3}{2}]_1$	+1.479	−1.700	+9.701	+1.045
$3s[\frac{3}{2}]_1- 3p[\frac{3}{2}]_1$	+1.439	−1.640	+9.422	+1.101
$3s[\frac{3}{2}]_2-3p[\frac{3}{2}]_1$	+1.436	−1.637	+9.415	+1.101
$3s[\frac{3}{2}]_1-3p[\frac{5}{2}]_2$	+0.636	−0.930	+7.146	+1.300
$3s[\frac{3}{2}]_2-3p[\frac{5}{2}]_2$	+0.639	−0.933	+7.156	+1.298

[a] To the fit $c_1T^3 + c_2T^2 + c_3T + c_4$

Table 2: Fits to line shifts (in units of $\times 10^{-21}$ MHz m^3/atom).

Transition	Shift coefficients[b]				
	$c_5 \times 10^{-12}$	$c_6 \times 10^{-9}$	$c_7 \times 10^{-7}$	$c_8 \times 10^{-5}$	$c_9 \times 10^{-2}$
$3s'[\frac{1}{2}]_0-3p'[\frac{1}{2}]_1$	+1.674	−1.755	+6.608	−5.546	−6.065
$3s[\frac{3}{2}]_1-3p'[\frac{1}{2}]_1$	+1.827	−1.910	+7.224	−6.804	−6.240
$3s[\frac{3}{2}]_2-3p'[\frac{1}{2}]_1$	+1.816	−1.902	+7.205	−6.786	−6.241
$3s[\frac{3}{2}]_1-3p'[\frac{3}{2}]_2$	+2.118	−2.240	+8.879	−13.00	−4.043
$3s[\frac{3}{2}]_2-3p'[\frac{3}{2}]_2$	+2.136	−2.257	+8.931	−13.07	−4.040
$3s'[\frac{1}{2}]_0-3p'[\frac{3}{2}]_1$	+2.973	−3.247	+13.20	−20.77	−5.271
$3s[\frac{3}{2}]_1- 3p'[\frac{3}{2}]_1$	+3.062	−3.343	+13.61	−21.36	−5.586
$3s[\frac{3}{2}]_1- 3p[\frac{1}{2}]_0$	+1.565	−1.644	+6.201	−4.340	−6.059
$3s'[\frac{1}{2}]_0-3p[\frac{1}{2}]_1$	−5.909	+6.542	−25.24	+37.54	−6.178
$3s[\frac{3}{2}]_1-3p[\frac{1}{2}]_1$	−4.288	+5.085	−21.39	+35.89	−6.615
$3s[\frac{3}{2}]_2-3p[\frac{1}{2}]_1$	−4.585	+5.396	−22.60	+37.98	−6.761
$3s[\frac{3}{2}]_1-3p[\frac{3}{2}]_2$	+1.294	−1.359	+5.119	−3.658	−6.195
$3s[\frac{3}{2}]_2-3p[\frac{3}{2}]_2$	+1.293	−1.358	+5.117	−3.655	−6.195
$3s'[\frac{1}{2}]_0-3p[\frac{3}{2}]_1$	+0.390	−0.512	+2.508	−1.064	−5.770
$3s[\frac{3}{2}]_1- 3p[\frac{3}{2}]_1$	+1.014	−1.214	+5.448	−6.621	−5.713
$3s[\frac{3}{2}]_2-3p[\frac{3}{2}]_1$	+1.014	−1.213	+5.441	−6.599	−5.716
$3s[\frac{3}{2}]_1-3p[\frac{5}{2}]_2$	+1.401	−1.385	+4.952	−3.940	−6.816
$3s[\frac{3}{2}]_2-3p[\frac{5}{2}]_2$	+1.405	−1.388	+4.965	−3.958	−6.815

[b] To the fit $c_5T^4 + c_6T^3 + c_7T^2 + c_8T + c_9$

Regularities in the Behaviour of Stark Widths of Argon Spectral Lines

E. V. Sarandaev and M. Kh. Salakhov

*Department of Physics, Kazan State University,
Kazan, Tatarstan, 420008, Russia.*

Abstract. The dependence for estimating the Stark widths of neutral argon lines have been proposed. The line widths calculated using the proposed dependence with the literature data are compared.

It was shown in papers [1,2] that there are dependencies of the Stark parameters of spectral lines on the respective effective principal quantum numbers (n*) for alkali metals and some other atoms and ions. These atomic systems are mainly characterized by L-S coupling.

In this paper, we attempt to obtain analogous dependencies for ArI atoms with a strong deviation from L-S coupling. The necessity of obtaining these dependencies stems from the fact that there are many measurements of the Stark parameters of the ArI lines [3,4,5,6] without theoretical calculations [7]. Moreover, these regularities are useful for the critical analysis of the theoretical and experimental data.

As previously, in order to find the Stark widths of the energy levels and to obtain the dependencies the semiclassical calculations of the line widths [7] were used. We have used the widths of 49 ArI lines at $N_e = 10^{17} cm^{-3}$ and T = 10000K because experimental measurements for ArI atoms were mainly carried out in the temperature range $(9-13) \cdot 10^3 K$ [3,4,5,6]. The analysis of the theoretical calculations and experimental data for ArI shows that the line widths slightly depend on the total moment (j), and thus the widths of the levels were found using averaging over j. As a result, the widths of 22 levels, including the states with two excited states were obtained.

As distinct from regularities found previously, in this case an essential dependence of the level widths on the orbital quantum number (l) was found. Its character is similar to the dependence of the squared matrix element for the dipole moment in Coulomb approximation on l

$$M = C(n^*)^2[5(n^*)^2 + 1 - 3l(l+1)] \qquad (1)$$

Therefore in order to improve the correlation between the widths of levels and their effective principal quantum numbers the dependence was plotted in the coordinates lg(DE) - lg(M) for C=1. For these coodinates a linear dependence of lg(DE) on lg(M) with the correlation coefficient R = 0.997 is observed. Thus this dependence may be written as follows

$$\Delta E = 3.58 \cdot 10^8 M^{1.255}, (rad/s) \qquad (2)$$

The line width (HWHM) in frequency units was determined as

$$\omega = \Delta E_1 + \Delta E_2, (rad/s) \qquad (3)$$

where ΔE_1 and ΔE_2 are the widths of the lower lying and upper levels, respectively.

It should be noted that the dependencies (1), (2), (3) are close to the semiempirical Griem formula [8]. In order to estimate the accuracy of this approximation we have compared the widths of ArI lines obtained according to (1),(2) and (3) with the accurate semiclassical calculations [7] which have been used to obtain this approximation. The mean discrepancy for 49 lines did not exceed 15%.

In order to demonstrate the validity of the dependence obtained the widths of ArI lines measured experimentally [3,4,5,6] were estimated. The mean discrepancy for 36 lines given in [3,4,5,6] did not exceed 12%.

REFERENCES

1. Sarandaev,E.V., Fishman,I.S., and Salakhov,M.Kh., "A New Approach to a Search for Regularities in the Behavior of the Stark Parameters of Atomic and Ionic Spectral Lines," presented at the *XX ICPIG*, Piza, Italy, Contr.Papers 1, pp.115-116, 1991.
2. Salakhov,M.Kh., Sarandaev,E.V., and Fishman,I.S., *Opt. Spektrosc.* **71**, 882-887 (1991).
3. Konjevic,N., and Wiese,W.L., *J.Phys.Chem.Ref.Data* **19**, 1307-1385 (1990).
4. Konjevic,N., Dimitrijevic,M.S., and Wiese,W.L., *J.Phys.Chem.Ref.Data* **13**, 619-647 (1984).
5. Musielok, J., *Acta Physica Polonica* **A 86**, 315-326 (1994).
6. Valognes,J.C., Bardet, J.P., and Vitel, Y., J.Phys.B: 26, 4751-4768 (1993).
7. Griem,H.R., *Spectral Line Broadening by Plasmas*, New York and London: Academic Press, 1974.
8. Griem,H.R., *Phys. Rev.* **165**, 258-266 (1968).

Quantum defect theory for momentum space wave functions

H. VAN REGEMORTER* and D. HOANG BINH*

Observatoire de Paris - 92190 MEUDON (France)
Fax : 33-1 45 07 74 69 - e-mail : HVR@obspm.fr

In the momentum space representation, of current use in many studies on atomic Rydberg states, the wave function $G_\alpha(\vec{q})$ is defined as the Fourier transform of the position space wave function $\Phi_\alpha(\vec{r})$ of a one electron system. In the case of a central field, the radial function $g_{vl}(q)$ can be written as an Hankel transform integral

$$g_{vl}(q) = q^{-1/2} \int_0^\infty r^{1/2} J_{l+1/2}(qr) P_{vl}(r) dr \qquad (1)$$

This form is very suitable for generalizing the quantum defect theory (QDT)[1] to the calculation for $g_{vl}(q)$ for non hydrogenic states, when $P_{vl}(r)$ is given by

$$P_{vl}(r) = \frac{1}{v} \frac{1}{[\Gamma(v+l+1)\Gamma(v-l)]^{1/2}} W_{v,l+1/2}\left(\frac{2r}{v}\right) \qquad (2)$$

where v is the effective quantum number, derived from the observed energy data. As known, the Whittaker function W gives the correct behavior of $P_{vl}(r)$ for $r>r_0$. For low l states, corresponding to penetrating orbits, r_0 is of the order of the ionic core boundary, usually much smaller than the mean radius $\bar{r} \approx v^2$ of the Rydberg electron.

Using a contour integral representation for the Whittaker function, we have rederived the hydrogenic analytical expression[2,3,4] which is not valid in the case of large quantum defect.

On the other hand, simple semi analytical expressions can be obtained for $g_{vl}(q)$ in using the well known Bates and Damgaard[5] truncated asymptotic expansion of W in expression (2)

$$P_{vl}(r) = \frac{1}{v} \frac{1}{[\Gamma(v+l+1)\Gamma(v-l)]^{1/2}} e^{-r/v} \left(\frac{2r}{v}\right)^v \sum_{t=0}^{t_0} b_t r^{-t} \qquad (3)$$

which terminates at $t=t_0$, the integer satisfying $v-l-1 \le t_0 < v-l$.

© 1997 American Institute of Physics

These wave functions $P_{vl}(r)$ have been extensively used for calculating bound bound and bound free radial transition integrals. Indeed, good results are obtained when small values of $r<r_0$, where the Coulomb approximation breaks down, give a negligible contribution to the integrals.

In the same way, the momentum radial function $g_{vl}(q)$ given by integral (1) will be accurately known when small r do not play any role. The ion boundary r_0 is the distance where the radial electron density falls to 10 % of its maximum value. Values of r_0 are derived from self consistent field calculations[6]. For alkali-like ions in their ground state, r_0 is about twice the mean radius.

To the failure of the Coulomb approximation for $P_{vl}(r)$, at small r, may correspond its failure for $g_{vl}(q)$, at large q. From the semi-classical JWKB expression of the momentum q(r), one knows that, for small values of $r \leq r_0 << v^2$, the actual momenta $q \approx (2/r)^{1/2}$ are much larger than the expectation value $\overline{q} \approx v^{-1}$.

Application of QDT, using (1) and (3), is particularly suitable for Rydberg states in alkali. But we have also applied the method to the $6s^3S$ and $6p^3P$ states in neutral magnesium -with $r_0 \approx 5,40$ a.u. for Mg^+ in its ground state. The contribution of $r<r_0$ to integral (1) is negligible, except for values arround the nodes. The momentum space wave functions $g_{vl}(q)$ are obtained with good accuracy for all values of q of interest in many applications involving intermediate and high Rydberg states in collision with neutral atoms.

REFERENCES :

1. M.J. SEATON (1958) Mon. Not. R. Astr. Soc., 118, 504
2. H.A. BETHE and E.E. SALPETER (1957) Quantum Mechanics of one and two electron atoms, Springer Verlag
3. B. PODOLSKY and L. PAULING (1929) Phys. rev., 34, 109
4. V. FOCK (1935) Z. Physik, 98, 145
5. D.R. BATES and A. DAMGAARD (1949) Phil. Trans. A., 242, 101
6. C.W. ALLEN (1963) Astrophysical quantities-Athlone Press

High Resolution Studies of Infrared and Raman Line Shapes at the University of Toronto

R. Berman, J. R. Drummond, P. Duggan, S. Hamid Fakhr–Eslam, J. W. Forsman, A. D. May, Guy D. Sheldon and P. M. Sinclair

Department of Physics, University of Toronto, Toronto, Canada, M5S 1A7

Abstract. We present some preliminary measurements of broadening, shifting and line mixing in HD, D_2, CO and CO_2 over a range of densities. We discuss problems with extracting the appropriate binary collision impact limit parameters from precise spectroscopic data when a number of physical processes contribute to the spectral profile. A specific spectral profile and fitting paradigm is recommended and a way to revise the profile to incorporate speed dependent effects, is proposed.

INTRODUCTION

Calculations of line shapes for neutral systems are almost exclusively confined to the dilute gas case i.e pressures approaching zero, where the usual binary collision impact approximations are valid. The problem facing the experimentalist is how to be certain that the correct binary collision impact parameters are being extracted from the data containing "spectral contaminants" or gathered over a range of finite pressures. This may be put another way. Many years ago one was ecstatic if theory and experiment agreed within a factor of two. Basically one was only identifying the forces that determined the time between collisions. Today, theory, computation and experimental techniques have advanced to the point where one is looking for agreement at the percent level. We are now in an era where one can examine many aspects of molecular collisions and their influence on spectral profiles. The question now becomes: What are the signatures that allow the experimentalist to disentangle them to get at the underlying physics? At the University of Toronto we are concerned with that question.

The study of the fundamentals of molecular dynamics in low density gases is just one of the two threads of our research program. The second thread is the improvement to the data base and the development of physically and computationally realistic spectral models for atmospheric (infrared) absorption. One is driven to develop spectral models simply because it is unrealistic even to contemplate ab initio calculations with so many species, so many lines and such a range of pressures and temperature as are found in the atmosphere. We illustrate these threads by briefly summarizing some previously published results and some new but yet incomplete research, along with comments.

EXPERIMENTAL SYSTEM

In the five years since the first measurement with our cw Raman gain spectrometer [1], we have made several improvements to the system and we have developed the infrared difference frequency capabilities. The present dual purpose spectrometer is highly automated, runs under 10,000 lines of computer control code, and is capable of generating infrared and Raman line shapes in dilute gases with a resolution on the order of 2 MHz and signal to noise ratios in excess of 2000. Figure 1 shows a schematic of the system.

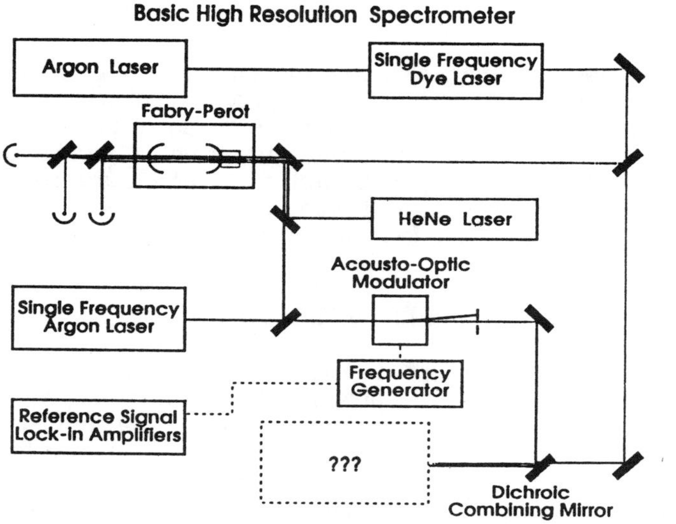

Figure 1. Schematic of the combined Raman and IR spectrometer. The single frequency dye and argon lasers both have a frequency jitter of approximately 1–2 MHz. Changes in the mean frequency are monitored with the Fabry–Perot interferometer and stabilized single frequency He–Ne laser. The optical system occupies a (1.2 m)x(3.6 m) "floating" table.

For Raman measurements the box with "???" is simply the Raman cell and detector system. For infrared work the "???" is a pair of $LiIO_3$ crystals for the generation of tunable IR radiation, followed by a conventional absorption setup. An important detail omitted from the schematic is the automated feedback system that controls the overlap of the axes of the two beams to about 1 μm. Rapid sampling of the various intensities is carried out for proper normalization of the signal.

A sample Raman line measured with the system is shown in Fig.2. Note that the error limits have been multiplied by 100. The experimental points match the fitted theoretical curve (solid line) within the error bars. It is clear that data like that shown in Fig.2 allow one to examine much finer details of spectral line shapes than have ever been possible in the past. For example, in [2] we examined departures from the soft collision model for the translational motion.

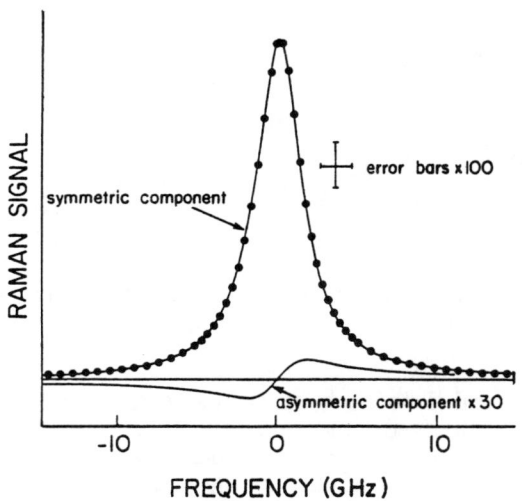

Figure 2. A sample Raman profile, the Q(2) line of D_2 at 302.4 K. The points are experimental results and the solid line a theoretical curve. Due to line mixing, the line has an asymmetric component equal in amplitude to about 1% of the peak amplitude of the symmetric component.

EXPERIMENTAL RESULTS AND COMMENTS.

Some Raman Measurements

The Raman Q branch of D_2 is well suited to test fully quantal calculations of the broadening and shifting etc. of spectral lines. Calculations exist for D_2 infinitely diluted in He [3] and it should soon be possible to calculate the self broadening case. We have recently completed a study of the broadening and shifting in He mixtures [4] and have also measured the line mixing [5] in pure D_2, all at room temperature. In the following paragraph we discuss a systematic error which exists in the older data. The example also illustrates how finer details can now be determined from spectral profiles.

One of the difficulties in determining accurate widths for the polarized Q branch is the presence of the underlying depolarized component. To measure this component we crossed the polarization of the probe and pump beam and resonated the latter in a ring cavity containing the Raman cell [6]. Figure 3 illustrates this novel configuration. We obtained a factor of 10 increase in the Raman signal with the resonator and were able to measure both the broadening and shifting for the Q(1)–Q(4) lines in pure D_2 with an accuracy of better than 1%. The broadening of the depolarized lines reported in [6] was about twice as large as for the isotropic component. A large difference was not unexpected since reorientation broadens the depolarized component, but not the isotropic part of the lines. Surprisingly, the shifts of the isotropic and anisotropic components were slightly different. What we choose to stress here is that by re-analyzing the spectrum with the two

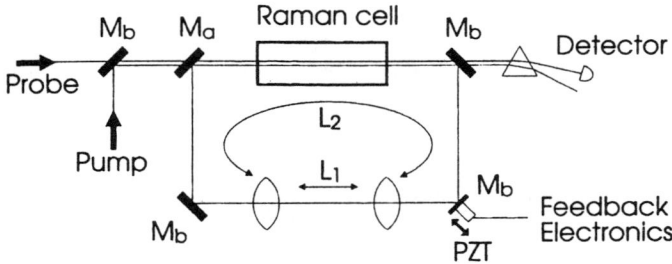

Figure 3. Ring resonator for enhancing a weak Raman signal. Only the argon pump beam is resonated.

lasers polarized parallel to each other, we were able to demonstrate that neglecting the depolarized contribution to the polarized spectrum had contributed a systematic error of 1% to the broadening of the isotropic part of the Q branch. This is significant in the present case as the random error had been estimated at only 0.2%. Furthermore, since the strength of the depolarized component varies with rotational quantum number, J, neglecting the depolarized part in a fitting routine produces a false J dependence to the broadening coefficients. Figure 4 shows the reconstruction of the line shape for the case of parallel polarizations.

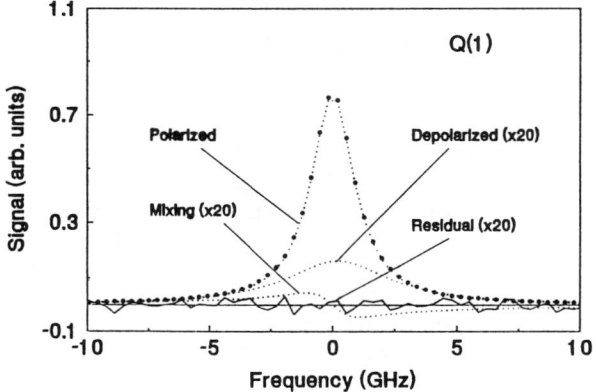

Figure 4. The Q(1) line in D_2 at 25.8 amagat showing the various contributions to the total profile for parallel polarization of the two lasers.

As shown in the figures above, the lines are asymmetric. This could arise from the finite duration of collisions [7] or from line mixing [8]. Neglecting the asymmetry in a fitting routine will also generate a systematic error in the extracted numbers. There is a clear signature that the asymmetry in the Q branch of D_2 is due to line mixing. The asymmetry is such that the Q(0) and Q(2) line appear to pull together as do the Q(1) and Q(3) lines, i.e. the mixing parameters have opposite signs for Q(0) and Q(2) and for Q(1) and Q(3). The asymmetry would be of the same sign for all the lines if it arose from departures from the impact approximation.

One of the consequences of the room temperature studies of line mixing was a new paradigm for fitting spectral lines [9]. In hindsight it should be attributed to Baranger [8], Gordon [10] and Royer [11]. Neglecting for the moment effects associated with the motion of the active molecule every band is the sum of "lines", S_i, each of which is the sum of a symmetric and an asymmetric component. S_i can be written in the form

$$S_i = X_i \Gamma_i / [(\omega - \omega_i - \Delta_i)^2 + \Gamma_i^2] + Y_i (\omega - \omega_i - \Delta_i) / [(\omega - \omega_i - \Delta_i)^2 + \Gamma_i^2]$$

The important point is that the relative line strength, X_i, the frequency shift, Δ_i, the width (HWHM), Γ_i and the mixing parameter, Y_i, for each line, <u>are all polynomials in the density.</u> At low density we have $\Delta_i = (\delta_1)_i \, \rho$, $\Gamma_i = (\gamma_1)_i \, \rho$. Rosenkranz [12] has given an expression for Y_i correct to first order in the density, ρ. If one ignores the asymmetry and assumes that the peak position is linear in density then one introduces a systematic error [9] into the measurements of $(\delta_1)_i$. In fact, due to line mixing, the peak of each line has a shift that is quadratic with density [13].

The process we use to compare theory and experiment is first to collect data with the highest possible signal to noise ratio and with a resolution high compared to the scale of the width of the lines and second to fit the data for the entire band to an expression which is the sum of line contributions, $S_i(\omega)$. From the fit we extract at each density the values of Δ_i, Γ_i etc. These are then fitted to a polynomial in density to determine $(\delta_1)_i$, $(\gamma_1)_i$ etc. Using this fitting paradigm generates what we believe to be the correct dilute gas coefficients that are to be compared to calculations. Furthermore, our hope is that having determined the terms linear in density we can use the information to begin to sort out the nonlinear terms. These terms have contributions from line mixing, from the finite duration of collisions and from three body terms in the interaction. There is very little theoretical literature on spectral properties that are nonlinear in the density, beyond that for line mixing.

Besides density studies, measurements with different isotopic species and at different temperatures help provide critical tests of theoretical calculations and in developing an understanding of the underlying physics. Figure 5 illustrates some preliminary measurements of line mixing in HD at

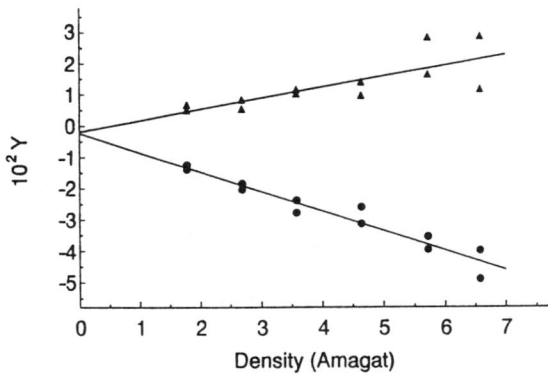

Figure 5. Line mixing parameter, Y, for the Q(0) and Q(1) as a function of density, for HD at 304.6 K; dots, Q(0) and triangles, Q(1).

room temperature. The mixing parameters in HD for the Q(0) and Q(1) lines have opposite slopes with respect to density. In contrast, in D_2, they have the same slope. This was expected because spin conservation no longer prevents $\Delta J = \pm 1$ transitions and mixing now occurs between the J = 0 and the J = 1 line.

Figs.6–8 show plots of widths, shifts and line mixing, as a function of density, in pure D_2 at 100.7 K. Approximate broadening, shifting and line

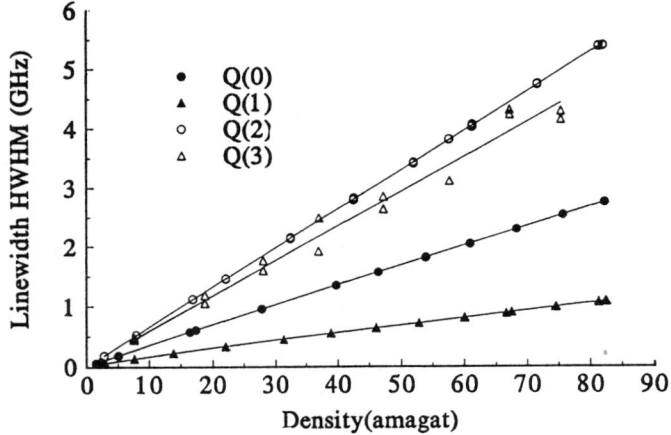

Figure 6. Width versus density for the Q branch lines in D_2 at 100.7 K.

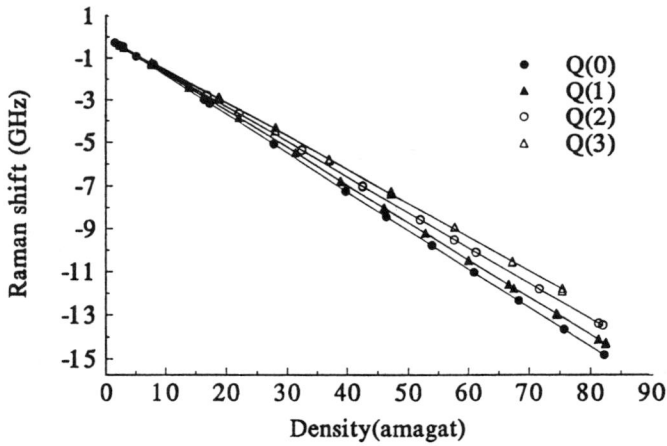

Figure 7. Shifts in the Q branch components in D_2 at 100.7 K.

mixing coefficients may be determined from the graphs. The low temperature data are so precise that we detect terms nonlinear in density up to the fourth power or even higher. As the value of the linear term extracted from the

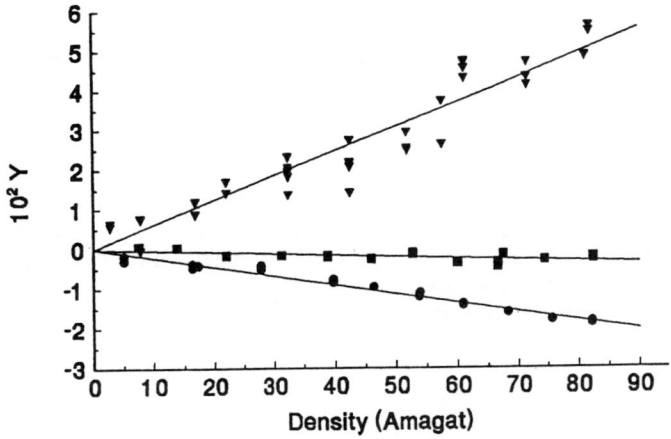

Figure 8. Line mixing parameter for the Q branch components in D_2 at 100.7 K; dots, Q(0), squares, Q(1) and triangles, Q(2).

measurements is sensitive to the specific terms included in the polynomial in the density one needs theoretical guidance as to which are the dominant and/or zero nonlinear terms. This is particularly true of the broadening coefficient where we find a variation of 10 % in the fitted value, depending upon the degree of the polynomial. There are only quantal calculations at a number of temperatures for the linear coefficients [3] for D_2 infinitely diluted in He. In progress are measurements at other temperatures and in D_2–He mixtures to check these calculations and to sort out, experimentally, the nonlinear dependence on density.

Some Infrared Measurements

Figure 9 illustrates what goes into the "???" box of Fig.1 when we

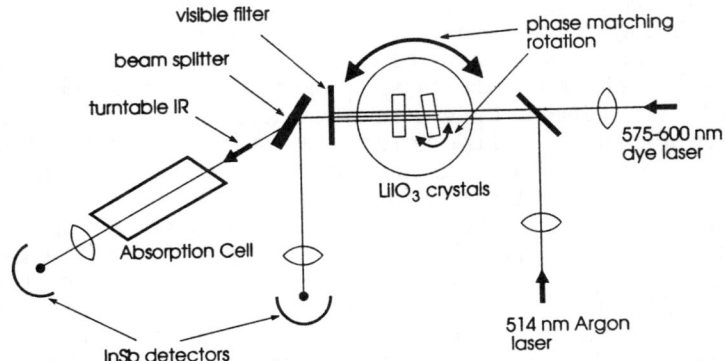

Figure 9. Schematic of the absorption cell and the system used to generate tunable infrared radiation.

operate the system as a difference frequency spectrometer. We are able to generate 10's of nanowatts of IR radiation with continuous coverage from approximately 2.5 μm to 5.5 μm with a resolution of 1–2 MHz. By paying considerable attention to ratioing the transmitted intensity to the input intensity and to etalonning effects in the optics we are able to obtain signal to noise ratios that sometimes reach 3000. We have carried out several room temperature studies in CO and CO_2.

Figure 10 shows a sample of the absorption coefficient for the $(\nu_1 + \nu_2)$ Q branch at 4.96 Torr and 301 K. The lower trace is the residual to our spectral model. The lines are so sharp that near line centre the noise is dominated by uncertainties in the frequency. Away from line centre the dominant noise arises from uncertainties in the amplitude.

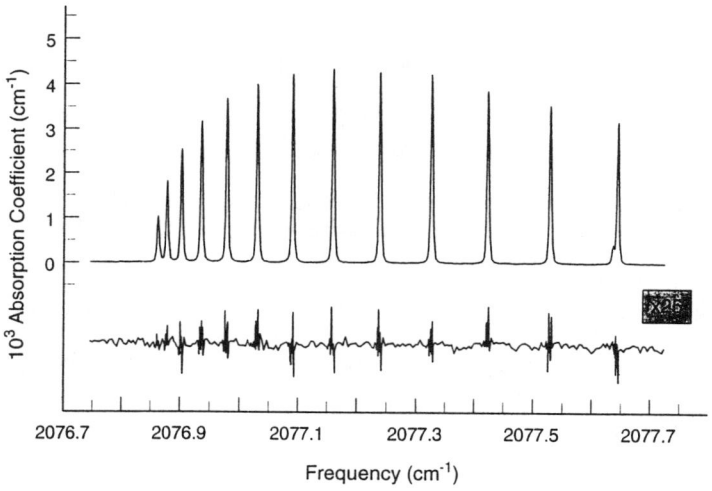

Figure 10. The absorption coefficient for the $(\nu_1 + \nu_2)$ Q branch in CO_2.

We use results in CO and CO_2 to illustrate some other fine details of spectral line shapes that we have omitted from the presentation up to this point. There are two aspects of the translational motion that contribute to the spectral profile. First, the phase of the field(s) at the molecule, $\mathbf{k}\cdot\mathbf{r}$ is modified by any change in the molecules position, $\mathbf{r}$. At low density this leads to Doppler broadening. At high density when $k\lambda$ is small compared to 1 (k is the wave vector and λ the kinetic mean free path) the molecule diffuses through the optical field(s) and we have the well known effect of Dicke narrowing. Second, the effect of collisions on the relaxation of the optical coherence may depend upon the speed of the molecule after averaging over the relative velocity of the perturbers. This gives a speed dependent shift and width to a line. The general properties of a line shape are dominated by which of the two degrees of freedom, the translational or internal, relaxes the most quickly. As the density increases, the translational degree relaxes more slowly and the internal degree relaxes more rapidly. Consequently, there is always a "crossing point".

In D_2, where the optical cross–section is small compared to the kinetic cross–section the "crossing point" is at a sufficiently high density that the translational motion is well into the diffusion regime before its behaviour is masked by the ever increasing rate of relaxation of the internal degrees of freedom. It for this reason that we were able to study details of the soft collision model in this system [2].

In CO and CO_2 the optical cross–section is larger than the kinetic cross–section and one expects to be sensitive only to the nearly free streaming aspects of the translational motion. It was with some surprise that fitting some P and R rovibrational lines in CO [14], using the standard model for the combined effects of translational and internal motion, produced a (Dicke) narrowing parameter that agreed with the mass diffusion constant at low pressures but increased dramatically beyond a certain pressure and varied from line to line. We find the same behaviour in CO_2, an example of which is shown in Fig.11. At low pressures the narrowing parameter

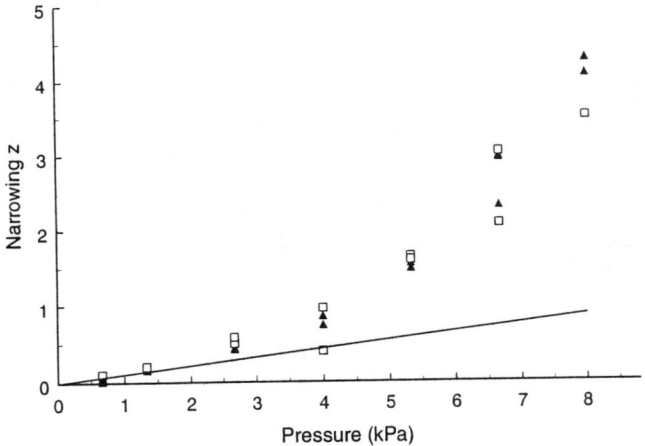

Figure 11. Dependence of the narrowing parameter on density for two lines in the $(\nu_1 + \nu_2)$ Q branch in CO_2 at room temperature; squares, Q(6) and triangles, Q(16). The solid line is calculated using the mass diffusion constant.

asymptotically approaches the narrowing parameter determined from the mass diffusion constant. At the very highest pressures one cannot obtain a fit even if the narrowing parameter is set to infinity (zero translational width). Clearly the model which combines a soft collision model for the translational motion and a Lorentzian profile for the optical coherence is not correct. The key to the problem became apparent when it was realized that the difficulty occurred, with increasing pressure at the "crossing point". Above the "crossing point" the spectrum is dominated by the evolution of the internal degrees of freedom and it is the description of these that was at fault. In [15] we identified the departure from a Lorentzian profile as being due to a speed dependent broadening. In [15] we used a speed dependent model for the broadening. For the translational motion we used a model that was speed independent, recognizing at the time that it was an undesirable approximation. Recently Ciurylo and Szudy [16] have given a fuller treatment.

The shifting in these systems is very small. This sets the speed dependence aspect apart from those observed by Farrow et al [17] where speed dependent shifts produce an inhomogeneous broadening. Here we have a superposition of Lorentzians with the same frequency but different widths. The result is a non–Lorentzian shape. One would think, by simply increasing the pressure to the point where the contribution of the translational motion to the width was negligible ($k\lambda \ll 1$), that one could study the non–Lorentzian profile. Unfortunately, in CO and CO_2 one quickly enters the line mixing region. In the case of a speed dependent broadening it would appear to be easy to include this in the mixing profile given above by an extension of the Duggan–Ciurylo calculation. This would represent a refinement to the band profile recommended above. We have already shown that simply modeling each line in the band as the sum of two line mixing profiles, S_i, but with different widths, fits the data very well.

Figure 12 shows the mixing coefficients, y_i, for the R and P branch in CO highly diluted in N_2. In the figure we have compared the measurements (points) to theoretical calculations (solid line) for CO molecules in a bath of He [18]. Experimentally, the broadening coefficients for CO–N_2 and CO–He are very similar. If we believe in any kind of "gap law", whereby one can predict the off–diagonal elements of the relaxation matrix form the diagonal elements, then it follows from a similar broadening that the mixing will be similar. The agreement between the points and the solid lines supports a scaling law approach [19, 20].

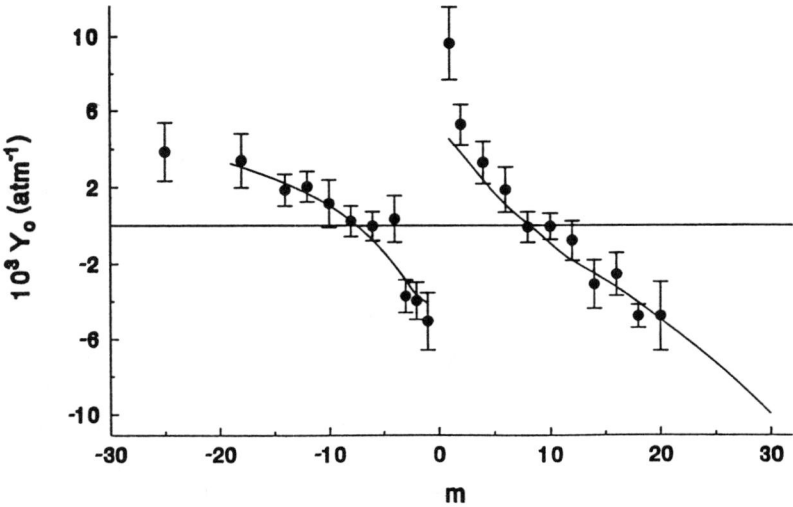

Figure 12. Plot of mixing coefficients versus m. Solid points, experimental mixing coefficients for the P ($m = -|J|$) and R branch ($m = J + 1$) in CO diluted in N_2 at 301 K. Solid line is a theoretical calculation for CO–He.

Figure 13 shows the mixing coefficients, y_i, for the $(\nu_1 + \nu_2)$ Q branch in CO_2. Here we have fitted the broadening coefficients to a "gap law" [19, 20] and then calculated the mixing coefficients using the the low density approximation of Rosenkranz [12]. Again, one can interpret the agreement between the experimental point and the solid lines as experimental support for gap laws.

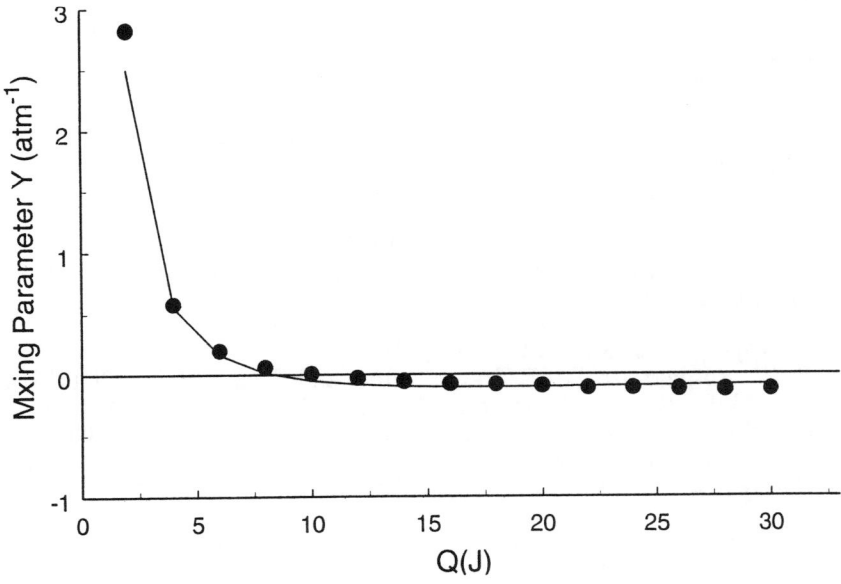

Figure 13. Plot of the mixing coefficients for the $(\nu_1 + \nu_2)$ Q branch in CO_2 at 301 K. The solid line was derived from an exponential gap law fitted to the broadening coefficients.

SUMMARY

In this paper we have given an overview of line shape studies at the University of Toronto and we have presented some preliminary data for studies in HD, D_2, CO and CO_2. We have recommended a fitting procedure that permits one to extract accurate low density collision parameters from high quality data when line mixing is important. It is suggested how effects nonlinear in density, such as three body interactions, might at least begin to be examined. A refinement to the spectral band model is suggested when speed dependent broadening or perhaps shifting, are to be included.

ACKNOWLEDGEMENTS

This work was supported by the Natural Sciences and Engineering Research Council of Canada and the Ontario Government through the Ontario Laser and Lightwave Research Centre.

REFERENCES

1. J. W. Forsman, P. M. Sinclair, P. Duggan, J. R. Drummond and A. D. May, Can. J. Phys. **69**, 558 (1991).
2. J. W. Forsman, P. M. Sinclair, A. D. May, P. Duggan and J. R. Drummond, J. Chem. Phys. **97**, 5355 (1992).
3. R. Brezina, W–K. Liu and S. Green, Phys. Rev. **A51**, 3645 (1995).
4. P. M. Sinclair, P. Duggan, M. Le Flohic, J. W. Forsman, J. R. Drummond and A. D. May, Can. J. Phys. **72**, 885 (1994).
5. P. M. Sinclair, J. W. Forsman, J. R. Drummond and A. D. May, Phys. Rev. **A48**, 3030 (1993).
6. P. M. Sinclair, P. Duggan, J. R. Drummond and A. D. May, Can. J. Phys. **73**, 530 (1995).
7. Ph. Marteau, C. Boulet and D. Robert, J. Chem. Phys. **80**, 3632 (1984).
8. M. Baranger, Phys. Rev. **111**, 494 (1958).
9. P. M.Sinclair, P. Duggan, J. W. Forsman, J. R. Drummond and A. D. May, Can. J. Phys. **72**, 891 (1994).
10. R. G. Gordon, J. Chem. Phys. **49**, 2455 (1968).
11. A. Royer, Phys. Rev. **A22**, 1625 (1980).
12. P. W. Rosenkranz, IEEE Trans. Anten. and Prop. **AP 23**, 498 (1975).
13. F. Thibault, J. Boiselles, R. le Doucen, R. Farreng, M. Morillion–Chapey and C. Boulet, Europhys. Lett. **12**, 319 (1990).
14. P. Duggan, P. M. Sinclair, M. P. Le Flohic, J. W. Forsman, R. Berman, A. D. May and J. R. Drummond, Phys. Rev. **A48**, 2077 (1993).
15. P. Duggan, P. M. Sinclair, A. D. May and J. R. Drummond, Phys. Rev. **A51**, 218 (1995).
16. R. Ciurylo and J. Szudy, submitted to J.Q.S.R.T.
17. R. Farrow, L.a. Rahn, G. O. Sitz and G. J. Rosasco, Phys. Rev. Lett. **63**, 746 (1989).
18. J. Boiselles, C. Boulet, D. Robert and S. Green, J. Chem. Phys. **90**, 5392 (1989).
19. B. Gentry and L. L. Strow, J. Chem. Phys. **86**, 5722 (1987).
20. A. E. DePristo and H. Rabitz, J. Mol. Spec. **70**, 476 (1978).

Collisional Coupling between Components of Molecular Spectral Lines

O. Tarrini, S. Belli and G. Buffa

Dipartimento di Fisica dell'Università
Piazza Torricelli 2, I-56126 Pisa, Italy

Abstract. A theoretical treatment is presented for collisional coupling, based on a calculation of the off-diagonal relaxation matrix in the Liouville space of the lines. An orthonormal line vector basis is adopted, allowing to prove that the relaxation matrix is symmetrical and to obtain insight into the use of sum rules. Particular attention is devoted to the case of a band of completely overlapped lines showing that collisional transitions between lines inside the band have a reduced relaxation effect respect to transitions to lines outside the band. This explains the observed reduction of the line width for the overlapping case with respect to the one of the resolved components. Detailed results, allowing a comparison with the existing measurements, are given for the case of collisional broadening and coupling of hyperfine and Stark components of spectral transitions.

INTRODUCTION

Line coupling effect received growing attention in the last decade because of its importance in the interpretation of atmospheric and astrophysical spectra. Indeed, the overlap of the wings of the lines can make the shape of the absorption profile very different from the simple sum of Lorentzians (or Voigtians) shapes; this can be particularly relevant for the weak absorption in the windows between the lines.

Another manifestation of line-coupling can be found in the case of full overlap of the lines: when a line can be observed as unresolved or resolved into several components, the collisional linewidth parameter of the unresolved line can be very different from the weighted average of those of its resolved components.

Both effects are due to the off-diagonal part of the collisional relaxation matrix. Indeed, the determination of the profile for a group of lines requires two steps: first, evaluation of the Liouville relaxation matrix; second, diagonalization of the non-Hermitian time evolution matrix in the space of the lines.

First of all we deal with the first problem. An orthonormal line vector basis is adopted, allowing to prove that the relaxation matrix is symmetrical. An explicit expression is given for the dependence of this matrix on the collisional amplitudes for the transitions from the lines. Our theoretical framework allows a discussion of some general properties of the off-diagonal relaxation matrix.

As far as the second problem, diagonalization of the relaxation matrix, is concerned, particular attention is devoted to the case of a band of completely overlapping lines showing that collisional transitions between lines inside the band have a reduced relaxation effect with respect to transitions to lines outside the band. This explains the observed reduction of the line width for the overlapping case with respect to the one of the resolved components.

OFF-DIAGONAL RELAXATION MATRIX

In the Liouville space formalism, the spectral line shape of a dipolar molecule, absorbing radiation at frequency ω and surrounded by the thermal bath of other perturbing molecules, is:

$$F(\omega) = -\frac{1}{\pi} \operatorname{Im} \langle\!\langle \boldsymbol{\mu} | \frac{1}{\omega - L_0 - i\Gamma(\omega)} | \boldsymbol{\mu} \rangle\!\rangle, \tag{1}$$

where $|\boldsymbol{\mu}\rangle\!\rangle$ is the dipole moment of the absorber, L_0 its unperturbed Liouvillian and Γ its collisional relaxation operator.

We define the inner product in Eq. (1) $\langle\!\langle a|b\rangle\!\rangle$ of two dynamical variables $|a\rangle\!\rangle$ and $|b\rangle\!\rangle$ as the equilibrium correlation product:

$$\langle\!\langle a|b\rangle\!\rangle = \int_0^1 d\lambda \, \left\langle e^{\lambda H_0/kT} \, a^\dagger \, e^{-\lambda H_0/kT} \, b \right\rangle_0. \tag{2}$$

The Hamiltonian H_0 and the equilibrium ensemble average $\langle\ldots\rangle_0 = Tr\{\rho_0 \ldots\}$ are defined on the internal degrees of the absorber.

$L_0 + i\Gamma(\omega)$ is the non-Hermitian Liouville operator describing the average time evolution of $|\boldsymbol{\mu}\rangle\rangle$ and acting on a complete set of orthonormalized line vectors $|\ell\rangle\rangle$ of dipolar transitions $\ell_i \to \ell_f$ between two internal states of the absorber. We assume:

$$|\ell\rangle\rangle = |\ell_i\rangle \, \rho_\ell^{-1/2} \, \langle\ell_f| \tag{3}$$

where the density matrix ρ in the Liouville space of the lines is defined by

$$\rho_\ell = (\rho_{\ell_i} - \rho_{\ell_f})kT/(E_i - E_f). \tag{4}$$

In this representation $|\boldsymbol{\mu}\rangle\rangle$ is given by:

$$|\boldsymbol{\mu}\rangle\rangle = \sum_\ell I_\ell^{1/2} |\ell\rangle\rangle, \tag{5}$$

where $I_\ell = |\langle\langle\ell|\boldsymbol{\mu}\rangle\rangle|^2 = \rho_\ell |\boldsymbol{\mu}_{\ell_i \to \ell_f}|^2$ is the line intensity. The line vector $|\ell\rangle\rangle$ is eigenvector of the unperturbed resonance frequency operator L_0 with eigenvalue $\omega_\ell = (E_f - E_i)/\hbar$.

The relaxation matrix $\Gamma(\omega)$ in Eq. (1) replaces the exact time dependent Liouville interaction operator $L_I(t)$ describing the fluctuating interaction between the absorber and the other perturbing molecules of the gas. In the low density limit the interaction is restricted to binary collisions and $\Gamma(\omega)$ is proportional to the number density n of perturbers:

$$(L_0 + i\Gamma)_{\ell\ell'} = \omega_\ell \delta_{\ell\ell'} + in\gamma_{\ell\ell'}, \tag{6}$$

where, in the short-memory approximation, $\Gamma(\omega) \simeq n\gamma$ is assumed independent on ω.

The diagonal part of the relaxation matrix, $n\gamma_{\ell\ell} = w_\ell + is_\ell$, gives the pressure width w_ℓ and shift s_ℓ of the single lines. The off-diagonal part $n\gamma_{\ell\ell'}(1 - \delta_{\ell\ell'})$ describes the cross relaxation between different lines.

The duration of collision effect [1] can be neglected when the the collision time is small with respect to the average time between two collisions and with respect to $1/(\omega - \omega_\ell)$. In this case the impact approximation [2] can be used and γ can be expressed in terms of a collisional cross-section σ:

$$\gamma_{\ell\ell'} = \langle v\,\sigma_{\ell\ell'}\rangle_v, \tag{7}$$

$$\sigma_{\ell\ell'} = \left\langle \pi \sum_l (2l+1) P_{\ell\ell'}(k,l,r)/k^2 \right\rangle_r. \tag{8}$$

$\sigma_{\ell\ell'}$ is the collisional cross-section for the transitions between the lines ℓ and ℓ'. $P_{\ell\ell'}(k,l,r)$ gives the contribution to $\sigma_{\ell\ell'}$ from a collision with given values of the bath variables, namely: internal state r of the perturbing molecule, wave vector $\mathbf{k}$ and orbital angular momentum $\mathbf{l}$ of the relative motion. $\langle\ \rangle_r$ in Eq. (8) and $\langle\ \rangle_v$ in Eq. (7) stand respectively for the thermal average on r and on the relative velocity $v = k\hbar/m$.

We define $P_{\ell\ell'}$ in terms of a Liouville scattering matrix $\mathbf{S}$ in the space of the lines.

$$P_{\ell\ell'}(k,l,r) = \delta_{\ell\ell'} - \mathbf{S}_{\ell\ell'}(k,l,r). \tag{9}$$

$\mathbf{S}$ is related to the usual S-matrix for the states by

$$\mathbf{S}_{\ell\ell'} = \rho_\ell^{1/2} \rho_{\ell'}^{-1/2} \frac{1}{2l+1} \sum_m \sum_{k',l',m',r'} \langle \ell_i, k, l, m, r | S | \ell'_i, k', l', m', r'\rangle$$
$$\times \langle \ell'_f, k', l', m', r' | S^\dagger | \ell_f, k, l, m, r\rangle, \tag{10}$$

where m and m' are the z components of $\mathbf{l}$ and $\mathbf{l'}$.

Rewriting Eq. (10) in terms of the transition matrix $T = \mathbb{1} - S$, from Eq. (9) we have:

$$P_{\ell\ell'}(k,l,r) = P_{\ell\ell'}^{(i)} + P_{\ell\ell'}^{(f)} + P_{\ell\ell'}^{(i,f)}, \tag{11a}$$

$$P_{\ell\ell'}^{(i)} = \delta_{\ell\ell'} \frac{1}{2l+1} \sum_m \langle \ell_i, k, l, m, r | T | \ell_i, k, l, m, r\rangle, \tag{11b}$$

$$P_{\ell\ell'}^{(f)} = \delta_{\ell\ell'} \frac{1}{2l+1} \sum_m \langle \ell_f, k, l, m, r | T | \ell_f, k, l, m, r\rangle, \tag{11c}$$

$$P^{(i,f)}_{\ell\ell'} = -\rho_\ell^{1/2}\rho_{\ell'}^{-1/2}\frac{1}{2l+1}\sum_m \sum_{k',l',m',r'} \langle \ell_i,k,l,m,r|T|\ell'_i,k',l',m',r'\rangle$$
$$\times \langle \ell'_f,k',l',m',r'|T^\dagger|\ell_f,k,l,m,r\rangle. \quad (11d)$$

The two "outer" terms $P^{(i)}_{\ell\ell'}$ and $P^{(f)}_{\ell\ell'}$ are diagonal in the space of the lines and are due to collisional transitions respectively from the initial and final states of the line ℓ.

Collisional coupling is due to the "middle" term $P^{(i,f)}_{\ell\ell'}$ given by the correlated product of the transition matrix elements between the initial and final states of the lines $\langle \ell_i|T|\ell'_i\rangle\langle \ell'_f|T^\dagger|\ell_f\rangle$ and accounting for the rate of amplitude transfer between the lines ℓ and ℓ'.

We note that the presence of the density factors $\rho_\ell^{1/2}\rho_{\ell'}^{-1/2}$ in Eqs. (10) and (11d) follows from our definition of orthonormal set (3) and inner product (2).

From our treatment it is clear that γ matrix is symmetric. Indeed, from Eqs. (7–10) we have

$$\gamma_{\ell\ell'} = \gamma_{\ell'\ell} \quad (12)$$

as a consequence of the time reversal symmetry $\langle a|S|b\rangle = \langle b|S^*|a\rangle$ of the S matrix and of the equality $\rho_\ell\rho_r\rho_v = \rho_{\ell'}\rho_{r'}\rho_{v'}$ due to energy conservation in the collision, valid within the width w of the lines.

Another remarkable property of the relaxation matrix γ is that its off-diagonal part, describing collisional interference between the lines, takes no contribution from "strong" collisions in which every coherence is lost. In this case the right hand side of Eq. (11d) vanishes, unless when $\ell_i = \ell_f$ and $\ell'_i = \ell'_f$.

As a consequence, the widely used sum rule

$$\gamma_{\ell\ell} = -\sum_{\ell\ne\ell'}\gamma_{\ell\ell'} \quad (13)$$

cannot be considered as exact. The simple case of an hard sphere intermolecular potential gives an example of full violation of Eq. (13). Its right hand side vanishes, while its left hand side is equal to the hard sphere cross section multiplied by the average velocity: $\gamma_{\ell\ell} = \bar{v}\sigma$. Indeed, Eq. (13) must be changed to:

$$\gamma_{\ell\ell} \geq - \sum_{\ell \neq \ell'} \gamma_{\ell\ell'} \tag{14}$$

the equal sign holding only for particular cases.

By the formal treatment given in Eqs. (7–11), one can derive the relaxation matrix $\gamma_{\ell\ell'}$ from the collisional transition matrix T for the initial and final states of lines ℓ and ℓ'.

Since an exact calculation of T is usually not easily realized, our theory can be adapted to the simplifying approximations commonly used in order to handle collision dynamics and obtain computable expressions.

The semiclassical approximation was widely used for line broadening calculations. It assumes that $T = T(b,v)$ is an operator on the internal degrees of the colliders while translation is described by classical trajectories with impact parameter b and relative velocity v. By Eqs. (7–11), we can extend the semiclassical approximation to the off-diagonal terms $\gamma_{\ell\ell'}$ [3]:

$$\gamma_{\ell\ell'} = \left\langle v \int_0^\infty 2\pi b \, db \, P_{\ell\ell'}(b,v,r) \right\rangle_{vr} \tag{15}$$

$$P_{\ell\ell'}(b,v,r) = \rho_\ell^{1/2} \rho_{\ell'}^{-1/2} \Big[\left(\langle \ell_i, r | T(b,v) | \ell_i, r \rangle + \langle \ell_f, r | T^\dagger(b,v) | \ell_f, r \rangle \right) \delta_{\ell\ell'} \\ - \sum_{r'} \langle \ell_i r | T(b,v) | \ell_i' r' \rangle \langle \ell_f' r' | T^\dagger(b,v) | \ell_f r \rangle \Big]. \tag{16}$$

The T matrix can be derived from the intermolecular interaction V by:

$$T = \mathbb{1} - \mathcal{O}\left(\frac{i}{\hbar} \exp \int_{-\infty}^{\infty} dt \, e^{iH_0 t/\hbar} V e^{-iH_0 t/\hbar} \right) \tag{17}$$

where $\mathcal{O}$ is the time-ordering operator.

STRONG OVERLAP

When the relaxation matrix γ is obtained, the problem arises of diagonalizing the non-Hermitian matrix $L_0 + in\gamma$. Perturbative approximations can be found in the literature for solving this problem in the case of weak overlap [4,5], that is to say, for lines whose frequency distance $|\omega_\ell - \omega_{\ell'}|$ is large in comparison to their coupling $n\gamma_{\ell\ell'}$. We treat here the opposite case of a group of lines $\bar{\ell}$ so overlapped that the off-diagonal part of the relaxation matrix exceeds their frequency spacing. If the interference with lines external to the overlapped band can be neglected, the bandshape $F_{\bar{\ell}}(\omega)$ can be obtained by solving Eq. (1) in the reduced set $\bar{\ell}$ of the overlapping lines of the band:

$$F_{\bar{\ell}}(\omega) = -\frac{1}{\pi} \operatorname{Im} \langle\!\langle \, \boldsymbol{\mu}_{\bar{\ell}} | \frac{1}{\omega - L_0 - in\gamma} | \boldsymbol{\mu}_{\bar{\ell}} \rangle\!\rangle, \tag{18}$$

where

$$|\boldsymbol{\mu}_{\bar{\ell}}\rangle\!\rangle = \sum_{\ell \in \bar{\ell}} I_\ell^{1/2} |\ell\rangle\!\rangle \tag{19}$$

is the dipole vector projected onto the subspace of the band.

The whole band shape can be described by a single Lorentzian $F_{\bar{\ell}}(\omega)$, whose intensity $I_{\bar{\ell}}$ and resonance frequency $\omega_{\bar{\ell}}$ are given respectively by the integrated intensity and by the centroid of the band:

$$F_{\bar{\ell}}(\omega) = I_{\bar{\ell}} n \gamma_{\bar{\ell}\bar{\ell}} / [(\omega - \omega_{\bar{\ell}})^2 + (n\gamma_{\bar{\ell}\bar{\ell}})^2], \tag{20}$$

$$I_{\bar{\ell}} = \sum_{\ell \in \bar{\ell}} I_\ell = \langle\!\langle \, \boldsymbol{\mu}_{\bar{\ell}} | \boldsymbol{\mu}_{\bar{\ell}} \rangle\!\rangle, \tag{21}$$

$$\omega_{\bar{\ell}} = \sum_{\ell \in \bar{\ell}} \omega_\ell I_\ell / I_{\bar{\ell}}. \tag{22}$$

Formally, this can be obtained by Eq. (18) if the Liouville operator $L_0 - in\gamma$ is approximated by its mean value:

$$\omega_{\bar{\ell}} - in\gamma_{\bar{\ell}\bar{\ell}} = \frac{\langle\!\langle \boldsymbol{\mu}_{\bar{\ell}} | L_0 - in\gamma | \boldsymbol{\mu}_{\bar{\ell}} \rangle\!\rangle}{I_{\bar{\ell}}}. \tag{23}$$

The deviation of the approximate single line representation, given by Eqs. (20–23), from the exact band shape given by Eq. (18), can be evaluated by the perturbative expansion of $1/(\omega - L_0 - in\gamma)$ with respect to its zero-order term $1/(\omega - \omega_{\bar{\ell}} - in\gamma_{\bar{\ell}\bar{\ell}})$. At the lowest order the deviation is proportional to the second moment $\langle\!\langle\boldsymbol{\mu}_{\bar{\ell}}|(L_0 - in\gamma - \omega_{\bar{\ell}} - in\gamma_{\bar{\ell}\bar{\ell}})^2|\boldsymbol{\mu}_{\bar{\ell}}\rangle\!\rangle$.

The broadening parameter $\gamma_{\bar{\ell}\bar{\ell}}$ of the line $\bar{\ell}$ can be separated into its diagonal and off-diagonal parts. From Eq. (23) we have:

$$\gamma_{\bar{\ell}\bar{\ell}} = \gamma_{\bar{\ell}\bar{\ell}}^{(d)} + \gamma_{\bar{\ell}\bar{\ell}}^{(od)} \tag{24a}$$

$$\gamma_{\bar{\ell}\bar{\ell}}^{(d)} = \sum_{\ell\in\bar{\ell}} I_\ell \gamma_{\ell\ell} / I_{\bar{\ell}} \tag{24b}$$

$$\gamma_{\bar{\ell}\bar{\ell}}^{(od)} = \sum_{\ell,\ell'\in\bar{\ell},\, \ell'\neq\ell} I_\ell^{1/2} \gamma_{\ell\ell'} I_{\ell'}^{1/2} / I_{\bar{\ell}} \tag{24c}$$

where $\gamma_{\bar{\ell}\bar{\ell}}^{(d)}$ is the weighted average of the widths of the individual lines, while $\gamma_{\bar{\ell}\bar{\ell}}^{(od)}$ represents the narrowing effect of line coupling and is an average of the off-diagonal relaxation matrix.

According to the single line profile for the fully overlapped lines, a normalized effective line vector [6] $|\bar{\ell}\rangle\!\rangle$ can be defined as an average of the individual line vectors, weighted with the line amplitude:

$$|\bar{\ell}\rangle\!\rangle = \frac{|\boldsymbol{\mu}_{\bar{\ell}}\rangle\!\rangle}{\langle\!\langle\boldsymbol{\mu}_{\bar{\ell}}|\boldsymbol{\mu}_{\bar{\ell}}\rangle\!\rangle^{1/2}} = \sum_{\ell\in\bar{\ell}} \frac{I_\ell^{1/2}|\ell\rangle\!\rangle}{I_{\bar{\ell}}}. \tag{25}$$

In this way the resonance frequency $\omega_{\bar{\ell}} = \langle\!\langle\bar{\ell}|L_0|\bar{\ell}\rangle\!\rangle$ and the broadening parameter $\gamma_{\bar{\ell}\bar{\ell}} = \langle\!\langle\bar{\ell}|\gamma|\bar{\ell}\rangle\!\rangle$ of the line $\bar{\ell}$ assume the form required by Eqs. (22) and (24).

In the frame of impact theory, $\gamma_{\bar{\ell}\bar{\ell}}$ can be expressed following the formal treatment given in Eqs. (7) and (8) for the case of not degenerated lines:

$$\gamma_{\bar{\ell}\bar{\ell}} = \langle v\, \sigma_{\bar{\ell}\bar{\ell}}\rangle_v \tag{26}$$

$$\sigma_{\bar{\ell}\bar{\ell}} = \left\langle \pi \sum_l (2l+1) P_{\bar{\ell}\bar{\ell}}(k,l,r)/k^2 \right\rangle_r, \tag{27}$$

where, according to Eqs. (24), the collisional efficiency function $P_{\bar{\ell}\bar{\ell}}(k,l,r)$ is given by an average on the lines of the band including both diagonal and off-diagonal parts of P:

$$P_{\bar{\ell}\bar{\ell}}(k,l,r) = \sum_{\ell,\ell' \in \bar{\ell}} I_\ell^{1/2} P_{\ell\ell'}(k,l,r) \, I_{\ell'}^{1/2} / I_{\bar{\ell}}. \qquad (28)$$

Eqs. (28) and (11) allow us to obtain $P_{\bar{\ell}\bar{\ell}}$ in terms of the scattering matrix elements describing the collisional transitions from the states ℓ_i and ℓ_f of the lines internal to the overlapped band.

Denoting by $\beta = k, l, m, r$ all quantum numbers other than the internal state of the absorber, and using the optical theorem

$$\mathrm{Re}\langle \ell_i, \beta | T | \ell_i, \beta \rangle = \frac{1}{2} \sum_{\ell_i', \beta'} |\langle \ell_i, \beta | T | \ell_i', \beta' \rangle|^2 \qquad (29a)$$

$$\mathrm{Re}\langle \ell_f, \beta | T | \ell_f, \beta \rangle = \frac{1}{2} \sum_{\ell_f', \beta'} |\langle \ell_f, \beta | T | \ell_f', \beta' \rangle|^2 \qquad (29b)$$

for Eqs. (11b) and (11c), we can express the real part of $P_{\bar{\ell}\bar{\ell}}$ as the sum of two terms:

$$\mathrm{Re}\, P_{\bar{\ell}\bar{\ell}}(\beta) = P_{\bar{\ell}\bar{\ell}}^{(\in)}(\beta) + P_{\bar{\ell}\bar{\ell}}^{(\notin)}(\beta). \qquad (30)$$

$P_{\bar{\ell}\bar{\ell}}^{(\in)}$ is given by transitions $\ell_i \to \ell_i'$ and $\ell_f \to \ell_f'$ connecting couples of lines internal to the band $\bar{\ell}$ while $P_{\bar{\ell}\bar{\ell}}^{(\notin)}$ is given by transitions $\ell_i \to \ell_i'$ and $\ell_f \to \ell_f'$ such that $\ell_i \to \ell_f$ is a line of the band, while $\ell_i' \to \ell_f'$ is not a line of the band:

$$P_{\bar{\ell}\bar{\ell}}^{(\in)}(\beta) = \sum_{\ell,\ell' \in \bar{\ell},\, \beta'} \left| I_\ell^{1/2} \langle \ell_i, \beta | T | \ell_i', \beta' \rangle - I_{\ell'}^{1/2} \langle \ell_f, \beta | T | \ell_f', \beta' \rangle \right|^2 / (2 I_{\bar{\ell}}), \qquad (31)$$

$$P_{\bar{\ell}\bar{\ell}}^{(\notin)}(\beta) = \sum_{\ell \in \bar{\ell},\, \ell' \notin \bar{\ell},\, \beta'} I_\ell \left(|\langle \ell_i, \beta | T | \ell_i', \beta' \rangle|^2 + |\langle \ell_f, \beta | T | \ell_f', \beta' \rangle|^2 \right) / (2 I_{\bar{\ell}}). \qquad (32)$$

Comparing Eqs. (31) and (32), one can see that the broadening effect is different for transitions to lines inside or outside the overlapped band. In the former case the contribution is given by the square of the absolute value of the difference between the scattering amplitudes of the transitions $\ell_i \to \ell_i'$ and $\ell_f \to \ell_f'$, while in the latter case the contribution comes from the sum of the squares of the absolute value of the same quantities. The reduction of the width $\gamma_{\bar{\ell}\bar{\ell}}$ of the overlapped band, with respect to the simple average of the resolved linewidths $\gamma_{\ell\ell}$, is hence explained as an effect of correlation between the transitions $\ell_i \to \ell'_i$ and $\ell_f \to \ell'_f$ for the lines inside the band.

A comparison between experimental measurements and theoretical results obtained within this framework, can be found in Ref. [6] for the case of non resonant Debye spectra of symmetric top molecules where a strong reduction of the line-width was observed with respect to the one of resonant lines.

PERTURBATIVE APPROXIMATIONS

In order to calculate the collisional broadening parameter $w_{\bar{\ell}} = \operatorname{Re} n \gamma_{\bar{\ell}\bar{\ell}}$ of group $\bar{\ell}$ of overlapping lines, the results of the previous section can be used within the approximations commonly used for line broadening calculations.

The semiclassical approximation, described by Eqs. (15–17), yields:

$$\gamma_{\bar{\ell}\bar{\ell}} = \left\langle v \int_0^\infty 2\pi b\, db\, P_{\bar{\ell}\bar{\ell}}(b,v,r) \right\rangle_{v,r}, \qquad (33)$$

where the efficiency function $P_{\bar{\ell}\bar{\ell}}(b,v,r)$ for the overlapped band $\bar{\ell}$ is given by the semiclassical equivalent of Eq. (28):

$$P_{\bar{\ell}\bar{\ell}}(b,v,r) = \sum_{\ell,\ell' \in \bar{\ell}} I_\ell^{1/2} P_{\ell\ell'}(b,v,r)\, I_{\ell'}^{1/2} / I_{\bar{\ell}}, \qquad (34)$$

and $P_{\ell\ell'}(b,v,r)$ is given by Eq. (16).

Slightly different semiclassical calculations methods can be found in the literature [7]. They use different approximations for collisional trajectories and for collisions occurring at intermediate values of the impact parameter b, that are neither strong nor weak. However all these theories derive the efficiency function from a second order perturbative expansion of T in terms of V.

The perturbative expansion of Eq. (17) and the use of optical theorem yields for the real part of the efficiency function:

$$\operatorname{Re} P_{\bar{\ell}\bar{\ell}}(b,v,r) = P_{\bar{\ell}\bar{\ell}}^{(i)} + P_{\bar{\ell}\bar{\ell}}^{(f)} + P_{\bar{\ell}\bar{\ell}}^{(i,f)}, \qquad (35a)$$

$$P_{\bar{\ell}\bar{\ell}}^{(i)} = \sum_{\ell \in \bar{\ell},\, \ell'_i, r'} \frac{I_\ell}{2\hbar^2 I_{\bar{\ell}}} |\langle \ell_i, r | V(\omega_{\ell_i \ell'_i} + \omega_{rr'}) | \ell'_i, r' \rangle|^2, \qquad (35b)$$

$$P_{\overline{\ell\ell}}^{(f)} = \sum_{\ell \in \overline{\ell},\, \ell'_f, r'} \frac{I_\ell}{2\hbar^2 I_{\overline{\ell}}} |\langle \ell_f, r|V(\omega_{\ell_f \ell'_f} + \omega_{rr'})|\ell'_f, r'\rangle|^2, \qquad (35c)$$

$$P_{\overline{\ell\ell}}^{(i,f)} = \sum_{\ell, \ell' \in \overline{\ell},\, r'} -\frac{\rho_\ell^{1/2} \rho_{\ell'}^{-1/2} I_\ell^{1/2} I_{\ell'}^{1/2}}{\hbar^2 I_{\overline{\ell}}} \langle \ell_i, r|V(\omega_{\ell_i \ell'_i} + \omega_{rr'})|\ell'_i, r'\rangle$$
$$\times \langle \ell_f, r|V(\omega_{\ell_f \ell'_f} + \omega_{rr'})|\ell'_f, r'\rangle \qquad (35d)$$

where $\omega_{aa'} = (E_a - E_{a'})/\hbar$ is the frequency of the molecular transition $a \to a'$ and $V(\omega)$ is the Fourier transform of the time dependent collisional interaction $V(t)$.

Collisional coupling effect is restricted to the subtractive "middle" part $P_{\overline{\ell\ell}}^{(i,f)}$ while the two "outer" terms $P_{\overline{\ell\ell}}^{(i)}$ and $P_{\overline{\ell\ell}}^{(f)}$ are just a weighted average on the overlapping lines.

COUPLING BETWEEN HYPERFINE AND STARK COMPONENTS OF A MOLECULAR TRANSITION

We consider now the case in which an hyperfine structure and/or a small electric field resolving the magnetic components can be present. Two questions arises. Is there, when the components are well resolved, an hyperfine and Stark dependence of collisional broadening? Is there, when the components are overlapped, collisional coupling between them?

The appropriate vector basis for treating these question is:

$$\ell = \ell_f \leftarrow \ell_i = \alpha_f J_f, F_f M_f \leftarrow \alpha_i J_i, F_i M_i. \qquad (36)$$

The magnetic quantum number M is $M = M_F$ if the hyperfine structure is present and $M = M_J$ in its absence; in this last case the quantum number F is not necessary. α summarizes all quantum numbers other than J, F, and M.

Eq.s (35) can be used in order to disentangle the dependence of the relaxation matrix from F and M quantum numbers and from α and J quantum numbers. Tis is possible when the hyperfine an Stark interactions are small perturbations of the rotational energy. In this case, it can be shown [8] that any hyperfine and magnetic dependence is restricted to the "middle" part $P^{(i,f)}_{\bar{\ell}\bar{\ell}}$ of the relaxation matrix, given by Eq. (35d). Denoting by $^N P^{(i,f)}_{\bar{\ell}\bar{\ell}}$ the "middle" collisional efficiency function due a molecular interaction which is a tensor τ_N of order N with respect to rotations of the absorbing molecule, the dependence can be written as:

$$^N P^{(i,f)}_{\bar{\ell}\bar{\ell}} = \langle J_i\alpha_i\|\tau_N\|J_i\alpha_i\rangle \langle J_f\alpha_f\|\tau_N\|J_f\alpha_f\rangle \, ^N O \, ^N R, \qquad (37)$$

where $^N O$ is invariant for rotations of the absorber and $^N R$ contains the rotational dependence and hence the dependence on F and M.

From Eq. (35d) explicit expressions for $^N O$ are obtained for the four different cases corresponding to presence or absence of hyperfine and of Stark structures.

We first consider the case in which hyperfine structure is absent. When the Stark components are resolved, the band $\bar{\ell}$ in Eq. (35d) is reduced to a single line and the rotational geometric factor $^N R$ is given by the simple product of two Clebsch-Gordan coefficients:

$$^N R = \langle J_i M_i|N0J_iM_i\rangle \langle J_f M_f|N0J_fM_f\rangle. \qquad (38)$$

If the field is removed, the magnetic components collapse into a degenerate line and their coupling can be calculated by Eq. (35d) yielding:

$$^N R = (-1)^{J_i+J_f+N+1}(2J_i+1)^{1/2}(2J_f+1)^{1/2}W(J_iJ_iJ_fJ_f;N1), \qquad (39)$$

where W is a 6-J Racah coefficient. As consequence of Eq. (38) in presence of Stark field the collisional broadening of the components will have a dependence on the quantum number M. When the field is removed a collisional coupling effect arises between the components producing a reduction of the line width to the value given by Eq. (39). This effect was observed for the spectrum of methyl fluoride [9–11] and the present treatment gives a satisfactory interpretation of these measurements. We point out that Eq. (38) was already given in Ref. [10] for the particular case $N = 1$ (dipole interaction) while Eq. (39) reproduces the well known and widely used result [12] for unresolved rotational lines, that can hence be considered as a particular case of our coupling theory.

When an Hyperfine structure is present, a dependence of the pressure broadening is obtained on both quantum numbers F and $M = M_F$. In presence of an electric field Eq. (35d) yields:

$$^NR = (-1)^{J_i+J_f+F_i+F_f+2I}(2F_i+1)^{1/2}(2F_f+1)^{1/2}\langle F_i M_i|N0F_i M_i\rangle$$
$$\times \langle F_f M_f|N0F_f M_f\rangle W(J_i J_i F_i F_i; NI)\ W(J_f J_f F_f F_f; NI). \qquad (40)$$

If the field is removed, the value of ^NR for an hyperfine line $\alpha_i J_i, F_i \to \alpha_f J_f, F_f$, is given by Eq. (35d) summing over all its Stark components:

$$^NR = (-1)^{J_i+J_f+2I+N+1}(2F_i+1)^{1/2}(2F_f+1)^{1/2}(2J_i+1)^{1/2}(2J_f+1)^{1/2}$$
$$\times W(F_i F_i F_f F_f; N1)W(J_i J_i F_i F_i; NI)\ W(J_f J_f F_f F_f; NI). \qquad (41)$$

The hyperfine dependence of pressure broadening was observed in Ref. [13] where Eqs. (40) and (41) were given for the particular case $N = 1$ and successfully compared to measurements performed on the hyperfine and hyperfine-Stark components of the rotational lines $J = 4 \to 5$, $K = 4$ of CH_3I and $J = 3 \to 4$, $K = 3$ of CH_3Br.

The present results can be shown in agreement with what found by Green [14] who studied the case of HCN perturbed by a noble gas atom and concluded that the hyperfine effects on the collisional lineshape are in that case null or very small.

In Ref. [3] the collisional lineshape of rotational transitions of CHF_2Cl asymmetric top molecule was studied. A negligible collisional coupling between hyperfine components of a single rotational transition was observed, while coupling was in some cases relevant between hyperfine components of different rotational transitions. Some hints were shortly given in that paper on the theoretical explanation of the phenomenon that is discussed in more details in the present work. Indeed, by Eq. (37) we can see that the reduced matrix elements $\langle J_i\alpha_i\|\tau_N\|J_i\alpha_i\rangle$ and $\langle J_f\alpha_f\|\tau_N\|J_f\alpha_f\rangle$ vanish for the case of an asymmetric top molecule interacting mainly by its dipole ($N = 1$).

REFERENCES

[1] C. Boulet, D. Robert, and L. Galatry, J. Chem. Phys. **72**, 751 (1980).

[2] M. Baranger, Phys. Rev. **111**, 481 (1958); **111**, 494 (1958; **112**, 855 (1958).

[3] G. Cazzoli, L. Cludi, G. Cotti, C. Degli Esposti, G. Buffa and O. Tarrini, J. Chem. Phys. **102**, 1149, (1995).

[4] G. Buffa, A. Di Giacomo and O. Tarrini, Nuovo Cimento B **20**, 281 (1974).

[5] P. W. Rosenkranz, I. E. E. E. Trans. Antennas Propagat. **23**, 498 (1975).

[6] G. Buffa and O. Tarrini, Phys. Rev. A **16**, 1612 (1977).

[7] G. Buffa and O. Tarrini, Appl. Opt. **28**, 1800 (1989) and references therein.

[8] S. Belli, G. Buffa and O. Tarrini, to be published.

[9] Ph. Brechignac, J. Chem. Phys. **76**, 3389 (1982).

[10] G. Buffa, A. Di Lieto, P. Minguzzi, O. Tarrini and M. Tonelli Phys. Rev. A **34**, 1065 (1986).

[11] V. Lemaire, L. Dore, G. Cazzoli, G. Buffa, O. Tarrini and S. Belli, "Collisional self-broadening, self-shifting and line-coupling in rotational transitions of CH_3F in presence of Stark fields", in 13th International Conference on Spectral Line Shapes, see the present book.

[12] P. W. Anderson, Phys. Rev. **76**, 647 (1949); C. T. Tsao and I. Curnutte, J. Quant. Spectrosc. Radiat. Transfer **2**, 41 (1962).

[13] G. Buffa, A. Di Lieto, P. Minguzzi, O. Tarrini and M. Tonelli Phys. Rev. A **37**, 3790 (1988).

[14] S. Green, J. Chem. Phys. **88**, 7331 (1988).

Velocity Selective Coherent Transient Study of Depolarizing Collisions in Molecular Gas

L.S. Vasilenko, N.N. Rubtsova, E.B. Khvorostov

*Institute of Semiconductor Physics, RAS,
prosp. acad. Lavrentyev, 13, Novosibirsk 630090, Russia*

Abstract. Method of stimulated photon echo with specially chosen linear polarizations of exciting pulses of coherent resonant radiation is applied for investigation of collisional decay of orientation and alignment in molecular gas SF_6 and its mixtures with buffers He and Xe. These relaxation rates are measured as a function of longitudinal velocities. Within experiment accuracy, no remarkable velocity dependence of depolarizing collisions rates was detected. This result confirms conventional approach of depolarizing collisions theoretical treatment.

Coherent transient phenomena like photon echo and its modifications are widely used for investigation of relaxation processes in various media [1]. This technique proved very fruitful being applied to studies of different types of molecular collisions and for extraction of data concerning details of the interaction potential.

Characteristic feature of our experiment is the application of continuous wave laser of good quality radiation as a source of excitation. Necessary sequence of exciting pulses of linear polarization is cutting out of this CW radiation by a set of electro-optical shutters. If necessary, additional electro-optical crystal could rotate polarization of definite exciting pulse by ninety degree. Line width of CW radiation of about 1 kHz ensures the mutual coherency of exciting pulses up to one millisecond time delay between them.

Another significant feature of our experiments is the possibility to excite gaseous sub-ensemble of definite translational velocities. The width of such sub-ensemble for longitudinal velocities is controlled by the parameters of exciting pulses (their time lengths and intensities), while the central velocity may be chosen by the frequency tuning of exciting radiation over the Doppler contour of a gas under investigation. This gives possibility to investigate velocity dependence of collisional relaxation by coherent transient phenomena and to extract from these data information concerning both mechanisms of relaxation and the steepness of the interaction potential [2,3].

In our experiments we use stable CO_2 laser with its frequency locked at the center of SF_6 transition as a reference. The radiation frequency of exciting laser was locked to this laser at definite adjustable frequency detuning, $\Omega = kv_z$, which determined the central velocity of molecular ensemble emitting coherent response. Our previous results [2,3] have demonstrated the ability of coherent transients to detect translational velocity selective collisional decay of the induced dipole moment of molecular transition, in others words, the homogeneous line width Γ, and to make definite conclusions about the steepness of the interaction potential.

Coherent transients are also very sensitive to another type of relaxation — to the depolarizing collisions. It's well known since the performance of first experiments on the optical pumping, that polarized light is able to create specific non-equilibrium populations distribution between magnetic sub-levels of excited or ground quantum state. Such polarized state is used to be described in the terms of polarization moments. The lowest of them are null moment, corresponding to the level population, first moment — orientation — corresponding to the creation of macroscopic magnetic moment of gaseous ensemble, second moment — alignment — corresponding to the macroscopic electric quadruple moment. The decay of polarization moments is called by collisions, which are very sensitive to the asymmetry of the interaction potential. Corresponding decay rates $\gamma^{(\alpha)}$, $\alpha = 0, 1, 2$ are decay rates of population, orientation and alignment.

In this work we used stimulated photon echo technique, suggested for the first time by russian theorists [4]. An idea of this method is the following: two first exciting pulses of resonant radiation, travelling along the axis "Y", with their linear polarizations forming angles ψ_1 and ψ_2 relative third pulse polarization, and separated by small time delay T_{12}, are employed to create the population, orientation, alignment and higher polarization moments in gaseous ensemble. These polarization moments undergo destroying by collisions during time delay T_{23} between second and third pulses. Third pulse converts the coherency of multiple moments of the resonant levels into the coherency of the resonant transition between these levels. Finally, after time delay T_{12} after the third pulse, this results in the generation of stimulated photon echo signal.

In the limit of small areas of exciting pulses, the expressions for "X" and "Z" components of electromagnetic field of stimulated photon echo depend on the angles ψ_1 and ψ_2 and contain only three types of decay rate constants $\gamma_i^{(\alpha)}$ ($\alpha = 0, 1, 2$; i denotes upper and lower levels), i.e. six decay rates. For infrared vibrational-rotational transitions corresponding decay rates of upper and lower levels are usually close to each other, which allows simplification of formula (in spite of six decay rate we have only three of them). Further simplification may be introduced by large dipole moment of transition $J \gg 1$, which is typically for SF_6 transition under investigation. Finally, special choice of angles ψ_1 and

ψ_2 reduces till two the number of decay rates, controlling stimulated photon echo amplitudes [5].

The most simple way to realize this idea experimentally is to rotate by ninety degree the polarization of one of three exciting pulses. In our previous work [6] we have realized the measurements of the difference $\delta = \gamma^{(2)} - \gamma^{(0)}$ between alignment and level population decay rates at $0 \to 1$ ν_3 Q(38) SF$_6$ transition by using the rotation of third pulse polarization by angle of ninety degree. The relaxation rate of alignment was estimated by addition of δ with populaton decay rates, known from another experiments [7].

In this work the measurement of collisional decay of orientation and alignment is performed at the transition $0 \to 1$ ν_3 P(33) A_2^1 SF$_6$. The polarizations of first and second pulses, rotated by ninety degree were used, which should warrant the contribution of only two decay rates to the stimulated photon echo amplitudes — $\gamma^{(1)}$ and $\gamma^{(2)}$. In this case, stimulated photon echo signals are very small, and we have attracted optical heterodyne technique to register the "X" component of stimulated photon echo. Simplified theoretical analysis shows, that $\gamma^{(1)}$ and $\gamma^{(2)}$ may be obtained by simple one-exponential fit of the sum and difference of stimulated photon echo amplitude A_{010}, corresponding second rotated pulse and A_{100}, corresponding to first rotated pulse.

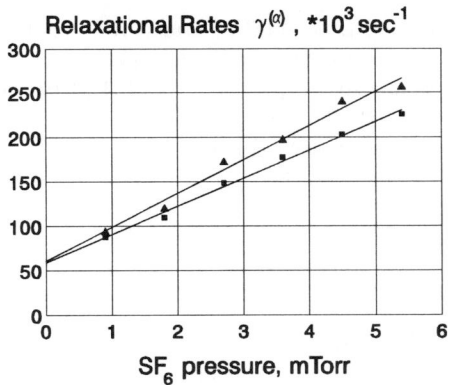

Figure 1: Relaxation rates (squares — $\gamma^{(1)}$, triangles — $\gamma^{(2)}$) vs SF_6 pressure. Straight lines are drawing by method of least squares.

Results are represented by Fig.1. In pure SF$_6$, the alignment decay rate proved some larger than orientation collisional decay rate and population decay rate. Our measurement gave following results: $\gamma^{(1)} = (32\pm5)\cdot 10^6$ sec$^{-1}\cdot$Torr^{-1}, $\gamma^{(2)} = (38\pm 5)\cdot 10^6$ sec$^{-1}\cdot$Torr^{-1}. Value of $\gamma^{(0)}$ we can extract from our earlier experiment [7]: $\gamma^{(0)} = (33 \pm 3)\cdot 10^6$ sec$^{-1}\cdot$Torr^{-1}. In the presence of buffers practically no difference between alignment and orientation collisonal decay rate was measured.

Velocity dependent measurements were performed for several values of frequency detuning of exciting laser radiation from the line center — 9 MHz, 4.5 MHz and at the center of line. No significant velocity dependent decay rate was detected neither in pure SF_6 nor in its mixtures with light (He) and heavy (Xe) buffer. Results are in qualitative agreement with theoretical predictions for collisions with buffers.

Authors are indebted to Russian Foundation for Basic Research for financial support of this work (Grant 95-02-04603).

REFERENCES

1. Shoemaker R.L., 1978, *Laser and Coherence Spectroscopy*, edited by J.I.Steinfeld (New York, London: Academic Press)
2. Vasilenko L.S., Rubtsova N.N, Chebotayev V.P., 1983, *Pis'ma v ZhETF*, **38**, pp. 391-394
3. Vasilenko L.S., Rubtsova N.N., Hvorostov E.B., 1996, *Laser Physics*, **6**, No 1, pp. 165-168
4. Yevseyev I.V., Yermachenko V.M., Reshetov V.A., 1980, *ZhETF*, **78**, 2213-2221
5. Yevseyev I.V., Yermachenko V.M., Samartsev V.V., 1992, *Depolarizing Collisions in Quantum Electrodynamics*, (Moscow, Nauka)
6. Belousov N.S., Vasilenko L.S., Matveyenko I.D., Rubtsova N.N., 1987, *Optika i Spektroskopiya*, **63**, 34-38
7. Vasilenko L.S., Rubtsova N.N., 1985, *Optika i spektroscopiya*, **58**, 697

Self and Foreign Broadening and Shift Versus Temperature of Ammonia Lines

G. Baldacchini,[*] G. Buffa,[†] F. D'Amato,[*] M. De Rosa,[‡]
F. Pelagalli,[‡] and O. Tarrini[†]

[*] *ENEA, INN-FIS, Via E. Fermi 27, 00044 Frascati (Roma)*
[†] *Dip. di Fisica, Università di Pisa, P.zza Torricelli 2, 56126 Pisa*
[‡] *ENEA guest*

Abstract. The effects of collisions with foreign gases (N_2, O_2, H_2, He, Ar) on the line parameters have been investigated in a wide range of temperature for the aQ(9,9) line of ammonia. When possible, the results have been compared with theory.

The effects of collisions on the molecular lineshapes are of great importance due to both fundamental and applicative aspects. Indeed the lineshapes contain useful information on the nature of interactions among molecules, and at the same time put some constraints to the design parameters of analytical instruments based on spectroscopical techniques.

The behaviours with temperature of broadening and shift coefficients are described by calculations based on a slightly modified ATC theory (1,2). Such calculations are suitable when the long range interaction is so effective that the broadening cross-section is far larger than the kinetic one. This is the case of self broadening, because of the large dipole-dipole interaction, and, at a lesser extent of N_2-broadening, because of the large quadrupole moment of nitrogen molecule. As far as the experimental measurements are concerned, we have realized an apparatus based on tunable diode lasers and variable temperature cell in the range 180÷380 K, in order to test the validity of the mentioned calculations.

Self- broadening and shift parameters have been measured for several lines in the v_2 band of ammonia around 900 cm^{-1} (3-5). Foreign gas effects have been measured with five gases (N_2, O_2, H_2, He, Ar) for the aQ(9,9) line (921.2550 cm^{-1}).

One of the results of our measurements is that the empirical power laws describing the temperature behaviours of broadening and shift coefficients are sometimes wrong, especially when wide temperature ranges are involved. Moreover the Figure shows the shift coefficients vs. temperature and it is clear how the results of our measurements can be applied for determining the behaviour

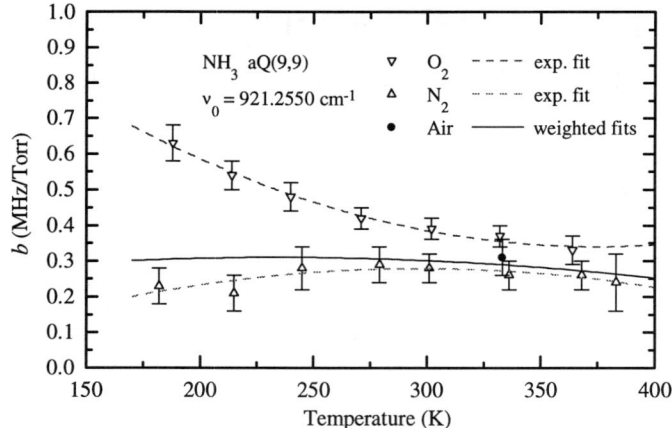

of absorption lines in different atmospheres. The dashed curves are the best fits of the experimental points.

Also we have verified that the calculation method used here, describes very well self and N_2 broadening of ammonia, has some limitations for self shift, while improvements are nedeed for other cases.

On the other side we have been able to produce very fine measurements with errors as low as 2÷3% for broadening parameters and very often less than 10% for shift parameters. These results, which are very important also for an improvement of the theoretical model, have been obtained thanks to a modest upgrading of the experimental apparatus, to a semiautomatic analysis of the data, and last but not least to surprisingly big foreign shifts compared to the foreign broadening parameters. Indeed, for foreign gas collisions, we found a shift/width ratio by for larger than for self-collisions.

REFERENCES

1) Anderson, *Phys. Rev.* **76**, 647 (1949).
2) Tsao and I. Curnutte, *J. Quant. Spectrosc. Radiat. Transfer* **2**, 41 (1962).
3) Baldacchini, A. Ciucci, F. D'Amato, G. Buffa and O. Tarrini, *J. Quant. Spectrosc. Radiat. Transfer* **53**, 671 (1995).
4) Baldacchini, F. D'Amato, M. De Rosa, G. Buffa and O. Tarrini, *J. Quant. Spectrosc. Radiat. Transfer*, to appear (1996).
5) Baldacchini, F. D'Amato, F. Pelagalli, G. Buffa and O. Tarrini, *J. Quant. Spectrosc. Radiat. Transfer*, to appear (1996).

Rotational contribution to Raman line shape of liquid carbonyl sulphide: MD simulation

Z. Gburski*, H. Stassen[+] and A. Kachel*

*Department of Physics, University of Silesia, Uniwersytecka 4, 40-007 Katowice, Poland,
+Inst. di Quimica, Universidade do Rio Grande, CEP 91540-000, Porto Alegre-RS, Brasil

Abstract. We apply the molecular dynamics (MD) simulation technique to calculate the angular velocity $G_\omega(t)$ and the second-order reorientational $G_2(t)$ correlation functions for liquid carbonyl sulphide. The relation between these correlation functions, predicted by the cumulant approximation, is analysed.

The relation linking the correlation functions $G_\omega(t)$ and $G_2(t)$ can be simplified by using a cumulant expansion and cutting it off after the second stage. A molecular dynamic simulations enable to calculate separately $G_\omega(t)$ and $G_2(t)$. Having both correlation functions from MD we are in a position to test the validity of the second cumulant approximation.

Molecular rotations can be observed by the analysis of Raman line shapes. Applications of the standard method furnishes the rotational spectrum $I_{rot}(\omega)$, which is the Fourier transform of the second order reorientational correlation functions $G_2(t) = < P_2(\cos\theta(t)) >$, $\theta(t)$ is the angle between **u**(t) and **u**(0), **u** is a unit vector along the axis for a linear molecule, P_2 is the second order Legendre polynomial. It has been shown (1) that, within the second cumulant approximation, there exists the relation between $G_2(t)$ and the angular velocity correlation function $G_\omega(t) = < \omega_\alpha(t) \omega_\alpha(0) >$ (summation over $\alpha = x,y,z$). Namely,

$$G_{2,c}(t) = \exp\{-6kT / I \int_0^t dt' \int_0^{t'} G_\omega(t'') dt'' \} \qquad (1)$$

where I is the moment of inertia of a given molecule, T is the temperature, k is the Boltzman constant. We have performed molecular dynamics simulations of 256 molecules of liquid carbonyl sulphide at T=250 K and the density ρ=1.125 g/cm^3. The usual periodic boundary conditions were applied, and the simulations extended up to 150 ps (integration step 3 10^{-15} s). The thermodynamic state has been simulated with a three-centre atom-atom Lennard-Jones (LJ) potential, the values of LJ parameters were taken from (2). We have simulated the correlation functions $G_\omega(t)$ and $G_2(t)$ then we calculated the second cumulant approximation to $G_2(t)$, i.e. the function $G_{2,c}(t)$ by numerical double integration of $G_\omega(t)$, following the eq.(1). The result is shown in Fig.1, where we compare $I_{rot}(\nu)$ with the Fourier transform of $G_{2,c}(t)$ denoted by $I_{rot,c}(\nu)$.

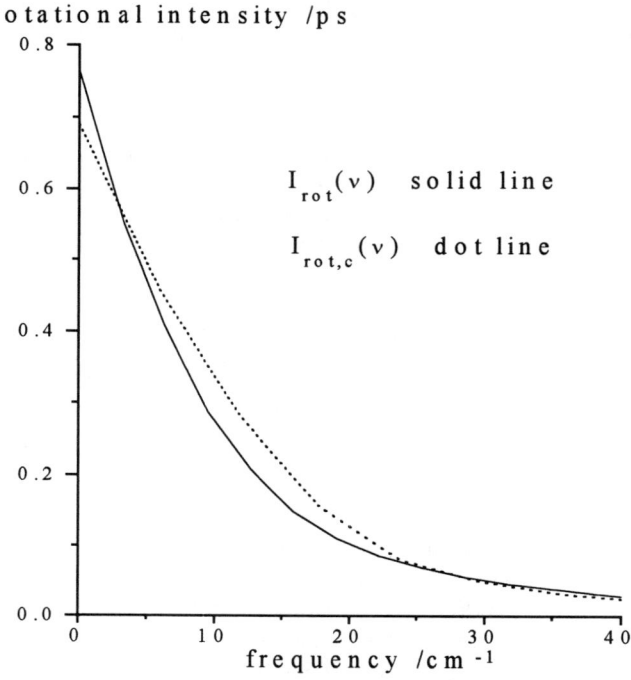

FIGURE 1. Comparison of the simulated intensities $I_{rot}(\nu)$ and $I_{rot,c}(\nu)$ for liquid carbonyl sulphide OCS at T=250 K and ρ=1.125 g/cm^3.

The simulation shows that the second cumulant approximation is good at high frequencies (short times), becoming less satisfactory for lower frequencies (longer times). One should be aware about the increasing uncertainty of a long time tail of the correlation function $G_\omega(t)$, if extracted from Raman line shape analysis.

REFERENCES

1. Lyndenn-Bell, R. M., *Molecular Liquids*, New York: Reidel, 1985, ch. 5, pp. 89-94.
2. Stassen, H., and Gburski, Z., Chem.Phys. Lett. **217**, 325-332 (1994).

Collisional self-broadening, shifting and line-coupling of rotational transitions of CH_3F in presence of Stark fields

Véronique Lemaire[*], Luca Dore[*], Gabriele Cazzoli[*],
Giovanni Buffa[!], Ottavio Tarrini[!], Simonetta Belli[!]

[*] Dip. di Chimica "G. Ciamician", Via Selmi 2, 40126 Bologna, Italy
[!] Dip. di Fisica, P.za Torricelli 2, 56126 Pisa, Italy

Abstract. Self-broadening and self-shifting of methyl fluoride, $^{12}CH_3F$, have been investigated in the 50-408 GHz range for the rotational transitions $J=0\rightarrow1$, $J=1\rightarrow2$, $J=2\rightarrow3$, $J=5\rightarrow6$ and $J=7\rightarrow8$ and for their $\Delta M_J=0$ and $\Delta M_J=\pm1$ Stark components, as they were well resolved by an appropriate electric field. The experimental M_J-dependence of the collisional relaxation parameters is in very good agreement with the theoretical one, developed within semiclassical perturbative approximations. In case where Stark components are not isolated anymore, discrepancy of profiles with sums of Voigt profile is observed. This is explained in term of collisional line-coupling effect.

Broadening measurements of CH_3F have been being inverstigated for some decades in different frequency ranges, for rotational and ro-vibrational transitions. More over, the first M_J-dependence of pressure-broadening has been reported for this molecule (1,2). Such a study contributes to a better understanding of the effect on lineshape of collisional relaxation, which for strong dipolar molecules, as is the case of CH_3F, is dominated by dipole-dipole interaction.

The M_J-dependence of relaxation parameters is also very useful, since Stark fields are commonly used to tune molecular laser and sometime in the particular application of collisional relaxation investigation.

In the present work, we show that the width of an M_J-unresolved transition can be quite different and usually smaller than the weighted average of the widths of the different Stark-resolved lines, since contributions to the broadening of collisions-induced transitions among M_J levels, resolved by an external electric field, have to be taken into account. As can be seen in the example of the $\Delta M_J=0$ Stark components of the $J=7\rightarrow8$ (fig. 1 and table 1), the experimental M_J-dependence of collisional parameters is very well reproduced by the theory, which is obtained by an extension of the semiclassical impact approximation (3,4) in order to calculate the relaxation matrix in the Liouville subspaces of lines (5).

We also present what would be the first M_J-dependence of the frequency-shift parameters. More over, in case of small electric field and for suitable pressure conditions, so that Stark components are not isolated anymore, we observed discrepancy from sums of Voigt profiles. This can be explained by line-coupling. The analysis of these profiles is in progress.

© 1997 American Institute of Physics

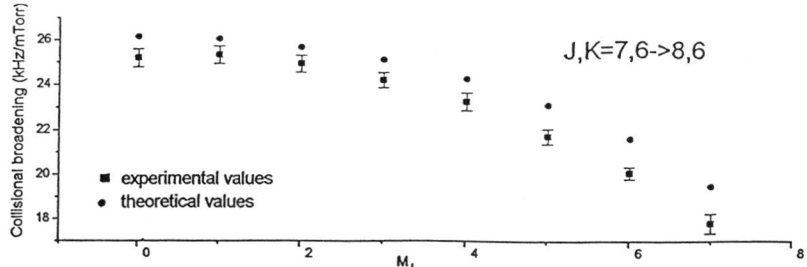

FIGURE 1: theoretical and experimental M_J-dependence of the broadening of the $\Delta M_J=0$ components of $J,K=7,6 \rightarrow 8,6$.

Transition	self-broadening (kHz/mTorr)		self-shifting (kHz/mTorr)	
	experim.	theory	experim.	theory
$J,K=7,6\rightarrow 8,6$	13.63 (16)*	16.28	0.84(5)	0.41
$M= 0\rightarrow 1$ / $-7\rightarrow -8$	17.80 (29)	19.90	0.52(7)	0.24
$M= 1\rightarrow 2$ / $-6\rightarrow -7$	21.21 (40)	22.24	-	0.16
$M= 2\rightarrow 3$ / $-5\rightarrow -6$	22.78 (40)	23.61	-	0.13
$M= 3\rightarrow 4$ / $-4\rightarrow -5$	23.46 (52)	24.27	-	0.12
$M= 0\rightarrow 0$	25.25 (40)	26.20		
$M= 1\rightarrow 1$	25.38 (41)	26.09		
$M= 2\rightarrow 2$	25.00 (39)	25.75		
$M= 3\rightarrow 3$	24.28 (35)	25.17		
$M= 4\rightarrow 4$	23.31 (39)	24.32		
$M= 5\rightarrow 5$	21.77 (35)	23.16		
$M= 6\rightarrow 6$	19.96 (30)	21.60		
$M= 7\rightarrow 7$	17.62 (45)	19.48		

The given errors correspond to one standard deviation. * determined from $^{13}CH_3F$.

TABLE 1: Experimental and theoretical self-broadening and self-shifting coefficients (in kHz/mTorr) of $\Delta M_J=0$ and $\Delta M_J=\pm 1$ Stark components for the $J,K=7,6\rightarrow 8,6$ transition of CH_3F.

REFERENCES

1 - Brechignac Ph., J. Chem Phys. **76**, 3389 (1982)
2 - Buffa G., Di Lieto A., Minguzzzi P., Tarrini O., Tonelli M., Phys. Rev. A **34**, 1065 (1986)
3 - Anderson P.W., Phys. Rev. **76**, 647 (1949), Tsao C.J. and Curnutte B., J. Quant. Spectrosc. Radiat. Transfer **2**, 41 (1962)
4 - Robert D. and Bonamy J., J. Phys. (Paris) **10**, 923 (1979)
5 - Belli S., Buffa G., Tarrini O., to be published

Collisional Broadening and Shift of Acetylene and Oxygen Overtones

A. Lucchesini, M. De Rosa, A. Ciucci
C. Gabbanini, S. Gozzini

Istituto di Fisica Atomica e Molecolare del CNR
Via del Giardino, 7 I-56127 Pisa, ITALY

Abstract. Overtone absorption lines of acetylene and oxygen around 12000 and 13000 cm^{-1} have been examined by the use of tunable diode lasers and the wavelength modulation (WM) spectroscopy with 2nd harmonic detection technique. The collisional broadening and shift coefficients have been obtained. The correct interpretation of the absorption features when detecting the second harmonic signal in the presence of a sloping background is discussed.

Overtone and combination of overtone ro-vibrational spectra in the visible and in the near-infrared have been studied more than sixty years ago and recently, by using tunable laser sources, there has been an increasing interest on the spectroscopy of such resonances[1]. Weak absorptions can be observed by using diode lasers and the WM technique[2]. This technique can be applied to the diode lasers without any special problem: the modulation of the injection current brings directly to a corresponding emission wavelength variation. The experimental apparatus adopted for the WM technique is explained in our previous paper[3].

The radiation $I(\nu)$ transmitted through the gas sample is well described by the Lambert-Beer equation:

$$I(\nu) = I_\circ(\nu)\exp[-\alpha(\nu)x] \qquad (1)$$

where $I_\circ(\nu)$ is the incoming radiation intensity at frequency ν, x is the optical density, that is the product of the gas density by the radiation path-length, $\alpha(\nu)$ is the absorption coefficient. Its shape is heavily influenced by the Doppler-broadening and pressure-broadening. In the diode lasers an amplitude modulation, which is associated with the emission frequency modulation, results in a sloping background of the direct absorption signal, and in an asymmetry of the 2nd derivative shape in the 2nd harmonic signal. Such an asymmetry involves an apparent displacement of the center of the line, which depends on both the sloping background and the width of the line itself. In presence of pressure-broadening, an apparent pressure-shift can occur. In order to avoid this "pseudo-shift" and to determine the correct value of the pressure-shift by fitting the experimental data, a proper function that takes into account this emission characteristic of the laser must be used. In case of weak absorptions, that is $\alpha(\nu)x \ll 1$, eq. (1) can be approximated by the expansion: $I(\nu) \simeq I_o(\nu)[1 - \alpha(\nu)x]$. Thus, in case of a linear background: $I_o(\nu) = \bar{I}_o(1 + s\nu)$, where $\bar{I}_o$ is the background intensity at $\nu = 0$ and s is the fractional change in the background. Then 2nd derivative of the total signal will be: $I''(\nu) = -\bar{I}_o x[(1+s\nu)\alpha''(\nu) + 2s\alpha'(\nu)]$ where $\alpha'(\nu)$ and $\alpha''(\nu)$ are the first and the second derivative of the absorption function. It can be verified that the two terms within square brackets have opposite effects on the asymmetry and the second one is predominant. Indeed, when fitting the experimental data, one can be inclined to use simply the 2nd derivative of the absorption function, multiplying it by a sloping linear function in order to match the asymmetry of the experimental signal. Although at first sight the quality of the fit could seem satisfactory, the resulting line parameters (intensity, width and line-center position) are incorrect. In particular the position of the center of the line can be strongly affected by the pseudo-shift. By using the 2nd derivative of the whole signal, the fit process gives the correct value for the absorption line center, even if the linear ramp is not well known.

The results of collisional broadening and shift measurements for acetylene around 12000 cm^{-1} are reported in our specific paper[4]. Preliminary results on pressure-shift have been obtained for oxygen at 13156.27 cm^{-1}:

$$S_{\text{self}} = (-0.3 \pm 0.1) \text{ MHz/Torr}$$

Measured pressure-broadening FWHM coefficients for oxygen are shown in Table 1.

TABLE 1. Pressure-broadening coefficients for oxygen at 13156.27 cm^{-1}

γ_{self}	γ_{N_2}	γ_{He}
(3.2 ± 0.1)MHz/Torr	(3.4 ± 0.1)MHz/Torr	(2.9 ± 0.1)MHz/Torr

REFERENCES

1. A. Campargue, M. Chenevier, and F. Stoekel, *J. Mol. Spectros.* **151**, 275–281 (1992)
2. F.S. Pavone and M. Inguscio: *Appl. Phys. B* **56**, 118–122 (1993)
3. A. Lucchesini, I. Longo, C. Gabbanini, S. Gozzini and L. Moi *Appl. Opt.* **32**, 5211–5216 (1993)
4. A. Lucchesini, M. De Rosa, D. Pelliccia, A. Ciucci, C. Gabbanini, S. Gozzini, *Appl. Phys. B*, to be published.

Vibrational Line Width and Line Shift in (Binary) Systems at High Pressures

J.A.Schouten and J.P.J.Michels

Van der Waals-Zeeman Institute, University of Amsterdam, Valckenierstraat 65, 1018 XE, Amsterdam, the Netherlands.

Abstract. In a dense fluid or solid the intermolecular forces can cause a considerable change in the molecular vibrational frequency and in the profile of the spectrum. In a pure substance the local environment of a molecule varies due to fluctuations in the density, while in a mixture also composition fluctuations may be important.
We will present experimental results concerning the frequency shift and the line width as a function of pressure (up to several GPa) for pure nitrogen. In the case of fluid nitrogen these results will be compared with computer simulations and theoretical calculations. Special attention will be paid to the so-called 'attractive' contribution.
The influence of the environment has been studied in dilute solutions of nitrogen in helium and hydrogen in helium, neon, argon and deuterium. The results will be discussed and compared with computer simulations in the case of nitrogen.
As we will show, in the above examples the modulation of the frequency is in the fast regime. Slow modulation can occur in the case of critical fluctuations. The system helium-nitrogen exhibits fluid-fluid demixing at high pressures. Raman measurements have been performed in the critical region. The inhomogeneous line broadening due to these concentration fluctuations will be discussed.

INTRODUCTION

Raman spectroscopy is an important tool for studying the physics of molecular systems at high pressure. In a dense fluid or solid the intermolecular forces may result in a considerable change in both the molecular vibrational frequency and in the profile of the spectrum. Since the local environment of the molecule can be probed with Raman spectroscopy, it allows for investigating the elementary molecular processes that are responsible for the formation of various solid phases as well as for the dynamical behavior in fluids. In a pure substance at constant pressure and temperature the local environment of a molecule, and thus the influence of the intermolecular forces, varies due to fluctuations in the density, while in a mixture also microscopic fluctuations in the composition may play a role. These fluctuations result in a broadening of the vibrational spectrum. Of course, changes in the interactions can also be effectuated by varying the pressure, the temperature or the composition. This results in a line shift as well as in a change of the line width. On the other hand these arguments suggest that detailed information about the intermolecular forces, in particular the unlike interaction in mixtures, can be obtained by performing Raman measurements. It will be shown that this is at most partly true.

Nitrogen is a favorite model system, since it is stable over a wide temperature and pressure range. The intermolecular interaction can be well represented by a site-site potential and its simple shape makes it perfectly suitable for studying the orientational effects. Moreover, the influence of various mechanisms, such as resonance coupling, change of anharmonicity, and vibrational population relaxation, on the vibrational properties can be neglected.

In this paper we will discuss the experimental results of Raman measurements on fluid nitrogen and nitrogen in helium as a function of pressure, temperature and composition. The investigation of nitrogen diluted in helium is very interesting, because it shows that the existing theories concerning the influence of attractive forces on the vibrational frequency and line width is not correct. We will also discuss some other binary mixtures. Computer simulations have been performed to calculate the frequency shift and line width in nitrogen and nitrogen diluted in helium. The results will be compared with experiments and the various contributions to the line width will be discussed.

The influence of concentration fluctuations is expected to be largest near the critical point of the mixture. However, such an effect was not observed in low pressure experiments. We will show that measurements at very high pressure are more appropriate to detect these concentration fluctuations.

EXPERIMENTAL DATA

The experimental results discussed in this paper have been obtained by performing Raman spectroscopy in a diamond anvil cell (DAC). Unless otherwise specified, the experiments have been carried out at room temperature. In Fig.1 the experimental shift of the vibrational frequency of nitrogen has been plotted as a function of pressure (1,2,3) up to the melting line. The very accurate results below 0.65 GPa have not been obtained in a DAC (1). At lower pressures the frequency first decreases as a function of pressure with respect to that of the isolated molecule while at higher pressures it increases to values even above the zero pressure value. Often the characteristic red shift in the lower pressure region is loosely attributed to the attractive forces (3) without specifying the exact meaning of this

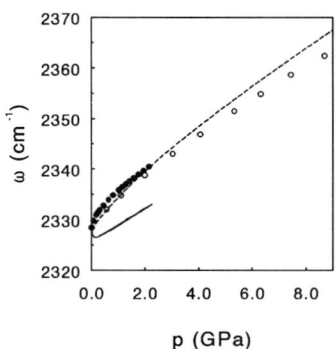
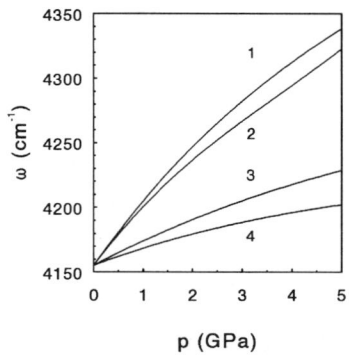

FIGURE 1A. Shift of Raman frequency of pure nitrogen and nitrogen in helium as a function of pressure. Full line: nitrogen experiment; dashed line nitrogen in helium; dots: MD results nitrogen; open circles: MD results nitrogen in helium.

FIGURE 1B. Shift of Raman frequency as a function of pressure. Hydrogen (2 mole percent) in helium (curve 1), in neon (curve 2), in deuterium (curve 3) and pure hydrogen (curve 4). The data for hydrogen in argon are not shown since they are only a little bit higher than those for hydrogen in neon.

statement; at higher pressures the contribution of the repulsive forces is assumed to be dominant. In order to check these assumptions we have also carried out measurements on nitrogen diluted in helium. Since the attractive forces in helium are very small it is to be expected that the red shift would disappear. Indeed, as shown in Fig.1A, the frequency immediately increases as the pressure is increased. It should be noted that there might be a small red shift below 0.1 GPa but that pressure range is not suitable for measurements with a DAC. Although the results seem to confirm the existing theories we will show that this is partly due to a cancellation of errors.

In this respect it is also interesting to look at the results of Loubeyre et al. (4) for hydrogen diluted in various fluids. They have measured the Raman shift of 2 mole percent hydrogen in helium, neon, argon, and deuterium, and compared their data with the frequency of pure hydrogen. An overview is given in Fig.1B. In this case it is very clear that the blue shift of the frequency in the other fluids is considerably higher than in pure hydrogen. The authors state that "A reasonable explanation is that the guest-host interaction has moved from its attractive region to its repulsive one. Therefore, the present data could be used to test and to refine the repulsive wall of the interactions between H_2 and rare-gas atoms (He, Ne, Ar)". We will show that this explanation is not correct.

For many physical properties density is the important variable apart from temperature. It turns out that for the vibrational frequency pressure is more important than density and the interpretation of isobars is much easier than that of iso-

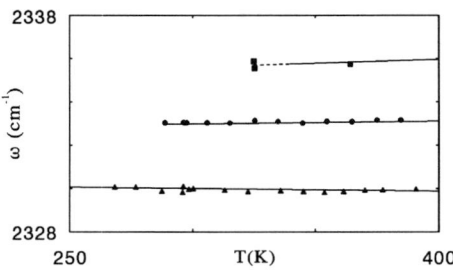

FIGURE 2. Experimental Raman shift of pure nitrogen as a function of temperature along an isobar. Triangles: 1.3 GPa; dots: 2.25 GPa; squares: 3.1 GPa. The short dashed extension of the 3.1 GPa line indicates that the fluid is in the metastable range.

chores. Therefore, we have also carried out a number of measurements as a function of temperature along a few isobars. Measurements along an isobar are unusual in high pressure physics due to experimental problems. The details of the measurements can be found in ref.2. The result has been plotted in Fig.2. Within the experimental accuracy the data points fall on a straight line and the frequency is almost independent of the temperature.

Apart from the frequency also the spectral line width is very important. Unfortunately it is very difficult to measure the line shape at very high pressures, in particular in dilute solutions, due to the low signal to noise ratio. In many cases, however, the line shape is characterized by the full width at half maximum (FWHM). We have determined the FWHM of pure nitrogen and nitrogen in helium at room temperature and a range of pressures. In Fig.3A the results of pure nitrogen have been plotted together with the low pressure data of Lavorel et al. (1). Kroon et al. (5) have determined dephasing times of pure nitrogen. The data above 0.8 GPa are from ref.5 and 6, while those at lower pressures have been taken from ref.7. We have converted those results to FWHM and added to Fig.3A. It is shown that our results are consistent with those of the time-resolved techniques.

The line width in pure nitrogen initially decreases as a function of pressure but via a minimum at about 0.5 GPa the width increases again at higher pressures up to the melting line. The initial decrease was generally explained by the increase of the collision rate as pressure increases. The increase of the width at higher pressures shows that the situation is more complicated. An explanation will be given in the next section

In Fig.3B the results for nitrogen in helium have been plotted. The general picture is the same as for pure nitrogen: an initial decrease as a function of pressure and via a minimum an increase at higher pressure. There are two important aspects that should be mentioned. First of all the line width of nitrogen in helium is at the same pressure considerably smaller than that of pure nitrogen. Secondly,

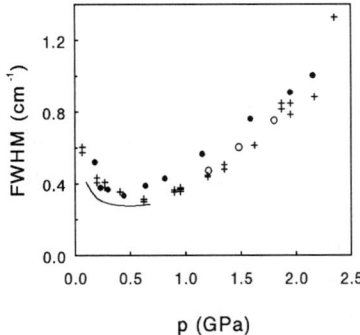 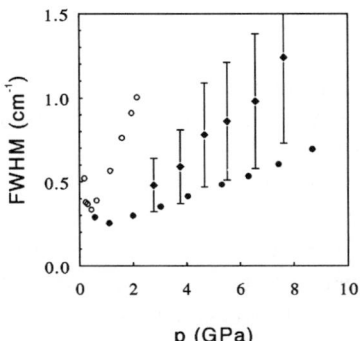

FIGURE 3A. Line width of nitrogen as a function of pressure. Full line: exp. ref.1; crosses: exp. refs. 5-7; open circles: exp. ref.2; dots: MD.

FIGURE 3B. Line width of nitrogen in helium and nitrogen as a function of pressure. Dots: exp.N_2 in He ref.2; open circles exp.N_2 ref.2; diamonds: MD results ref.8 and 9.

the slope of the width as a function of pressure is also much smaller than in pure nitrogen.

INTERPRETATION

In the case of nitrogen some simplifying assumptions hold. First of all the resonance coupling (RC) in the fluid is very small. An experimental study on liquid N_2 at ambient pressure showed that, if RC is present at all, the contribution to the line shift would be less than 10% (10). A MD study showed that it is negative and less than 10% (11). So there is essentially no RC effect on the Raman shift and it is very likely that the same holds for the width. The second assumption is that the change in the anharmonicity is negligible, as shown in shock wave experiments (12). The third assumption is that the influence of the vibrational energy relaxation on the line width is negligible, since the relaxation time has been measured to be orders of magnitude larger than the pure dephasing time (6,7,13). Finally it is assumed that the vibrational and the molecular motions act on different time scales, such that the potential energy of a momentary configuration of the molecules can be treated as a time-independent potential for the calculation of the vibrational energy.

Due to this favorable situation, agreement exists on many aspects of the influence of the intermolecular interaction on the vibrational frequency. Generally this influence is determined by the derivatives of the intermolecular potential energy with respect to the molecular bond length (r). The frequency distribution over the molecules and the correlation time behavior is reflected in the profile of the spectrum. The main contribution to the bondlength derivatives is considered to be the change of the site-site energy due to the translational motion of the atoms (14,15). In the calculations the spatial gradients of the intermolecular potential along the bond are determined or, alternatively, the molecule is considered as a rigid dumbbell and the reaction forces are determined, necessary to keep the distance between the atoms fixed. This contribution is usually referred to as the repulsive one, because the attractive part of the intermolecular potential in the Weeks, Chandler and Anderson (WCA) sense has a negligible spatial gradient (14,15). If only this contribution is taken into account, one achieves too large values for the Raman shift both in the solid and in the fluid, while it is impossible to explain the characteristic minimum in the frequency as a function of pressure. Therefore, it has been proposed that the site-site potential function itself is dependent on the bondlength r (15-18). For instance, the part of the potential being due to the dispersive interaction is dependent on the molecular polarizability, which is known to be linearly dependent on r. An increase of r thus decreases the potential energy via a larger negative dispersive contribution of the interaction potential. Because it was assumed that this dispersive mechanism dominates the contribution of the r dependence of the site-site potential, this was called attractive (15).

More than a decade ago Schweizer and Chandler (15) developed the hard fluid (HF) theory for the calculation of the vibrational frequency and the halfwidth. The repulsive contribution to the Raman shift was calculated by modeling the nitrogen molecule as a hard dumb-bell in a surrounding of hard spheres. The attractive contribution, resulting from the bond length dependence of the site-site interaction parameters, was taken to be a negative, term linear in the density. Because of the fact that the proportionality constant was an adjustable parameter the question about the contribution of the r-dependence of the repulsive part of the potential was avoided. The calculated value for the line width of N_2 was not in agreement with experiment (15).

Pratt and Chandler (16) showed that the red gas to liquid shift for Br_2 dissolved in argon cannot be explained on the basis of an atom-atom additive potential. Herman and Berne (17) performed Monte Carlo calculations for Br_2 in argon with a bondlength dependent potential and obtained a significant red shift at increasing density. De Souza et al. (18) derived theoretical expressions for the solvent configuration averaged force on a diatomic solute molecule and also performed Monte Carlo calculations. They have shown that, by introducing a three atom potential, for any diatomic solute and spherical solvent system with realistic parameters a red frequency shift can be obtained.

Recently Ben-Amotz and coworkers (3,19) have modified the theory. One aspect is the theoretical determination of the value of the proportionality constant for the attractive contribution, mentioned above. They state that the constant depends on the differential attractive solvation energy of the solute in the ground and excited vibrational states (19). It turned out that the calculated value was far off for acetonitril, which was attributed to errors in the values for the permanent multipole interactions. However for nitrogen, where these interactions do not play a role, the constant was calculated with a purely dispersive model and this led to a good agreement with experiment (3). This would imply that the bondlength dependence of the repulsive part of the potential is negligible.

We have performed molecular-dynamical (MD) simulations of the frequency shift as well as of the line width for both pure nitrogen and nitrogen in helium. The method is described in ref.8. A diatomic model has been used in which the intermolecular forces act on two sides that coincide with the two constituent atoms. First we consider three mechanisms that influence the vibrational frequency and the line width: the first and second order effect by the external forces and the vibration-rotation coupling. The external forces, exerted by the surrounding molecules on the sites, result in an axial force, relevant for the vibrational motion. The resulting change in the vibrational frequency is proportional to this force. This force is determined by the simulations while the proportionality constant can be calculated from the values for the harmonic and cubic potential parameters of the intramolecular forces. Because the external forces will not be homogeneous, the axial force will depend on the bond length; this is called the second order effect. A rotation of the molecules results in a centrifugal force that would stretch the bond length and consequently will change the harmonic frequency. Besides, the momentary change in the bondlength, due to vibrational motion, will also change the rotational frequency. This well-known vibration-rotation coupling results in a change of the frequency that is proportional to the rotational energy. Using the potential parameters mentioned above, this effect can straightforwardly be calculated by classical mechanics for the two-site model.

We have calculated the mean value for the vibrational frequency $<\omega_{vib}>$, the width of the distribution of the momentary individual values of the frequency Δ (the amplitude of modulation), and the time dependence of the momentary frequency in terms of the self-correlation function $\Omega(t)$, from which the correlation time t_{corr} can be calculated. It turns out that the second order effect of the external forces has the same sign as the first order effect. Its value is about 20% of the first order. Of course, the contribution of the vibration-rotation coupling is independent of pressure. In Fig.1A the results for the frequency have been plotted for pure nitrogen as well as nitrogen in helium. It is seen that in both cases the total change in the frequency increases monotonously with pressure. In the case of nitrogen the agreement with experiment is very poor since in the calculated results the red shift is completely absent. In the case of nitrogen in helium the agreement with experiment is reasonable. It is noteworthy that the calculated results for both cases nearly coincide. This means that the direct effect of the intermolecular forces is nearly independent of the medium but that it depends only on pressure.

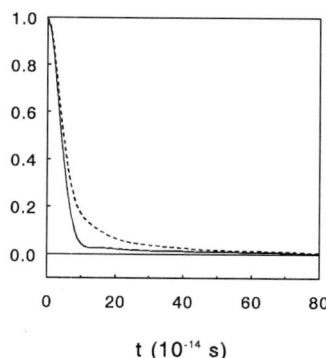

FIGURE 4. The MD results for the time correlation function of the frequency of nitrogen at 1.15 GPa. Full line: total; dashed line: discarding vibration-rotation coupling.

A serious simplification in our model is the assumption that the intermolecular potential is the same for excited molecules as for molecules in the ground state. It is known that the polarizability increases at excitation, but it is hard to quantify this effect. Therefore, we determined the change in potential energy at excitation directly from the difference in frequency, that is the energy gap between the simulated and experimental frequency. Remarkably, the calculated frequency for nitrogen in helium is lower than the experimental value, in contrast to pure nitrogen. Therefore, the bond length dependence is not only due to the attractive part of the potential or in other words the differential solvation energy is not only attractive. We determined an intermolecular site-site potential function of the form $\Delta\varphi = a/r^{12} - b/r^{6}$. This energy jump $\Delta\varphi$ results in a frequency jump $\Delta\omega_{vib} = \Delta\varphi/(2\pi c)$ which is added to the shift. In both cases good agreement with experiment can be obtained by adjusting the parameters a and b. It should be noted that although the energy jump has a marked influence on the shift, it barely affects the total potential energy of a molecule. At site distances where the attractive forces dominate the value for $\Delta\varphi$ is always less than 0.6% of the total potential for pure nitrogen and is an order of magnitude smaller for nitrogen in helium. In the repulsive region the contribution is negligible. In the calculation of the line width the 'dispersive' contribution is also taken into account.

Since it is often suggested that the influence of vibration-rotation coupling (VRC) on the line width is negligible above a few thousand bar, we have paid special attention to the time behavior of the frequency correlation function $\Omega(t)$. MD allows for a separate investigation of the influence of each mechanism that influences the frequency or the line width. In Fig.4 this function is depicted for nitrogen at 1.15 GPa but it is representative for other pressures. The function rapidly decays within 10^{-13} s after which a long tail persists. A quantitative demon-

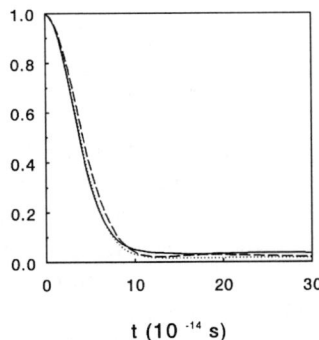
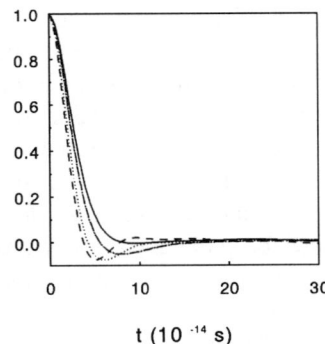

FIGURE 5. A) the autocorrelation function of nitrogen at 0.4, 1.1, and 2.0 GPa. b) the autocorrelation function of nitrogen in helium at 1.1, 2.0, 5.3, and 8.7 GPa.

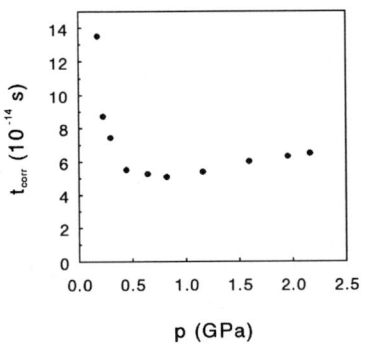
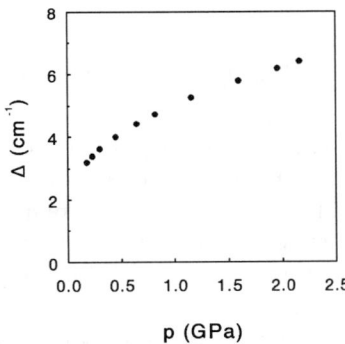

FIGURE 6. A) MD results for the correlation time t_{corr} of nitrogen as a function of pressure. b) *idem dito* for the amplitude Δ.

stration of the importance of VRC at high pressures is given in Fig.4 were the function $\Omega(t)$ has also been plotted for the case that VRC is discarded. Without VRC the behavior of the time correlation function is more regular and the resulting correlation time is longer. VRC has also influence on the amplitude of modulation.

The correlation time and the amplitude have been plotted for pure nitrogen in Fig.6. At increasing pressure t_{corr} falls off rapidly until at about 0.8 GPa a minimum is reached. At still higher pressures t_{corr} slightly increases. This increase is due to the fact that at higher pressures the long tail in $\Omega(t)$, predominantly caused by the first order effect, is more pronounced (Fig.5a). The amplitude of modulation gradually increases as a function of pressure but levels off at higher pressure. For nitrogen in helium the results are qualitatively the same. In the pressure range of this investigation the product of Δ and t_{corr} is considerably smaller than 1 (about 0.06). Therefore, the system is in the fast modulation regime and consequently the line width can be calculated by FWHM = $2\Delta^2 \cdot t_{corr}$. The calculated values for the line width are compared in Fig.3 with the experimental data. In the case of pure nitrogen the agreement is very good, while for nitrogen in helium the calculated data points are just outside the indicated uncertainty region of the experimental results. One should realize, however, that the experimental results have been obtained from extrapolations to zero nitrogen concentration, while the pressure dependence on composition is very strong for low nitrogen concentrations. Moreover, at this low concentration the experiments are less accurate. As mentioned before, the line width of nitrogen in helium is much smaller than that of pure nitrogen. The amplitude of modulation is nearly the same in both systems but the correlation time decreases to much lower values for nitrogen in helium. The reason for this is found in the shape of the autocorrelation function $\Omega(t)$ (See Fig.5). This function reveals even a negative part, an effect not seen in the simulations on pure nitrogen.

CRITICAL CONCENTRATION FLUCTUATIONS

When approaching a critical point, large fluctuations of the order parameter arise. Up to now, the only way to determine experimentally the amplitude of the critical fluctuations has been by means of Raman spectroscopy, which probes the local environment of the molecule. For pure substances an increase of the line width occurs (20), from which an order parameter distribution (density) can be extracted (21). However, a similar experiment on the liquid mixture of 3-methylpentane and nitroethane, which is a model system for critical phenomena, showed no critical broadening (22). Recently (23), the width and shift of the vibrational Raman spectra of mixtures of helium and nitrogen have been measured in the homogeneous fluid phase as a function of pressure, temperature and composition. The authors have shown that the broadening is due to critical concentration fluctuations.

The results have been displayed in Fig.7A. This figure was obtained in the following way. The instrumental profile can be represented by a Gaussian with FWHM=1.5 cm^{-1}. The spectra can also be represented with Gaussians rather than with Lorentzians, so the FWHM of the deconvoluted spectrum was determined by subtracting the instrumental width quadratically from that of the Gaussians. In fact data have been collected for eight different compositions in a temperature range from 200 to 400 K and pressures up to 10 GPa. The data have been fitted for each composition separately and from these fits Fig.7A has been constructed. At the higher pressures the points for pure nitrogen have been obtained by extrapolating into the metastable region. The choice for the proper order parameter is often

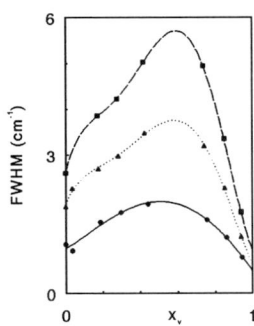
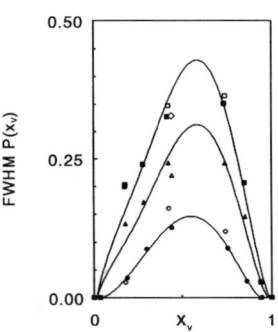

FIGURE 7A. FWHM of nitrogen in helium versus volume fraction of helium. Squares: 6.5 GPa; triangles: 4.5 GPa; dots: 2.5 Gpa.

FIGURE 7B. FWHM of the distribution function $P(x_v)$ versus volume fraction. Closed symbols as in Fig.7A for T=296 K. Open symbols, circles: 4.5 GPa and 360 K (ε=0.42); squares: 4.5 GPa and 260 K (ε=0.026); triangles: 2.5 GPa and 244 K (ε=0.17); diamond: 2.5 GPa and 213 K (ε=0.026).

made by using the criterion of a symmetric coexistence curve. In the case of helium-nitrogen our choice for the volume fraction instead of, for instance, the mole fraction does not only result in a more symmetric coexistence curve, but also in a nearly symmetric FWHM curve, while the frequency shift as a function of the volume fraction (at constant T and p) is then almost a straight line.

The isobars show a distinct maximum at a volume fraction of helium of about 0.55, which corresponds to the critical volume fraction (24). The value of the FWHM in the maximum is about twice that for pure nitrogen, the ratio being highest for the highest pressures. The FWHM decreases rapidly for the higher volume fractions and its value in dilute solutions is smaller than in pure nitrogen, as mentioned before.

In order to be sure that the line broadening is due to an anomaly at criticality other explanations have to be excluded, for example an anomalous density effect in the critical region. Such a density effect will have a direct influence on the vibrational frequency as well and thus on the line shift as a function of volume fraction. Therefore, it is interesting to look at the behavior of the line *shift* as a function of temperature, pressure and composition. Along all isobars (and constant composition) the shift is constant as a function of temperature; only not far from the critical line a very small increase has been measured. At constant composition and temperature the Raman shift increases regularly with pressure and the data can be properly fitted (standard deviation σ=0.15 cm^{-1}). Using these fits the composition dependence of the Raman shift at a given pressure can be determined.

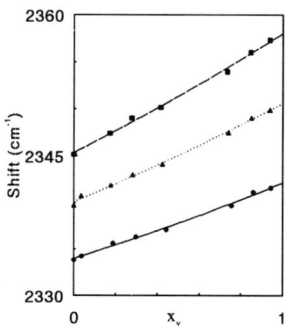

FIGURE 8. The Raman shift of nitrogen in helium versus volume fraction of helium at 6.5 GPa (squares), 4.5 GPa (triangles), and 2.5 GPa (dots).

In Fig.8 the results at room temperature have been plotted for 2.5, 4.5, and 6.5 GPa. The dependence is almost linear. This regular behavior versus x_v and versus T means that along the isobar the *average* interaction of a nitrogen molecule with its environment only gradually changes as a function of composition; it is constant as a function of T and so it does not show any anomaly in the critical region. Therefore, there is no anomalous density effect on the line width (25).

The results can be used to make an estimate of the amplitude of the critical concentration fluctuations as a function of p, T, and x. It is assumed that the decay time of the critical fluctuations is much larger than the relevant time scale, given by the inverse line width, so this contribution has no significant motional narrowing. This is a reasonable assumption, because the time scale given by our spectra is about 2ps while that of the decay of the fluctuations in the mixture 3MP-NE is several orders of magnitude larger (26) and in mixtures far from the critical point the lifetime of the concentration fluctuations is about 5 ps (27). Moreover, the profiles of our spectra can be fitted better with Gaussians than with Lorentzians. We now will use a finite-size method analogous to that of the pure substances (28). Thus we can divide the sample space in small volumes with a length scale larger than the interaction range but much smaller than the wavelengths of the critical fluctuations. For a given *overall* value the *local* value of the order parameter then shows a certain distribution which (as we will show) is directly reflected in the spectral shape. We denote such a distribution by P (28). With x_v as order parameter $P(x_v)$ for mixtures will be almost symmetrical and a comparison can be made with the symmetrical $P(\rho)$ in pure substances (28). Note that $P(0)$ and $P(1)$ need not to be zero. Because of the almost linear behavior in Fig.8, the relation between the local frequency $\omega(x_v)$ and the frequency corresponding to the overall concentration $\omega(x_{vo})$ in the sample is:

$$\omega(x_v) = \omega(x_{vo}) + (\partial\omega/\partial x_v)_{p,T} (x_v - x_{vo})$$

This relation can be inverted to yield an expression for x_v as a function of the frequency. The critical contribution to the spectrum at given p and T is then:

$$I_{cr}(\omega) \sim (1-x_v(\omega)).P(x_v(\omega))$$

where $P(x_v(\omega))$ is the distribution of the local composition. The factor $(1-x_v(\omega))$ expresses the fact that intensity of the signal is proportional to the number of nitrogen molecules in the part of the sample under consideration. Because of this factor a small correction has to be made (1-5%) to obtain the width of $P(x_v(\omega))$ from that of $I_{cr}(\omega)$. We calculated this correction by modeling $P(x_v(\omega))$ with a Gaussian: it turns out that the resulting asymmetry is too small (<1.5%) to be detected, but the resulting negative shift of the peak position can be a few tenths of a cm^{-1}. We have probably observed this effect. The FWHM of $I_{cr}(\omega)$ was obtained by subtracting the non-critical contribution quadratically from the total width. The main source of uncertainty in the critical width originates from this procedure.

The results for the FWHM of $P(x_v)$ at room temperature have been plotted in fig.8 for three pressures (2.5, 4.5 and 6.5GPa). If we define a reduced temperature ε by $\varepsilon(T,p)=T/T_{cr}(p)-1$, where $T_{cr}(p)$ is the critical temperature at pressure p, the values of ε for these isobars are respectively 0.42, 0.17, and 0.026. Also results for other temperatures and pressures have been determined. Fig.8 shows that within the accuracy the results seem similar for the same value of ε, nearly independent of p. Thus at a given composition one can approximately measure the distance to the critical line by regarding only the temperature. This is probably due to the steepness of the critical line and compensation by a small change in critical composition. The isobar at 6.5 GPa comes close to the critical line. Here the FWHM of $P(x_v)$ is estimated to be 0.5, which is 0.9 times the critical value of x_v (0.57). The density (ρ) fluctuations in pure N_2 give a FWHM of $0.55\rho_{cr}$. The experimental data show that the critical enhancement is still noticeable up to $\varepsilon=0.8$; in pure N_2 this is $\varepsilon=0.4$. In the mixture the effect seems to be larger than in the pure compound.

From the above equations (1) and (2) it appears that there will be no critical broadening if $(\partial\omega/\partial x_v)_{p,T} = 0$. For the density fluctuations a similar correlation has been found between $\partial\omega/\partial\rho$ and the magnitude of the broadening effect. If $(\partial\omega/\partial x_v)_{p,T} = 0$ the critical fluctuations will not affect the local interactions, because these are then accidently equivalent for both components. In the mixture 3MP-NE this is true for all the Raman lines explaining the absence of the broadening effect.

REFERENCES

1. B.Lavorel, B.Oksengorn, D.Fabre, R.Saint-Loup, and H.Berger, Mol.Phys. **75**, 397 (1992).
2. M.I.M.Scheerboom, J.P.J.Michels, and J.A.Schouten, J.Chem.Phys. (1996), in press.
3. G.S.Devendorf and D.Ben-Amotz, J.Phys.Chem. **97**, 2307 (1993).
4. P.Loubeyre, R. LeToullec, and J.P.Pinceaux, Phys.Rev.B **45**, 12844 (1992).
5. R.Kroon, R.Sprik, and A.Lagendijk, Chem.Phys.Lett. **161**, 137 (1989).
6. R Kroon, M.Baggen, and A Lagendijk, J.Chem.Phys. **91**, 74 (1989).
7. R.Kroon, thesis, University of Amsterdam, 1993.
8. J.P.J.Michels, M.I.M.Scheerboom, and J.A.Schouten, J.Chem.Phys. **103**, 8338 (1995)
9. J.P.J.Michels, M.I.M.Scheerboom, and J.A.Schouten, J.Chem.Phys., submitted.
10. M.J.Clouter, H.Kiefte, and R.K.Jain, J.Chem.Phys. **73**, 673 (1980).
11. D.Levesque, J.J.Weis, and D.W.Oxtoby, J.Chem.Phys. **72**, 2744 (1980).
12. S.C.Schmidt, D.Schiferl, A.S.Zinn, D.D.Ragan, and D.S.Moore, J.Appl.Phys. **69**, 2793 (1991).
13. B.Khalil-Yahyavi, M.Chatelet, and B.Oksengorn, J.Chem.Phys. **89**, 3573 (1988).
14. L.R.Pratt, C.S.Hsu, and D.Chandler, J.Chem.Phys. **68**, 4202 (1978).
15. K.S.Schweizer and D.Chandler, J.Chem.Phys. **76**, 2296 (1982).
16. L.R.Pratt and D.Chandler, J.Chem.Phys. **72**, 4045 (1980).
17. M.F.Herman and B.J.Berne, J.Chem.Phys. **78**, 4103 (1983).
18. L.E.S.de Souza, C.B.E.Guerin, D.Ben-Amotz, and I.Szleifer, J.Chem.Phys. **99**, 9954 (1993).
19. D.Ben-Amotz and D.R.Herschbach, J.Phys.Chem. **97**, 2295 (1993).
20. M.J.Clouter and H.Kiefte, Phys.Rev.Lett. **52**, 763 (1984).
21. M.J.Clouter, H.Kiefte, and C.G.Deacon, Phys.Rev.A **33**, 2749 (1986).
22. K.A.Wood and H.L.Strauss, J.Chem.Phys. **74**, 6027 (1981).
23. M.I.M.Scheerboom and J.A.Schouten, Phys.Rev.E **51**, R2747 (1995).
24. W.L.Vos and J.A.Schouten, Physica A **182**, 365 (1992).
25. D.W.Oxtoby, Annu.Rev.Phys.Chem. **32**, 77 (1981).
26. H.Burnstyn and J.V.Sengers, Phys.Rev.A **25**, 448 (1982).
27. L.J.Muller, D.Vanden Bout, and M.Berg, J.Chem.Phys. **99**, 810 (1993).
28. M.Clouter, Annu.Rev.Phys.Chem. **39**, 69 (1988).

Vibrational Raman Spectrum of O_2 in Ar matrix at room temperature and pressure to 60 GPa.

Federico Gorelli,# Lorenzo Ulivi,+ and Marco Zoppi+

#*Dipartimento di Fisica, Università di Firenze, Largo Enrico Fermi 2, Firenze, Italy*
+*Istituto di Elettronica Quantistica, CNR, Via Panciatichi 56/30, 50127 Firenze, Italy*

Abstract. Low concentration oxygen impurities in a solid argon matrix, have been studied by Raman scattering at high pressure and room temperature. The vibron peak shows a splitting, due to vibrational coupling interaction between two nearest neighbour oxygen molecules, which increases with pressure. The pressure dependence of the vibrational coupling parameter, derived from these measurements, demonstrates a strong difference between oxygen and hydrogen, with respect to intermolecular interaction.

Studies of molecular systems at high pressure give valuable information on the interaction between neighbouring molecules. The system used in the present experiment is a solid binary mixture of argon and oxygen. The low oxygen concentration (3–5%) has permitted to study the interaction between only two molecules isolated as substitutional impurities in an argon crystal. We have performed Raman measurements of the vibrational transition of O_2 at pressures up to 60 GPa, using a Diamond Anvil Cell, and a triple monochromator equipped with a liquid nitrogen cooled detector. Pressure calibration has been obtained by the ruby fluorescence technique.[1]

The spectra show a splitting of the vibron peak, that starts to be appreciable at 10 GPa, as a shoulder on the low frequency side, and increases with pressure (fig. 1a). The peaks were fitted using two gaussian lineshapes and the parameters obtained for different concentrations are almost the same. The presence of the second peak, beside that of the O_2 molecule isolated in argon, arises from the interaction of vibrating nearest neighbour pairs of molecules. The relevant O_2-O_2 interaction for this effect is that part of the potential which depends on the internal coordinates r_1, r_2 of the two molecules, and on the intermolecular distance R. Neglecting the terms of the intermolecular potential which depend also on the relative orientation of the molecules, this may be written as $F_2(r_1,r_2,R)$.[2] Expanding this term for small deviation from the equilibrium bond distance r_e we can write $F_2(r_1,r_2,R) = -g(R)u_1u_2$ where $u_i=(r_i-r_e)$. With simplifying assumption,[3] we obtain

$$g(R) = 4\pi^2\mu\left(v_1^2 - v_2^2\right) \quad (1)$$

where μ is the reduced mass for the intramolecular vibration.

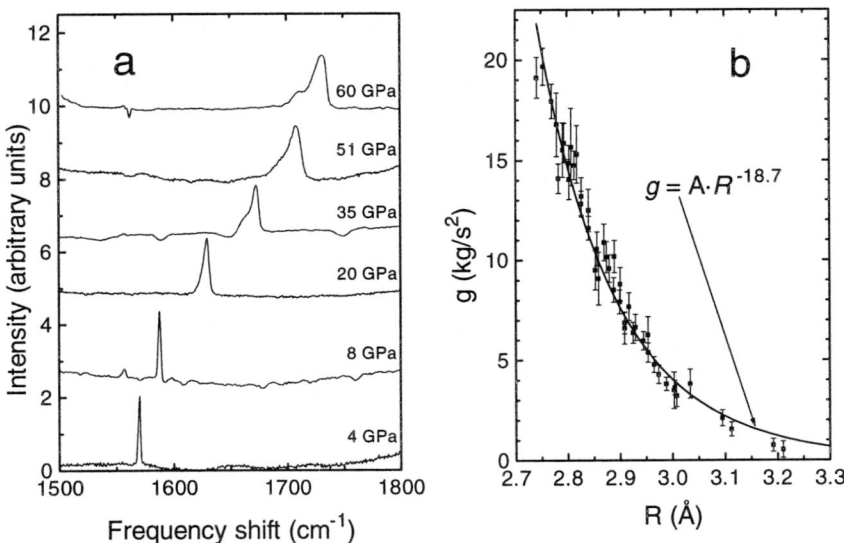

Figure 1. (a): Raman spectra at different pressures; (b): vibrational coupling parameter as a function of nearest neighbour distance.

The dependence of the vibrational coupling parameter $g(R)$ on the intermolecular distance is much stronger (R^{-19}) than what is predicted by Van Kranendonk[4] assuming a dipole-dipole interaction (R^{-6}), and which has been observed in the case of hydrogen molecules.[3] This strong dependence indicates that the dominant interaction between two oxygen molecules at his densities is probably a charge transfer type. The present measurement may therefore give valuable information for the interpretation of the metallization of solid oxygen, observed recently around 96 GPa.[5]

REFERENCES

1. H.K. Mao, and P.M. Bell, Science, **200**, 1145 (1978).
2. J. Van Kranendonk, *Solid Hydrogen*, Plenum press New York, London, (1984), ch.2, p. 33.
3. P. Loubeyre, R. LeToullec, and J.P. Pinceaux, Phys. Rev. **B45**, 12844 (1992).
4. See ref. 2, ch. 3, p.62.
5. Y. Akahama, H. Kawamura, D. Häusermann, M. Hanfland, and O. Shimomura, Phys. Rev. Lett. **74**, 4690 (1995).

ROUND TABLE DISCUSSION
HIGHLIGHTS AND OPEN PROBLEMS ON LINE SHAPES

INTRODUCTORY TALK BY
Józef Szudy
Institute of Physics, Nicholas Copernicus University,
Grudziądzka 5/7, 87-100 Toruń, Poland

Highlights! There are so many exciting results of both theoretical and experimental studies from various groups obtained during the last two years that it is practically impossible to incorporate all of them into one round table discussion. Therefore, in order to start the discussion I would like to outline only a few recent developments that yielded results which in my opinion are of importance for the whole line shape community.

1. CLOSE-COUPLED CALCULATION of the SELF-BROADENING of HELIUM TRIPLET LINES and NEON $2p^53s-2p^53p$ LINES.

Gillian Peach at the University College, London and her coworkers at James Cook University, Townsville, Australia [1,2] have performed fully quantum-mechanical calculations of the collisional self-broadening and shift of twelve triplet lines of helium corresponding to the 2^3S-3^3P, 2^3P-n^3S, and 2^3P-n^3D transitions with n = 3, 4, 5 and seven neon lines corresponding to the transitions between configurations $2p^53p$ and $2p^53p$. Their calculations were based on the impact theory of Baranger and the interatomic interaction was presented by adiabatic molecular potentials computed using model potentials for the electron-atom and atomic core-core interactions. Close-coupled equations were solved using a R-matrix method. Comparison of theoretical results and measurements is restricted by the paucity and quality of currently available experimental data.

An OPEN PROBLEM *addressed to the experimentalists* is: New measurements of self-broadening of the 2^3P_0-3^3D_1, 2^3P_2-3^3D_3, and 2^3P-3^3S_1 lines should be performed using *absorption spectroscopy*, a technique which eliminates the uncertainties of not well understood excitation mechanisms which plague the measurements based on emission spectroscopy.

2. NON-LORENTZIAN BEHAVIOUR and SPEED-DEPENDENT COLLISIONAL EFFECTS on INFRARED MOLECULAR LINES

A. David May and his group at University of Toronto have carried out high-resolution and high-signal-to-noise ratio measurements of infrared line shapes of CO perturbed by Xe, N_2 and He in the intermediate regime from Doppler to collision broadening. They have found that the collisionally-broadened component of the line shape cannot be fit by a Lorentzian line shape, and the largest deviations occur in the core of the line [3]. It was shown that a speed-dependent collisional width is required to fit the spectra perturbed by Xe. A line shape model based on the theory of Robert *et al.* [4], folding in the Dicke narrowing, has been derived by the Toronto group [5] in a dilute CO:N_2 mixture, in which non-Lorentzian collision broadening scheme was suggested. In another study at Toronto [6] the role of line mixing and finite duration of collision in the formation of profiles of infrared lines of D_2 was investigated.

3. RAINBOW SATELLITES on the WINGS of the LYMAN-α LINE due to the H-H and H-H$^+$ COLLISIONS in the SPECTRA of WHITE DWARF STARS and the FIRST LABORATORY OBSERVATION of these SATELLITES

The shape of the Lyman-α line of hydrogen still remains one of the most fundamental problems in the field of collisional broadening of atomic spectral lines. Sando et al. [7] as the first have shown that the contributions from the *free-free* $B^1\Sigma_u^+ - X^1\Sigma_g^+$ transitions during the H-H collision should give rise to a feature termed a *rainbow satellite* located on the red wing of the Lyman-α line at the wavelength 162.3 nm. This satellite corresponds to the *singularity* of the quasistatic profile due to the minimum of the difference potential $\Delta V(R) = V(B^1\Sigma_u^+) - V(X^1\Sigma_g^+)$ for the H-H interaction. The first quantum-mechanical calculation of the shape of this satellite has been made by Sando and Wormhoudt in 1973 [8]. Also in 1973 Stewart et al. [9] have predicted theoretically that another rainbow satellite should appear on the red wing of the Lyman-α line of H at the wavelength 140.5 nm if the perturbations by protons are taken into account.

In spite of these theoretical predictions there have been no observations of such satellites in laboratory conditions until 1995 when John F. Kielkopf at University of Louisville, Kentucky completed his experimental study on the shapes of Lyman-α line emitted by a *laser-produced plasma* in hydrogen [10]. In the September 1995 issue of *The Astrophysical Journal* [11] J. F. Kielkopf and Nicole F. Allard reported results of their measurements of the profiles of satellite features on the red wing of the Lyman-α line emitted by a laser-produced plasma in pure hydrogen. To my knowledge this was the first laboratory observation of such satellites (26 years after the H-H satellites on the red wing of Lyman-α were theoretically predicted by Sando, Doyle and Dalgarno !).

The wavelength, relative intensity and shape of satellites have been found to agree with new theoretical calculations based on *ab initio* H-H and H-H$^+$ potentials performed by Allard et al. [19].

It should be emphasized, however, that in an astromical context, satellite features on the red wing of Lyman-α were first observed as early as in 1980 in the spectra of the DA *white dwarf stars* obtained with the *International Ultraviolet Explorer* (IUE)[13]. In 1994 Koester et al.[12] have found these satellite features in the UV spectra of the DA white dwarfs obtained by the *Hubble Space Telescope* (HST). The main constituent of the atmospheres of the DA white dwarfs is hydrogen. Greenstein [13] as the first reported the existence of an absorption feature at 140 nm in the IUE spectrum of a DA white dwarf. Now it is generally accepted that all DA white dwarfs with effective temperatures T_{ef} between 10000 and 20000 K show the absorption at 140 nm. For cooler white dwarfs (T_{ef} ~ 11000 - 13000 K) another absorption feature was detected at 160 nm [14]. In 1985 Holm et al. [14], Koester et al. [15] and Nelan and Wegner [16] have finally identified these absorption features as *red rainbow satellites* of the Lyman-α line perturbed by protons (140 nm) and by neutral hydrogen atoms (160 nm).

It should be noted that in the atmospheres of white dwarfs the densities of perturbers (H, H$^+$) are usually so high that the effects of the simultaneous interactions of the absorbing hydrogen atom with many perturbers may be important. To include such effects in calculations of the profiles of the wings of the Lyman-α line Allard and Kielkopf [17] and Allard and Koester [18] have used a unified theory of Anderson and Talman.

A new much more general analysis of the shape of the wings of Lyman-α has recently been published in the December 1994 issue of *Astronomy and Astrophysics Supplement Series* by Allard et al. [19]. In their treatment degeneracy of atomic states perturbed due to H-H and H-H$^+$ collisions is taken into account. The calculated profile shows a number of satellite features, two of them (162.3 and 140.5 nm) have been observed in the IUE and HST absorption spectra of white dwarfs and also *very recently* in laboratory conditions in the Lyman-α line emitted by a laser-produced plasma in pure hydrogen [11]. In particular, these new calculations [19] predicted the existence of a satellite at 123.4 nm due to H-H$^+$ interaction which has been observed in the laboratory by Kielkopf and Allard [11] but has not been found in the spectra of white dwarf stars.

Two OPEN PROBLEMS should be noted:

(i) The calculations of Allard et al. [19] were performed assuming the dipole transition moment to be constant. It is interesting to establish the effect of the variation of the transition probability with interatomic separation on the resulting profile.

(ii) The contributions from the asymptotically forbidden *bound-free transitions* $B'^1\Sigma_u^+ \leftarrow X^1\Sigma_g^+$ and $b^3\Sigma_u^+ \leftarrow a^3\Sigma_g^+$ which lead to the well-known continuous spectrum in the region 170 – 500 nm have not been taken into account. These contributions may be important in the spectral region above 160 nm.

During the last few months new exciting results have been obtained on astronomical side. Detlev Koester and his coworkers have been able to extend their work on the spectra of white dwarfs to the profile of the Lyman-β line. In a space experiment (*Hopkins Ultraviolet Telescope* on board the Space Shuttle) a white dwarf with a temperature around 20000 K was oberved in the region between 90 and 180 nm. It showed a very unusual profile of Lyman-β with clear absorption satellite feature on the red wing. Koester et al. [20] have explained this as a satellite due to the perturbation by protons, and obtained very nice fits between theoretical spectra and observations.

4. STUDY of POTENTIAL BARRIERS in Σ STATES of the Li*(3P,3D)-Ne and Li*(3P,3D)- He SYSTEMS by means of LASER EXCITATION SPECTROSCOPY

Wolfgang Behmenburg and his group at Duesseldorf have measured excitation spectra of the transitions 2PΛ →3DΛ and 2PΛ→3PΛ in LiNe and LiHe collision molecules in the spectral range 15800-17600 cm^{-1} about the Li 2P → 3D line by means of two-step laser excitation [21]. Rainbow satellites were observed and have been identified as due to maxima in the difference potentials 2PΣ-3DΣ, 2PΠ-3DΣ, and 2PΛ-3PΣ that are in turn related to potential barriers in the upper 3PΣ and 3DΣ states. Approximate calculations of interaction potential curves and dipole transition moments for Li-He and Li-Ne based on the Fermi-Omont model have been performed. The results lead to a quantitative understanding of the origin and shapes of potential barriers in the 3PΣ and 3DΣ states.

5. LINE-SHAPES of ALKALI-METAL ATOMS in SUPERFLUID HELIUM

In 1993 Fujisaki *et al.* [22] at Kyoto University have developed a new method for implanting a high density of neutral atoms in superfluid helium (He II); laser sputtering within liquid helium. Using this method Kinoshita *et al.* [23,24] have performed measurements of excitation and emission spectra of alkali-metal atoms (Cs and Rb) in liquid helium. The helium presssure was changed from the saturated vapour pressure to about 25 atm at 1.6 K, under which conditions the liquid helium is in a superfluid state. Earlier studies of ions (e^- and He^+) in liquid helium revealed that some structures may be created: *"bubbles"* for electrons and *"snowballs"* for He^+ ions. Spectroscopic researches for neutral atoms immersed in superfluid helium have shown that these atoms form a similar bubble structure (*atomic bubble*). Kinoshita *et al.* [23,24] have found that the pressure shift and broadening of the observed spectra corresponding to the D lines of Cs and Rb can be explained qualitatively by theoretical calculations based on the spherical atomic bubble model. The Cs D_2 excitation spectra were found to have double peaks, indicating the existence of anisotropic oscillations of the bubble surface. In a very recent paper Kinoshita *et al.* [25] have developed a *"deformed"* bubble model such that the bubble structure of surrounding helium is not spherical but is deformed instanteneously by a quadrupole oscillation. The calculated energy level of the $P_{3/2}$ state of the deformed atomic bubble is then split into two branches due to the dynamic Jahn-Teller effect giving the D_2 excitation line shape consisting of two components with different peak intensities and widths.

REFERENCES

1. Leo, P.J., Peach, G., and Whittingham, I.B., *J.Phys.B* 28, 591 (1995).
2. Leo, P.J., Mullamphy, D.F.T., Peach, G., and Whittingham, I.B., *J.Phys.B* 28, 4449 (1995).
3. Duggan, P., Sinclair, P.M., May, A.D., and Drummond, J.R., *Phys. Rev.A* 51, 218 (1995).
4. Robert, D., Thuet, J.M., Bonamy, J., and Temkin, S., *Phys. Rev.A* 47, R771 (1993).
5. Duggan, P., Sinclair, P.M., Le Flohic, M.P.,Forsman, J.W., Berman, R., May, A.D., and Drummond, J.R., *Phys. Rev.A* 48, 2077 (1993).
6. Sinclair, P.M., Duggan, P., Forsman, J.W., Drummond, J.R., and May, A.D., *Can. J. Phys.* 72, 891 (1994).
7. Sando, K., Doyle, R.D., and Dalgarno, A., *Astrophys.J.* 157, L143 (1969).
8. Sando,K., and Wormhoudt,J.C.,*Phys.Rev.A* 7, 1889 (1973).
9. Stewart, J.C., Peek, J.M., and Cooper, J., *Astrophys.J.* 179, 983 (1973).
10. Kielkopf, J.F., *Phys. Rev.E* 52, 2013 (1995).
11. Kielkopf, J.F., and Allard, N.F., *Astrophys.J.* 450, L75 (1005).
12. Koester, D., Allard, N.F., and Vauclair, G., *Astron. Astrophys.* 291, L9 (1994).
13. Greenstein, J.L., *Astrophys.J.* 241, L87 (1980).
14. Holm, A.V., Panek, R.J., Schiffer III,F.H., Bond, H.E., Kemper, E., and Grauer, A.D., *Astrophys.J.* 289, 7741 (1985).

15. Koester, D., Weidemann, V., Zaidler-K.T.,E.M., and Vauclair, G., *Astron. Astrophys.* 142, L5 (1985).
16. Nelan, E.P., and Wegner, G., *Astrophys. J.* 289, L31 (1985).
17. Allard, N.F., and Kielkopf,J.F., *Astron. Astrophys.* 242, 133 (1991).
18. Allard, N.F., and Koester, D., *Astron. Astrophys.* 258, 464 (1992); Koester, D., and Allard, N.F., in: *Spectral Line Shapes*, Vol.7, eds. R. Stamm and B. Tallin (Nova Science Publ., Commack, NY, 1993).
19. Allard, N.F., Koester, D., Feautrier, N., and Spielfiedel, A., *Astron. Astrophys. Suppl. Ser.* 108, 417 (1994).
20. Koester,D. et al., *Astrophys. J. Letters* (accepted for publication, 1996).
21. Makonnen, A., Kaiser, A., and Behmenburg, W., *Z. Phys. D* 36, 325 (1996); Behmenburg,W., Makonnen, A., and Findeisen, M., *Z. Phys. D* 25, 315 (1993).
22. Fujisaki, A., Sano, K., Kinoshita, T., Takahashi, Y.,and Yabuzaki,T., *Phys. Rev. Lett.* 71, 1039(1993).
23. Kinoshita, T., Fukuda, K., Takahashi, Y., and Yabuzaki, T., *Phys. Rev. A* 52, 2707 (1995).
24. Kinoshita, T., Fukuda, K., Takahashi, Y., and Yabuzaki, T., *Z. Phys. B* 98, 387 (1995).
25. Kinoshita, T., Fukuda, K., and Yabuzaki, T., *Phys. Rev. B* (accepted for publication, 1996).

APPLICATIONS: ASTROPHYSICS, ATMOSPHERE, AND ENVIRONMENT

Line Profiles as Diagnostic and Actor: Abundance Anomalies in Stars

Georges Michaud and Jacques Richer

Département de physique, Université de Montréal, Montréal, Canada, H3C 3J7
and CERCA, 5160, boul. Décarie, bureau 400, Montréal, Canada, H3X 2H9

Abstract. While most stars have a chemical composition similar to that of the Sun, some 10% of those more massive than the Sun have large anomalies of some chemical elements. Many of those anomalies are now believed to be caused by diffusion in the presence of selective radiative acceleration. Those chemical species that absorb more of the outgoing photons are dragged by them to the stellar surface.
 Line widths play a double role. One needs a good knowledge of them to determine the abundance anomalies from the observed stellar spectra. One also needs them to evaluate precisely the radiative accelerations of all chemical species that are abundant enough to have saturated lines and to determine how blends modify the radiative accelerations.
 The effect of line widths will be quantitatively analyzed in a number of stellar objects. The approximations currently used will be reviewed and the need for improvements discussed.

I- IMPORTANCE OF ATOMIC DIFFUSION IN STARS

Most stars have relative abundances of the chemical elements similar, within a factor of 1.5, to those in the Sun. However, particle transport within stars plays a major role (factors of 10 or more) in the surface abundance of some 10% of stars more massive than the Sun. In those, atomic diffusion is often important and atomic physics then becomes a major player. We here only give a few examples. A more complete review of the importance of abundance anomalies and of particle transport may be found in Michaud and Tutukov (1).

The largest abundance anomalies are observed in the ApBp stars (including the HgMn, the magnetic Ap and the AmFm stars). Together they represent some 20 % of stars with $T_{eff} > 7000$ K (T_{eff} is the *effective temperature* defined as the black body temperature needed to emit the radiative flux emitted by the star). They are also the more slowly rotating stars of their T_{eff} group. At $T_{eff} < 7000$ K the superficial convection zones become deeper, so that diffusion leads only to smaller anomalies. In the HgMn stars (10000 < T_{eff} < 15000 K) for instance, overabundances by up to factors of 10^6 (i. e. 10^6 larger abundances than seen in the Sun and most other stars) have been observed for Hg (2), (3). These come at the same time as isotope abundance anomalies. While in the solar system, the isotope with the largest abundance is ^{202}Hg (29.8%), the most abundant isotope in many HgMn stars is ^{204}Hg. Most other anomalies are smaller, but they still are often by a few orders of magnitude (e. g. Mn, P, ...).

In the Sun, atomic diffusion is expected to lead to much smaller anomalies because of the deep superficial convection zone. Given the particle transport rate below the superficial convection zone and the mass of that zone, the abundance anomalies are expected to be limited to some 10 % (4), (5), (6). Such small effects could not be confirmed observationally until recently. Through solar pulsations they become measurable. Most recent calculations of the solar pulsation spectrum suggest that the observed spectrum is better reproduced by those evolutionary models that include atomic diffusion (7), (8), (9), (10).

Atomic diffusion then appears to play a role in most stars. In some it modifies the superficial abundance by orders of magnitude. Even when the astrophysical context does not allow such large effects, smaller ones occur such as in the Sun.

Particle transport by atomic diffusion is a fundamental physical process that in principle should always be included in the modelling of stellar evolution. In practice, it has been largely neglected in the past because it was believed that large scale hydrodynamical processes rendered its effect negligible (11). The model that will be described in § III neglects the competing hydrodynamical processes that are yet to be more completely understood from first principles: meridional circulation, turbulence, mass loss. It includes convection, atomic diffusion and all physical processes that currently can be calculated from first principles. It requires a large amount of atomic data in order to arrive at reliable estimates of surface chemical composition. These estimates are then compared to observed spectra and atomic data play a role in the interpretation (§II).

II. LINE PROFILES AND ABUNDANCE DETERMINATIONS

There are at least five contributors to the observed line widths that are suggested to play a role in abundance determinations in stars: *natural* broadening, Doppler broadening, collisional broadening, micro-turbulence and magnetic desaturation. I will not discuss the effects of magnetic fields here, though Ap stars are frequently observed to have magnetic fields of more than 1000 Gauss. Their effect should be included through a proper treatment of the radiative transfer of polarized radiation (12). Micro-turbulence is in principle caused by turbulence in the light forming region. It is not clear that strong enough turbulence is present in these stars for micro-turbulence to play any significant role in the broadening of spectral lines. The micro-turbulence parameter is often used however as a *fudge factor* to reduce the scatter in the abundance determined by different spectral lines of a given chemical element. It will not be discussed here.

The natural damping can be evaluated precisely once the various radiative decay possibilities are included, leading to a Lorentz profile of full width at half maximum $\Delta v_R = [\tau_L^{-1} + \tau_U^{-1}]/(2\pi)]$, where τ_L and τ_U are respectively the timescales of the lower and the upper levels of the transition. For the quadratic Stark effect to be discussed below, the broadening is also Lorentzian of width Δv_C. This leads to a Lorentz profile of width $\Delta v_R + \Delta v_C$. The thermal motion leads to a superposition of individual profiles. One performs a convolution of a Gaussian profile of Doppler width $\Delta v_D = v_0 (2kT/m_a)^{0.5} c^{-1}$, where v_0 is the central frequency of the transition, with a Lorentz profile to obtain the final profile given by:

$$\phi_{ij}(v) = \frac{1}{\pi^{0.5} \Delta v_D} H(a, v) \qquad (1)$$

where H(a,v) is the Voigt function and

$$a = \frac{\Delta v_R + \Delta v_C}{2 \Delta v_D}. \qquad (2)$$

All quantities are reasonably well known except for Δv_C which is frequently approximated in astrophysics by:

$$\Delta v_C = \gamma \frac{(n_L^4 + n_U^4)}{(Z_i + 1)^2} \frac{N_e}{\sqrt{T}} \qquad (3)$$

where n_L and n_U are the principal quantum numbers of respectively the lower and upper levels of the transition.

The run of the values of Δv_R, Δv_C and Δv_D in a main sequence star of 2.5 $M_\odot$ is shown in Fig. 1 for a *resonance* line, defined as a line between the ground state of the dominant state of ionization and an excited state at 50 % of the ionization energy of the ground state. A line starting from that level and ending at 60% of the ionization potential of the ground state is also shown. Those two lines are assumed to have respectively f = 1 and f = 0.01. The value of γ was estimated following Griem (13):

$$\gamma = 8.0 \times 10^{-6} cm^3 K^{0.5} s^{-1}. \qquad (4)$$

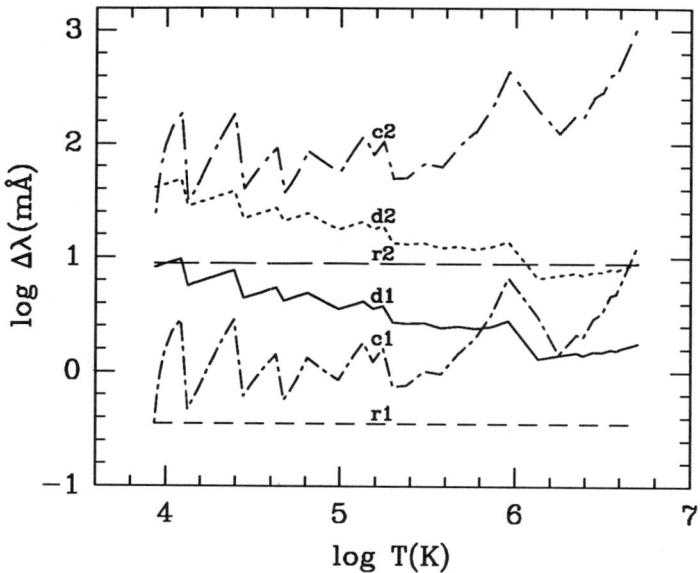

FIGURE 1. Radiative (r), Doppler (d), and collisional (c) line widths of two hypothetical spectral lines of a Fe-like atom in its dominant state of ionization. The widths are calculated in the envelope of a $M = 2.4 M_\odot$, $T_{eff} = 10^4$ K stellar model.

The choice of γ will be further discussed below; see also Gonzalez et al. (14).

The main points to note are:

a) The collisional width (c1) dominates in the stellar interior for log T > 5.5 and c2 dominates throughout the interior.

b) Close to the stellar surface, the Doppler width is expected to dominate for most lines. The natural width can become larger than the pressure width close to the surface where the rapid reduction of c1 and c2 is due to the decrease of pressure close to the surface.

The widths are then expected to be uncertain mainly in the interior where the pressure broadening is the main contribution. In the atmosphere, where the observed lines are formed and where abundances are determined, the pressure width only comes in for some of the more abundant elements when the pressure broadening ends up dominating far from line center even if the pressure width is smaller than the Doppler width. Pressure broadening is however important for hydrogen and helium whose lines are important to determine stellar gravity and T_{eff}.

Isotope Determinations

One recent example of the importance of the detailed shape of spectral lines is given by the Space Telescope observations of the HgMn star χ Lupi (15). These are high resolution spectra from which it is possible to determine the relative isotope composition, for instance, of thallium. However using the available line frequencies, it was not possible to fit the two Tl II hyperfine lines with any abundances. Either the identification was incorrect or the available line frequencies were inaccurate.

Using the Godard High Resolution Spectrograph, it is possible to determine individual wavelengths to an accuracy of 1 mA. The laboratory accuracy of the wavelengths must be at least of the same quality. Fourier transform spectra were obtained at the University of Lund (16) to an accuracy better than 0.3 mA for measured wavelengths. Isotope shifts were obtained. They are given in the following table for the ^1S-^3P intercombination line to the ground state of Tl II.

TABLE 1

Isotope	F_l-F_u	Wavelength (A)
203	0.5-1.5	1908.5632
203	0.5-0.5	1908.6982
205	0.5-1.5	1908.5725
205	0.5-0.5	1908.7087

The isotope shift is needed to determine the isotope composition which turns out to be significantly different from the terrestrial mix.

In the stellar spectrum (15), the line width is dominated by a thermal broadening of approximately .04 A for the 1908 A line. The two hyperfine components are clearly resolved but the two isotope lines are not resolved and only appear as a shift which was determined to an accuracy of 0.0005 A or 1/20 of the isotopic separation (see Fig. 3 and 4 of (15)). By a careful spectrum synthesis of the the two hyperfine lines, one then determines that the abundance of Tl is a factor of $10^{3.8}$ larger in that star than in the Sun. The abundance was determined using oscillator strengths obtained through *ab initio* Multi-Configuration Dirac-Fock calculations with an estimated uncertainty of 20%. Using the accurate wavelength determinations, one further obtains that the lines could be caused by the heavier of the two isotopes, ^{205}Tl with no discernable contribution from the lighter one, ^{203}Tl. This puts strict constraints on the possible models to explain the abundance anomalies in this star.

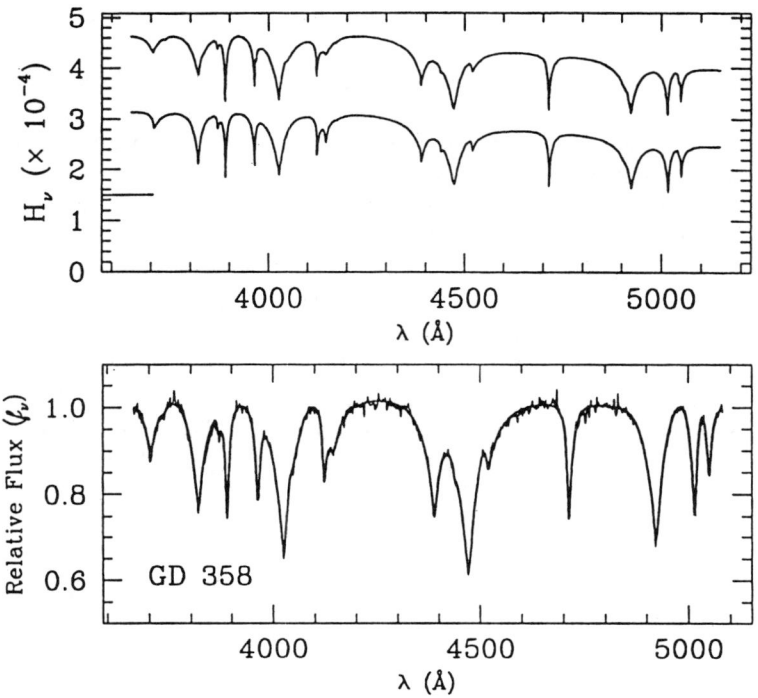

FIGURE 2. On the upper part, is shown the profiles of He lines in a DB white dwarf atmosphere with (top line) and without (bottlom line) new broadening data. On the bottom part are shown synthetic and observed spectra of GD 358.

Line Broadening and Forbidden Components

White dwarfs are one of the late stages of stellar evolution, after stars have exhausted their nuclear fuels. The He white dwarfs (DB stars) show the effect of nuclear reactions. However particle transport also plays a role because of the extreme purity of H or He often shown by their spectra. In those stars whose atmospheric composition is dominated by He, the determination of T_{eff} and gravity must come from an analysis of the He line profiles. The knowledge of T_{eff} and g is essential to understand the evolutionary status of these objects and in particular the separation in He rich and H rich white dwarfs.

In Fig.. 2 we show profiles of He lines calculated both without (bottom line) and with (upper line) the new broadening data. The changes are mainly significant for $\lambda < 4400$ A where the transition from the impact to the quasi-static regime for electrons, the transition from quadratic to linear Stark broadening and the presence of forbidden components are important and had not been included in the past. The presence of forbidden components is caused by the large perturbing electric fields in the dense plasma. They had been calculated for the 4471 line however. On the bottom part of the figure, a synthetic spectrum is compared to observations of GD 358 (17). In this case, the broadening leads to the appearance of components, some 50 A from line centre. The modelling of that part of the spectrum can only be successful if it includes a detailed calculation of the broadening for many He I lines.

III. THE DIFFUSION MODEL

In the diffusion model, stars are assumed to arrive on the main sequence with a chemical composition similar to that of the Sun. There they burn hydrogen in their centre for their energy source. The modelling shows that there are convection zones in the superficial region. These are dominated by rapid mass motion and are well mixed. Atomic diffusion is relatively slow and can play a significant role to modify abundances only outside of convection zones. Solar type stars spend some 10^{10} yr on the main sequence with relatively little movement of their convection zones. One needs to calculate the particle transport throughout the region below the convection zone.

More massive stars spend a shorter period of time on the main sequence but have much smaller convection zones than the Sun so that the overall effect is that much larger abundance anomalies are expected in the more massive than in the solar type stars.

The particle transport equation is given by:

$$\rho \frac{\partial c}{\partial t} = \frac{1}{r^2} \frac{\partial}{\partial r}[\rho r^2 (D_{12} + D_T) \frac{\partial c}{\partial r} - \rho v_g c] - S(\rho, T) c, \quad (5)$$

where D_{12} is the atomic diffusion coefficient, v_g is the advective part of the atomic diffusion velocity containing the contributions of gravitational settling, thermal diffusion, and radiation driven diffusion (18). Through the radiative acceleration, v_g depends on c. Mass loss could also be included. S is a source/sink term which can for instance be caused by nuclear reactions in the deep stellar interior.

The turbulent diffusion coefficient, D_T, reduces the effect of atomic diffusion where turbulence is important, for instance in the more rapidly rotating stars. The interaction of turbulence and atomic diffusion is an area of active current research but will not be discussed here.

The advective part of the atomic diffusion velocity, v_g, generally dominates the transport and is given by:

$$v_g = D_{12}\{[-(A-\frac{Z}{2}-\frac{1}{2})g - Ag_R]\frac{m_p}{kT} - k_T\frac{\partial \ln T}{\partial r}\}. \quad (6)$$

The thermal diffusion term is included in the calculations but does not dominate, so that the particle transport in stars is in practice dominated by the competition between gravity, g, and g_R, the radiative acceleration. A detailed discussion of the different coefficients appearing in the diffusion equation and of its form in ternary mixtures may be found in Michaud and Proffitt (19).

Radiative Accelerations

The radiative acceleration, g_R, comes from the selective absorption of the outstreaming photons. Bound-bound and bound-free transitions, including auto-ionization features, play a significant role but lines usually dominate (see for instance (14)). g_R depends on the atomic properties of each state of ionization.

In stellar interiors, the radiative transfer is well represented by the *diffusion approximation* and the contribution of some frequency interval (v, v+dv) to radiative acceleration may be written (20):

$$g_{Rv}(A)dv = \frac{\pi T_{eff}^4 k^4}{2c^3 h^3} \frac{R^2}{r^2} \frac{K_v(A)K_R}{X(A)K_v} P(u)du \quad (7)$$

where

$$u = \frac{hv}{kT} \quad (8)$$

and

$$P(u) = u^4 \frac{e^u}{(e^u - 1)}. \quad (9)$$

K_v the total opacity at frequency v, $K_v(A)$ is the contribution of element A to K_v, $X(A)$ is the mass fraction of element A and K_R is the Rosseland averaged opacity which is defined by:

$$K_R^{-1} = \frac{\pi}{4\sigma_R T^3} \int_0^\infty K_\nu^{-1} \frac{\partial B_\nu}{\partial T} d\nu \qquad (10)$$

From Eq. 7, atomic data are needed to compute the ratio:

$$\frac{K_\nu(A) K_R}{K_\nu} \qquad (11)$$

which depends on both the oscillator strengths and line widths. The atomic data banks such as those developed by the Opacity project (21) and Kurucz (22) are now used with an aimed accuracy of a factor of 1.1 on g_R.

When one computes evolutionary models one needs K_R to compute the radiative energy transfer. In *standard* stellar evolution, this is done by computing tables of K_R beforehand and interpolating in the tables during the evolution. When Fe, C, He, or any of the important contributors to the Rosseland averaged opacity diffuses during the evolution, one needs to recalculate K_R at the same time as the evolution proceeds. One then needs to integrate over frequencies at a sufficient number of points to follow the important frequency variations. Currently 4000 frequency points are used. All values of $K_\nu(A)$ must be kept in memory for computing time to be reasonable.

In order to obtain radiative accelerations, the ratio given by Eq. (11) must be calculated for each of the important chemical elements. This ratio depends on atomic data. When the opacity at ν is not dominated by element A, the contribution of the frequency interval (ν, $\nu+d\nu$) to g_R depends on the oscillator strength (for lines; and on σ for the continuum). When most of K_ν comes from element A however, the contribution of (ν, $\nu+d\nu$) does not depend on f but the width of the region where the opacity of one line dominates is determined by both f and the line width. For pressure broadened lines the contribution of one line to g_R is proportional to:

$$\frac{(f \Delta \nu_c N)^{0.5}}{N} \qquad (12)$$

where N is the number density of the absorbing state. The contribution of one line to the number of absorbed photons is proportional to the square root of the oscillator strength, $\Delta \nu_c$ and N. However, since the push must be divided by the number of ions sharing it, N appears in the denominator and the contribution to radiative acceleration is proportional to $N^{-0.5}$ as the abundance is increased.

One has the largest effect of collision broadening for abundant elements. The dependence of g_R on the width is shown on Fig 3 for carbon (from (14) where we refer the reader for more details). One finds in the astrophysical literature, collisional widths calculated with values of γ ranging from $1.6\ 10^{-6}$ to $2.0\ 10^{-5}$ (see Eq. 3 and 4 above). Radiative accelerations were calculated for C with these two extreme values, the long and short dashes respectively. The maximum uncertainty on g_R is by a factor of 3 but is closer to a factor of 2 over most of the T range. This is however much larger than the acceptable uncertainty. The detailed g_R calculations for carbon were carried out using a more elaborate *modified semi-empirical approximation* (MSE) developed by Dimitrijevic & Konjevic (23) for the quadratic Stark effect along with the needed matrix elements computed using the oscillator strengths provided in the Opacity Project data base. The results are given by the full line. These are acceptably reproduced by using Eq. 4 with $\gamma = 8\ 10^{-6}$ (dotted line on Fig. 3) instead of the MSE formulae as a comparison of the full line with the dotted line shows. Using this value of γ, calibrated using the MSE formulae, probably would be sufficient for many applications. For greater accuracy, the MSE theory is currently being used for radiative acceleration calculations using Opacity Project data such as (24) for iron.

For the linear Stark effect of hydrogenoid states, one currently uses an approximation suggested by Cox (25), involving Eq. (3) but with n_L and n_U raised to the fifth instead of the fourth power. This may not be a very good approximation and the results of Stehlé (26) will be used in future studies. This may affect radiative accelerations over the temperature range where hydrogenoid states are the more important ones for g_R.

Blends

Line widths are also important for blends and for the effects these have on radiative acceleration. The accuracy with which one needs to know the wavelength of lines to calculate the effect of blends may be evaluated from Fig. 1. Since line widths of strong lines are equal to 3 to 5 times the pressure width, one needs to know the transition wavelength within some 50 mA to know if two lines overlap. The accuracy with which the Opacity Project determines energy levels does not satisfy this criterion. One may calculate blends only if a large number of lines are involved so that the blending is statistical.

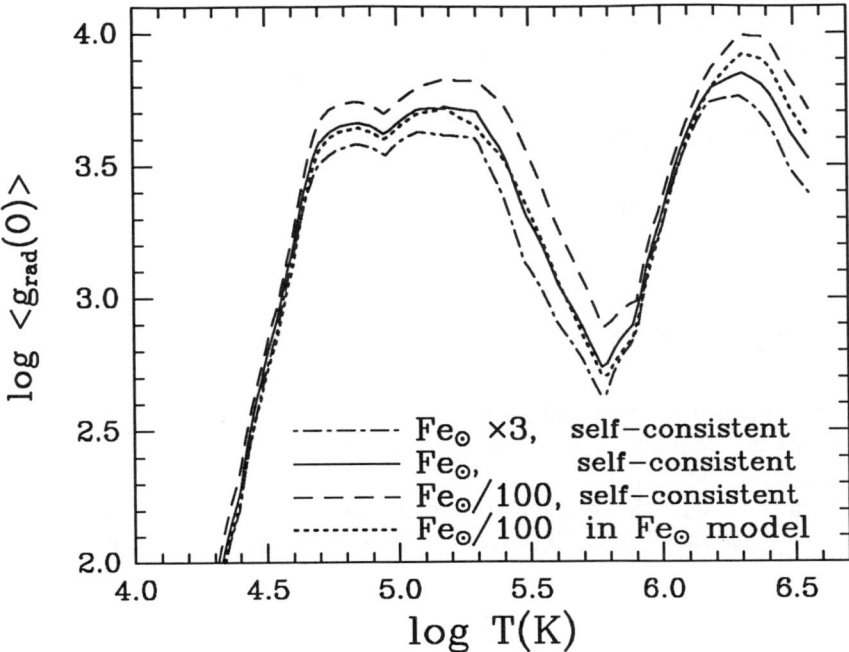

FIGURE 3. Radiative acceleration of oxygen in the envelope of a star with a fixed core (region with T>3.6 10^6 K), and different Fe abundances in the envelope (other elements, including O, kept in solar proportions). In the self-consistent cases, the same abundances are used to compute the model and g_R. In the fourth case (dotted), Eq. (7) is used to compute g_R with K_R computed with reduced Fe while the K_R used in the model was calculated with solar composition.

On Fig. 3 is shown the radiative acceleration on oxygen in mixtures of solar composition (continuous), where Fe has been reduced by a factor of 100 (dash) and where it has been increased by a factor of 3 (dash-dot). The Fe abundance in the star was arbitrarily modified for T < 3.6 10^6 K and kept solar at higher T. The dot curve is not a completely self consistent calculation in that the Rosseland opacity used for g_R corresponds to $Fe_\odot/100$ while the Rosseland opacity used to calculate the stellar model had a solar Fe abundance. The luminosity is forced not to be affected and the T to be continuous. This shows the importance of blends and of the contribution of Fe to the average opacity. These results show the importance of calculating simultaneously as the evolutionary model the settling of those chemical elements that contribute significantly to K_R and to recalculate K_R continuously as the abundances change. The radiative acceleration of O may be changed by a factor of 1.5 either way by abundance changes of Fe. There are effects of model adjustment as can be seen from Table 2. Changes in $g_R(O)$ are *not* simply due to the effect of K_R in Eq. (7).

TABLE 2

X(Fe)/X(Fe$_\odot$)	T$_{eff}$ (10^4 K)	log g	R (10^{11} cm)
3	1.04	4.24	1.38
1	1.08	4.31	1.28
0.01	1.12	4.37	1.12

The T_{eff} and K_R both change and so does K_v. Effects partly cancel and the global effect is difficult to predict without doing the calculation. These models should be viewed as preliminary; they show the interest of obtaining self consistent evolutionary models.

Similarly the reduction of the He abundance by a factor of 10^3 has been shown to cause an increase of $g_R(Li)$ by a factor of more than ten (27) at some depths in stars.

Comparison to other Work

I received recently a draft version of a paper by Seaton (28) describing his use of the Opacity Project data to calculate radiative accelerations. It contains a detailed comparison to the calculations described above. There is general agreement when one compares calculations for similar compositions and done with the same atomic data. This is specially true for CNO. He however had a more complete set of transitions available from the OP project than available to us for Fe and this leads to a factor of 3 increase in the radiative acceleration on Fe at some temperatures. His results are available for more chemical species than available previously but do not contain all the corrections described in (14). This is important mainly for T < 50000 K. In order to use them in evolutionary calculations one needs to find a way to include the effect of the diffusion of one specie on K_R and on the radiative acceleration of other species as described above. He makes tests showing that the use of 10^4 frequency points to calculate the integrals over frequencies leads to an accuracy of about 3% on g_R.

Auto-ionization Levels

The lines involving auto-ionization levels saturate slowly as the abundance increases because of their width. They can consequently play a large role for the radiative acceleration of abundant elements. At the same time they can show their signature in stellar spectra and so allow an observational confirmation of the presence of states that are otherwise very difficult to measure in the laboratory.

One recent example is given by Si^+ (29). The progress made possible by the recent atomic structure calculations can be estimated by comparing to a previous study on the same subject in 1981 (30). Fig. (6) of Ref. (29) shows a synthetic spectrum calculated with the autoionization resonances of Si^+ obtained with an R-matrix code for its first 50 energy levels. The importance of autoionization levels is clearly demonstrated. The observations constitute a confirmation of the calculations that cannot be obtained from the laboratory for most of the transitions.

The radiative acceleration to be expected from those autoionization features has not yet been calculated. Since some of the features are the main opacity source over 100 A, the contribution is expected to be large. It had been suggested by (18) that in order to explain the observed overabundance of Si in some Ap stars, autoionization features with widths of some 100 A and f ~ 1 were needed. No such features were known to exist in Si^+ in 1970. They have now been calculated.

REFERENCES

1. Michaud, G. and Tutukov A. 1991, IAU Symposium 145, **Evolution of Stars: the Photospheric Abundance Connection**, Golden Sands, Bulgaria, 25-31 August, ed. G. Michaud and A. Tutukov (Dordrecht: Kluwer).
2. Leckrone, D. S., Wahlgren, G. M., and Johansson, S. G. 1991, ApJL, 377, L37.
3. Wahlgren, G. M., Leckrone, D. S., Johansson, S. G., Rosberg, M., Brage, T. 1995, ApJ, 444, 438.
4. Aller, L. H., and Chapman, S. 1960, ApJ, 132, 461.
5. Michaud, G. 1977, Nature, 266, 433.
6. Proffitt, C. R., and Michaud, G. 1991, ApJ, 380, 238.
7. Cox, A. N., Guzik, J. A., and Kidman, R. B. 1989, ApJ, 342, 1187.
8. Christensen-Dalsgaard, J., Proffitt, C. R., and Thompson, M. J. 1992, ApJL, preprint.
9. Bahcall J. N., & Pinsoneault, M. H. 1995, RMP, 67, 781.
10. Richard, O., S. Vauclair, C. Charbonnel, W.A. Dziembowski 1996, A&A, submitted.
11. Eddington, A. S. 1926, The Internal Constitution of the Stars (New York: Dover [1959] reprint), § 199.
12. Babel, J., and Michaud, G. 1991, A&A, 241, 493.
13. Griem, H. R. 1960, Ap. J., 132, 883; Griem, H. R. 1968, Phys. Rev. 165, 258.
14. Gonzalez, J.-F., LeBlanc, F., Artru, M.-C., and Michaud, G. 1995, A&A, 297, 223.
15. Leckrone, D. S., Johansson, S. G., Kalus, G., Wahlgren, G. M., Brage, T., and Proffitt, C. R. 1996, ApJ, 462, 937.
16. Johansson, S. G., Kalus, G., Brage, T., Leckrone, D. S., and Wahlgren, G. M., 1996, ApJ, 462, 943.
17. Beauchamp, A., Wesemael, F., Bergeron, P., and Liebert, J. 1995, ApJ, 441, L85. Beauchamp, A., Wesemael, F., Bergeron, P., Saffer, R. A., and Liebert, J. 1996, in Proceedings - 9th European Workshop on White Dwarfs, ed. D. Koester & K. Werner (Springer: Heidelberg), in press.

18. Michaud, G. 1970, ApJ, 160, 641.
19. Michaud, G., and Proffitt, C. R. 1993, in Inside the Stars, IAU COLLOQUIUM 137, Vienna, April 1992, ed. W. W. Weiss and A. Baglin, ASP Conference Series, 40, 246.
20. Michaud, G., Charland, Y., Vauclair, S., and Vauclair, G. 1976, ApJ, 210, 447.
21. Seaton, M. J. in Inside the Stars, IAU COLLOQUIUM 137, Vienna, April 1992, ed. W. W. Weiss and A. Baglin, ASP Conference Series, 40, 222.
22. Kurucz, R. L. 1991, in Stellar Atmospheres: Beyond Classical Models, ed. by L. Crivellari, I. Hubeny and D. G. Hummer, NATO ASI Series (Dordrecht, Kluwer), p. 441.
23. Dimitrijevic, M. S., and Konjevic, N. 1980, J. Quant. Rad. Transfer 24, 451; Dimitrijevic, M. S., and Konjevic, N. 1986, A&A, 163, 297; Dimitrijevic, M. S., and Konjevic, N. 1987, A&A, 172, 345.
24. LeBlanc, F., and Michaud, G. 1995, A&A, 303, 166.
25. Cox, A. N., 1965, Stellar Structure. In Aller L. H., and McLaughlin D. B. (eds.) Stars and Stellar Systems, vol. 8. University of Chicago Press, Chicago, p.218.
26. Stehlé, C. 1996, in preparation; Stehlé, C. 1994, A&AS, 104, 509.
27. Richer, J., Michaud, G., and Massacrier, G. 1996, A&A, in press.
28. Seaton, M. J. 1996, preprint.
29. Lanz, T., Artru, M.-C., Le Dourneuf, M., and Hubeny, I. 1996, A&A, 309, 218.
30. Artru, M.-C., Jamar, C., Petrini, D., and Praderie, F. 1981, A&A, 96, 380.

Data Bases for diagnostic of high temperature astrophysical plasmas

Massimo Landini

*Department of Astronomy and Space Science,
University of Florence, Italy*

Abstract The spectral region below 2000 Å is crowded of lines from the most important elements in the universe and the x-ray and EUV emission of astrophysical plasmas is an extremely powerful tool to investigate temperature and density models of celestial sources. In the last two decades a number of space missions has been devoted to investigate the X-ray and EUV sky, and, even before, space born spectrograph, measured detailed spectra of the solar corona. Two high spectral resolution instruments, CDS and SUMER , on the SOHO mission, are producing a lot of high quality spectra of the solar corona between 150 and 1600 Å and high resolution observations are planned for the near future also from stars and galaxies. To properly exploit the huge amount of information supplied by the observations and to plan new observations, the most updated sets of atomic data are necessary. Models of neutral atoms and ions ,details of the most important atomic processes,(rates of collision and radiative ionizations and excitation, radiative decays , recombinations) are being collected in extended databases, to be accessed by the scientific community. A brief description is given of some of them that are in the way of upgrading.

Introduction

The Extreme UltraViolet spectral region supplies a lot of information for the diagnostic of temperature, density, velocity and chemical composition of astrophysical plasmas. The EUV (100-912 Å) and UV (912 - 3000 Å) spectral regions contain a large number of spectral lines emitted by ions which exist in plasmas where the electron temperature ranges from 10^4 to 10^8 K and electron density ranges from 10^4 to 10^{15} cm^{-3}. Hot and low density plasmas are present in the Universe in a variety of objects: solar and stellar coronae, hot stars, planetary nebulae, supernovae remnants, interstellar medium, quasar, thin galactic halos.

The EUV-UV emission has been measured during the last 25 years by a large number of space programs: Skylab, OSO, IUE, Einstein, EXOSAT, ROSAT, HST, HUT, SERT, EUVE and many other experiments are now flying or are planned for the future.

The SOHO mission, successfully launched at the beginning of last December, will study the solar corona with high sensitivity spectroscopic instruments (CDS, SUMER, EIT, UVCS). Many space missions are planned so far aiming to achieve larger spectral and spatial resolution: UVSTAR, FUSE.

To exploit completely the huge amount of information supplied by the observations, the most updated and extended data-bases on atomic data are necessary.

In this paper a review will be given on the information necessary to do proper diagnostic of temperature and density of optically thin plasmas using the Extreme Ultraviolet radiation

Some Available Data Bases

A few atomic data-bases have been compiled and may be used to do theoretical predictions of lines emission in the EUV.

I will mention :

ADAS (Atomic Data and Analysis Structure) developed at Rutherford Appleton Laboratory (1) as a set of atomic data collections and computer codes to evaluate radiation properties of atoms and ions and help users to interpret spectral measurements.

The ARCETRI Spectral Code, originally developed in 1970 and several time updated during the last 25 years, to evaluate continuum and line emission from high temperature and low density plasmas (a poster supplies more details) (2).

The CHIANTI Data Base which is being assessed by H. Mason, K. Dere, and E. Landi, and includes detailed evaluations of the most important ions of Li-like, Be-like B-like, C-like, N-like isoelectronic sequences.

The IRON Project which is producing a number of papers concerning new evaluation of electron collision excitation cross sections for low ionization degrees of iron (FeII to FeVIII)and other elements pertaining to the same isoelectronic sequences.

The radiation processes.

Most of the astrophysical plasmas are hot (T$\geq 10^4$ K) and low density ($N_e \leq 10^{13}$ cm^{-3}) for optically thin conditions the most important radiative processes to be included in the evaluation of the radiative power emitted by the plasma are *continuum radiation*, that includes *free - free continuum*, important for log T ≥ 6.5, *free - bound continuum*, important for log T ≤ 7, *two - photons continuum*, important for $N_e \leq 10^8$ cm^{-3} and *line radiation*, usually much more important than continuum emission.

Line emission

Most of the power radiated by optically thin plasmas concerns line emission. The *bound - bound* emissivity is given by:

$$P_{j,i} = N_j(X^{+m})\, A_{j,i}\, h\nu_{j,i} \qquad erg\ cm^{-3} sec^{-1}$$

$A_{j,i}(sec^{-1})$ is the radiative transition probabilities; $N_j(X^{+m})$ is the number density of the level j of the ion:

$$N_j(X^{+m}) = \frac{N_j(X^{+m})}{N(X^{+m})} \frac{N(X^{+m})}{N(X)} \frac{N(X)}{N(H)} \frac{N(H)}{N_e} N_e \qquad cm^{-3}$$

N_e is the electron number density; $\frac{N_j(X^{+m})}{N(X^{+m})}$ is the population of level j relative to the total $N(X^{+m})$ number density of ion X^{+m}; it is function of the electron temperature and density; $\frac{N(X^{+m})}{N(X)}$ is the ionization ratio of ion X^{+m} which is predominantly a function of temperature; $\frac{N(X)}{N(H)}$ is the element abundance relative to Hydrogen which varies in different astrophysical plasmas and also in different solar features; $\frac{N(H)}{N_e}$ is the Hydrogen abundance relative to electron density (~ 0.8).

From above it is clear that the theoretical evaluation of $P_{j,i}$, requires the knowledge of proper atomic models, radiative transition probabilities and oscillator strength evaluation, ionization balance and level population computation.

The Atomic Model

The most updated an extended atomic models should be included; extremely useful are compilations of observed and theoretical energy levels (see f.i. Kelly compilations available at NIST) (3). A lot of numerical computations have been performed during the last twenty years and a large amount of data may be found in the literature. (see f.i references given in Atomic Data and Nuclear Data Tables 57,numbers 1 and 2 , 1994)

All the Data-Bases include files containing for each level of each ion configuration, term, multiplicity, statistical weight, energy level both observed and predicted.

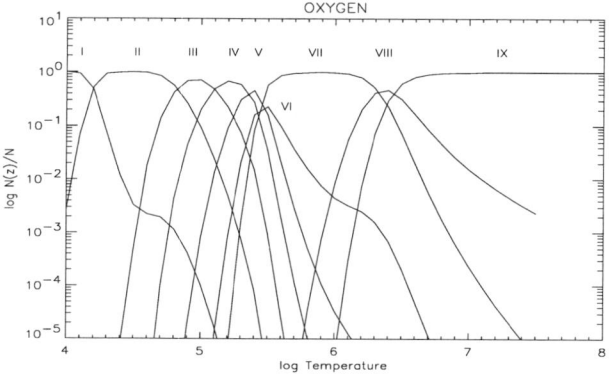

Figure 1: Ionization equilibrium of Oxygen ions.

The Ionization Balance

The ion abundance is evaluated assuming stationary balance among all the important ionization and recombination processes.

The main processes for the ionization balance of most ions are *electron collision ionization* and *radiative recombination*. But also *autoionization, dielectronic recombination* and *charge exchange* are taken into account.

Usually numerical computations are fitted by simple analytical expressions and used to solve the detailed balance among recombinations and ionizations in a stationary state:

$$N_e S^{(z-1,z)} N^{(z-1)} = N_e \alpha^{(z,z-1)} N^{(z)} + N_H C^{(z,z-1)} N^{(z)}$$

where $S^{(z-1,z)}$ is the electron ionization and autoionization rate, $\alpha^{(z,z-1)}$ is the radiative and dielectronic recombination rate, $C^{(z,z-1)}$ is the charge exchange. The ion balance may be supplied externally. Frequently used computations are from (4),(5). In fig. 1 the equilibrium balance of Oxygen is plotted versus temperature.

Level population

In low density plasmas the collisional excitation processes are generally faster than ionization and recombination timescales, therefore *the collisional excitation is dominant over ionization and recombination* in producing excited state.

The number density population of level j must be calculated by solving the *statistical equilibrium* equations for all the levels and including collisional and radiative excitation and de-excitation mechanisms; the most important are electron collision and radiative decay, but in many cases the *proton collisional excitation and de-excitation* rates must also be included. They become comparable with electron collisional processes only for transitions where $\Delta E_{i,j} \ll kT_e$, usually for transitions between fine structure levels at high temperatures.

$$N_j(N_e \Sigma_i C^e_{j,i} + \Sigma_{i<j} A_{j,i}) = \Sigma_i N_i N_e C^e_{i,j} + \Sigma_{i>j} N_i A_{i,j}$$

$C^e_{i,j}$,is the collisional excitation rate coefficient for a Maxwellian electron velocity distribution:

$$C^e_{i,j} = \frac{8.63 \times 10^{-6}}{T_e^{1/2}} \frac{\Upsilon_{i,j}(T_e)}{\omega_i} \exp\left(\frac{-\Delta E_{i,j}}{kT_e}\right)$$

ω_i is the statistical weight of level i, k, the Boltzmann constant $\Upsilon_{i,j}$, the thermally-averaged collision strength

$$\Upsilon_{i,j}(T_e) = \int_0^\infty \Omega_{i,j} \exp\left(-\frac{E_j}{kT_e}\right) d\left(\frac{E_j}{kT_e}\right)$$

Ω , the collision strength a dimensionless quantity, related to the *electron excitation cross-section*

E_j , the energy of the scattered electron relative to the final energy state of the ion.

$A_{j,i}$ the *spontaneous radiation transition probability* (sec^{-1}).

In the early stages of astrophysical plasmas computations (7-10), extended use has been made of the so called "coronal model" approximation, where the assumptions are made that the population of the upper level of transition j occurs mainly via collisional excitation from the ground state g and that the radiative decay overwhelms any other depopulation process. The solution of the equilibrium equations is straightforward and much less atomic data are necessary.

To evaluate $\Upsilon_{i,j}(T_e)$a in the general case one needs to know Ω over the full energy range from 0 to ∞.

Electron-ions scattering computations are performed using different approximations such as *Distorted Wave, Coulomb Bethe* and *Close-Coupling* approximation. Their accuracy ranges usually between 25% and 10%. They are very time consuming and frequently $\Omega(E_j)$ is evaluated only for a few energy values and interpolation/extrapolations techniques are used .

A useful scaling law has been suggested by Burgess and Tully (11). It allows both to evaluate Υ and to store infinite data into a few mesh points.

The collision rates and the radiative rates are used to evaluate population for any level of each ion as a function of temperature and density . Usually the density dependence is much more important than the temperature one

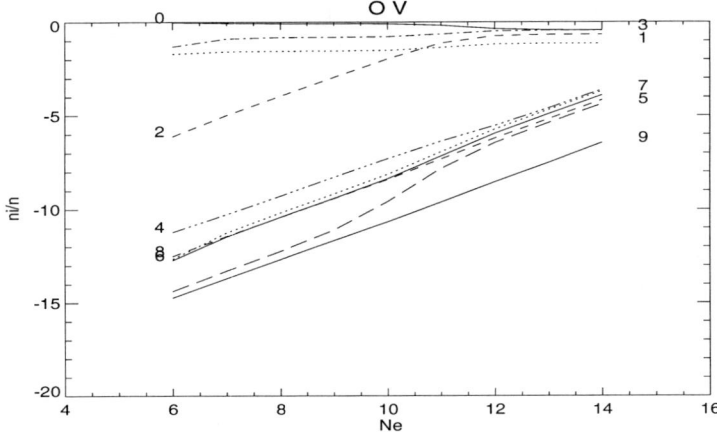

Figure 2: The population of the first ten levels of OV for different densities.

An example for OV is given in fig. 2.

Plasma diagnostic

For optically thin lines the flux at the Earth of a spectral line j is given by:

$$I(\lambda_{j,i}) = \frac{1}{4\pi R^2} \int_V P_{j,i}\, dV \qquad erg\ cm^{-2} sec^{-1}$$

where V is the volume of the source and R is the earth-to-object distance. Following the " coronal model approximation " the flux is usually written as

$$I(\lambda_{j,i}) = \frac{1}{4\pi R^2} \int_V G_{j,i}(T, n_e) n_e^2\, dV = \frac{1}{4\pi R^2} \int_V G_{j,i}(T, n_e) f(T) dT$$

where $G(T, n_e)$ is the *Contribution Function*, the *Differential Emission Measure* is defined

$$DEM = n_e^2 \frac{dV}{dT} = f(T)$$

and assumption is made that it is function of temperature only.

Selecting a set of non density dependent lines, the inversion of a system of equations $I(\lambda_{j,i})$ allows to evaluate the $f(T)$ which supplies the temperature diagnostic, i.e. the information on the matter distribution at different temperatures in the source.

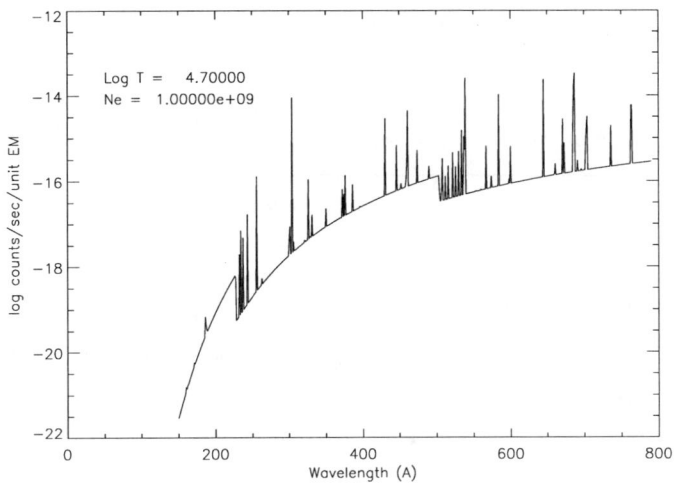

Figure 3: The number of photons emitted per unit emission measure and 1 Å band for log T= 4.7 and $n_e = 10^9$ from 150 Å to 800 Å.

Density diagnostic may be performed using ratios of line pairs of the same ion.

Fig. 3 shows an example of $G(T, n_e)$ for logt= 4.7 and $n_e = 10^9$ from 150 Å to 800 Å.

Some Data-Base supply also auxiliary computer programs that allow to evaluate DEM and to generate synthetic spectra.

One example is given in fig. 4 where the observed spectrum of the active star EQ Pegasi, measured by the SW Spectrometer of Extreme Ultraviolet Explorer, is compared with the synthetic spectrum (12).

Conclusions

The necessity of extended and critically assessed Atomic Data Bases for the diagnostic of temperature and density of optically thin astrophysical plasmas is stressed.

A big effort has been done in the last few years, to collect updated and critically evaluated data on atomic models, radiative probabilities and excitation collision rates useful to interpret the spectra measured in the UV from the celestial sources.

A few Data-Bases are mentioned ; some of them are available for public release. Some result from the Arcetri Spectral Code, that has been used for data

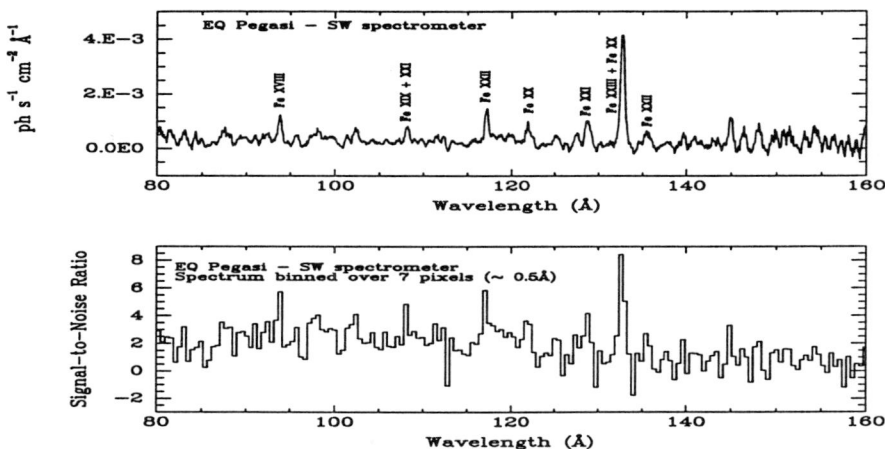

Figure 4: Observed and simulated spectra of EQ Pegasi. The synthetic spectrum of EQ Pegasi between 80 and 160 Å, evaluated using the procedure shown by Monsignori Fossi et al.(1995) (*top*). The Short Wavelength (80 - 160 Å) spectrum of EQ Pegasi observed by the Extreme Ultraviolet Explorer (EUVE) spectrometer (*bottom*).

reduction and analysis of the Extreme Ultraviolet Explorer may be browsed via Internet (http://www.arcetri.astro.it and http://cea.berkeley.edu/ego)

The CHIANTI Data-Base will be released next July, the ADAS system must be requested to the Rutherford Appleton Laboratory, OPACITY Project is available at the Strasbourg Data Center.

REFERENCES

1. H.P. Summers, D.H. Brooks, T.J. Hammond, A.C Lanzafame, J. Lang,1996,Tech Rep RAL-TR-96-017
2. Monsignori-Fossi,B.C. and Landini,M. 1996,*Proceedings* I.A.U. Coll. 152, 543-552
3. Wiese, W.L., Smith, M.W., Glennon, B.M., 1966, Atomic Trans. Prob., 1 NBS
4. Arnaud,M. and Rothenflug,R. 1985,*A&AS*,60,425
5. Arnaud,M. and Raymond,J.C. 1992,*A&AS*, 398,394
6. Landini,M. and Monsignori Fossi,B.C. 1970, *A&AS*,6,468
7. Landini,M. and Monsignori Fossi,B.C. 1990, *A&AS*,82,229
8. Mewe,R. 1972 , *Sol Phys*, 22, 459
9. Mewe,R. Groneshild,E.H.B.M., van den Oord, G.H.J. 1985, *A&AS*, 62, 197
10. Raymond J.C., and Smith, B.W. 1978,*ApJS*, 35, 419
11. Burgess,A. and Tully,J.A. 1992,*A&A*, 254, 436
12. Monsignori Fossi,B.C., Landini, M., Fruscione, A., Dupuis, J. 1995, *ApJ*, 449, 376-385

Density diagnostic of astrophysical plasmas

Massimo Landini and Enrico Landi

Dipartimento di Astronomia e Fisica dello Spazio
Università di Firenze, Italy

Abstract. The Arcetri Spectral Code, that evaluates XUV line and continuum emission of thin plasmas in the range 1-2000 Å, has been recently updated and includes detailed computation of levels populations for the Iron lines from Fe IX to Fe XXIII and the most important Be-like, C-like and N-like ions (1). A new technique is developed to evaluate the electron density, comparing observations and theoretical predictions for a set of density dependent lines of selected ions. Examples are given for Fe XIII and Si IX using recent observations of the solar corona with the S.E.R.T.S. spectrometers.

The Method

Electron density diagnostics from low density hot plasmas is usually performed comparing observed and predicted ratios of selected line pairs emitted by the same ion; this method assumes that the emitting region is isothermal and furthermore very frequently it gives not consistent results when applied to different pairs of lines and it is not always simple to understand which lines are responsible for the disagreement. We suggest a new method which allows to make use at the same time of all the available density dependent (and not dependent) lines and to compare them with the theoretical predictions.

The observed intensity of a spectral line can be put in the form

$$I_{obs} = K \int G(T, N_e) \phi(T) dT$$

We make the basic assumption that $G_{ij}(T, Ne) = K' f_{i,j}(T, N_e) \, g(T)$ so that we may define the L-Function for fully resolved lines as follows:

$$L(N_e) = \frac{I_{obs}}{G(T_{max}, N_e)}$$

and a similar form for unresolved multiplets of the same ion. It may be shown that, for $f_{i,j}$ slowly varying function of the temperature, all the $L(N_e)$ will meet for $N_e = N_e^o$, where N_e^o is the density of the emitting source.

A plot of the L-Functions of all the observed spectral lines of a given ion allows to evaluate N_e. It is also possible to identify transitions not agreeing with the others (blends, wrong intensity and wavelength calibration, wrong identifications).

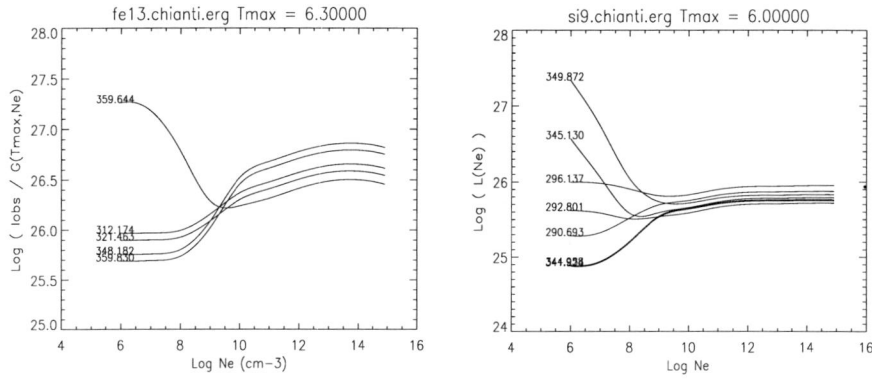

Figure 1: The Fe XIII and Si IX L-Functions.

Analysis of Fe XIII and Si IX solar spectra

We have studied the spectrum of a solar Active Region taken the May, 5th 1989 with the S.E.R.T.S. spectrometer(2). The observation covers the spectral range from 170 Å to 449 Å. The spectral resolution achieved is $\simeq 50-80\ m\text{Å}$. The high sensitivity of the instrument has allowed to detect a large number of lines emitted by high temperature plasma, from $T \simeq 10^{4.7}$ K to $\simeq 10^{6.8}$ K. The theoretical data used to analyze the observed spectrum are included in the Arcetri Spectral Code and are the most updated available in the literature.

The analysis of the Si IX spectrum provides a lower limit for the electron density of the emitting plasma: $N_e \geq 10^{9.1} cm^{-3}$. Moreover we are able to identify an unresolved contribution of Fe XVII to the intensity of line 296.137, which justifies the problems of this line.

The analysis of the Fe XIII spectrum with $\lambda \geq 300 \text{Å}$ allows to measure $N_e = 10^{9.3 \pm 0.2} cm^{-3}$.

Five more lines are observed with $\lambda \leq 300 \text{Å}$; when blending with other ions is taken into account, four of them meet on the same density, while the last ($\lambda = 240.723 \text{Å}$) largely disagree and needs further analysis.

REFERENCES

1. Monsignori Fossi B.C. and Landini M.: 1996, *IAU Coll n. 152*, 543, Eds S. Bowyer and R.F. Malina
2. Thomas R.J. and W.M. Neupert: 1994, *Astrophysical Journal Supplement Series 91*, 461-482

The XUV spectral code of Arcetri

Massimo Landini and Enrico Landi
Dipartimento di Astronomia e Fisica dello Spazio,
Università di Firenze, Italy

Brunella Monsignori Fossi[1]
Osservatorio Astrofisico di Arcetri, Firenze, Italy

Abstract The Arcetri Spectral Code evaluates lines and continuum emission of thin thermal plasmas in the temperature range from 10^4 K to 10^8 K and for electron number density lower than 10^{12} cm^{-3}, in the spectral range 1-2000 Å. For each ion of the most common elements the ionization balance is evaluated and the population level is computed assuming statistical equilibrium between excitation and decay processes. A sampling of the code, that is in the way of upgrading. is available on Mosaic(http://www.arcetri.astro.it).

The numerical code and the Data Base

The X and EUV region contains a large amount of emission lines pertaining to ions of the most abundant elements in the universe. In the last 25 years the Xray-UV emission from optically thin astrophysical plasmas has been measured by a large number of space programs, with the aim to study both solar and celestial sources. To fully exploit the observations, theoretical evaluations of plasma radiation must be performed using the most updated atomic bases. The Arcetri Spectral Code dates back to 1970 , when it covered the spectral range from 1 to 100 Å and included about 100 emission lines, and it has been repeatedly updated during the last twenty years. The present version of the code evaluates both continuum and line emission for any specified chemical composition , for temperature ranging from 10^4 K to 10^8 K and for electron number density lower than 10^{12} cm^{-3}, in the wavelength interval from 1 to 2000 Å. The continuum emission includes : free-free , free-bound from the ions of the most abundant elements, two-photons continuum from H-like ions. The line emission includes a large number of radiative transitions of all the ions of the most important elements. The power emitted both in the continuum and in the lines depends on the knowledge of the ionization balance that includes electron collision ionization, autoionization, charge exchange, radiative and dielectronic recombination. The number density population of each level is calculated by solving the statistical equilibrium equations among electron collision excitation and decay and radiative decay. An extensive review of the atomic data available in the literature has been performed to find the most

---

[1] We express our dear memory of Brunella Monsignori Fossi, who suddenly died last January; her deep engagement in continuously updating this Data Base and promoting its use will be always acknowledged.

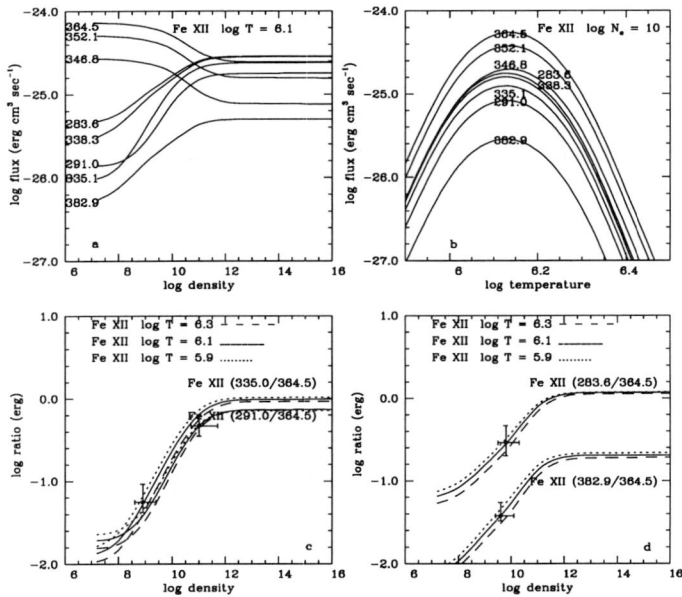

Figure 1: Examples of temperature and density effects on some line emission of Fe XII. *[a]* Power emitted per unit emission measure versus density for the peak temperature, *[b]* Power emitted per unit emission measure versus temperature for $N_e = 10^9$, *[c] and [d]* ratio of four density sensitive lines and comparison with observations of SERTS 89 (*crosses*).

updated atomic models, the radiative Einstein coefficients and the collision strengths, and the effective collision strengths over maxwellian distributions have been computed. The Data Base includes the Iron ions from Fe IX to FeXXIII the following *Be-like ions*, C III, O V, Ne VII, Mg IX, Ca VII, Si XI, Ni XXV, the *C-like ions*, O III, Ne V , Mg VII, Si IX, S XI, Ca XV, Zn XXV, the *N-like ions*, Mg VI, Si VIII, S X, Ar XII, Ca XIV. For the ions that are not included in the list, the coronal model assumption has been made and level population is evaluated via the excitation from the ground level only.

EFFECT OF THE CONTINUUM AND SELECTIVE ABSORPTION RATIO ON CALCULATED SYNTHETIC SOLAR SPECTRAL LINE PROFILES

Galal *A. , Yousef ** N. , Behery ***M. and Hamid* R .

* National Research Institute of Astronomy and Geophysics , **Astronomy Department, Cairo University and *** Astronomy Department , Al Azhar University , ARE

Empirical and synthetic ratio - values (r_λ) of continuum and selective absorption coefficients have been calculated at various displacements from the center of OI solar infrared triplet lines (7771.69 - 7774.18 - 7775.4 Å) . The firstly approximated empirical ratio - values are defined under the assumption of the constancy of this ratio throughout an infinitismal interval of the continuum optical depth in the solar atmosphere, but it varies on the average from one interval of depths to another . On the other hand the synthetic ratio values are determined under the LTE assumption .

As shown from Figure 1, the LTE synthetic ratio values (solid line) are systematically increasing while the empirical ones (various symbols) are decreasing with the decrease of the continuum optical depth in the solar photosphere . The numbers 1, 2 and 3 correspond to displacements 0.0, 0.5 and 1.0 Å from the line centers respectively .

The best fit empirical ratio-values (Dashed line) do not deviate much from the LTE synthetic ones at deeper layers of the photosphere . It seems that LTE assumption leads to overestimated values of the continuum and selective absorption coefficient ratios in the top layers of the photosphere . For this reason , the calculated LTE line profiles at limb positions are shallower in comparison to the observed ones (1) .

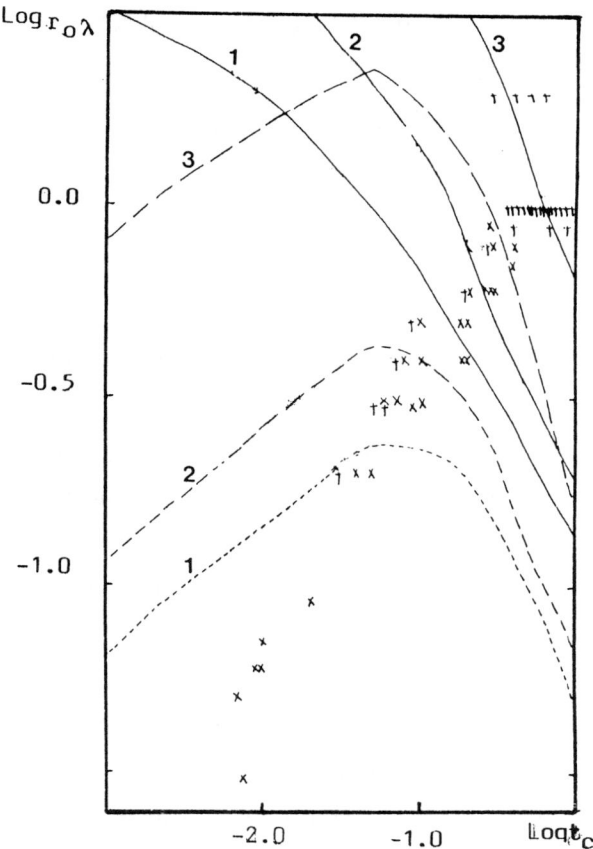

FIGURE 1. Depth-variations of continuum and selective absorption-ratio in the OI line 7771.96Å.

REFERENCES

1. Galal, A. A., Youssef, N. H. and Behery, M. M., "Accurate profiles of solar Infrared OI triplet lines", in AIP conference proceedings 328 on Spectral Line Shapes, Vol. 8, Toronto, Canada, 1994, pp 254-255.

Laboratory Studies of Infrared Features Observed in the Atmospheres of the Outer Planets [†]

Prasad Varanasi and Vassilii Nemtchinov
Institute for Terrestrial and Planetary Atmospheres
The University at Stony Brook
Stony Brook, NY 11794-5000, U.S.A.

Abstract

High-resolution measurements of the half-widths of several hydrogen-broadened lines in the ν_2^s-fundamental and ν_2^a-fundamental bands of $^{14}NH_3$ and $^{15}NH_3$ at temperatures relevant to the Jovian atmosphere are presented as an illustration of studies in our laboratory of the spectroscopic features observed in the atmospheres of the outer planets. The variation of the linewidths upon temperature and rotational quantum numbers is deduced.

Introduction

The continuing observation of the Jovian atmosphere and the atmospheres of the other giant planets by various orbiting probes and ground-based astronomers creates a sustained need for laboratory data on the spectral lines of CO, CH_4, CH_3D, C_2H_2, C_2H_4, C_2H_6, $^{14}NH_3$, and several other infrared-active molecules at appropriate temperatures using hydrogen and helium as broadening gases. In particular the analysis of the infrared spectra of the Jovian atmosphere in the 10 μm region is influenced greatly by the availability of accurate laboratory data on the spectral lines of $^{14}NH_3$ and $^{15}NH_3$. The infrared spectra of CO, CH_4, CH_3D, and hydrocarbons produced by photodissociation of methane in the respective planetary atmospheres are also the subject of investigations in our laboratory but are beyond the scope of this presentaion. We present in this talk some of our most recent measurements of the hydrogen-broadened half-widths, and line shapes of lines in the fundamental bands of CO, $^{12}CH_4$, $^{13}CH_4$, and $^{12}CH_3D$ in the (3-5 μm) region and the data on lines in the fundamental bands of $^{14}NH_3$ and $^{15}NH_3$ at 10 μm. However, for brevity, this paper limits itself to data on a few lines in the 10 μm bands of $^{14}NH_3$ and $^{15}NH_3$ only.

[0†]Supported by the Planetary Atmospheres Branch of NASA's Solar System Exploration Division under Grant-in-Aid No. NAGW-1894.

Linewidths of $^{14}NH_3$ and $^{15}NH_3$

A Bruker IFS-120HR Fourier-transform spectrometer with spectral resolution of 0.01 cm^{-1} ($^{14}NH_3$ and $^{15}NH_3$ bands) was employed. H$_2$ has been used as the broadening gas. The data have been measured at several temperatures appropriate to the Jovian atmosphere. The collison-broadened half-widths of the spectral lines were determined using a non-linear least-squares line-fitting computer routine. By using data obtained at 200, 255, and 295 K the temperature dependence of the linewidths γ_{JK} (cm^{-1} atm^{-1}) has been determined according to the simple relation

$$\gamma_{JK}(T)/\gamma_{JK}(T_0) = (T_0/T)^n,$$

where J and K are the rotational quantum numbers, T_0 is a reference temperature, and n is an exponent to be determined from linewidth data at various temperatures T. The functional dependence of the linewidths on the rotational quantum numbers J and K is attempted. A few examples are given in Tables 1-3. A selective comparison with the data of Margolis and Poynter[2] is presented in Table 3. The data are useful not only in the analyses of the $^{14}NH_3$ and $^{15}NH_3$ lines in the planetary spectra but also in determining the influence of the trough absorption of ammonia lines on the overlapping lines of other molecular species. Planetary observations of the same lines belonging to $^{14}NH_3$ and $^{15}NH_3$ should lead astronomers to an unambiguous determination of the ^{15}N:^{14}N in the planetary atmosphere depending upon the avaialablity of accurate laboratory spectroscopic information.

References

1. J. S. Margolis and R. L. Poynter, *Appl. Opt.* **30**, 3023 (1991).

Table 1. Hydrogen-broadened half-widths of the P(4,K) Lines in the ν_2^a and ν_2^s bands of ^{14}NH$_3$.

Line	$\nu(J,K)$ (cm^{-1})	γ_L^0 (cm^{-1} atm^{-1})			n
		296 K	255 K	200 K	
$aP(4,3)$	851.3251	0.0870	0.0989	0.1186	0.79
$sP(4,3)$	887.8730	0.0921	0.1020	0.1208	0.69
$aP(4,2)$	852.7233	0.0824	0.0903	0.1081	0.70
$sP(4,2)$	887.9970	0.0875	0.0917	0.1131	0.68
$aP(4,1)$	853.5473	0.0736	0.0798	0.0937	0.62
$sP(4,1)$	888.0750	0.0765	0.0833	0.0940	0.52
$aP(4,0)$	853.8174	0.0721	0.0776	0.0891	0.54

Table 2. Hydrogen-broadened half-widths of the P(4,K) Lines in the ν_2^a and ν_2^s bands of ^{15}NH$_3$.

Line	$\nu(J,K)$ (cm^{-1})	γ_L^0 (cm^{-1} atm^{-1})			n
		296 K	255 K	201 K	
$aP(4,3)$	847.6011	0.0919	0.1016	0.1206	0.70
$sP(4,3)$	882.8530	0.0899	0.0978	0.1177	0.70
$aP(4,2)$	849.0060	0.0828	0.0908	0.1104	0.72
$sP(4,2)$	883.0079	0.0826	0.0934	0.1064	0.65
$aP(4,1)$	849.8340	0.0762	0.0777	0.0884	0.58
$sP(4,1)$	883.1046	0.0771	0.0830	0.0962	0.58
$aP(4,0)$	850.1080	0.0670	0.0729	0.0824	0.53

Table 3. Comparison of our hydrogen-broadened half-width data at 200 K of lines in the ν_2^a and ν_2^s bands of $^{14}NH_3$ with those measured by Margolis and Poynter.[1]

Line	$\nu(J,K)$ cm^{-1}	γ_L^0 (cm^{-1} atm^{-1}) This Work	Ref. 1
$aP(4,3)$	851.3251	0.1186	0.1231
$sP(4,3)$	887.8730	0.1208	0.1312
$aP(4,2)$	852.7233	0.1081	0.1120
$sP(4,2)$	887.9970	0.1131	0.1072
$aP(4,1)$	853.5473	0.0937	0.0960
$sP(4,1)$	888.0750	0.0940	0.1004
$aP(4,0)$	853.8174	0.0891	0.0918
$sP(1,0)$	948.2316	0.1335	0.1130
$aR(5,5)$	1049.3454	0.1120	0.1146
$aR(5,4)$	1051.5107	0.1045	0.1083
$aR(5,3)$	1053.1292	0.0924	0.0963
$aR(5,2)$	1054.2515	0.0810	0.0843
$aR(5,1)$	1054.9114	0.0742	0.0755

MD Simulations of the RT CIA of CO_2 for the Atmosphere of Venus

M. Gruszka*, and A. Borysow*,**

* Department of Physics, Michigan Technological University, Houghton, MI 49931, USA
** Niels Bohr Institute, Copenhagen University Observatory, Juliane Maries vej 30, DK-2100 Copenhagen, Denmark

Abstract: The Roto – Translational (RT) collision – induced absorption (CIA) spectra of gaseous CO_2 were computed using molecular dynamics (MD) simulations. By adjusting an overlap contribution to the multipolar induction, a good agreement between MD simulations and existing experiments (233 K to 400 K) was achieved. Based on these results an analytical model of the CIA spectral profile from 200 K to 800 K for modeling of Venus' atmosphere is proposed.

Theory: MD is the only currently available method of computing CIA spectra for relatively heavy and large molecules like CO_2, where the anisotropy of the intermolecular potential can not be neglected, and at relatively high temperatures, where the quantum effects can be disregarded.
In the classical approximation and at the low density limit the absorption coefficient $\alpha(\omega)$ of a gas of the number density n, is given by:

$$\alpha(\omega) = \frac{2\pi^2}{3} \frac{n^2}{kTc} \omega^2 \, V \, I(\omega),$$

where the spectral density $I(\omega)$ is a Fourier transform of the correlation function $C(t) = <\vec{\mu}(0) \cdot \vec{\mu}(t)>$. The $\vec{\mu}(t)$ is the intermolecular induced dipole.

Computer Experiment: The NVE ensemble of 256 CO_2 molecules at the density of 25 amagats (binary limit) was simulated using *Moliq − Dynamo* program [1] written by David Fincham. A rigid rod representation of a CO_2 molecule was assumed together with the 5q(CO_2) [2] intermolecular potential. The trajectories were collected for a total time of over 750 ps, with a time step of 0.005 ps. The reliability of the averaging procedure was confirmed by less than 10% difference between the values of the two lowest spectral moments obtained from simulations and the results from analytical formulas [3, 4].

Induced dipole: Purely electrostatic origins of the induced dipole were found to be insufficient to satisfactorily reproduce the experimental data. For each simulated temperature (233 K, 273 K, 300 K and 400 K), the obtained results were about 40% too small around the maximum of the absorption band. This discrepancy was corrected by introducing the quantum overlap to the quadrupole and the hexadecapole induction terms, through the trace and the anisotropy of the molecular polarizability. In Fig. 1 the simulated spectra with and without the overlap correction are compared with the experimental results [5, 6] at 300 K. The values of the maximum intensity of the computed, and the experimental spectra are within the experimental uncertainty (10%). Similarly good agreement was achieved at other simulated temperatures [7].

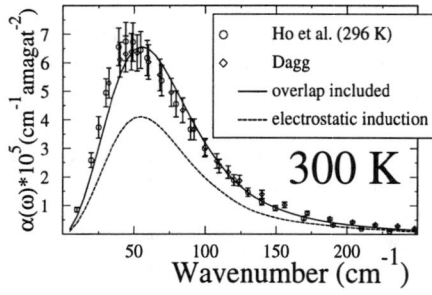

Figure 1: Simulated RT CIA of CO_2 at 300 K using two different sets of induction mechanisms.

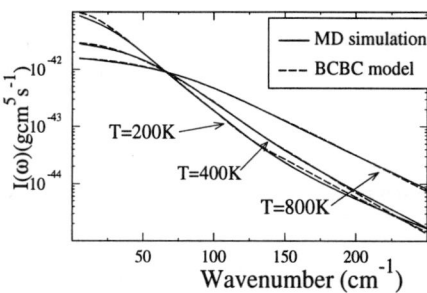

Figure 2: Spectral density $I(\omega)$ obtained from MD simulations and from the analytical model.

Model: The same anisotropic intermolecular potential and the induced dipole parameters were used in MD simulations performed at higher temperatures, up to 800 K to cover the range of existing atmospheric conditions on Venus. Based on these new results we propose an analytical BCBC model [8] of the CIA spectral profile from 200 K to 800 K for modeling of Venus' atmosphere. In Fig. 2 the comparison between simulated and modeled $I(\omega)$ at 200 K, 400 K and 800 K is shown. At all temperatures and over the span of four orders of magnitudes the discrepancy between the simulation and the model is below 10%.

Acknowledgments: The support of the Planetary Atmospheres Division of NASA is gratefully acknowledged.

REFERENCES:
1. D.Fincham. *Dynamo - A program for molecular dynamics simulation.*
2. C.S.Murthy, S.F.O'Shea and I.R.McDonald. *Mol. Phys.*, 50:531, 1983
3. A.Borysow and M.Moraldi. *Phys. Rev. Lett.*, 68:3686, 1992
4. M. Gruszka and A. Borysow. *Mol. Phys., In press.*
5. W.Ho, G.Birnbaum and A.Rosenberg. *J. Chem. Phys.*, 55:1028, 1971
6. I.R.Dagg. *private comm.*
7. M. Gruszka and A. Borysow. *Mol. Phys., submitted*
8. M. Gruszka and A. Borysow. *Icarus, in preparation*

Millimeter Wave Farwings of Water Vapor in N_2 and CO_2 Atmospheres

A. Bauer*, M. Godon*, J. Carlier* and R.R. Gamache[†]

*Laboratoire de Spectroscopie Hertzienne. URA CNRS. CERLA.
Université de Lille I. 59655 Villeneuve d'Ascq Cedex, France
and † Center for Atmospheric Research, University of Massachusetts Lowell,
450 Aiken Street, Lowell, MA 01854, USA

Abstract. Absolute absorption has been measured in the laboratory for H_2O-N_2 and H_2O-CO_2 mixtures at 239 GHz, in the atmospheric window of the H_2O rotational spectrum. Both data are compared to existing models.

The region of 240 GHz, in the gap between two strong water vapor lines, consitutes a window for the Earth atmospheric millimeter wave spectrum. As CO_2 has no pure rotational spectrum, this region is also a large window for planets with CO_2 atmosphere like Venus and Mars.

As it has long been known from Earth atmospheric observations that in these farwing regions of the H_2O spectrum, absorption cannot be accounted for by impact profiles, measurements in the laboratory have been carried out for a better understanding of the farwing behavior.

A high sensitivity Fabry Perot interferometer has been used. The method relies on measurements of the power loss factor α through the gas when it is irradiated by a wave at frequency ν. The high quality factor Q of the resonator (10^6) entails an equivalent path $l = cQ/2\pi\nu$ of about 200 m at 239 GHz.

Pressure and temperature dependences of absorption could then be obtained around $\alpha = 10^{-8}$ cm^{-1} per torr.

Following measurements for pure water vapor, where absorptions of one order of magnitude larger than those predicted by impact profiles had been observed, measurements for atmospheric mixtures of $H_2O + N_2$ and $H_2O + CO_2$ were carried out, and compared to available line profiles in the millimeter wave range.

Besides the usual Van Vleck-Weisskopf (VVW) and Gross, rewritten by Zhevakin-Naumov (ZN) lineshapes, profiles involving an empirical or semi empirical continuum have been used. The Clough-Kneizys-Davies(CKD) model, included in the FASCODE atmospheric transmission computer code, has been used in our calculations ; it involves a multiplying factor applied to a VVW line shape far from the center. Our absorption data have also been compared to those obtained from the Liebe (L) model, involving a totally empirical continuum added a VVW contribution.

Collisional line broadening data for rotational lines up to a high frequency limit (12 000 GHz) are necessary for the calculation of the absorption predicted by the VVW and ZN models. Whereas these data were available for H_2O-N_2, calculations had to be carried out for the $H_2O - CO_2$ mixture, using the Robert-Bonamy theory.

At 239 GHz, a comparison between experimental data and models gives the following results.

. Collisional line widths $\Delta\nu$ are about 1.5 times higher for H_2O-CO_2 than for H_2O-N_2, leading to the same calculated absorption rate for VVW and ZN models.

. For both mixtures, experimental absorption α_{exp} is much higher than α_{vvw}. But at 296 K the discrepancy is 6.2 times larger for H_2O-CO_2 than for H_2O-N_2.

. A quadratic term b(T) in the H_2O pressure dependence of α appears, negligible in the VVW model when $p_{H2O} \gg p_{N2,CO2}$. Although this term should correspond to H_2O-H_2O collisions, it is larger in the H_2O-CO_2 mixture.

. The temperature dependence of absorption is usually described by $\alpha(T) = \alpha(T_0)(T/T_0)^n$. The strong temperature dependence observed (n = - 4.9 and - 5.3 respectively) with respect to the VVW model (n = - 3.7) also reflects the "continuum effect".

. The continuum models for H_2O-N_2 reproduce reasonably experimental absorption. No continuums are available for H_2O-CO_2.

A. Bauer, M. Godon, J. Carlier and Q. Ma, J.Q.S.R.T. **53**, 411-423 (1995).
A. Bauer, M. Godon, J. Carlier and R.R. Gamache, J. Mol. Spectrosc. **176**, 45-57 (1996).

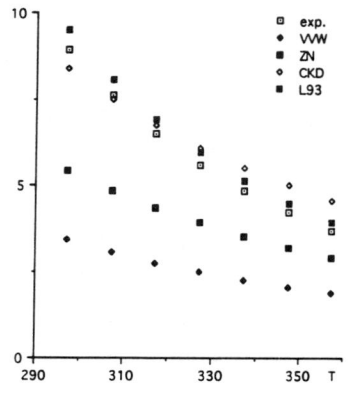 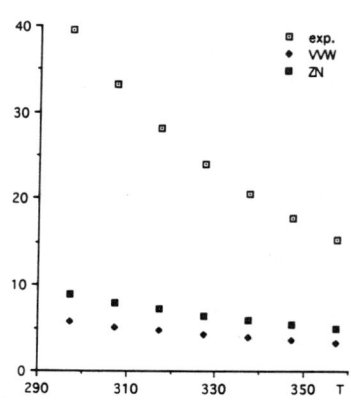

H_2O-N_2 and H_2O-CO_2 absorption ($10^{-6} cm^{-1}$) at 239 GHz, for p_{H2O} = 10 torr, $p_{N2,CO2}$ = 750 torr.

Raman scattering as a probe for remote sensing measurement of water temperature

Giovanna Cecchi(*), Piero Mazzinghi(+), Valentina Raimondi(*), Marco Zoppi(+)

(*) Consiglio Nazionale delle Ricerche, Istituto di Ricerca Onde Elettromagnetiche, Via Panciatichi 64, I-50127 Firenze (Italy) [cecchi@iroe.iroe.fi.cnr.it].

(+) Consiglio Nazionale delle Ricerche, Istituto di Elettronica Quantistica, Via Panciatichi 56/30, I-50127 Firenze (Italy) [mazzinghi@ieq.fi.cnr.it; zoppi@ieq.fi.cnr.it].

The determination of the sea-water temperature is a fundamental step in the investigation of the marine environment. However, while the surface temperature is monitored by passive remote sensors, the temperature profile is obtained using the traditional *in situ* technique which implies longer measurement time. The development of a remote sensing technique, to detect water temperature profiles, represents a remarkable improvement to monitor the sea underwater temperature.

Lidar systems are remote sensing instruments capable of detecting underwater temperature. The technique is based on the analysis of the light scattering spectrum. Laboratory experiments have demonstrated that the OH-stretching Raman spectrum of water changes significantly as a function of temperature and, in a lesser part, by varying salinity [1, 2, 3]. Raman spectra of water were investigated in our laboratory both on pure and sea-water samples. The experimental set-up consists of an Argon ion laser tuned on the 514.5nm line, a thermostatized container for the water sample, and a triple spectrometer coupled to

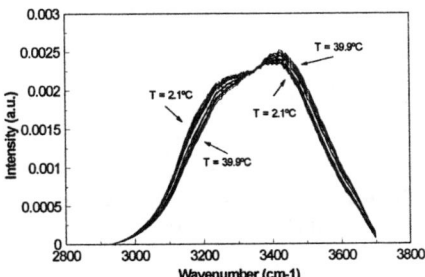

Figure 1. Sea-water Raman spectra vs. temperature.

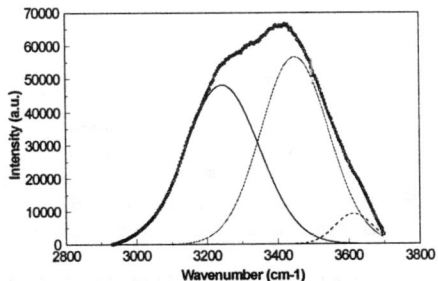

Figure 2. Raman spectrum at T=22.9°C and its three-Gaussian fit.

© 1997 American Institute of Physics

an Optical Multichannel Analyser. The temperature was measured directly, with an accuracy of 0.1°C, by a thermocouple placed in the sample close to the scattering region. The Raman spectra, obtained by varying the temperature from 2 to 40°C, are reported in Fig.1. The spectra are normalised to costant areas and show different components whose intensities change with temperature. The aim of the laboratory experiment was to parametrize the spectra and to calibrate the temperature dependence of the parameters [4]. We found that the experimental spectra could be described by the sum of three Gaussian components. The fit of the Raman spectrum at T=22.9°C is reported in Fig.2, while in Fig.3 we show the intensity ratio between the two main components as a function of temperature. The slope of the straight-line, resulting from a least square fit of the data, gives the sensitivity of the method: i.e. a relative variation of about 0.9% for $\Delta T=1°C$.

Three field experiments were carried out between 1991 and 1994 aboard the National Research Council oceanographic ships. The whole Lidar equipment was placed on the main deck of the ship and the laser beam was deflected into the water by a 45° mirror, extending out of the ship. Here, the excitation source was an excimer laser (XeCl, λ=308nm). At this wavelength the penetration depth is still good while the Raman signal, being proportional to λ^4, is enhanced. The Raman spectra were accumulated over several laser pulses while the background was automatically subtracted during the dead time. The temperature and salinity profiles were measured showing that both quantities were constant down to 15m. In addition, the temperature was detected every 10min, by a sensor ($\delta T=0.1°C$) at the depth of 3m. The same Gaussian fitting procedure was adopted for these spectra obtaining the results shown in Fig.4. The linear fit now gives a sensitivity figure of 4.7% for $\Delta T=1°C$.

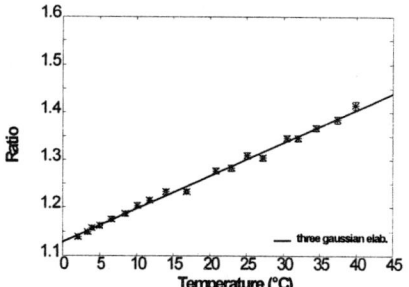

Figure 3. The intensity ratio between the two main Gaussians vs. temperature.

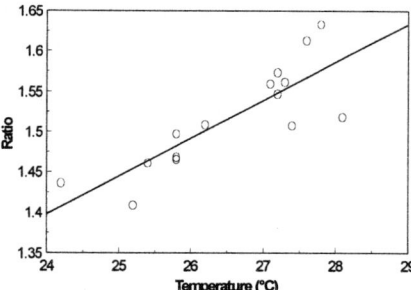

Figure 4. Field experiment: the intensity ratios vs. temperature.

[1] Walrafen E., J. Chem. Phys. **47**, 114 (1967).
[2] Angell C.A., Rodgers V., J Chem. Phys. **80**, 6245 (1984).
[3] Chang C.H., Young L.A., Leonard D.A., *Remote measurements of fluid temperature by Raman scattered radiation,* U.S. Patent 3.986.775 (1974).
[4] Raimondi V., Breschi B., Cecchi G., Pantani L., Tirelli D., Valmori G., Mazzinghi P., Zoppi M., EARSeL Advances in Remote Sensing **1**, n.2, 131 (1992).

Implementation of the Voigt Line-Shape Calculation in the Forward Model for Operational MIPAS Retrievals

M. Ridolfi[(*)], M. Höpfner[(+)], P. Raspollini[($)]

[(*)] Istituto di Ricerca sulle Onde Elettromagnetiche 'Nello Carrara' (IROE - CNR),
via Panciatichi, 64 - Firenze, Italy (Fax: +39 55 4222475, E-Mail: ridolfi@iroe.iroe.fi.cnr.it)
[(+)] IMK - FZK/Universität, Karlsruhe, Germany
[($)] Fondazione per la Meteorologia Applicata, via Caproni, 8 - Firenze, Italy

An ESA-supported study, for the development of an optimised algorithm for routine p, T and VMR retrievals from emission spectra measured by MIPAS, is in progress at IROE-CNR, in collaboration with the University of Bologna and the Institut für Meteorologie und Klimaforschung (IMK).

MIPAS (Michelson Interferometer for Passive Atmospheric Sounding) is an ESA-developed instrument that will operate on board ENVISAT-1 as part of the first Polar Orbit Earth Observation Mission program (POEM-1). MIPAS will perform limb-sounding observations of the atmospheric emission spectrum in the middle infrared region.

Altitude profiles of atmospheric pressure and temperature (p,T), and of the volume mixing ratio (VMR) of five high-priority species (O_3, H_2O, HNO_3, CH_4 and N_2O) will be routinely retrieved from MIPAS measurements in near real time (NRT), in the 8 - 53 km altitude range. The retrieval of these parameters is performed by fitting synthetic spectra, calculated by using a model of radiative transfer through an inhomogeneous atmosphere (forward model), to the observations.

The aim of the study is to develop a code for MIPAS NRT data analysis, optimised in accordance with the requirements of accuracy and speed. Since synthetic spectra calculations are responsible for the largest part of the retrieval time, both mathematical and physical optimisations of the forward model have been considered. Within this framework, calculation of the line-profile is a critical step because of the large number of floating point operations that are needed and due to the accuracy requirements that have to be satisfied. Even if MIPAS does not have a very high spectral resolution (0.025 cm^{-1}), the spectral features have to be simulated using a fine frequency grid (0.0005 cm^{-1}, i.e. smaller than the smallest atmospheric line-width), so that the radiative transfer in an optical path that spans the altitudes from the tangent altitude up to the upper limit of the atmosphere is correctly modelled. Three different algorithms have been compared for the computation of the Voigt line-shape: Humlicek [1], AFGL [2] and Drayson [3]. The Humlicek algorithm has been selected as the fastest procedure.

© 1997 American Institute of Physics

Table 1: Run-time comparison of different methods for the calculation of the Voigt line-shape.

Method	Relative run time
LOR-HUM	1
Humlicek [1]	5.4
AFGL [2]	7.1
Drayson [3]	8.4

However, a further optimisation has been considered. The optimised procedure (LOR-HUM) consists of the calculation of the line-profile using the Humlicek approach inside a ± 30 Doppler half-width broad region around the line-centre and simply using Lorentz profile beyond 30 half-widths. The relative computing times (only for the calculation of the line-profile) of the different algorithms are compared in Table 1. A significant time saving is observed when the LOR-HUM method is used. Panel (a) of Fig.1 shows a spectrum simulated by means of the Humlicek algorithm: this simulation is considered as the reference for the comparison with the other methods.

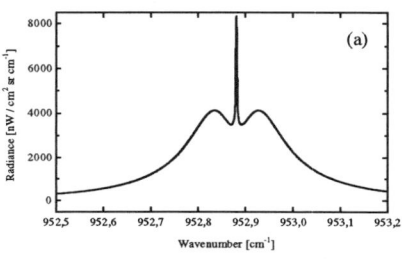
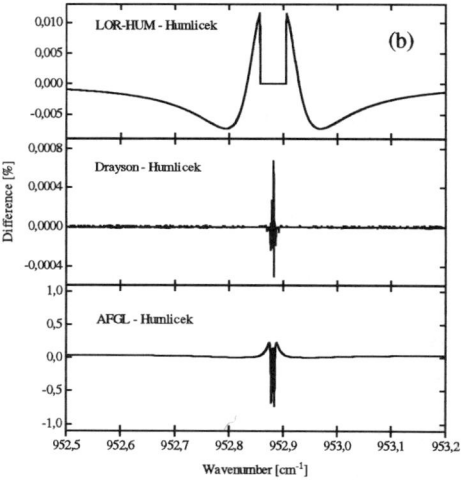

Fig. 1. (a): Radiative transfer calculation of a CO_2-line for a tangent altitude of 8 km. Humlicek algorithm has been used.
(b): Deviations of a limb forward calculation (8 km tangent altitude) for different line-shape algorithms. The reference calculation has been performed by using the Humlicek algorithm (see panel (a)).

In the (b) panel of Fig. 1, the percentage differences between the spectra obtained with the different algorithms and the reference spectrum shown in panel (a) are plotted. It results that the LOR-HUM method can be used within the frame of our accuracy requirements.

REFERENCES

[1] Humlicek, J., 'Optimised computation of the Voigt and complex probability functions', J. Quant. Spectr. Radiat. Transfer, 27, 437-444, 1982.
[2] Clough, S. A., F. X. Kneizys, L. S. Rothman, and W. O. Gallery, 'Atmospheric spectral transmission and radiance: FASCOD1B', SPIE 277, 152-166, 1981.
[3] Drayson, S. R., 'Rapid computation of the Voigt profile', J Quant. Spectr. Radiat. Transfer, 16, 611-614, 1976.

Airborne Infrared Diode Laser Spectrometer for *in Situ* Measurement of Stratospheric Trace Gas Concentration

Guido Toci, Piero Mazzinghi and Matteo Vannini

Istituto di Elettronica Quantistica del Consiglio Nazionale delle Ricerche
Via Panciatichi 56/30, 50127 Firenze, Italy

Abstract. This paper describes a spectrometer working with a laser diode for the measurement of concentration of molecular components traces by absorption spectroscopy in the mid infrared. The instrument currently under development is to be employed on the stratospheric aircraft *Geophysica M-55* for the *in situ* measurement of the nitric acid concentration in the Polar Stratospheric Clouds (PSC).

We present the design and operating criteria of a spectrometer employing as exciting source a diode laser for the quantitative measurement of molecular trace amounts in air by direct absorption measurement of a spectral line of the compound under study.

The instrument under development will be specifically devoted to the direct measurement of the concentration of the nitric acid (HNO_3) in the Polar Stratospheric Clouds (PSC) found in the lower stratosphere (18-22 km of altitude) at a temperature of 180-220 °K. The PSCs composition plays a key role in the stratospheric ozone depletion process (1, 2), but it is still under debate. Binary (H_2O and HNO_3) or ternary (H_2O, HNO_3 and H_2SO_4) liquid or solid aerosol particles are possible, but, up to date, direct experimental data do not exist.

The spectrometer is composed by a narrow band (1×10^{-4} cm^{-1}), single mode, CW diode laser emitting in the mid infrared, cooled in liquid nitrogen, tunable over a range of 25 cm^{-1}. The laser beam is sent into a multipass (182 passes) astigmatic Herriott cell (3), providing for 36 mt absorption path length with a small longitudinal size (20 cm mirror separation) and a small active volume (0.3 lt.) filled with the gas sample providing a sufficiently fast sampling time. The signal is detected by a cooled photoconductor. The absorption level is measured by means of second harmonic modulation spectroscopy (4), providing an enhanced detection sensitivity and an efficient rejection of the electronic and mechanical noise originated by the aircraft. Tuning and absorption are calibrated *on line* by means of a calibration cell filled in with a reference absorbing gas. The optomechanical design has been optimized against mechanical and thermal stress. For this reason the optical layout uses mainly transmission optics as the folding mirrors employed in laboratory measurements are very sensitive to misalignment, being the path necessary for the absorption detection very long.

The gas sample is provided by a Counterflow Virtual Impactor (CVI) (5) aerodynamic probe which selectively collects and concentrates by a known factor (5-50) the PSC aerosol particles by rejecting the ambient air. The particles are then evaporated at about 5 °C in the same synthetic air of the CVI and fed in the absorption cell by the gas stream. An HNO_3 amount in the cell of about 2×10^{11} molec/cm^3 is expected when in a PSC.

The HNO_3 molecule shows two strong roto-vibrational absorption bands near 1320 cm^{-1} (v_3, v_4) and near 1710 cm^{-1} (v_2) (5). In the experimental conditions (pressure 50-80 mbar, temperature 5 °C) the pressure line broadening is the dominant effect (γ_p=0.11 cm^{-1}/atm). The rotational fine structure of the P and R branches of the bands is well resolved (line separation $\approx$ 0.5 cm^{-1}), with a peak cross section of about 2.5×10^{-18} cm^2/molec (v_3 band) and 5.0×10^{-18} cm^2/molec (v_2 band). The absorbance in the cell is expected to be about 1.7×10^{-3} (v_3 band) and 3.4×10^{-3} (v_2 band). The operating wavelength has to be chosen to match a well resolved, unambiguously identifiable spectral line with large peak absorption cross sections in a spectral window with reduced interference from other chemical species (i.e. H_2O).

ACKNOWLEDGMENT

This research has been funded from the European Union, Environment and Climate Program, contract # ENV 4-CT 99-0039 in the frame of the Airborne Polar Experiment (APE).

REFERENCES

1. Toon, O. B. and Turco, R. P., Scientific American **264**, p. 68 (1991)
2. Wayne, R. P. *Chemistry of Atmospheres*, Oxford Science Publications, 1993
3. J. B. McManus, P. L. Kebabian, M. S. Zahniser, Appl. Opt. (1991)
4. C. R. Webster, R. T. Mentzies, E. D. Hinkley, in *Laser Remote Chemical Sensing*, Ed. M. R. Measures, New York, Wiley-Interscience, 1988, cap. 3, p. 163
5. K. J. Noone, J. A. Ogren, J. Heintzenberg, R. J. Charleston, D. S. Covert, Aerosol Sci. Technol. **8**, p. 235 (1988)
6. HITRAN 96; (by courtesy of Dr. Rothman, Ontar Co.) original data by R. D. May, C. R. Webster, J. Mol. Spectrosc. **138**, p. 383 (1989) and A. Maki, J. Mol. Spectrosc. **127**, p.104 (1988)

**COLLISION INDUCED SPECTROSCOPY:
SYMPOSIUM IN HONOR OF GEORGE BIRNBAUM**

George Birnbaum and the Collision-Induced Raman and Infrared Spectroscopies

Lothar Frommhold

Physics Department, University of Texas at Austin, Texas 78712-1081.

George Birnbaum received his Ph. D. degree in physics from the George Washington University in 1956. Prior to this he worked at the National Bureau of Standards in Washington, D. C., and then in Boulder, Colorado, in microwave spectroscopy and explored nonresonant absorption in gases [1-5]. Later he moved to the research laboratories of Hughes Aircraft Co. and Rockwell International, where he expanded his studies of collision-induced absorption to the far infrared [6-12]. Most notably, he also proposed and actually demonstrated collision-induced depolarized light scattering spectra in remarkable detail [13-16]. He returned to the National Bureau of Standards as a senior scientist in 1975, where he continued his research in the collision-induced spectroscopies [17-58], as well as in other areas, such as the line shapes of pressure broadened molecular bands [59-61]. He is a research professor at the Catholic University of America, where he directs an infrared laboratory. He has been an invited lecturer at the University of California at Los Angeles and the University of Texas at Austin, and a Visiting Professor and Guest Researcher at various branches of the University of Paris. Worldwide, he is regarded an innovator and leader in the fields of molecular and supramolecular infrared and Raman spectroscopy. In the past forty years or so, I think he must have reviewed most of the papers published in these fields in U.S. journals, besides a fair fraction of such papers submitted to the journals abroad. His understanding of the field, his constructive and imaginative comments and wise counsel as a referee, friend, and/or scientific consultant have advanced our field in a multitude of subtle, immeasurable ways. He has given numerous invited talks and written many review articles in the fields mentioned [62-73]. In 1983, as the chairman, he was the principal designer of that historical workshop in Bonas, France, where the entire field of collision-induced Raman and infrared spectroscopy, both of rarefied gases and dense matter, was reviewed. He was also the editor of the Proceedings of that conference [67]. That volume has served our community as an indispensable compendium and it will likely continue to be just that for years to come. He is consulted by scientists concerned with the basic processes of molecular interactions, collision-induced properties and the many important applications, especially in liquid state physics and astrophysics.

NON-RESONANT ABSORPTION AT MICROWAVE AND FAR-INFRARED FREQUENCIES [1–12]

In 1954 Birnbaum and Maryott discovered collision-induced absorption (CIA) of infrared-inactive gases in the microwave region [1, 2]. They found that the absorption varied roughly as density squared. Using Mizushima's theory of quadrupole-induced dipoles, Birnbaum was able to show that the observed microwave absorption in carbon dioxide was indeed collision-induced absorption caused mainly by such dipoles [1]. Subsequent work also demonstrated collision-induced microwave absorption in other gases, specifically oxygen [2, 3], nitrogen and ethylene [5]. This work was the first demonstration of the collision-induced translational absorption. We note that the absorption continua of oxygen and nitrogen are of course of considerable interest in connection with studies of the earth's atmosphere; gaseous carbon dioxide is important for the study of the atmosphere of Venus which is composed mostly of CO_2.

The microwave absorption studies were followed by a theory of collision-induced absorption line shapes in rare gas mixtures in the far infrared [6]. Based on classical radiation theory, Levine and Birnbaum computed the profiles of such emission and absorption spectra, using a simple analytical function to model the variation of the collision-induced dipole moments with increasing internuclear separation. Assuming straight-line collision paths, an analytic expression, which nowadays is usually referred to as the Levine-Birnbaum (LB) model, was obtained for the spectral profile which was shown to agree quite well with the measured spectra. The work proved to be a mainstay for the interpretation of collision-induced spectra. Birnbaum and collaborators also reported a number of collision-induced absorption measurements in various gases, namely gaseous carbon tetrafluoride [8], methane and fully deuterated methane (CD_4) [9], and carbon dioxide [11,12]. These measurements have withstood the test of time and still are indispensable in important ways; for example, as representative examples of such spectra for detailed analyses. Such measurements are also needed for the planetary sciences, e.g., for determining vertical temperature profiles of planetary atmospheres. Gas and liquid spectra of sulfur hexafluoride were compared and a collision-induced band was identified near 50 cm^{-1} [10].

COLLISION-INDUCED DEPOLARIZED LIGHT SCATTERING [13–16]

In a theoretical paper in 1968 Birnbaum and Levine predicted that dense, monatomic gases scatter light in form of a depolarized continuum. The width of the spectrum is of the order $\pm 1/\tau$, where τ is the characteristic duration of a collision. These authors predicted the shape of such spectra at low gas densities, specifically the characteristic, near exponential wings and the quadratic density dependence of the scattered intensities [13]. The spectra were thought to be caused by the fluctuations of the incremental polarizability of atomic pairs — that is the variation with time of the *excess* polarizability during collisional interactions over the sum of polarizabilities of the unperturbed (i.e., non-interacting) atoms. The polarizability of atomic pairs is a tensor with a non-vanishing anisotropy — a remarkable feature of monatomic gases composed of isotropic particles! — which introduces the depolarization of the scattered light. The anisotropy may be estimated by the classical electrostatic theory, the so-called dipole-induced dipole (DID) model, but quantum corrections due to the rearrangement of the electronic clouds of the atoms during collisions were also thought to be more or less important.

Shortly after the prediction of collision-induced depolarized light scattering (CIDLS) spectra, in papers by Birnbaum and McTague, the first experimental observations of such spectra were reported in argon and krypton [14] and xenon gas [15]. At relatively low gas density, the spectral shape and the quadratic density dependence of the scattered intensities were found in remarkable harmony with the theoretical predictions.

At the higher densities, interestingly, deviations from the quadratic density dependence were observed. These were thought to be due to ternary spectral components, arising from collisional complexes of exactly three interacting atoms, whose polarizability tensor and the spectra of course differ from the pair polarizability and pair spectra in characteristic ways. Extending the idea of the virial expansion in powers of density that was so successful in collision-induced absorption studies to the case of the collision-induced Raman spectra, the binary and ternary Raman spectra have actually been separated in this work on the basis of their differing density dependences. The ternary spectra were shown to be of a *negative* intensity, a fact that is related to the destructive interference characteristic of three-body induced spectra.

The fundamental properties of collision-induced Raman spectra, the ever present exponential spectral wings, the density squared dependence of the intensities at fixed frequencies at low densities, and the cancellation effect characteristic of complexes of three interacting atoms were clearly demonstrated in this seminal work [15].

The significance of this work was recognized immediately. This phenomenon has attracted much attention by theoreticians as well as experimentalists because the effect results from molecular interactions that are manifest during collisions and, therefore, provide a powerful tool for investigating in detail various aspects of intermolecular interactions. For example, such studies provide a unique method for determining the higher-order polarizabilities of spherical molecules. Moreover, they permit basic studies of many-body interactions in compressed fluids and much effort has been expanded in this direction since that pioneering work appeared. Since its inception the field of supramolecular Raman scattering has attracted a number of outstanding researchers. To date over 800 original papers have been published in this area[1] which confirm just about every aspect and the basic ideas of Birnbaum's classic discovery.

In fact, the idea of the collision-induced depolarized spectra as presented by Birnbaum and associates was so powerful, so crisp, so compelling that most of us seemed to almost forget that besides the depolarized collision-induced scattering process, there exists a collision-induced *polarized* scattering process, arising from the *trace* of the pair polarizability increment. That process, of course, is weak and a result strictly of quantum mechanics; no good classical model exist which would provide a reasonable, quantitative estimate. However, polarized collision-induced spectra carry valuable information on molecular interactions as well and should be as rewarding to study as its depolarized counterpart. To this day the bulk of the research efforts, however, has been and in essence still is in the nearly classical, depolarized scattering process predicted by the imaginative, influential discoverer and master of the field...

Birnbaum's early work already suggests that collision-induced light scattering is ubiquitous. In the purest and most frequently studied form, collision-induced light scattering spectra appear as forbidden spectra in systems composed of atoms or spherical molecules. However, polarizability fluctuations induced by collisions between spherical particles also occur in collisions between particles with anisotropic polarizabilities that do have allowed rotational Raman spectra. Thus, the contribution due to collisional induction is found virtually in all Raman spectra of dense matter, including the liquid and solid states. This suggestion is fully confirmed in the 800 or so original papers that appeared since the discovery of collision-induced depolarized light scattering.

THE FAR-INFRARED SPECTRA OF NONPOLAR GASES [17–47]

At the National Bureau of Standards Birnbaum and associates measured a number of very important collison-induced absorption spectra in the far-infrared

---

[1]A. Borysow and L. Frommhold, Collision-induced Light Scattering — a Bibliography, *Adv. Chem. Phys.* **75**, 439 (1989); update 1993 by L. Ulivi and L. Frommhold.

of gases composed of nonpolar molecules [6–12]. Such studies have continued to recent times [17–53]. He explored this effect in gases composed of molecules of various symmetries, illustrating the roles in the induction and cancellation processes of the various molecular multipoles [62].

Birnbaum also reported the celebrated, extra-ordinarily important measurements of the complete rototranslational spectra of H_2–H_2 and H_2–He pairs in the far infrared. These data have been indispensable for the planetary scientists for their studies of the thermal emission from the atmospheres of the outer planets that are composed primarily of H_2 and He [21]. Besides such measurements, especially those of the just mentioned spectra of H_2–H_2 and H_2–He pairs [21,25,27], were of crucial importance for detailed tests of accurate quantum chemical calculations from first principles. Agreement was found between measurement and the (parameter-free!) *ab initio* calculations of these spectra, within the small uncertainties of the measurement [30, 37]. This remarkable fact not only demonstrates the deep, quantitative understanding obtained of the physics involved in the spectroscopic collisions of such systems. It also made possible the most reliable predictions of such spectra at temperatures at which measurements do not exist, for example for the applications to very cold planetary atmospheres and to hot gases and their emission spectra (shockwaves and stellar atmospheres) for these and other systems. Birnbaum's measurements were among the most reliable in our field and have thus been of immeasurable value for the many uses they have found.

Like collision-induced light scattering, collision-induced absorption occurs in virtually all molecular gases, regardless of whether the molecules are infrared active or not. In molecular liquids collision-induced spectra provide a sensitive probe of molecular interactions and dynamics. In systems composed of spherical particles, these spectra are due to the collisional interactions. In short, the supramolecular spectroscopies are almost always important if the density of matter is sufficiently high, e.g., in all planetary atmospheres. Scientists at institutions such as NASA and JPL realized this soon after the discovery of collision-induced absorption. Over the years Birnbaum has served the community of planetary scientists as a most valuable advisor, coordinator and provider of reliable spectroscopic data as needed for the significant progress that ensued.

LINE SHAPE [6,37–45,49–56]

Throughout his career, Birnbaum was interested in the line shape problem. We have mentioned above the analytical LB model [6], a classical model that was physical, intuitive and invaluable for the development of our field. While it was not (yet) an *ab initio* computation, it was a useful two-parameter model which describes the experimental profiles quite well. It would thus serve useful purposes for reasonable extrapolations of measured spectral profiles, for example for improved determinations of spectral moments, and for the prediction of spectra at temperatures where measurements did not exist.

In 1976 Birnbaum and Cohen [39] developed a theory of collision-induced absorption line profiles that has continued to this day to be of great importance for the interpretation and prediction of collision-induced spectra. Detailed computational and experimental studies have shown that the so-called Birnbaum- Cohen (BC) model profile, which is analytic and easy to compute, approximates real spectral profiles of the rototranslational bands amazingly well, over a peak- to wing- intensity ratio that is often much greater than 100:1 [42] — even under nonclassical, strongly quantal conditions, e.g., induced hydrogen absorption spectra at low temperatures. For the rotovibrational bands, simple modifications of the BC shape have been proposed [44]. In the collision-induced spectroscopies, the BC profile is as important as the lorentzian, gaussian and voigt profiles of the ordinary spectroscopies: it represents the most useful analytical model profile for the supramolecular spectroscopies. It is demonstrably superior to several other models that have been proposed in our field over the years. In fact it is so good that — if used correctly — quantum line shape calculations are almost obsolete in a number of cases of great practical interest, unless the spectral features due to van der Waals molecules need to be modeled accurately.

In collaboration with a number of individuals, Birnbaum participated in the quantum formulation of line shapes of collision-induced absorption spectra [41, 37]. These efforts made possible detailed comparisons of measurements of binary absorption spectra with the fundamental theory. The perhaps most important conclusion from that work was that for the simpler systems (e.g., H_2–He or H_2–H_2 pairs), theory is perfectly capable to produce spectra of an accuracy that is comparable to that of the finest measurements. Furthermore, such work facilitated the most reliable temperature extrapolations of measured spectra which are necessary, especially for the astrophysical applications.

It is also noteworthy that thus the spectroscopic signatures of the $(H_2)_2$ van der Waals dimer could be predicted in the induced H_2 $S_0(0)$ and $S_0(1)$ lines from theory, for a clear identification of such features in the Voyager spectra of Jupiter [43]. These dimer features are potentially most significant for the determination of the all-important He/H_2 abundance ratio in the atmospheres of the outer planets and cool stars — that is one of the most important parameters for the quantitative understanding of the formation of the solar system, or any other planetary or stellar system that was formed from primordial matter.

In collaboration with Guillot and Mountain, Birnbaum calculated spectral profiles of the intercollisional interference process of dense systems [54]. Important comclusions, such as that of the significance of irreducible ternary dipole components and specifically the exchange quadrupole-induced dipole have been drawn from these efforts [55] which we will briefly address below.

MANY-BODY DIPOLES AND SPECTRA [54,55]

Collision-induced spectra are observable in virtually all dense, predominantly neutral fluids and solids. Intensities of collision-induced spectra typically increase with increasing density. The idea of expanding the measured intensities in terms of powers of density — a virial expansion — has been used early-on by Birnbaum for intriguing analyses of such measurements to separate the two- and three-body spectral contributions [7,15,10,12,etc.] Birnbaum and associates have also advanced the theoretical and computational aspects of the many-body spectral profiles [48–55,73].

In a significant extension of the studies of ternary Raman spectra, Birnbaum and Guillot investigated the effects of three- and four-body spectroscopic interactions in compressed atomic fluids [71]. In particular, the role of the irreducible three-body interactions, i.e., interactions that cannot be described by pairwise interactions [54, 55], was investigated. In such studies the modeling of the intercollisional dips observed in various dense fluids was attempted. The authors conclude that purely pairwise-additive intermolecular interactions cannot explain the observed absorption dip. Rather, an irreducible ternary dipole component of the exchange-induced quadrupole-induced dipole (EQID) type must be assumed in order to reproduce the measurements — a conclusion that has received substantial support since.

COLLISION-INDUCED EMISSION [57–58]

It is clear that collision-induced dipoles manifest themselves in various ways. Among other effects, collision-induced emission (CIE) must be expected in hot, dense, predominantly neutral gases as encountered in certain types of flames, rocket jets, shockwaves, and the atmospheres of cool stars. Yet it took many years after collision-induced absorption was discovered to actually point out a case of collision-induced emission under fairly common laboratory conditions, namely in linear shock wave experiments [57, 58]. Of course, George Birnbaum was there to demonstrate the significance of CIE under such conditions: the induced fundamental band of H_2 was prominently exhibited in emission of a mixture of hydrogen and argon heated by the shock wave. Collision-induced emission in the visible and infrared of the molecular gases is of a special interest in connection with its anticipated significance under the conditions mentioned. Like all emission spectra, CIE is much more complex than its counterpart, collision-induced absorption, because at the high temperatures highly vibrationally excited molecular states are significant whose induced 'hot' bands are at present not known, but for certain systems may be predicted by the fundamental theory that should be important for numerous applications. A vast area of future experimental and theoretical studies was thus pointed out by George Birnbaum.

PRESSURE BROADENING [59–61]

The publications listed at the end are highly selected; we list only papers related to his light scattering and absorption work mentioned above. He has written many more papers in other fields, too numerous to list here. We just mention as one example that he is considered a leading expert in pressure broadening of molecular lines, a field that has preoccupied his thinking for many years. We mention just a few recent review articles [59, 60, 61] which are frequently quoted in the literature on pressure broadening of molecular gases. Such work has also influenced his activities in the collision-induced spectroscopies.

CONCLUSION

Dr. Birnbaum gained worldwide recognition for his pioneering work in the spectroscopy of collision-induced depolarized light scattering spectra and collision-induced absorption in the infrared, particularly for his contribution to the theory of the shape of such spectra, and for his important work concerning the far infrared spectra of the nonpolar, molecular gases found in the atmospheres of the outer planets. He published many review articles [62–72], usually by invitation from the organizers of advanced research workshops and specialists' conferences. He organized the historic NATO Advanced Science Institute (ASI) in 1983 dealing with all aspects of collision-induced spectroscopy, light scattering and infrared absorption which occurs in all dense states of matter. Related collision-induced phenomena were also considered, e.g., the effects of collision-induced dipoles and polarizabilities on the dielectric and refractive properties of dense fluids (Clausius- Mosotti and Lorentz- Lorenz equations and their virial expansions). He was the editor of the book [66] that resulted from this conference. Immediately, that volume became the standard reference in these fields and remains to be just that to this day. Most of us have sought his counsel on more than one occasion and over the many years we were privileged to know him, and this is not likely to change any time soon. He must be one of the most influential leaders in our field, with all due respect to the great pioneers we were blessed to know in the fields of the interaction-induced infrared and Raman spectroscopies.

G. Birnbaum — SELECTED PAPERS

[1] G. Birnbaum, A. A. Maryott, and P. F. Wacker. Microwave absorption by the nonpolar gas CO_2. *J. Chem. Phys.* **22**, 1782 (1954).

[2] A. A. Maryott and G. Birnbaum. Microwave absorption in compressed oxygen. *Phys. Rev.* **99**, 1886 (1955).

[3] A. A. Maryott and G. Birnbaum. Microwave absorption in compressed oxygen. *J. Chem. Phys.* **32**, 686 (1960).

[4] A. A. Maryott and G. Birnbaum. Collision induced microwave absorption in compressed gases. I Dependence on density, temperature, and frequencies in CO_2. *J. Chem. Phys.* **36**, 2026 (1962).

[5] G. Birnbaum and A. A. Maryott. Collision-induced microwave absorption in compressed gases. II. Molecular electric quadrupole moments. *J. Chem. Phys.* **36**, 2032 (1962).

[6] H. B. Levine and G. Birnbaum. Classical theory of collision-induced absorption in rare gas mixtures. *Phys. Rev.* **154**, 86 (1967).

[7] G. Birnbaum, H. B. Levine, and D. A. McQuarrie. Determination of two- and three-body relaxation times from collision-induced absorption. *J. Chem. Phys.* **46**, 1557 (1967).

[8] A. Rosenberg and G. Birnbaum. Far infrared absorption in gaseous CF_4. *J. Chem. Phys.* **48**, 1396 (1968).

[9] G. Birnbaum and A. Rosenberg. Collision-induced absorption in gaseous CH_4 and CD_4 in the far infrared region. *Physics Letters* A **27**, 272 (1968).

[10] A. Rosenberg and G. Birnbaum. Far infrared spectra of gaseous and liquid SF_6. *J. Chem. Phys.* **52**, 683 (1970).

[11] W. Ho, G. Birnbaum, and A. Rosenberg. Far infrared collision-induced absorption in CO_2. I. Temperature dependence. *J. Chem. Phys.* **55**, 1028 (1971).

[12] G. Birnbaum, W. Ho, and A. Rosenberg. Far-infrared collision-induced absorption in CO_2. II. Pressure dependence in the gas phase and absorption in the liquid. *J. Chem. Phys.* **55**, 1039 (1971).

[13] H. B. Levine and G. Birnbaum. Collision induced light scattering. *Phys. Rev. Letters* **20**, 439 (1968).

[14] J. P. McTague and G. Birnbaum. Collision induced light scattering in gaseous Ar and Kr. *Phys. Rev. Letters* **21**, 661 (1968).

[15] J. P. McTague and G. Birnbaum. Collision induced light scattering in gases: I. Rare gases Ar, Kr and Xe. *Phys. Rev.*, A **3**, 1376 (1971).

[16] H. B. Levine and G. Birnbaum. Determination of models for collision induced polarizability by the method of moments. *J. Chem. Phys.* **55**, 2914 (1971).

[17] G. Birnbaum. Far infrared collision-induced spectrum in gaseous methane. I. Band shape and temperature dependence. *J. Chem. Phys.* **62**, 59 (1975).

[18] G. Birnbaum and E. R. Cohen. Far infrared collision-induced absorption in gaseous methane. II. Determination of the octopole and hexadecapole moments. *J. Chem. Phys.* **62**, 3807 (1975).

[19] G. Birnbaum and E. R. Cohen. Determination of molecular multipole moments and potential function parameters of non-polar molecules from far infrared spectra. *Molec. Phys.* **32**, 161 (1976).

[20] E. R. Cohen and G. Birnbaum. Influence of the potential function on the determination of multipole moments from pressure-induced far-infrared spectra. *J. Chem. Phys.* **66**, 2443 (1977).

[21] G. Birnbaum. Far-infrared absorption in H_2 and H_2-He mixtures. *J.Q.S.R.T.* **19**, 51 (1978).

[22] G. Birnbaum and T. K. Bose. Comparison of dielectric and refractive virial coefficients and collision-induced absorption bands. *J. Chem. Phys.* **71**, 17 (1979).

[23] G. Birnbaum and H. Sutter. Collision induced absorption in a highly symmetric molecule: SF_6. *Molec. Phys.* **42**, 21 (1981).

[24] E. R. Cohen, L. Frommhold, and G. Birnbaum. Analysis of the far-infrared H_2-He spectrum. *J. Chem. Phys.* **77**, 4933 (1982). Erratum: *ibid.*, **78**, 5283 (1983).

[25] P. Dore, L. Nencini, and G. Birnbaum. Far infrared absorption in normal H_2 from 77 to 298 K. *J.Q.S.R.T.* **30**, 245 (1983).

[26] G. Birnbaum, L. Frommhold, L. Nencini, and H. Sutter. The collision-induced far-infrared absorption band of gaseous methane in the region 30-900 cm^{-1}. *Chem. Phys. Letters* **100**, 292 (1983).

[27] G. Bachet, E. R. Cohen, P. Dore, and G. Birnbaum. The translational rotational absorption spectrum of hydrogen. *Can. J. Phys.* **61**, 591 (1983).

[28] G. Birnbaum, M. Krauss, and L. Frommhold. Collision-induced dipoles of rare gas mixtures. *J. Chem. Phys.* **80**, 2669 (1984).

[29] C. Chapados and G. Birnbaum. The forbidden far infrared ν_6 band of SF_6. *J. Molec. Spectrosc.* **105**, 206 (1984).

[30] G. Birnbaum, G. Bachet, and L. Frommhold. Experimental and theoretical investigation of the far-infrared spectrum of H_2-He mixtures. *Phys. Rev. A* **36**, 3729 (1987).

[31] G. Birnbaum, A. Borysow, and H. G. Sutter. Measurement and analysis of the far infrared absorption spectrum of the gaseous mixture H_2-CH_4. *J.Q.S.R.T.* **38**, 189 (1987).

[32] J. Borysow, L. Frommhold, and G. Birnbaum. Collision induced rototranslational absorption spectra of H_2-He pairs. *Astrophys. J.* **326**, 509 (1988).

[33] C. Chapados and G. Birnbaum. Infrared absorption of SF_6 from 32 to 3000 cm^{-1} in the gaseous and liquid states. *Molec. Phys.* **132**, 323 (1988).

[34] P. Dore, A. Filabozzi, and G. Birnbaum. Measurements and analysis of rototranslational absorption spectra of low density H_2-Ar mixtures. *Can. J. Phys.* **66**, 803 (1988).

[35] P. Dore, A. Filabozzi, and G. Birnbaum. Rototranslational absorption in gaseous H_2-Ar mixtures at intermediate densities. *Can. J. Phys.* **67**, 599 (1989).

[36] P. Dore, M. Moraldi, J. D. Poll, and G. Birnbaum. Analysis of rototranslational absorption spectra induced in low-density gases of non-polar molecules: The methane case.

Molec. Phys. **66**, 355 (1989).

[37] W. Meyer, L. Frommhold, and G. Birnbaum. Rototranslational absorption spectra of H_2–H_2 pairs in the far infrared. *Phys. Rev.* A **39**, 2434 (1989).

[38] M. S. Miller, D. A. McQuarrie, G. Birnbaum, and J. D. Poll. Constant acceleration approximation in collision induced absorption. *J. Chem. Phys.* **57**, 618 (1972).

[39] G. Birnbaum and E. R. Cohen. Theory of line shapes in pressure induced absorption. *Can. J. Phys.* **54**, 593 (1976).

[40] G. Birnbaum, M. S. Brown, and L. Frommhold. Lineshapes and dipole moments in collision-induced absorption. *Can. J. Phys.* **59**, 1544 (1981).

[41] G. Birnbaum, Shih-I Chu, A. Dalgarno, L. Frommhold, and E. L. Wright. Theory of collision-induced translation-rotation spectra: H_2-He. *Phys. Rev.* A **29**, 595 (1984).

[42] J. Borysow, L. Trafton, L. Frommhold, and G. Birnbaum. Modelling of pressure-induced far infrared absorption spectra: Molecular hydrogen pairs. *Astrophys. J.* **296**, 644 (1985).

[43] L. Frommhold, R. Samuelson, and G. Birnbaum. Hydrogen dimer structures in the far-infrared spectra of Jupiter and Saturn. *Astrophys. J.* **283**, L79 (1984).

[44] G. Birnbaum and A. Borysow. On the problem of detailed balance and model line shapes in collision-induced rotovibrational bands. *Molec. Phys.* **73**, 57 (1991).

[45] M. S. Brown, L. Frommhold, and G. Birnbaum. About an information theoretical spectral shape proposed for the collision induced spectroscopies. *Molec. Phys.* **62**, 907 (1987). See also *ibid.* **64**, 1001 (1988).

[46] G. Birnbaum. Collision-induced vibrational spectroscopy in liquids. *Vib. Spect. Mol. Liq. and Sol.* **56**, 147 (1980).

[47] W. A. Steele and G. Birnbaum. Molecular calculations of moments of the induced spectra for N_2, O_2, and CO_2. *J. Chem. Phys.* **72**, 2250 (1980).

[48] B. Guillot, S. Bratos, and G. Birnbaum. Theoretical study of spectra of depolarized light scattered from dense rare-gas fluids. *Phys. Rev.* A **22**, 2230 (1980).

[49] B. Guillot, S. Bratos, and G. Birnbaum. Theory of collision induced absorption in dense rare gas mixtures. *Molec. Phys.* **44**, 1021 (1981).

[50] B. Guillot, S. Bratos, and G. Birnbaum. Theoretical study of collision induced far infrared absorption of dense rare gas mixtures. *Phys. Rev.* A **25**, 773 (1982).

[51] B. Guillot and G. Birnbaum. Theoretical study of the far infrared absorption spectrum of dense nitrogen. *J. Chem. Phys.* **79**, 686 (1983).

[52] G. Birnbaum and R. D. Mountain. Molecular dynamics study of collision-induced absorption in rare gas liquid mixtures. *J. Chem. Phys.* **81**, 2347 (1984).

[53] R. D. Mountain and G. Birnbaum. Molecular dynamics study of intercollisional interference in collision induced absorption in compressed fluids. *J. Chem. Soc. Faraday Trans. 2*, **83**, 1791 (1987).

[54] B. Guillot, R. D. Mountain, and G. Birnbaum. Theoretical study of the 3-body absorption spectrum in pure rare-gas fluids. *Molec. Phys.* **64**, 747 (1988).

[55] B. Guillot, R. D. Mountain, and G. Birnbaum. Triplet dipoles in the absorption spectra of dense rare gas mixtures: I Short range interactions. *J. Chem. Phys.* **90**, 650 (1989).

[56] C. J. Montrose, T. A. Litovitz, G. Birnbaum, and R. Mennella. Viscoelastic relaxation

in simple liquids, an interaction-induced phenomenon. *J. Non-cryst. Solids* **130**, 177 (1991).

[57] R. Krech, G. Caledonia, S. Schertzer, K. Ritter, T. W. Wilkerson, L. Cotnoir, R. Taylor, and G. Birnbaum. Laboratory observation of collision induced emission in the fundamental vibration rotation band of H_2. *Phys. Rev. Letters* **49**, 1913 (1982).

[58] G. E. Caledonia, R. H. Krech, T. D. Wilkerson, R. L. Taylor, and G. Birnbaum. Collision-induced emission in the fundamental vibration-band of H_2. *Phys. Rev.* A **43**, 6010 (1991).

[59] G. Birnbaum. Microwave pressure broadening and its application to intermolecular forces. *Adv. Chem. Phys.* **12**, 487 (1967).

[60] G. Birnbaum. The shape of collision broadened lines from resonance to the far wings. *J.Q.S.R.T.* **21**, 597 (1979).

[61] G. Birnbaum. A kinetic approach to the shape of pressure broadened molecular bands. In L. Frommhold and J. W. Keto, eds., *Spectral Line Shapes* **6**, p. 337, Am. Inst. Physics, N.Y., 1990.

[62] G. Birnbaum. Determination of molecular constants from collision-induced far-infra-red spectra. In J. van Kranendonk, ed., *Intermolecular Spectroscopy and Dynamical Properties of Dense Systems — Proc. Int. School of Physics "Enrico Fermi"*, p. 111 (1980).

[63] G. Birnbaum. Collision induced vibrational spectroscopy in liquids. In S. Bratos and R. M. Pick, eds., *Vibrational Spectroscopy of Molecular Liquids and Solids*, p. 147. Plenum, New York, 1980.

[64] G. Birnbaum, B. Guillot, and S. Bratos. Theory of collision-induced line-shapes: Absorption and light-scattering at low density. *Adv. Chem. Phys.* **51**, 49 (1982).

[65] G. Birnbaum. Study of atomic and molecular interactions from collision-induced spectra. In J. V. Sengers, ed., *Thermophysical Properties of Fluids*, p. 8. Am. Soc. Mech. Engineers, New York, 1981.

[66] G. Birnbaum, ed. *Phenomena Induced by Intermolec. Interactions*. Plenum Press, New York, 1985.

[67] S. Bratos, B. Guillot, and G. Birnbaum. Theory of collision induced light scattering and absorption in dense rare gas fluids. In Ref. [66], p. 363.

[68] B. Guillot and G. Birnbaum. Theoretical interpretation of the far infrared absorption spectrum in molecular liquids: Nitrogen. In Ref. [66], p. 437.

[69] G. Birnbaum. Comments on the spectra of the halogens and halogen complexes in solution. In Ref. [66], p. 775.

[70] G. Birnbaum, L. Frommhold, and G.C. Tabisz. Collision induced spectroscopy: Absorption and light scattering. In J. Szudy, ed., *Spectral Line Shapes* **5**, p. 623. Ossolineum, Warsaw, 1989.

[71] B. Guillot and G. Birnbaum. Interaction-induced absorption in simple to complex liquids. In Th. Dorfmüller, ed., *Reactive and Flexible Molecules in Liquids*, p. 1. NATO ASI, Kluwer Academic Publ., 1989.

[72] G. Birnbaum and B. Guillot. Cancellation effects in collision-induced phenomena. In G. C. Tabisz and M. N. Neuman, eds., *Collision- and Interaction-Induced Spectroscopy*, p. 1, Kluwer, Dordrecht, 1995.

Collision Induced Light Scattering in Metal Vapors: the Mercury Spectra

F. Barocchi*+, M. Sampoli#+, L. Ulivi‡

* Physics Dept., Universita' di Firenze, Italy;
\# Energetic Dept., Universita' di Firenze, Italy;
+Istituto Nazionale di Fisica della Materia, Italy;
‡ Istituto di Elettronica Quantistica, CNR, Italy

I. Introduction

The depolarized part of the collision induced light scattering (CILS), the so called depolarized induced light scattering (DILS), has been studied extensively in the last two decades in molecular gases and liquids[1] in order to understand the details of the mechanism that is responsible for the phenomenon and to use this phenomenon as a probe of the molecular and fluid properties. In atomic fluids the depolarized spectrum for frequencies greater than $\sim 1 cm^{-1}$ is due only to polarizability anisotropies created in the medium by both interacting atoms and stable complex like dimers and trimers. When the density is sufficiently low, this spectrum is due only to interacting transient pairs and dimers[2,3] and is related to the transitions between the energy states of a pair of atoms, which for convenience can be divided between the "free" states of the translational continuum and the "bound" states of the dimers. Since the DILS spectrum of an atomic pair is related directly to the induction mechanism, which is responsible of the polarizability anisotropy, and to the interaction potential of the pair; both these properties can in principle be studied by DILS. This has been extensively done for the case of noble gases, for which the pair potentials are well known, and valuable information is derived on the details of the induction mechanism.[2,3] Indeed, it has been established that for noble gas systems, the long-range induced polarizability anisotropy, the so-called dipole-induced-dipole (DID) part, dominates the behavior of the low frequency portion of the spectrum (90÷100% of the total intensity) while only at high frequencies ($\nu > 50$ cm^{-1}) the short-range induced anisotropies are important.

A very interesting category of atomic systems are the metal vapors. For those systems the pair potential is not very well established. This is mostly due to experimental difficulties, because high-temperature corrosive vapors must be produced, handled and studied.

Here we review recent experiments in mercury vapors.[4-8] Mercury was chosen among the various metals, as a starting point, because of its microscopic properties connected partly to the relativistic behavior of its electrons[9] and partly to its pair interaction which is stronger than in van der Waals systems. From the experimental point of view, mercury vapor does not absorb light in the visible region of the spectrum for the densities regimes we are interested in and can be produced, with some care, at moderately high densities and pressures, suitable for our experiments, at temperatures below 1200 °C.

The main motivation of our work has been to give precise experimental results for the mercury vapor DILS spectra in order to study the form of the pair interaction potential together with the nature of the induction mechanism responsible for the light scattering process of pairs of atomic metals.

II. The depolarized spectra.

To carry out a detailed study of the DILS spectra of mercury, as it has been done for the case of noble gases,[2,3] the density behavior of the DILS spectrum in isothermal condition has to be measured for obtaining the pair-spectrum. The scattering cell has been devised for holding a metal vapor up to temperatures of the order of T=1000 °C and pressures up to P=100 MPa. A schematic drawing of this fused quartz cell together with the heating system and pressure compensation apparatus is given in Fig.1.

The cell is heated by a surrounding oven and inserted in a steel autoclave. The pressure of the mercury vapor inside the cell is compensated by the pressure outside the cell of an inert gas which fills the internal body of the autoclave. As is sketched in Fig.1, the pressure circuit fills at the same time the scattering cell (with mercury vapor) and the autoclave (with the inert gas), so that the pressure is continuously equalized inside and outside the scattering cell. The heating system of the scattering cell has a temperature controller which stabilizes the temperatures within ±1 °C, while the pressure is measured with a precision of ±0.5%. The thermodynamic conditions of mercury vapor in our experiments are reported in Table 1. The mercury densities ρ were derived by means of the equation of state.[10] The spectra have been measured with an argon ion laser of 0.5 W of power and the laser beam was focused into the sample with the polarization in the direction of the scattered light. The scattering angle was 90° and both polarizations were collected. The power transmitted through the cell was continuously monitored in order to detect any possible mercury deposition on the windows.

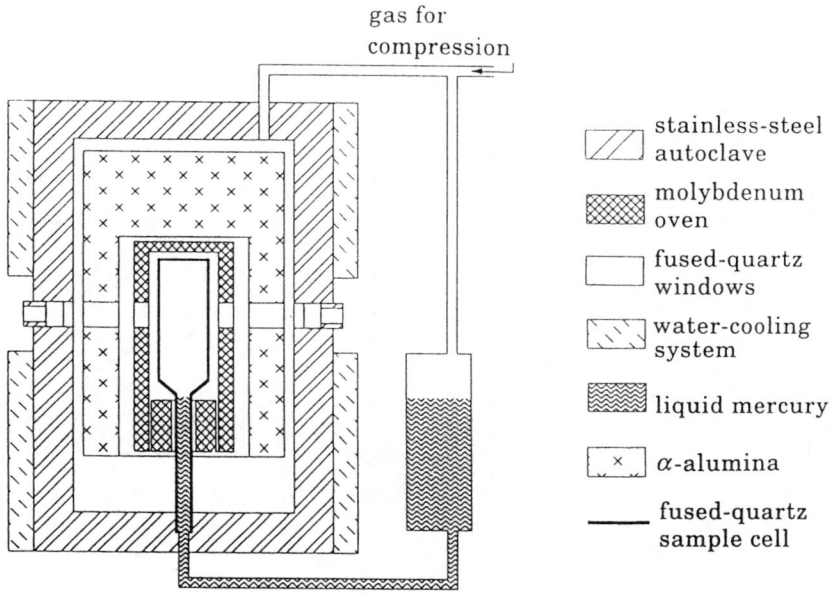

Figure 1: Schematic drawing of the high pressure high temperature light scattering cell together with pressure compensation apparatus.

TABLE 1. Thermodynamic states of mercury vapor investigated

T °C	P MPa	ρ At./nm^3	T °C	P MPa	ρ At./nm^3
460	0.40	0.040	700	0.87	0.065
480	0.53	0.051		1.87	0.141
520	0.28	0.026		2.87	0.218
	0.40	0.037		3.87	0.296
	0.53	0.049	800	0.92	0.062
	0.68	0.063		2.02	0.138
	0.82	0.075		3.13	0.215
550	0.86	0.077		4.24	0.293
	1.0	0.091		5.35	0.372
580	1.3	0.11		6.5	0.46
600	1.3	0.11		7.5	0.53
				8.5	0.60
			900	~15	~1

The spectral response of the overall spectrometer plus the detection apparatus was carefully calibrated by means of a black body radiation source.

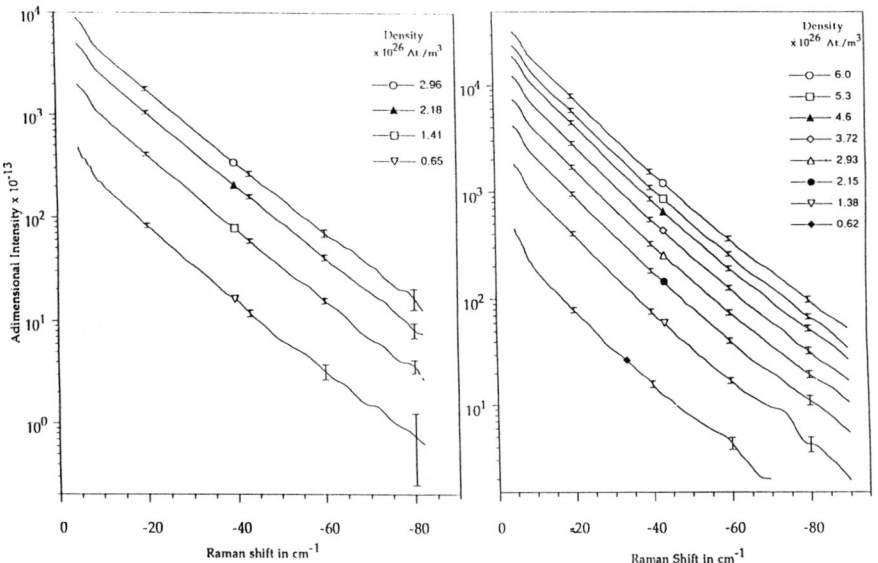

Figure 2: DILS pair spectra (adimensional units) of mercury vapor at 700°C (left) and 900°C (right).

For the absolute calibration, a small amount of hydrogen gas, of the order of 0.2 MPa, was mixed with the mercury and both spectra, i.e. the rotational S(1) line of hydrogen and the DILS spectrum of mercury, were measured at the same time. The value of the anisotropic polarizability of hydrogen we have used for the calculation of the S(1) line intensity, is $\beta = 0.320$ Å^3 at the 488 nm wavelength.[11,12] Figure 2 shows examples of experimental spectra. By means of the calibration of the mercury spectra, we have been able to determine their density dependence. As an example, this density dependence is given in Fig.3 for three different frequencies at the temperature T=800 °C. For the sake of convenience the quantity $Y(\rho) \propto I(\nu,\rho)/\rho + cost(\nu)$ is plotted. We can see that, within the experimental uncertainties, our measurements were performed in the density squared regime. Therefore, we have determined the spectrum of an isolated colliding pair for frequencies higher than 5 cm^{-1} by extrapolating the spectral intensity to zero density.

With good approximation we can now derive the zeroth and second spectral moments of the pair spectra at the measured temperatures. This has been accomplished by extrapolating the spectral intensity from 5 cm^{-1} down to zero frequency with a polynomial shape and at high frequency with an exponential decay. An estimate of the uncertainty due to both extrapolation procedures has been taken into account in the evaluation of moment error bars. The absolute values of the experimental moments are defined by

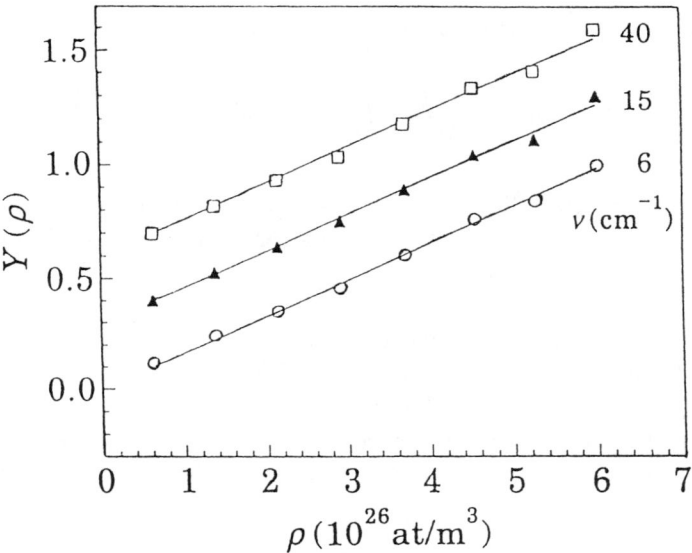

Figure 3: Density sependence of the DILS spectral intensity at 800°C and three different frequencies. For the sake of convenience the function $Y(\rho)$ is reported; $cost(\nu)$ yields 0, 0.3, 0.6 for $\nu = 6, 15, 40$ cm^{-1} respectively.

$$M_n^{pair} = \int_{-\infty}^{+\infty} \nu^n D_\|(\nu)\, d\nu \qquad (1)$$

where $D_\|(\nu)$ is related to the double differential scattering cross section for the pair and is given by the usual theoretical expression:

$$\begin{aligned} D_\|(\nu) &= V \frac{d^2\sigma}{d\Omega\, d\nu} \\ &= \tfrac{2}{15} V k_o k_s^3 \int_{-\infty}^{+\infty} dt\; \exp(-i\, 2\pi\nu t)\; \langle \beta(t)\beta(0)\, P_2[\cos(\theta(t))]\rangle \end{aligned} \qquad (2)$$

Here V stands for the volume of the scattering system, $\beta[r(t)]$ for the polarizability anisotropy of the interacting pair at time t, $P_2(x)$ for the second Legendre polynomial and θ for the angle between the pair axis at time $t = 0$ and $t = t$. The frequency and wave vector of the incoming and scattered light are denoted by ν_o, k_o and ν_s, k_s respectively; $\nu = \nu_o - \nu_s$ stands for the frequency shift and angular brackets for the equilibrium ensemble average.

Once a model form for the pair potential and pair polarizability anisotropy are given, theoretical calculations of the zeroth and second moment can easily be performed within the framework of classical mechanics, which is appropriate for this case at high temperature, by means of the standard expressions[3]:

$$M_o^{pair} = \int_{-\infty}^{+\infty} D_\|(\nu) \, d\nu \simeq \frac{2}{15} k_o^4 \int_V g(r) \, \beta^2(r) \, 4\pi r^2 \, dr \qquad (3)$$

$$M_2^{pair} = \int_{-\infty}^{+\infty} \nu^2 D_\|(\nu) d\nu \simeq \frac{4\,k_B T}{15\,m} k_o^4 \int_V g(r) \left[\left(\frac{d\beta(r)}{dr}\right)^2 + 6\frac{\beta^2(r)}{r^2}\right] 4\pi r^2 dr \qquad (4)$$

Here $\beta(r)$ is the polarizability anisotropy as a function of the interatomic distance r, $g(r)$ is the pair distribution function given by $g(r) = \exp\left[-U(r)/k_B T\right]$ where $U(r)$ is the pair potential, m is the atomic mass and we have approximated $k_s \sim k_o$.

For the sake of comparison and discussion, in the reported evaluations, we have used two models for the pair polarizability anisotropy: the first order DID model which is given by: $\beta_{DID}(r) = 6\alpha^2/r^3$, where α is the atomic polarizability and the all-order DID which was derived by Silberstein[13] by the resummation of the infinite number of terms of the dipolar series expansion and reads:

$$\beta_{all}(r) = \frac{6\alpha^2 r^3}{r^6 - \alpha r^3 - 2\alpha^2} \qquad (5)$$

The use of this form for the polarizability anisotropy is justified by the large ratio of α/r_{min}^3 for mercury compared to noble gases, i.e. $\alpha/r_{min}^3 \simeq 0.12$, 0.048, 0.038, 0.031 for mercury, xenon, krypton, argon respectively. Indeed this ratio determines the importance of the successive terms in the dipolar series which give substantial contributions in the case of mercury. Table 2 gives the comparison between the experimental moments and the calculation performed with the first order DID approximation and the all order DID anisotropy by using the following three potentials, i.e. the empirical Lennard-Jones model[14](LJ), the theoretical model derived by Baylis[15](BA), and the more recent Morse model derived from spectroscopic measurements[16](MSKAK).

First we notice, in both cases, that the differences in the values of the moments evaluated with the different potentials are minor. Therefore, even though the mercury potentials we have used, cannot be considered refined ones, nevertheless the large discrepancies between the theoretical and experimental values cannot be attributed to some deficiency in the potentials.

The inadequacy of both the models of the polarizability anisotropy for the correct description of the experimental results is clear from the comparison, even though the all order DID brings the values of the zeroth moments at the various temperatures closer to the experimental ones. In particular we notice that the ratio between the second and zeroth moments can be determined experimentally with good precision; its temperature behavior can therefore give a good test of the models we have used.

TABLE.2 Comparison between experimental and theoretical moments. For each potential model, the top value refers to DID approximation at first order, while the bottom to the all-order DID.

T(°C)	600	700	800	900
LJ				
M_0^{pair} (10^{-47}cm^5)	279	270	262	257
	370	358	349	343
$\frac{M_2^{pair}}{M_0^{pair}}$ (cm^2)	197	218	242	2.65
	275	310	345	381
BA				
M_0^{pair} (10^{-47}cm^5)	290	280	273	268
	388	376	368	363
$\frac{M_2^{pair}}{M_0^{pair}}$ (cm^2)	200	224	247	270
	286	324	363	336
MSKAK				
M_0^{pair} (10^{-47}cm^5)	264	258	253	250
	353	345	340	336
$\frac{M_2^{pair}}{M_0^{pair}}$ (cm^2)	199	222	245	268
	286	323	361	401
EXP				
M_0^{pair} (10^{-47}cm^5)	580±10	620±20	580±20	-
$\frac{M_2^{pair}}{M_0^{pair}}$ (10^2cm^2)	233±20	252±4	290±3	348±13

Both polarizability models give values for this ratio well outside the experimental uncertainties in all the explored temperature range. This result indicates that in the case of mercury an additional positive polarizability anisotropy at long and intermediate range must be added to the DID one in order to meet the experimental results for both zeroth and second moments and their temperature behavior. In fact, both moments must be largely increased to fit the experimental results within experimental uncertainties, while their ratio and temperature dependence have to be changed only slightly. This pair polarizability behavior is very different from that of noble gases where an empirical negative short range contribution, beside the DID one, was necessary, mostly, in order to give the correct theoretical values of the higher-order moments and high-frequency lineshapes.[3]

Following the previous considerations also in the case of mercury we derive an empirical pair polarizability anisotropy model. For the sake of simplicity we use the same empirical model suggested by Barocchi and Zoppi[17] which can be written as:

$$\beta_{emp}(r) = \frac{6\alpha^2}{r^3} + \frac{A}{r^6} + B\exp(-r/r_0) \qquad (6)$$

where B and r_0 are two free parameters and A is the coefficient of the second order DID approximation corrected for the hyperpolarizability contribution as indicated by Buckingham[18] and used by Meinader et al.[19] for the case of noble gases: $A = 6\alpha^3 + \frac{\gamma C_6}{3\alpha}$, where γ is the hyperpolarizability and C_6 is the first dispersion force coefficient. Since the hyperpolarizability γ and the parameter C_6, which enter in the determination of A, are not known for mercury, we have derived them by following a simple observation and extrapolation,[20–23] as explained by Barocchi et al..[8] For mercury this leads to a value of $A \simeq 1300$ Å^9. Within our hypothesis we can now test the empirical polarizability anisotropy model for mercury by means of the comparison between experimental and calculated values of the first two even moments of the DILS spectral intensity at various temperatures.

For the calculation of the theoretical values of the moments with the empirical polarizability anisotropy, we have used only the most recent pair potential, i.e. the MSKAK, because the details of the shape of the different reliable potential affects the values of the zeroth and second moments only slightly. In contrast, we have to reconcile large discrepancies presumably due to the polarizability model. Table 3 gives the results of the comparison between the experimental and theoretical moments where the values of the parameters of the empirical polarizability model are: $A = 1300$Å^9, $B = 37$Å^3 and $r_0 = 1.2$ Å.

TABLE 3. Comparison between experimental and theoretical moments evaluated with the empirical polarizability model.

	T (°C)	600	700	800	900
MSKAK	M_0^{pair} (10^{-47}cm^5)	601	594	583	575
	$\frac{M_2^{pair}}{M_0^{pair}}$ (cm^{-2})	242	267	297	325
EXP	M_0^{pair} (10^{-47}cm^5)	580 ± 10	620 ± 20	580 ± 20	–
	$\frac{M_2^{pair}}{M_0^{pair}}$ (cm^{-2})	233 ± 20	252 ± 4	290 ± 3	348 ± 13

As one can see from Table 3 the agreement between the experimental values and the theoretical ones is quite good and we have verified that it remains acceptable for $30 < B < 50$ Å^3 and $r_0 = 1.2 \pm 0.2$ Å. Also the temperature behavior of the moment ratio is satisfactorily reproduced.

This indicates the breakdown of the simple point dipole approximation for mercury at intermediate range and the onset of specific interactions in the electronic properties. This makes mercury pair interaction properties not completely van der Waals in nature and therefore quite different from noble gases and globular molecular systems.

III. The isotropic spectrum.

Most investigations, like the one reported in the previous section, have been devoted to study the anisotropic part of the pair Interaction Induced (I-I) polarizability tensor.[1,24] The study of the I-I isotropic polarizability, i.e. of the trace of the I-I polarizability tensor, is a much more difficult task, both experimentally and theoretically. Indeed experimentally the I-I polarizability trace contributes to the CILS polarized spectrum only marginally, the corresponding anisotropic part being usually responsible for most of the spectral integrated intensity. The first experimental determination of the spectrum due to the I-I isotropic polarizability was performed in noble gases by Proffitt, Keto and Frommhold[25] and in molecular system by De Santis and Sampoli.[26]

Experimentally, the signal due to the polarizability trace, which from now on we call 'isotropic spectrum',[27] is derived from a small difference of two already weak signals, which have to be calibrated relatively to each other with high precision. This brought large uncertainties from 25% to more than 100% on the spectral intensity of noble gases.

Theoretically, the difficulties in the calculation come mainly from large cancellations between the various contributions to the trace. In turn, this fact makes the calculations of the trace both a theoretical challenge and a stringent test of the employed wave functions respectively. The empirical and theoretical models for the I-I polarizability trace of noble gases from helium to xenon have extensively been discussed by Dacre and Frommhold.[28]

Here we report the measurements of the collision induced isotropic light scattering spectrum of mercury diatom in the range $6 \div 70$ cm^{-1}, and we describe an empirical model for the I-I polarizability trace of mercury. This type of high temperature spectroscopy can be applied to several metal vapors and is the only experimental method suitable to study the I-I trace of metal diatom. The experiment has been performed with the same apparatus described before. The only differences worth to be mentioned are: *i)* the use of an image rotator in order to rotate the polarization of the incoming beam in the two required geometries, i.e. either normal to the scattering plane or along the collecting optical axis; *ii)* the addition inside the sample cell of a small amount of nitrogen for calibrating both CILS spectra against the rotational spectrum of nitrogen.

The scattered light was selected always with vertical polarization by means of a polaroid sheet. The isotropic spectrum, $I_{iso}(\nu)$, is derived from the measured spectra by:

$$I_{iso}(\nu) = I_{VV}(\nu) - \frac{4}{3}I_{HV}(\nu) \tag{7}$$

where $I_{VV}(\nu)$ and $I_{HV}(\nu)$ are the scattered intensities when the laser beam is polarized vertically and horizontally respectively. The VV and HV intensities of the mercury have been measured separately at T=700±2 °C and three different densities, namely 0.065, 0.113, 0.141 atoms/nm³; the intensities are found to be quadratic in density within experimental uncertainties indicating that all the spectra are to be attributed to I-I diatomic polarizability.

The main difficulty for precise measurements of the isotropic signal is caused by the change in the optical alignment when the incoming polarization is rotated from vertical to horizontal direction. These changes may alter spuriously the weak isotropic contribution. In order to control as much as possible these modifications we have chosen to calibrate both spectra with an internal standard. To this purpose we have recorded in the same run the induced spectra of mercury and the rotational spectrum of nitrogen. In this way we have been able to calibrate the VV and HV spectra from the knowledge of the absolute intensity of the nitrogen rotational lines (the polarizability anisotropy of nitrogen[29] being $\beta_{N_2} = 0.69$ Å³) and to ascertain that no alignment change was detectable with our apparatus, since the ratio of the integrated intensity of each rotational lines between the two polarizations was equal to the theoretical value 3/4 within 1%.

Fig. 4 shows the Stokes side of the $I_{VV}(\nu)$ and $I_{HV}(\nu)$ measured spectra, together with the evaluated $I_{iso}(\nu)$ experimental spectrum, in logarithmic scale for $\rho_{Hg} = 0.113$ atoms/nm³ and $\rho_{N_2} \approx \rho_{Hg}/3$ molecules/nm³. For clarity I_{HV} has been multiplied by 4/3. It is evident that at high frequency, where the nitrogen contribution is dominant in both spectra, I_{VV} and $\frac{4}{3}I_{HV}$ superimpose each other within the experimental uncertainties and the I_{iso} spectrum is practically vanishing at frequencies higher than 70 cm⁻¹. For frequencies below 70 cm⁻¹ the clear difference between I_{VV} and $\frac{4}{3}I_{HV}$ manifests the presence of a collision induced isotropic contribution. The rotational lines disappears in the obtained I_{iso} spectrum between 6 and 70 cm⁻¹ within the error bars.

The precise absolute calibration of pair spectrum I_{iso} has been determined by comparison with the depolarized spectrum[7] and controlled with the intensity of the high frequency rotational lines of nitrogen.

The values of the experimental zeroth (M_o^{iso}) and second (M_n^{iso}) moments of I_{iso} have been calculated in the same way as for the DILS spectra and are given in table 4. $I_{iso}(\nu)$ is related to the double differential scattering cross section for the diatom and is given by the theoretical expression[2]:

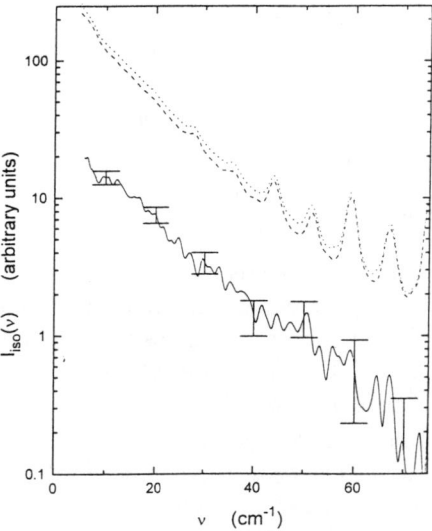

Figure 4: Experimental spectra: I_{VV} (dotted line), I_{VH} (dashed line), I_{iso} (solid line). In I_{iso} the estimated error bars are reported.

$$I_{iso}(\nu) = V\frac{d^2\sigma_{iso}}{d\Omega\,d\nu}$$
$$= V k_o k_s^3 \int_{-\infty}^{+\infty} dt\,\exp(-i\,2\pi\nu t)\,\langle a\,(r\,(t))\,a\,(r\,(0))\rangle \quad (8)$$

where $a(r(t))$ stands for the trace of I-I polarizability of the pair at distance r and time t.

Once a model form for the pair potential and pair polarizability trace are given, theoretical calculations of zeroth and second moments can easily be performed by:

$$M_o^{iso} = \int_{-\infty}^{+\infty} I_{iso}(\nu)\,d\nu \simeq k_o^4 \int_V g(r)\,a^2\,(r)\,4\pi r^2\,dr \quad (9)$$

$$M_2^{iso} = \int_{-\infty}^{+\infty} \nu^2 I_{iso}(\nu) d\nu \simeq \frac{2\,k_B T}{m} k_o^4 \int_V g(r) \left(\frac{da\,(r)}{dr}\right)^2 4\pi r^2 dr \quad (10)$$

where also in this case we have approximated $k_s \sim k_o$.

For the sake of comparison and discussion, we have first calculated the zeroth and second moments of $I_{iso}(\nu)$ by using eqs.(9,10) and two elementary models of the pair polarizability trace, i.e. the lowest order and all-order DID approximation[13] plus the hyperpolarizability term[18] given by:

$$a(r) = \left(4\alpha^3 + 5\gamma C_6/9\alpha\right)/r^6 \tag{11}$$

$$a(r) = \frac{4\alpha^3}{r^6 - \alpha r^3 - 2\alpha^2} + \frac{5\gamma C_6}{9\alpha r^6} \tag{12}$$

Table 4 gives the comparison between the experimental moments of $I_{iso}(\nu)$ derived from the measured spectrum and the moments calculated by using the two expressions for the polarizability trace eqs.(11,12) and eqs.(9,10), with $\gamma \simeq 8 \times 10^{12}$Å^{6}erg^{-1} ($1 \times 10^{-60}$ C^{4}m^{4}J^{-3}) and $C_6 \simeq 4 \times 10^{-10}$Å^{6}erg ($4 \times 10^{-77}$ m^{6}J)[8] and the MSKAK pair potential.[16]

The inadequacy of both models of the trace for the correct description of the experimental results is clear from the comparison, mainly because these models do not reproduce the normalized second moment $M_{2,n} = M_2/M_0$, which can be determined experimentally with good precision, not being affected by calibration errors. Both models give $M_{2,n}$ values much higher than the experimental ones and well outside the experimental uncertainties. The experimental value of M_0 being about 20% smaller than the lowest order DID one, indicates that a negative short range term must be added to $a(r)$ to reconcile the calculations with the experiment. This is similarly to what happens to the induced anisotropy of noble gases.[19]

TABLE 4. Experimental and theoretical moments of the isotropic spectrum

	M_0^{iso} ($10^{-47}cm^5$)	M_2^{iso}/M_0^{iso} (cm^{-2})
EXP	200 ± 3	458 ± 10
$DID - first$	237	697
$DID - all$	322	904
$empirical$	199	457

In agreement with previous works[2] we have adopted the following empirical model of the trace of mercury diatom:

$$a(r) = \left(4\alpha^3 + 5\gamma C_6/9\alpha\right)/r^6 - B\exp(-r/r_0) \tag{13}$$

and determined the parameters B and r_0 by fitting the calculated M_0 and M_2 to the experimental values. The best fit gives: $B \approx 7 \times 10^6$ Å^3 and $r_0 \approx 0.18$

Å. The moments corresponding to the empirical polarizability trace are also reported in Table 4.

Similarly to what happens for the polarizability anisotropy, also the trace behaves quite differently from the traces of noble gases which all become negative at a distance $r \sim r_m$.[30] In the case of mercury the trace is large and positive down to $\sim 0.8\ r_m$, indicating the existence of a contribution due to strong electron correlation effects in the overlap region, which is known to be positive for both trace and anisotropy,[30] thus confirming what already found for the I-I polarizability anisotropy of mercury diatom.

IV. The dimer spectra

The bound dimer band in the DILS spectrum is recognized as a shoulder at relatively low frequency shift. Dimer features in DILS spectra have been revealed for different atomic and molecular gases. It is known since a long time that the importance of the contribution of bound dimers is critically dependent on the reduced temperature,[31] $T^* = k_B T/\varepsilon$, where ε is the interaction potential well depth, indeed this controls the population of the bound states of the atomic pair. For mercury vapor around 500 C the reduced temperature is small enough to make the dimer population relatively high. The dimer band should then be observable, provided a high resolution and high contrast instrument is used. To this aim, we have performed a series of measurements of the mercury DILS spectrum using the high resolution high contrast SOPRA monochromator.[32] Four series of measurements, at temperatures 480, 500, 520, and 580 °C have been done. For each temperature two or three different densities were investigated. This density analysis has permitted to demonstrate that all the measured intensity can be ascribed to the pair spectrum. Spectra has been recorded in the frequency range 0.2-30 cm^{-1}, with a resolution of 0.1 cm^{-1}. In all cases the dimer band is very well determined, and the DILS spectrum shows a maximum at about 4 cm^{-1}. The experimental spectra are shown in Fig. 5.

We have calculated the mercury DILS spectra assuming different pair potential and polarizability models, both in classical and quantum mechanical approximation. The results of these calculations show that the shape of the dimer band behaves differently from that of the translational band which, in general, is dependent on both the pair potential and polarizability in a way that the effects of the two physical properties cannot be separated. The shape of the dimer band is strongly dependent on the pair potential only, while the intensity depends essentially on the pair polarizability.

This fact is not surprising if one think that the dimer is bound at an almost fixed distance, at least if compared with the large distance range where

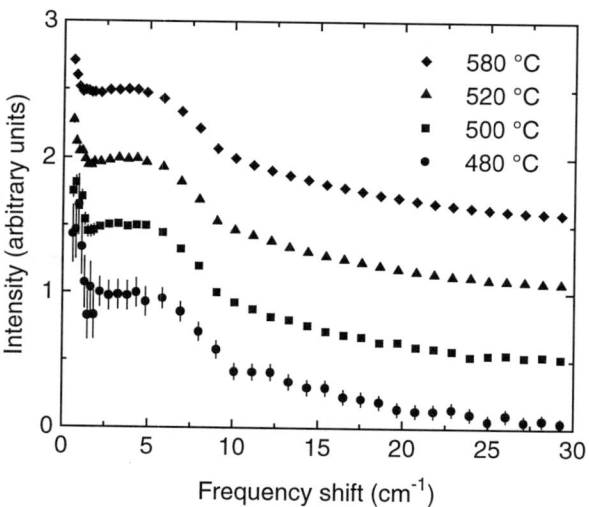

Figure 5: Experimental spectra showing the dimer band; for convenience the spectra are translated on the vertical scale and divided by the square of the density.

a free pair of atoms has a non vanishing variation of the polarizability. Therefore it is the constant value of the polarizability anisotropy at that distance that determines the intensity. As an example of this fact we report in Fig.6a the dimer band calculated with the same potential but different polarizability models. Once the two band are scaled by a suitable factor, they superimpose each other. On the other hand, the spectra obtained with different potentials are different mainly in the frequency extension of the pure rotational dimer band. The extension is very well determined, since it appears as an evident change of slope in the DILS experimental spectra. In principle, the precise knowledge of the shape of the band would allow a direct inversion, to obtain the potential well.[33] In our case the procedure cannot be so straightforward, since this band is superimposed to the translational spectrum, which can be calculated with enough precision only after having modelled both the pair potential and polarizability. To compare the shape of our spectra with the calculated ones, we have normalized these latter in order to match the experimental results in the frequency range 15-30 cm^{-1}. As an example, we report in Fig.6b the comparison performed at 580°C, using the DID model for the pair polarizability. The three potential used are referred in the figure legend as LJ,[14] BA,[15] MSKAK.[16]

As is apparent, all the investigated potentials seem not to reproduce the

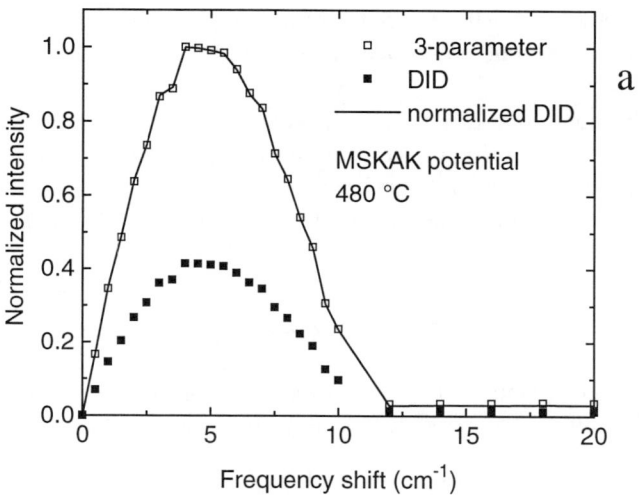

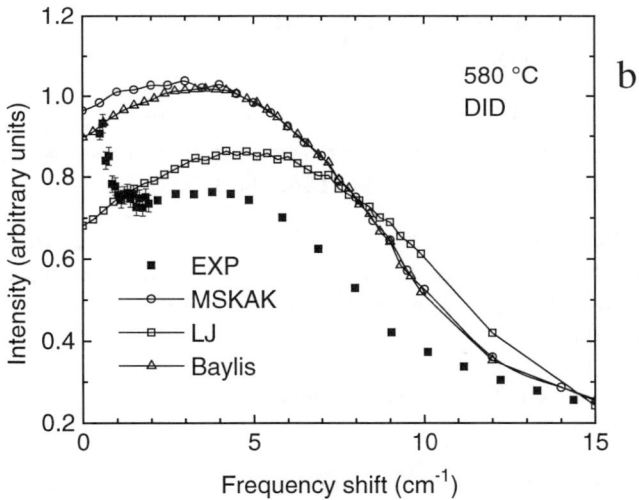

Figure 6: (a) Effect of canging the polarizability model on the dimer band: full square, DID model; open square, empirical model (eq.6); the solid line represents the DID results suitably scaled. (b) Comparison between theoretical and experimental spectra in the dimer band region.

shape of the dimer band with respect to its frequency extension, indicating a specific deficiency of present potentials, at least in the region of the potential well.

As a concluding remark, we want to stress that measurements at low reduced temperature and at high resolution of the dimer band is as a promising tool for the check, and perhaps determination, of potential models, especially in metal vapors, where information from other sources are difficult to obtain.

REFERENCES

1. See, for example, *Phenomena Induced by intermolecular Interactions*, ed. Birnbaum G. (Plenum, New York, 1985), and references quoted therein

2. Frommhold L. *Adv.Chem.Phys.*, **46**,1 (1981)

3. Barocchi F. and Zoppi M., in *Phenomena Induced by intermolecular Interactions*, ed. Birnbaum G. (Plenum, New York, 1985), p. 311 and references quoted therein

4. Sampoli M., Guasti A., Barocchi F., Winter R., Rathenow J., Hensel F., *Phys.Rev.* **A42**, 6910 (1992)

5. Barocchi F., Sampoli M., Hensel F., Rathenow J., Winter R., in *Proceeding of the Eighth International Conference on Liquid and Amorphous Metals*, J. Hafner editor, *J.Non-Crystal.Solids*, **156-158**, 663 (1993)

6. Barocchi F., Sampoli M., Hensel F., Rathenow J., Winter R., in Proceeding of the Nato Advanced Research Workshop on *Collision- and Interaction-Induced Spectroscopy* Ed. by Tabisz G.C. and Neuman M.N., (Kluwer Academic Publisher, 1995)

7. Sampoli M., Barocchi F., Grassi L., Hensel F., Rathenow J., *Europhys. Lett.* **28**, 483 (1994)

8. Barocchi F., Hensel F., Sampoli M., *Chem. Phys. Lett.* **232**, 445 (1995)

9. Pyykko P. and Desclaux J.P., *Acc.Chem. Res.*, **12**, 276 (1979)

10. Schoenherr G., *Dissertation*, University of Marburg, Germany (1978)

11. Bridge N.J. and Buckingham A.D., *Proc.Roy.Soc.* **A295**, 334 (1966)

12. Lewis Ford A. and Browne J.C., *Phys. Rev.* **A7**, 418 (1973)

13. Silberstein L., *Phil. Mag.* **33**, 521 (1917)

14. Epstein L.F. and Powers M.D., *J.Phys.Chem.* **336** , 57 (1953)

15. Baylis W.E., *J.Phys.B: At.Mol.Phys.* **10**, L583 (1977)

16. Koperski J., Atkinson J.B., Krause L., *Chem. Phys. Lett.* **219**, 161 (1994)

17. Barocchi F. and Zoppi M., *J.Chem.Phys.* **65** , 901 (1976)

18. Buckingham A.D., *Trans. Faraday Soc.* **52**, 1035 (1956)

19. Meinander N., Tabisz G.C., Zoppi M., *J. Chem. Phys.* **84**, 3005 (1986)

20. Dummur D.A., Hunt D.C., Jessup N.E., *Mol.Phys.*, **37**, 713 (1979)

21. Buckingham A.D. and Clarke K.L., *Chem. Phys.Lett.*, **57**, 321 (1978)

22. Tang K.T., Norbeck J.M., Certain P.R., *J. Chem. Phys.* **64**, 3063 (1976)

23. Thomas J.F. and Meath W.J., *Mol.Phys.*, **34**, 113 (1977)

24. For a recent review see *Collision- and Interaction-Induced Spectroscopy* Ed. by Tabisz G.C. and Neuman M.N., (Kluwer Academic Publisher, 1995)

25. Proffitt M.H. , Keto J.W., Frommhold L., *Phys. Rev. Lett.* **45**, 1843 (1980)

26. De Santis A. and Sampoli M., *Chem. Phys. Lett.* **96**, 114 (1983)

27. In previous works on noble gases,[25,28] the spectrum due to the I-I polarizability trace has been called 'polarized spectrum'. Here 'polarized spectrum' stands for the scattered intensity with polarization equal to the one of incoming beam.

28. Dacre P.D. and Frommhold L., *J. Chem. Phys.* **76**, 3447 (1982)

29. ·Penny C.M., Peters S.T., Lapp M., *J. Opt. Soc. America* **64**, 712 (1974); Bridge N.J. and Buckingham A.D., *Proc. Roy. Soc.* **A295**, 334 (1968)

30. Dacre P.D., *Mol. Phys.*, **45**, 1 (1982); *ibidem* **45**, 17 (1982), *ibidem* **47**, 193 (1982)

31. Levine H.B., *J. Chem. Phys.* **56**, 2455 (1972)

32. Mazzacurati V., Benassi P., Ruocco G., *J. Phys. E Sci.Instrum.* **21**, 798 (1988)

33. Maitland G.C., Rigby M., Smith E.B., Wakeham W.A., *Intermolecular Forces* (Clarendon Press, Oxford, 1981) p. 396

Collision-induced Emission and Sonoluminescence

Lothar Frommhold[a] and Wilfried Meyer[b]

[a]*Physics Department, University of Texas at Austin, Texas 78712-1081.*
[b]*Fachbereich Chemie, Universität, D-67661 Kaiserslautern, Germany*

It is well known that infrared inactive gases such as hydrogen and nitrogen are capable of absorbing infrared radiation efficiently if the gas densities are sufficiently high. The absorption is due to dipole moments induced by exchange, overlap, and dispersion forces, and/or multipolar induction, in pairs (triples, etc.) of interacting molecules. These collision-induced dipoles also *emit* radiation in the infrared and, at sufficiently high temperatures, in the visible and near ultraviolet as well. We estimate the shape and intensity of the binary emission spectra of nitrogen and nitrogen-argon mixtures, along with the opacities of such sources. At densities of several times 10^{21} molecules per cubic centimeter and temperatures in the ten thousand kelvin range, the computed spectra match the spectral profiles and emission intensities of sonoluminescence in air, nitrogen and nitrogen-argon mixtures, etc., at wavelengths from 200 to 700 nm. The work suggests that the radiation source is optically thin. Detailed *ab initio* calculation of collision-induced emission spectra are being undertaken and first results are shown.

COLLISION-INDUCED EMISSION

It is well known that in collisional interaction a system of two or more neutral particles $X, Y, \ldots$ may acquire new properties, in addition to the simple sum of properties of the non-interacting (i.e., well separated) particles. These new properties we will call the interaction-induced or supramolecular properties. One case in point is the collision-induced dipole moment observable in all molecular gases and even in mixtures of monatomic gases, which at high gas densities gives rise to the familiar collision-induced absorption (CIA) spectra in the infrared — even if the individual particles of the gas are infrared-inactive. A substantial body of experimental and theoretical work exists and a deep understanding of such supramolecular absorption has been obtained, certainly for interactions of the simpler atoms (He, Ar,...) and/or molecules (H_2, N_2, etc.) at not too high gas densities (e.g., up to a fraction of liquid densities) [1]. Collision-induced dipoles are weak compared to the dipoles of polar molecules such as HCl, typically one hundredth or less of an atomic unit $e \cdot a_0$ if thermal collisions are considered. Nevertheless, in gases composed of non-polar molecules, at the higher gas densities, CIA is most striking because the absorption increases as density squared (with cubic contributions if enough ternary interactions occur), while the absorption by polar molecules increases only linearly with density. High temperatures also enhance the interaction-induced dipole strengths significantly, because exchange force-induced dipoles are exponential functions of the separation of the collisional pair: with decreasing separation the dipoles increase rapidly — just like the intermolecular repulsive forces.

Without doubt, interaction-induced dipoles also emit radiation. For the nonpolar gases (e.g. Jupiter, with its atmosphere of hydrogen and helium), the thermal radiation in the far infrared is of a supramolecular origin. More relevant here is the emission of hot, dense and basically neutral matter into a cooler environment, e.g., the atmospheres of cool stars and planets, certain melts and types of flames, rocket jets, shockwaves, etc., must be more or less of the interaction-induced type, depending on the circumstances. Detailed laboratory investigations of collision-induced emission (CIE) from such sources are, however, not known — with a single exception: using linear shocktubes Caledonia et al. [2] observed strong emission in the H_2 fundamental band from the shockwave-heated mixture of hydrogen and argon. Since emission in the H_2 fundamental band is forbidden in the non-interacting H_2 molecule and a quadratic density dependence of the intensity was observed, the emission must be of a supramolecular (mostly binary) origin. The spectral profile was modeled successfully with the techniques commonly employed in CIA work.

The present work is concerned with more or less detailed computations of collision-induced emission spectra for cases of practical interest. Collision-induced absorption spectra have long been computed from first principles [1, 3]. The comparison of such calculations with the finest measurements of CIA spectra showed agreement within the small uncertainties of the measurements. In other words, ample evidence exists that the supramolecular spectra of simple pairs may dependably be predicted from theory [1]. The prediction of CIE spectra should be no exception, even if the computation of emission spectra is in general more complex because of the high temperatures.

SONOLUMINESCENCE

Sonoluminescence (SL) generates light from sound. It is a long familiar phenomenon which has recently excited renewed interest as a device for creating inexpensively very hot and dense environments. A single, stable speck of light may be maintained in water for many hours in acoustic cavitation fields of spherical symmetry (with the so-called single bubble arrangement [4]). When the pressure amplitude of sound waves in water somewhat exceeds one atmosphere, in the tensile phase of the sound wave, cavitation occurs. The bubbles are roughly spherical, of an initial radius of 50 μm or so, and are filled with air, but other gases may be substituted by suitable techniques [5, 6]. In the compression phase of the sound wave, the bubble collapses with such a speed that spherical, convergent shockwaves are thought to be driven in the gas [7]. At the center of the bubble the shock is reflected and a surprisingly strong light emission (sonoluminescence) is observed for a short time ($\approx 10^{-12}$ seconds) from a small region (< 1 μm; perhaps smaller) at the bubble center. Densities around 4×10^{21} cm^{-3} and temperatures up to millions of kelvin have been seriously proposed on various occasions, but more conservative estimates based on hydrodynamic modeling of the shockwave suggest temperatures around 7000 K. This high concentration of energy at the center of the bubble is experimentally attained using mostly standard, low-cost laboratory equipment [8].

In striking contrast to the ease of producing sonoluminescence in the laboratory is the difficulty of understanding the nature of the emission processes. There seems to be some consensus (at least in general terms) about the role of the imploding and reflected shockwaves, but the origin of the emitted light is not at all clear. The emission is widely considered to arise from bremsstrahlung in the (hypothetical) high-temperature plasma which is usually modeled as a black-body source. Several alternative mechanisms have been proposed in greater or lesser detail in recent years; see the concluding remarks below. We think that sonoluminescence is collision-induced emission from a basically neutral environment. If this is true, the source is not at all optically dense [9] so that Planck's radiation law, upon which most analyses of the sonoluminescence spectra were based, must not be used for analyses of the emission process.

Previous work [9, 10] suggests that the observed emission of light is of a supramolecular origin, arising from fluctuating dipoles induced by molecular interactions (collisions). An extensive body of work exists both on interaction-induced dipoles and the associated spectra, which has been collected in a recent monograph [1]. Collision-induced emission, as the process is called, is expected to be significant virtually in all hot gases and gas mixtures, if the gas densities are sufficiently high, e.g., tens or hundreds of amagats. (One amagat roughly equals a density of 2.68675×10^{19} molecules per cm^3.) Moreover, CIE is also known from studies of certain stellar and planetary atmospheres [11, 12].

CIE LINE SHAPE CALCULATIONS

If the dipole and interaction potential of a given molecular pair are known, absorption and emission spectra can in principle be computed. Most computations of binary spectra assume isotropic interaction, which makes quantum calculations of massive systems with many open channels possible. The absorption coefficient α at temperature T and angular frequency $\omega = 2\pi\nu$ may be written [1]

$$\alpha(\omega;T) = \frac{4\pi^2}{3\hbar nc} N_L^2 \, \varrho_1 \, \varrho_2 \, \omega \left[1 - \exp\left(-\frac{\hbar\omega}{kT}\right)\right] V g_a(\omega;T) \qquad (1)$$

where $N_L \, \varrho_1$ and $N_L \, \varrho_2$ are the number densities of the molecules N_2 and X; the ϱ_i are densities in amagat units; V is the volume; $N_L = 2.68675 \times 10^{19}$ cm^{-3} is Loschmidt's number; n is the refractive index; and $g_a(\omega;T)$ is the spectral density,

$$g_a(\omega;T) = \sum_{s,s'} P_s \sum_{t,t'} P_t \, |\langle t|\boldsymbol{\mu}_{s,s'}|t'\rangle|^2 \, \delta(\omega_{s,s'} + \omega_{t,t'} - \omega) \, . \qquad (2)$$

Here, $\boldsymbol{\mu}_{s,s'}$ is the rotovibrational matrix element given by $\langle s|\boldsymbol{\mu}|s'\rangle$; the subscript $s = \{v_1, j_1, v_2, j_2\}$ denotes the rotational and vibrational quantum numbers of the interacting molecules; the matrix elements $\langle t|\boldsymbol{\mu}_{s,s'}|t'\rangle$ of the electric dipole moment $\boldsymbol{\mu}$ between initial and final translational states $|t\rangle$, $|t'\rangle$ of the pair; Bohr's frequency conditions of the well separated collisional partners are $\hbar\omega_{s,s'} = E_{s'} - E_s$ and $\hbar\omega_{t,t'} = E_{t'} - E_t$, where E_t is the kinetic energy of relative motion of the collisional partners and E_s is the sum of the rotovibrational energies of the two molecules in collisional interaction; the P_t and P_s are the normalized Boltzmann factors of translational and rotovibrational state populations, respectively; thermal equilibrium is here assumed for lack of better information. The total energy is $E_s + E_t$ and for absorption the initial energy must be less than the final energy. Further details may be found elsewhere [1]. We also note that $1/\alpha$ is the mean free path for the absorption of a photon of frequency ω. If $1/\alpha$ is much greater than the size of the source, we may talk about an "optically thin" source, and *vice versa*.

Similarly, for emission we may write

$$I(\omega;T) \, d\omega = \frac{4\omega^4}{3c^3} N_L^2 \, \varrho_1 \, \varrho_2 \, V g_e(\omega;T) \, d\omega \, . \qquad (3)$$

The difference of the spectral density, $g_e(\omega;T)$, relative to Eq. (2), is that now the initial energy $E_s + E_t$ must be *greater* than the final energy $E_{s'} + E_{t'}$. In other words, frequencies now reverse sign and the δ function in (2) should now read $\delta(\omega_{s,s'} + \omega_{t,t'} + \omega)$; everything else remains the same. The expression (3) is valid if the source is optically thin.

The emission model considered previously [9, 10] is that of a nitrogen molecule N_2 interacting with X, an atom (e.g., Ar) or another molecule (e.g.,

N_2, O_2, H_2O, etc.),

$$N_2^* + X^{(*)} \to N_2 + X + \hbar\omega + \Delta E_t.$$

Such collisional systems are representative for shocks in air, pure nitrogen, nitrogen argon mixtures, etc. The interacting pair N_2–X (a supramolecular quantal system) possesses rotovibrational, translational, and perhaps electronic energies, which may be fully or partially converted into photons $\hbar\omega$ with the help of interaction-induced dipoles. CIE spectra occur as very diffuse bands at the energetically accessible rotovibrational (and electronic) transition frequencies of N_2 and X, and also at sums (differences) of such frequencies if X is a molecule [1].

The computation of the CIE spectra requires as input the intermolecular potential and dipole surfaces, $V(R, r_1, r_2)$ and $A(R, r_1, r_2; \lambda L)$, which are functions of the intermolecular separation (R) and the vibrational coordinates (r_1, r_2) of the molecules N_2 and X; the λ and L parameters reflect molecular and dipole symmetries [13, 3].

Quantal computer codes based on the isotropic interaction approximation exist for the computation of such profiles [1]. Simple, analytical functions, $\Gamma(\omega; T)$, are known which approximate these quantum profiles well [1] so that the complex quantum computations may often be avoided. The better model profiles have three parameters which may be readily related to the three spectral moments of lowest order [1]. The spectra are thus easy to compute if the dipole surface and the isotropic interaction potential are known.

A crude model. For N_2–N_2 and similar pairs, only a very limited subset of the required information is presently available. For example, instead of $V(R, r_1, r_2)$ only a few vibrational matrix elements $\langle v|V|v\rangle$ are known for the smallest vibrational quantum numbers $v = 0, 1 \ldots$ A similar statement can be made for the spherical dipole tensor components $B_{\lambda L}^{v'v} = \langle v'|A_{\lambda L}|v\rangle$. In other words, for a first estimate of the CIE spectra of N_2–X pairs, one must somehow estimate the vibrational matrix elements of the dipole and potential surfaces. Our previous work [9, 10] was based on the isotropic part of the N_2–N_2 interaction potential for molecules in the vibrational groundstate ($v = 0$), which we have used regardless of the actual vibrational excitation of the N_2 molecules. For the dipole matrix elements, we simply took a conservative average of the overlap-induced dipole components with $\lambda, L = 0, 1$ of other, better known systems,

$$B_{01}^{v'v}(R) = b\, \exp[-a(R - \sigma)], \qquad (4)$$

with $a = 10/\sigma$, $b = 1 \times 10^{-3}$ a.u., again assumed to be independent of v, v', and $\sigma = 7.70$ bohr, for lack of better data. Electronic excitations of the N_2 molecules are neglected. Spectral profiles of collision-induced emission may

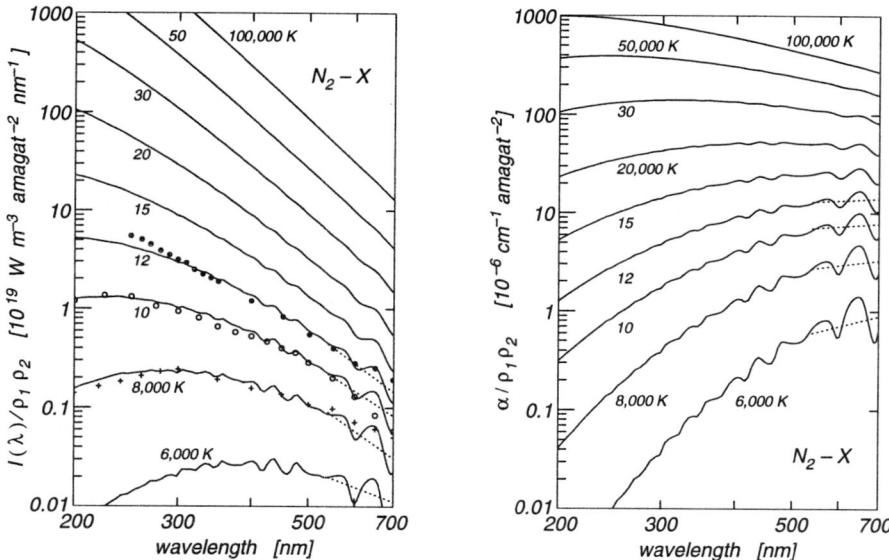

Figure 1: *Left:* Crude estimates of the normalized collision-induced emission spectra of the N_2–X system (single transitions) at nine temperatures, from 6,000 to 100,000 K (solid lines). Representative measured spectral profiles are indicated by dots and circles. *Right:* The normalized absorption coefficient at the same nine temperatures. The normalization factor is the product of densities ρ_1, ρ_2 of N_2 and X. (The structures seen at the lower temperatures and long wavelengths are artifacts of the calculation.)

thus be computed from Eq. (3). In our calculations we consider nine temperatures ranging from six thousand to hundred thousand kelvin, Fig. 1.

Figure 1 (left panel) shows that the profiles obtained (solid lines) reproduce existing measurements (dots and circles) of sonoluminescence in air and nitrogen bubbles [14, 6] quite well at temperatures around 10,000 K [9]. One may distinguish three types of profiles: type one has a clear maximum in the wavelength range shown; type two shows some curvature in the logarithmic plot but maximum intensity seems to occur at wavelengths smaller than 200 nm; and type three looks alsmost like a straight line in the grid chosen. Type 1 is the characteristic profile at the low temperature end and type 3 is found at the highest temperatures, with type 2 occuring at intermediate temperatures. The calculated profiles are compared with three measured profiles in Fig. 1: the pluses (+) are for nitrogen mixed with 2% xenon [15] and matches closely a type 1 profile at 8,000 K. The circles (o) are for air bubbles [14] and correspond more or less to the type 2 profile at 10,000 K. The dots (•) are for nitrogen bubbles [6] and resemble type 3 at 15,000 K or perhaps higher.

We note that the experimental profiles were shifted freely up or down for easy comparison with the calculated profiles of similar shape, without regard to the actual intensity of the measurements. We will show next that the intensities of the measurements actually are consistent with the calculated spectra.

In order to compare the computed intensities I (which are normalized by the product of densities $\varrho_1 \varrho_2$ of N_2 and X) with the measurements (dots, circles, pluses; $\approx 10^7$ photons per flash), we need to know the densities ρ_1 and $\rho_2 \approx 100$ amagats each; the duration of a flash (≈ 1 ps); the size of the source (≈ 10 nm); and the repetition rate given by the sound frequency (≈ 10 kHz). At temperatures of about 10,000 K agreement with the measurements is observed: a simple integration yields 10^7 photons per pulse at 10,000 K in the wavelength intervall shown.

The right panel of Fig. 1 shows the computed, normalized absorption coefficient α. The mean free path of a photon is given by $\ell = 1/\alpha$. Using the same densities ρ_1, ρ_2 and the temperature of 10,000 K, we obtain a mean free path for absorption of photons of several centimeters. That value must be compared with the dimension of the source (roughly 10 nm). We conclude that the source is optically thin at the wavelengths shown. In other words: Planck's black-body radiation law used elsewhere for the analyses of sonoluminescence spectra is not valid here, according to our estimate.

Figure 1 represents the case of particle X not undergoing any rotovibrational transitions. X may be a rare gas atom or another nitrogen molecule. In the latter case the collisional partner may undergo a simultaneous transition [1]. That possibility has been accounted for and the results are shown in Fig. 2. For double transitions we have assumed a somewhat smaller dipole strength parameter, $b = 10^{-4}$ a.u., for a reason to be explained below. Simultaneous transitions enhance the emission at short wavelengths (left panel), especially if strongly bound molecules are considered [9, 10]. The right panel shows again the absorption coefficient. The mean free paths for photons in the source is now smaller, but still orders of magnitude greater than the source dimensions.

Generalization. Crude estimates of sonoluminescence spectral profiles and intensities, such as the ones shown in the figures above, are based on two pieces of input: the exponential dipole model, Eq. (4), and a potential function. For the results shown above, we used an isotropic Maitland-Smith potential model [16] of welldepth $\epsilon/k = 104$ K, position of the minimum $R_m = 7.70$ bohr, and the exponent $13 + 8(R/R_m - 1)$, independent of the vibrational excitations v_1, v_2 of the two N_2 molecules. (This potential is quite good for N_2 pairs in the vibrational groundstates, $v_1 = v_2 = 0$.) The potential $V_0(R)$ goes through 0 for $R = \sigma = R_m/1.1217$. Besides these dipole and potential models, the reduced mass of the collisional pair affects the results somewhat.

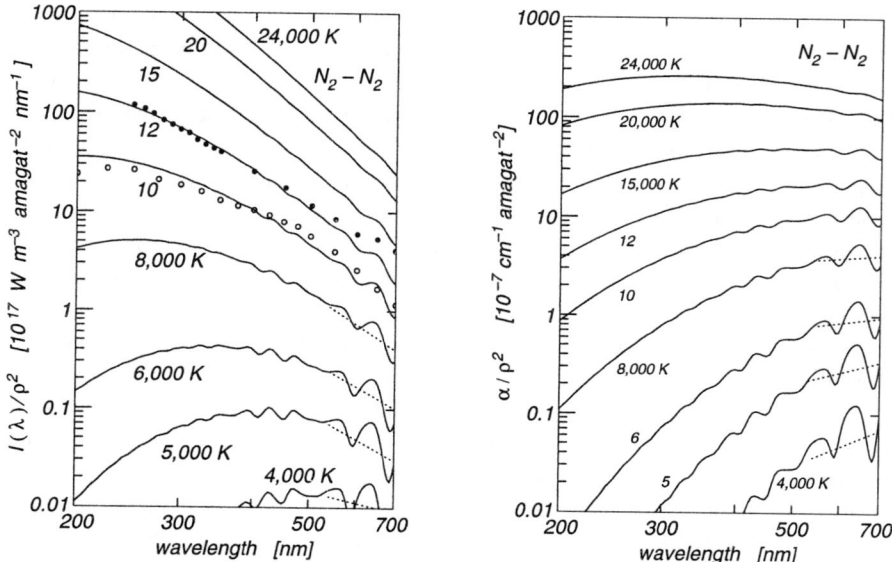

Figure 2: *Left:* Crude estimate of the normalized collision-induced emission spectra of the N_2–N_2 system (double transitions included) at nine temperatures from 6,000 to 100,000 K. Note the intensity scale which differs from that of Fig. 1, mostly because of the smaller dipole strength parameter b here assumed. Double transitions shift the maxima to higher frequencies. *Right:* The normalized absorption coefficient. The normalization factor is the square of the N_2 density, ρ^2.

These models of dipole and potential function are quite characteristic of collisional pairs commonly encountered in sonoluminescence studies, but the parameters are of course system specific. Of all the parameters used above, the dipole strength b and dipole range parameters a, Eq. (4), have the greatest effect on the computational results, while the reduced mass and potential parameters may be varied fairly widely before the results, Fig. 1, change dramatically. It is therefore worthwhile to point out that the estimates shown in the above figures may be considered representative for a variety of collisional pairs of interest in sonoluminescence studies, e.g., for X = Ar, N_2, ..., with one important correction to be discussed next.

In a certain sense, the dipole strength parameter b is a measure of the electronic asymmetry of the collisional pair. For example, the Ar–Ar pair has a zero dipole strength, $b = 0$, because of the inversion symmetry of the pair which is inconsistent with the existence of a dipole moment. Similarly, two nitrogen molecules in precisely the same rotovibrational (and electronic) states have a zero dipole moment of the type of Eq. (4) with $\lambda, L = 0, 1$.

(Such an N_2–N_2 complex may, however, have other dipole components; see [1] for details; present estimates show that the overall effects of these other dipole components is generally relatively weak at the high temperatures of interest here.) This situation of highly symmetric pairs must be contrasted with that of dissimilar pairs, e.g., N_2–Ar or N_2–Xe, where relatively strong dipoles must be expected under otherwise comparable conditions [1]. For that reason we have chosen a much smaller dipole strength parameter for Fig. 2, which is concerned with the 'similar' N_2–N_2 pairs (in different rotovibrational states.) This simple fact will also make plausible certain observations with gas mixtures in sonoluminescence studies that are difficult — if not impossible — to explain with the other competing theories of the emission process, e.g., black-body or quantum vacuum radiation. We note that the spectral intensities are proportional to the dipole strength parameter squared, $I \sim b^2$, so that the spectra of dissimilar pairs will always be more intense than the spectra of highly symmetric pairs, as long as the shockwave mechanism is not altered significantly by such additions — a fact which is most familiar in CIA, but was suppressed in our figures above that are computed with fixed b values.

Summarizing, one may say that the crude N_2–X CIE model is reasonably successful in reproducing the three types of observed spectral profiles, as well as the intensities, of sonoluminescence in air, nitrogen and nitrogen-rare gas mixtures. Qualitatively even the observed higher intensities in nitrogen with small admixtures of rare gases are plausible. At the same time, it is clear that a more quantitative treatment is desirable. Especially, our assumption of v, v'-independent dipole matrix elements must certainly be corrected eventually — along with the vibrational state dependence of the interaction potential that was neglected above.

Ab initio **work.** Improved treatments of the kind must resort to quantum chemical computations of van der Waals pairs. For the lighter systems, e.g., H_2–He, H_2–H_2, He–Ar, H_2–Ar, etc., very accurate potentials [17, 18] and interaction-induced dipole surfaces [19, 20, 3] exist. Such information was used to accurately predict collision-induced absorption spectra at low temperatures (< 300 K) [1]. We have extended our previous *ab initio* calculations of the H_2–He system so that higher temperatures may be considered, using a small basis set in a multireference-CI treatment. Specifically, we have computed potential and dipole moments at twenty different bond distances, from 0.6 to 4 bohr, eleven separations R from 2 to 6 bohr, and five orientational angles between 0 and 90°. Matrix elements $\langle v'|A_{\lambda L}|v\rangle$ and $\langle v|V_0|v\rangle$ are obtained and used for the line shape calculations. In this way, the crude assumptions of the N_2–X estimate above are avoidable. The assumption of isotropic intermolecular interactions is, however, still necessary at this time.

Figure 3 shows the results at two temperatures. In the figure, only the

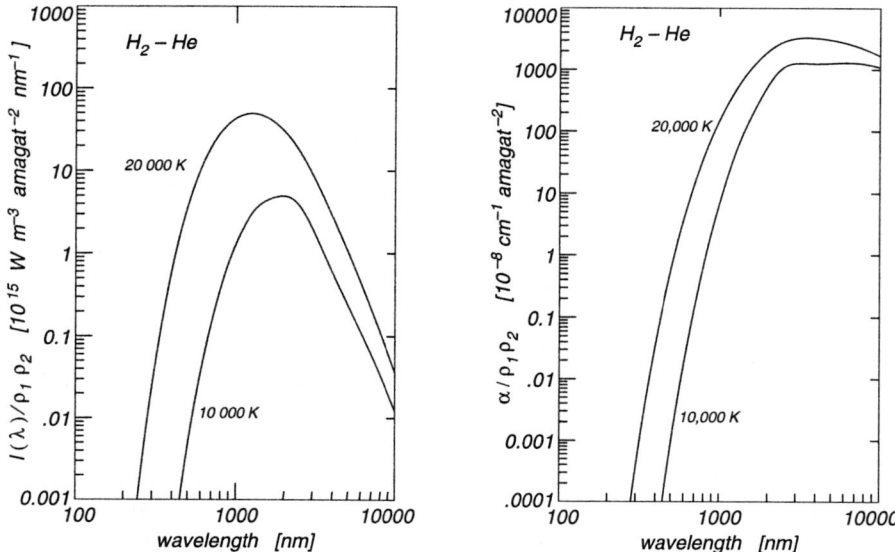

Figure 3: *Ab initio* computation of the collision-induced emission (*left*) and absorption (*right*) spectra of H_2–He pairs at three temperatures. We note that only the $\lambda, L = 0, 1$ dipole component is here accounted for.

main dipole component $\lambda, L = 0, 1$ was accounted for. Two other components ($\lambda, L = 2, 1$ and 2,3) have comparable intensities so that our theoretical prediction should actually be increased by roughly a factor of two, in order to reflect the over-all emission of that pair. Of the spectra shown, only the parts with wavelengths between 200 and 700 nm are observable for sonoluminescence in water, but larger portions of the spectrum should be observable in other liquids.

Unfortunately, as far as we know, no sonoluminescence spectra of bubbles filled with a pure hydrogen- helium mixture exist. However, we think that such data could be forthcoming in the future. Moreover, The H_2–He system may be considered a prototype of diatom-atom systems and we hope to obtain with this study some insight into the collision-induced emission process involving 'hot' rotovibrational bands, which are not yet very well known. We hope that the work will eventually be supplemented by a study of the H_2–H_2 system, thought to be a prototype of diatom-diatom systems. Furthermore, sonoluminescence of hydrogen bubbles is being studied elsewhere and we hope that something may be learned from the comparison of theory and measurement. Actually, the H_2–He system should exhibit some of the main features of the H_2–H_2 system, as well as of other systems, such as H_2–Ar, etc.

It is interesting to compare the results for N_2–X, Fig. 1, and H_2–He, Fig. 3.

For all the similarities the N_2–X and H_2–He systems may have, the main difference of interest here is the much greater dissociation energy of N_2 (≈ 10 eV) relative to that of H_2 (4.2 eV). For the emission of photons of 4 to 6 eV energy (i.e., 300 to 200 nm wavelength), vibrational excitations of such magnitude are needed but not available in H_2–He. This simple fact is the main reason for the small emission intensities at the short wavelengths of H_2–X systems. For H_2–H_2 and isotopes, some of that defect may be made up by double transitions (as in Fig. 2); detailed theoretical studies of H_2–H_2 pairs are however presently not available.

CONCLUDING REMARKS

We have argued that the basic emission process of sonoluminescence is probably collision-induced emission. If the spherical shockwave model is correct, on a picosecond time scale, conditions of high density and high temperature are created at the bubble center in a presumably mostly neutral environment, which strongly support collision-induced emission. At densities in the $\approx 5 \times 10^{21}$ cm^{-3} range, at a given temperature (say 10,000 K or so), CIE of optically thin sources is a most efficient emission (and cooling!) process in the visible region of the electromagnetic spectrum. In fact, under such conditions, the emission is much stronger than that of black bodies at the same temperature. This is possible because the sonoluminescence source is not in a thermodynamic equilibrium with the surroundings: for the duration of the flash kinetic energy flows from the translational and rotovibrational motions of the molecules into electromagnetic energy which is lost (radiated), but energy does not flow in the reverse direction so that no conflict exists with established thermodynamic principles.

It has occasionally been said that sonoluminescence in water is different from that in other liquids, e.g., organic liquids, even though a casual observer might not necessarily agree: the methods to generate sonoluminescence and the appearances are certainly very similar. The main difference is that in studies in water temperatures of 100,000 K or more were reported, but only 5,000 K in organic liquids [21, 22]. The temperature measurements in organic liquids are based on spectroscopic signatures of organic radicals and must be considered fairly reliable. We have presented arguments above that in sonoluminescence studies in water the temperatures amount to roughly 10,000 K — a number which is much more in line with the spectroscopic measurements in organic liquids. Moreover, according to the present work, the sonoluminescence source is optically thin at the frequencies of interest here, which rules out the use of Planck's radiation law. But above all it was the use of Planck's law which suggested the very high temperatures, which we thus consider uncertain. We

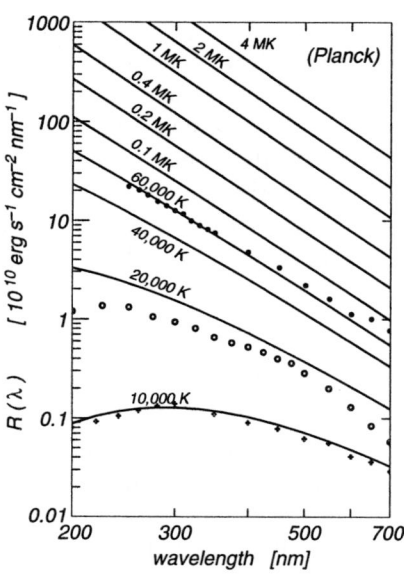

Figure 4: Planck's radiation law in the optical window of water. For comparison, the same measured profiles as in Fig. 1 are shown, as above shifted vertically so that the measurements more or less coincide with a similar Planck profile.

also mention that very simple hydrodynamic models of the shockwave mechanism typically suggest temperatures more in line with the lower ones just mentioned.

It is interesting to look at the part of Planck's black-body radiation law with wavelengths between 200 and 700 nm, the optical window of water, Fig. 4. Quite obviously, the three types of measured profiles are here, too, reasonably well represented by Planck's law, but again temperatures at the low end of the scale are necessary for the curved measured profiles, types 2 and 3. The high temperatures (>100,000 K) have been justified mainly on the basis of the known absolute intensities ($\approx 10^7$ photons per flash) — which is not correct if our CIE model of sonoluminescence indeed reflects the physics of the sonoluminescence spectra. We note that the slopes of those profiles that look nearly like straight lines in Figs. 1 (left panel) and 4 are very different, suggesting perhaps a possible experimental distinction between black-body radiation and CIE.

One of the more interesting alternative explanations of sonoluminescence is quantum vacuum radiation (QVR) by the accelerated water/air boundary

[23, 24, 25]. Eberlein shows that the QVR spectrum is given by an expression of the same form as Planck's radiation law. This fact reproduces most (but not all) of the observed spectral profiles reasonable well if an adjustable parameter γ, which describes the critical duration of significant acceleration, is set equal to 10^{-15} s; no other freely adjustable parameters are found in that theory. That is a very short time, perhaps unrealistically short for the dynamics of particles other than electrons — particularly since a coordinated motion of many particles is assumed. Moreover, we have mentioned experiments that show that the addition of a small amount of a rare gas to nitrogen enhances the emission by an order of magnitude or so. The single-parameter QVR hypothesis cannot account for this fact. (We have argued above that the addition of small amounts of a rare gas — or indeed any gas — to nitrogen will doubtlessly increase collision-induced emission if the shockwave process is basically unaltered.) Furthermore, the sonoluminescence bubbles, which without doubt are filled with some gas, must emit some light by CIE as estimated above by the widely accepted shockwave mechanism, which might outshine QVR.

Acknowledgments

The support of the Deutsche Forschungsgemeinschaft is gratefully acknowledged.

REFERENCES

[1] L. Frommhold. *Collision-induced Absorption in Gases*. Cambridge University Press, Cambridge, New York, 1993.
[2] G. E. Caledonia, R. H. Krech, T. D. Wilkerson, R. L. Taylor, and G. Birnbaum. Collision-induced emission in the fundamental vibration-band of H_2. *Phys. Review A*, 43:6010 – 6017, 1991.
[3] W. Meyer and L. Frommhold. Ab initio interaction-induced dipoles and absorption spectra. In G. C. Tabisz and M. N. Neuman, editors, *Collision- and Interaction-Induced Spectroscopy*, Dordrecht, 1995. Kluwer.
[4] D. F. Gaitan, L. A. Crum, C. C. Church, and R. A. Roy. Sonoluminescence and bubble dynamics for a single, stable cavitation bubble. *J. Acoust. Soc. Am.*, 91:3166 – 3183, 1992.
[5] L. A. Crum and R. A. Roy. Sonoluminescence. *Science*, 266:233–234, 1994.
[6] R. Hiller, K. Weninger, S. J. Putterman, and B. P. Barber. Effect of noble gas doping in single-bubble sonoluminescence. *Science*, 266:248–250, 1994.
[7] C. C. Wu and P. H. Roberts. A model of sonoluminescence. *Proc. Roy. Soc. (London)*, A 445:323 – 349, 1994.
[8] R. A. Hiller and B. P. Barber. Producing light from a bubble of air. *Sci. American*,

272:96 – 98, 1995.

[9] L. Frommhold and A. A. Atchley. Is sonoluminescence due to collision-induced emission? *Phys. Rev. Letters*, 73:2883 – 2886, 1994.

[10] L. Frommhold and A. A. Atchley. presented at [26], p. 3240.

[11] A. Borysow. Pressure-induced molecular absorption in stellar atmospheres. In U. G. Jørgensen, editor, *Molecules in the Stellar Environment*, volume 428 of *Lecture Notes in Physics*, pages 209 – 222. Springer, 1 edition, 1994.

[12] R. H. Tipping. Collision induced effects in planetary atmospheres. In G. Birnbaum, editor, *Phenomena Induced by Intermolec. Interactions*, pages 727 – 738. Plenum Press, New York, 1985.

[13] J. L. Hunt and J. D. Poll. Lineshape analysis of collision-induced spectra of gases. *Can. J. Phys.*, 56:950 – 961, 1978.

[14] J. T. Carlson, S. D. Lewis, A. A. A. Atchley, D. F. Gaitan, X. K. Maruyama, M. E. Lowry, M. J. Moran, and D. R. Sweider. In H. Hobaek, editor, *Advances in Nonlinear Acoustics*, page 406. World Scientific, 1993.

[15] S. Putterman. presented at [26].

[16] G. C. Maitland, M. Rigby, E. B. Smith, and W. A. Wakeham. *Intermolecular Forces*. Clarendon Press, Oxford, 1981.

[17] W. Meyer, P. C. Hariharan, and W. Kutzelnigg. Refined *ab initio* calculation of the potential energy surface of the H_2-He interaction with special emphasis to the region of the van der Waals minimum. *J. Chem. Phys.*, 73:1880 – 1897, 1980.

[18] J. Schaefer and W. E. Koehler. Quantum calculations of rotational and NMR relaxation, depolarized Rayleigh and rotational Raman line shapes for H_2(HD)–He mixtures. *Physica A*, 129:469 – 502, 1985.

[19] W. Meyer and L. Frommhold. Collision induced rototranslational spectra of H_2-He from an accurate *ab initio* potential surface. *Phys. Review A*, 34:2771 – 2779, 1986.

[20] L. Frommhold and W. Meyer. Collision induced rotovibrational spectra of H_2-He pairs from first principles. *Phys. Review*, A 35:632 – 638, 1987. Erratum: Phys. Rev. **A 41** (1990) 534.

[21] K. S. Suslick. Sonochemistry. *Science*, 247:1439–1445, 1990.

[22] E. B. Flint and K. S. Suslick. The temperature of cavitation. *Science*, 253:1397–1399, 1991.

[23] J. Schwinger. *Proc. Nat. Acad. Science (U.S.A.)*, 90:2105, 1993.

[24] C. Eberlein. Sonoluminescence as quantum vacuum radiation. *Phys. Rev. Letters*, 76:3842 – 3845, 1996.

[25] C. Eberlein. Theory of quantum radiation observed as sonoluminescence. *Phys. Review A*, 53:2772 – 2787, 1996.

[26] Program of the 128th Meeting of the Acoustical Society of America, *J. Acoust. Soc. America* **96**, 3239-3254 (1994), November 1994; Sonoluminescence and Sonochemistry sessions.

Bull's eye model for Collision-Induced Light Scattering

Umberto Balucani[1] and Renzo Vallauri[2]

[1]*Istituto di Elettronica Quantistica del CNR, I 50127 Firenze (Italy)*, [2]*Istituto Nazionale per la Fisica della Materia e Dipartimento di Fisica dell' Università di Trento, I 38050 Povo (Trento)*

Abstract. The description of CILS in the pairwise DID approximation, in terms of a model, implemented in collaboration with George Birnbaum, is presented. Computer simulation results for a LJ system at different densities are reported, which point out the merits and disadvantages of the model.

The process which leads to light depolarization from a system of molecules with an intrinsic isotropic polarizability, revealed in experimental investigations by George Birnbaum et al. (1), is the appearance of a transient non diagonal component in the polarizability tensor due to collisional interaction, so that the effect is known as *"Collision-Induced Light Scattering"* CILS. The simplest mechanism devised to account for the measured intensity is a dipole induced dipole (DID) effect i.e. the assumption that a molecule feels not only the external electric field of the incident e.m. wave, but even the field generated by the dipole moments induced on the surrounding molecules. In the so called first order DID approximation, the depolarized spectrum is proportional to the Fourier transform of the correlation function of the projection of the total dipole moment of the system onto the direction **u** orthogonal to that of the incident electric field, **v**, i.e

$$C(t) = \frac{V}{N} \left\langle \left[\sum_i \sum_{j \neq i} (\mathbf{T}(\mathbf{r}_{ij}(0)) \cdot \mathbf{v}) \cdot \mathbf{u} \right] \cdot \left[\sum_k \sum_{l \neq k} (\mathbf{T}(\mathbf{r}_{kl}(t)) \cdot \mathbf{v}) \cdot \mathbf{u} \right] \right\rangle \quad (1)$$

where $\mathbf{T}(\mathbf{r}_{ij}(t))$ is the dipole propagator tensor; $\mathbf{r}_{ij}(t) = \mathbf{r}_i(t) - \mathbf{r}_j(t)$, and V and N represent the volume and the number of molecules, respectively. Under this assumption the total correlation function and consequently the depolarized spectrum, contains terms which reflect the dynamics of separate pairs ($C_2(t)$), triplets ($C_3(t)$), and quadruplets ($C_4(t)$), of molecules. Eq. (1) can in fact, naturally be split into three contributions, namely $C(t) = 2C_2(t) + 4C_3(t) + C_4(t)$.

A direct evaluation of these contributions has been possible through molecular dynamics (MD) simulation, which has allowed a deeper insight into the dynamical processes responsible for the measured spectra. Namely: i) the relaxation times of the individual two, three and four-body terms are considerably longer than that pertaining to their sum, ii) a substantial cancellation occurs between the two and four-body positive contributions and the three-body part which turns out to be negative. Several results of these calculations have been reported for an argon like system simulated through a Lennard-Jones (LJ) potential, at different thermodynamic conditions from the compressed fluid to the liquid phase (2). A difficulty of computer simulation turns out to be that, whereas the two and three body contributions can be obtained with reasonably accuracy, the evaluation of the total correlation function (being a bulk property) needs extremely long runs before achieving a decent convergence (3).

A method to overcome these difficulties was devised in 1979 in collaboration with George Birnbaum during one of his frequent visits to Florence. From eq. (1) one realizes that by defining $S(\mathbf{r}_i(t)) = \sqrt{V/N} \sum_{i \neq j} (\mathbf{T}(\mathbf{r}_{ij}(t)) \cdot \mathbf{v}) \cdot \mathbf{u}$, i.e. a quantity proportional to the component of the induced electric field on molecule i, which gives rise to the depolarized spectrum, C(t) can be decomposed as

$$C(t) = \left\langle \sum_i S(\mathbf{r}_i(0)) \cdot S(\mathbf{r}_i(t)) \right\rangle + \left\langle \sum_i \sum_{j \neq i} S(\mathbf{r}_i(0)) \cdot S(\mathbf{r}_j(t)) \right\rangle = C_{self}(t) + C_{dist}(t) \quad (2)$$

It can easily be shown that $C_{self}(t) = C_2(t) + C_3(t)$ and $C_{dist}(t) = C_2(t) + 3C_3(t) + C_4(t)$. The picture of the CILS which emerges from this decomposition is the following: the spectrum is the sum of a term which comes from the self correlation of the induced electric field on each individual molecule plus a cross-correlation part which accounts for the interference of the electric fields induced on two distinct molecules, as illustrated in figure 1. Such a picture prompted George to baptize "**Bull's eye**" the proposed model.

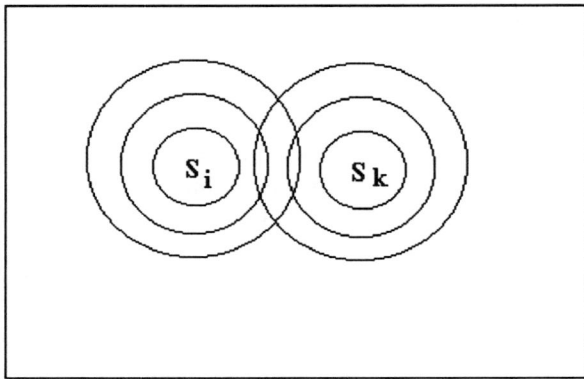

Figure 1. Illustration of the Bull's eye model.

The usefulness of the model has been tested in the case of LJ argon by studying the density variation of the values at t=0 of the various contributions. The results are reported in Table 1.

Table 1. Contributions to the total scattering intensity from pairs, triplets and quadruplets and their combination as resulted from the Bull's eye model. MD data for LJ argon at $T^* = 2.5$.

n^*	$C(0)$	$2C_2(0)$	$4C_3(0)$	$C_4(0)$	$C_{self}(0)$	$C_{dist}(0)$
0.157	0.95	1.49	-0.72	0.18	0.565	0.385
0.322	1.09	3.26	-3.0	0.83	0.88	0.21
0.422	1.01	4.23	-4.75	1.53	0.93	0.09
0.499	0.98	4.96	-6.04	2.06	0.97	0.01
0.533	0.97	5.56	-7.47	2.88	0.91	0.06
0.960	0.38	12.30	-21.76	9.84	0.70	-0.32

In figure 2 we present some MD results for the correlation functions. As is apparent, for the two reported densities, the total correlation function is very close to the self contribution $C_{self}(t)$, defined in eq.(2). Since this term is a single particle quantity, it can be more accurately evaluated due to the averaging process over all the molecules of the system.

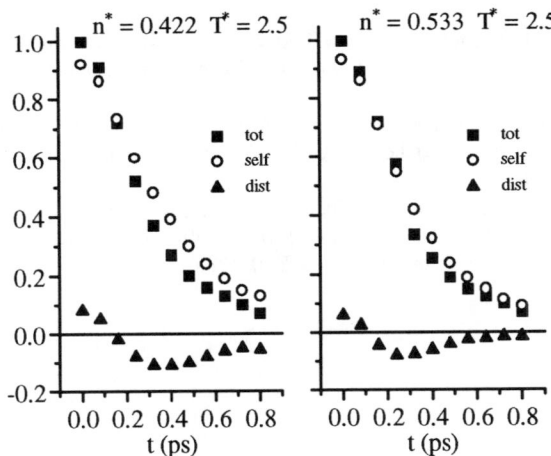

Figure 2. Computer simulation results for LJ argon. The density and temperature are given in reduced units, i.e. $n^* = n\sigma^3$, $T^* = k_B T/\varepsilon$, being σ and ε the LJ parameters and n the number density. Squares: the total correlation function C(t), circles and triangles: the self and distinct contributions defined in eq.(2), respectively. All the correlation functions have been normalized to C(t=0).

From this investigation it comes out that at very low density (lower than the one reported in the table) the dominant contribution comes from the pair term so that the model can give reasonable account for the collision induced intensity. By increasing the density the triplet term becomes important and the model doesn't give a good estimate of the process since the distinct part cannot be neglected and a full calculation of the interference term is necessary to reproduce the total intensity. This is true even for the time behaviour of the correlation function. However, there is a region of intermediate densities (approximately between $n^* = 0.422$ and 0.533) where a cancellation occurs between two, three and four body terms, such that the distinct part, defined through the Bull's eye model, becomes small and the self part gives the overwhelming contribution to the total intensity. From figure 2 one can infer that even the time dependence of the total and self part is quite similar. Therefore in this range reasonable results for the collision induced effect can be derived by evaluating the self term of the Bull's eye model, which, as already pointed out, turns to be very accurate due to the averaging process over all the particles used in the simulation. At higher density such a cancellation is no longer present and a full evaluation of the distinct part is necessary, thus making the Bull's eye model no longer efficient.

REFERENCES

1. Levine H.B., Birnbaum G., Phys. Rev. Lett. **20**, 439 (1968); Thibeau M., Oksengorn B. and Vodar B., J. de Physique **29**, 287 (1968); Mc Tague J.P., Birnbaum G., Phys. Rev. A**3**, 1376 (1971).
2. Alder B.J., Strauss H. L. and Weis J.J., J. Chem. Phys. **62**, 2328 (1975); Ladd A.J., Litovitz T.A. and Montrose C.J., J. Chem. Phys **71**, 4242 (1979); Balucani U., Vallauri R., Mol. Phys. **38**, 1099, (1979).
3. Clarke J.R.H. and Woodcock L.V., Chem. Phys. Lett., **78**, 121 (1981); Vallauri R., in *"Phenomena Induced by Molecular Interactions"* edited by G. Birnbaum, NATO ASI Series B; Physics Vol.127, Plenum Press, (New York 1985).

The Wings of the Rototranslational Raman Spectrum of Nitrogen at low density

Aleksandra Borysow *,†, Yi Fu * and Massimo Moraldi §

* Physics Department, Michigan Technological University, Houghton, Michigan 49931; † Niels Bohr Institute, Copenhagen University Observatory, Juliane Maries vej 30, DK-2100 Copenhagen, Denmark and § Dipartimento di Fisica dell' Universita' di Firenze and Istituto Nazionale di Fisica della Materia, Unita' di Firenze, Largo E.Fermi 2, I-50125 Firenze, Italy

Abstract. Calculations of the far wings of Raman rotational spectrum of N_2 have been performed and compared with existing measurements. The wings of both the allowed and the collision induced spectrum have been taken into account. The results show that the allowed contribution dominates the spectral intensity in the far wings.

This work is devoted to the analysis of the far wings of the rototranslational Raman spectrum of N_2 at room temperature. The rototranslational spetrum of N_2 has been measured recently at different densities and it has been found that the spectral intensity scales quadratically with density at all frequencies larger than 350 cm^{-1} [1]. This spectral region is what we call the far wings.

Our aim is to understand what is the particular microscopic mechanism which is responsible for the scattered intensity in that spectral region. The question is important because, depending on what particular mechanism is effective, we can extract different information about molecular properties and intermolecular interactions. Two different mechanisms can be considered as responsible for the scattered intensity at low density. One is due to the interaction of radiation with the permanent polarizability of each single molecule. Such an interaction gives rise to the allowed rotational spectrum which, in a first approximation, appears as a simple stick spectrum. The other mechanism is due to the polarizability induced in a pair of molecules and is responsible for the broad rototranslational collision induced spectrum. Higher order contributions in the polarizability are to be ruled out due to the low density and because they would be responsible for a dependence on density stronger than a power of two.

The far wings of the induced spectrum are determined by the correlation of the induced polarizabilities, which depends mostly on the isotropic part of the intermolecular potential: the anisotropic component of the potential is responsible only for small corrections, at least as long as the molecules are not too anisotropic. The collision induced spectral contribution is thus calculated according to usual methods that involve the solution of the Shroedinger equation for a system of two particles interacting through a purely isotropic potential [2]. We have included 14 inducing mechanisms [3] which arise from dipole induced dipole up to second order (DID), dipole induced octopole (DIO) and the term coupling the permanent quadrupole moment with the dipole-dipole-quadrupole hyperpolarizability. The total integrated intensity is contributed mostly by the DID term. However, due to the shorter range behaviour and to selection rules for transitions between rotational states, it is the DIO term that contributes mostly in the far wings. It is however found that such a contribution is about two orders of magnitude smaller than the measured scattered intensity: the collision induced contribution represents only a small fraction of the intensity in the far wings. As a consequence the far wings of rototranslational Raman spectrum cannot be expected to give valuable information on the polarizability induced by intermolecular forces.

We have thus calculated the wings of the allowed rotational spectrum as well. Such a calculation needs a theory that will account for the role of the anisotropic part of the intermolecular potential. Indeed, the simple assumption of a purely isotropic potential, which is reasonable for the collision induced spectrum, fails completely to reproduce the wings of the allowed one. Birnbaum [4] has contributed a paper at the 10th ICSLS where he develops some ideas proposed previously by Bulanin et al. [5]. He suggests that three main mechanisms are to be taken into account in order to reproduce the far wings of an allowed spectrum: line interference, finite duration of collisions and molecular torques incorporating initial orientational correlations. We propose a unified theory [6] that contains all the above mentioned ingredients but that treats the anisotropic part of the intermolecular potential as a perturbation. We assume that is a good approximation for N_2 due to the relative smallness of the anisotropic part of the intermolecular potential. Apart from that approximation, the relaxation matrix (memory function) is calculated rigorously, with the inclusion of all the contributions and treating the pair dynamics quantum mechanically. In the theory we use, the wings of the allowed spectrum are determined by the correlation of the various spherical components of the anisotropic part of the intermolecular potential. Such a correlation, because of the perturbative treatment, is to be calculated for a pair of molecules that interact through an isotropic potential. The calculation is performed with the same methods as those

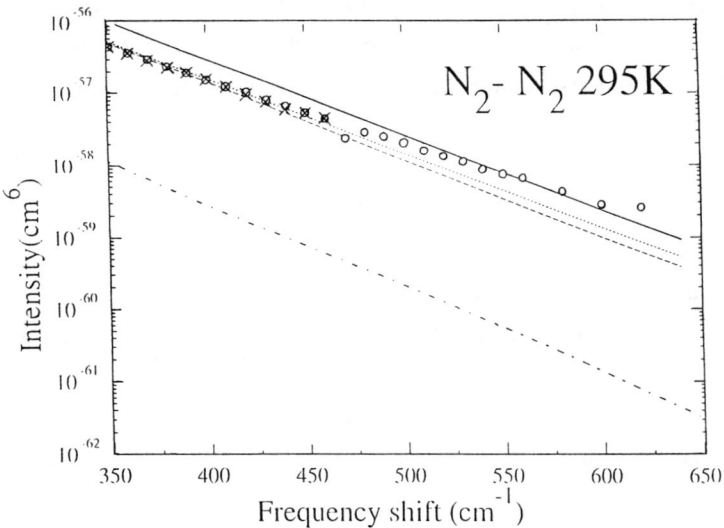

FIGURE 1. X and O: measured intensity divided by density squared at 41 and 169 amagats respectively (see ref.[1]). -·- : collision induced contribution. ··· : allowed contribution from V_{224} term. --- : allowed contribution from V_{202} term. The continuous line represents the total contribution of the allowed spectrum.

involved in the calculation of the collision induced spectrum apart for the replacement of the spherical components of the induced polarizability with the spherical components of the anisotropic contribution to the intermolecular potential. In our calculation we have used potential components from litterature [7]. The results are shown in the figure, in which circles and crosses represent the experimental results and the continuous line represents the contribution of the allowed spectrum. The allowed contribution is composed mostly of two parts corresponding to the two most important anisotropic components of the intermolecular potential, that is V_{224} and V_{202} [7]. As anticipated, the collision induced spectrum represents only a negligible contribution to the far wings of the rototranslational spectrum of nitrogen. The allowed spectrum, on the contrary, is able to explain, at least qualitatevily, the experimental results. A quantitative comparison shows discrepancies that can be as large as 50% though. Such discrepancies can be attributed either to the perturbative treatment of the anisotropic part of the intermolecular potential or, more likely, to inaccuracies of the intermolecular potential model. Once more tests are made on the theory, the experiemental spectra could thus be used to derive a more reliable intermolecular potential.

REFERENCES

1. Le Duff Y. and Teboul V.,Phys. Lett. **A 157**, 44 (1991)
2. Frommhold L., in Advances in Chemical Physics, edited by I. Prigogine and S. Rice (Wiley, New York, 1981), vol.46, p.1
3. Le Duff Y., Bancewicz T. and Glaz W., in *Collision and Interaction Induced Spectroscopy* , NATO Advanced Research Workshops, edited by G.Tabisz and M.N.Newman (Kluwer, Dordrecht, 1994), pp. 423-440
4. Birnbaum G., in *Proceedings of the 10th ICSLS,* edited by L.Frommhold and J.W. Keto (American Institute of Physics, New York 1990),pp.337-351
5. Bulanin M.O., Tonkov M.V. and Filippov N.N., Can. J.Phys. **62**, 1306-1314 (1984)
6. Fu Y., Borysow A. and Moraldi M., Phys. Rev. **A 53**, 201-205 (1996)
7. van der Avoird A., Wormer P.E.S. and Jansen P.J., J.Chem.Phys. **84**, 1629-1635 (1986)

Non-Resonant Microwave Absorption and Collisional Coupling in Infrared Q Bands

G. Buffa and O. Tarrini

Dipartimento di Fisica dell'Università, Piazza Torricelli 2, 56126 Pisa, Italy

Abstract. For symmetric top molecules a non-resonant absorption line was observed many years ago and shown to have a collisional broadening cross section far lower than the rotational absorptions lines of the same molecules. An "effective line model" explains this linewidth reduction as being due to collisional coupling between different $\Delta J = 0$ rotational lines. The same model can be used for describing collisional coupling in infrared Q bands.

The aim of this paper is to show that two seemingly rather different topics are strictly connected. The two topics are: reduced collisional broadening observed in the microwave region for non-resonant absorption of symmetric top molecules and collisional coupling observed in the infrared region for the lineshape of molecular vibrational Q bands.

For symmetric top molecules, the electric dipole moment has a static component parallel to the axis of rotation giving rise to a Debye type absorption at zero frequency. Due to collisions, the line is broadened and absorption can be observed in his wings, which are not at zero, but at very low frequency. The main result obtained by this kind of investigation is that the collisional broadening coefficient for this non-resonant Debye line is far smaller than for the resonant lines of the same molecule. This statement, given by Birnbaum (1) many years ago, was confirmed by more recent studies on rotational transitions of symmetric top-molecules. For instance, a recent detailed study of CH_3CN rotational transitions (2) has shown an average pressure broadening parameter about four times larger than what obtained for the Debye line.

A large amount of information is available nowadays also on the pressure broadening of infrared roto-vibrational P, Q and R bands, characterized by $\Delta J = -1, 0$, and $+1$, respectively. Apart from particular cases, such as ammonia, the general result is that while the rotational dependence of pressure broadening (for instance the J dependence) can be large, the vibrational dependence is small: different vibrational transitions have similar pressure broadening coefficients and also P, Q, and R bands give similar results. This makes the exception of the Debye line more remarkable. Indeed, if we consider the ground state just as a particular vibrational state, rotational absorption is an R band, stimulated rotational emission is a P band, and Debye absorption is a Q band with the peculiarity that while in a vibrational Q band rotational structure can usually be resolved, this is impossible for the Debye line. However, in a vibrational Q band the spectrum is usually very crowded. When pressure is raised, the rotational components begin to overlap and collisional coupling effect may occur. If in Debye line a collisional coupling effect between rotational components is introduced, the analogy with vibrational Q bands is restored.

Some years ago we proposed a way to handle this problem (3). Indeed, a group of overlapping lines can be treated by a multiline scheme or by an "effective line scheme". In the multiline scheme one has n lines: $|\ell_1\rangle, |\ell_2\rangle \ldots |\ell_n\rangle$, and the collisional lineshape is given by

$$F(\nu) = -\frac{1}{\pi} \text{Im} \langle\!\langle \mu | \frac{1}{\nu - L_0 - i\Gamma} | \mu \rangle\!\rangle, \tag{1}$$

where μ is the dipole moment of the molecule, L_0 its unperturbed Liouvillian and Γ its collisional relaxation operator. A diagonal term of Γ, Γ_{kk}, gives the collisional broadening of line ℓ_k, while an off-diagonal term, Γ_{kj} gives collisional coupling between lines ℓ_k and ℓ_j. In this scheme, in order to obtain the lineshape, a complete knowledge of the $n \times n$ relaxation matrix is needed and a diagonalization of the non Hermitian operator $L_0 + i\Gamma$ must be performed.

A far simpler description is obtained (3) by use of the "effective line" $\overline{\ell}$

$$|\overline{\ell}\rangle = \sum_k I_k^{\frac{1}{2}} |\ell_k\rangle / \overline{I}^{\frac{1}{2}}, \tag{2}$$

where I_k is the intensity of line ℓ_k. The effective line has an intensity $\overline{I} = \sum_k I_k$ given by the sum of line intensities and a frequency $\overline{\nu} = \sum_k I_k \nu_k / \overline{I}$ given by the weighted average of line frequencies. On the contrary, its collisional broadening $\overline{\Gamma}$ is not the weighted average of line broadening because a reduction is given by the, usually negative, line coupling terms Γ_{kj}:

$$\overline{\Gamma} = \sum_{k,j} I_k^{\frac{1}{2}} I_j^{\frac{1}{2}} \Gamma_{kj} / \overline{I} = \sum_k I_k \Gamma_{kk} / \overline{I} + \sum_{k,j, k \neq j} I_k^{\frac{1}{2}} I_j^{\frac{1}{2}} \Gamma_{kj} / \overline{I}. \tag{3}$$

As can be seen in Table 1, the effective line scheme gives good results for

TABLE 1. Collisional broadening radius (cm^{-8}) for non–resonant absorption of symmetric top molecules. Comparison between measurements and calculations taking into account, or not, collisional coupling between rotational components.

molecule	without coupling	with coupling[3]	measurements
CH_3F	14.2	7.4	6.8[1]
CH_3Cl	14.0	8.0	7.4–7.9[1,4,5]
CH_3I	14.5	7.9	8.6[1]
CH_3Br	16.5	8.1	8.0–9.0[1,4,5]
CH_3CN	28.0	13.8	12.5[1]
CHF_3	11.8	5.8	5.4[1]

TABLE 2. Collisional broadening radius (cm^{-8}) for non–resonant absorption of ammonia. Comparison between measurements and calculations not taking into account collisional coupling, or taking into account collisional coupling between parity components only, or taking also into account collisional coupling between rotational components.

isotopic species	without coupling	only parity coupling	also rotational coupling[3]	measurements
NH_3	14.4	7.9	7.1	7.7[5]
ND_3	15.1	7.9	6.6	6.9[6]

non resonant spectra of dipolar symmetric top molecules. Collisional coupling effect reduces the broadening radius by a factor two with respect to what one would obtain by omitting the coupling term in Eq. (3).

This kind of study can be extended to the ammonia molecule, whose nitrogen atom can cross the plane of the three hydrogens giving rise to the well known microwave inversion spectrum involving a transition between symmetric (+) and antisymmetric (−) parity states. Since the dipole moment oscillates in time, there is no Debye absorption at low pressures. However, when pressure is raised and the width of inversion lines becomes comparable to their frequencies, a Debye line appears as a consequence of collisional coupling between inversion absorption line $+ \to -$ and inversion emission line $- \to +$. In this case the effective line method can be used to describe collisional coupling both between the two parity components and between rotational components. As shown in Table 2, the narrowing effect is due for this molecule mainly to the coupling between components of different parity.

On the whole, from this overview it is clear that collisional coupling observed for vibrational Q bands and reduced linewidth observed for non resonant Debye lines are the same effect. The effective line model can account for the shape of Debye lines and hence can be used also for Q bands when they have so close frequencies, or are so broadened, that any structure disappears. It also allows an extension to collisional coupling effects of the models commonly used for calculation of line broadening.

Finally, we note that, as far as band wings are concerned, the effective line method gives a good lineshape also at low pressures, when the structure of the band is resolved. This can be seen from the scheme reported in Figure 1. The shape of a band, where collisional coupling is present, is plotted at different pressures near to the band center (left part of the figure) and also, suitably magnified, in the band wing (right part of the figure). Pressure, and all Γ matrix elements, are increased by a factor 5 changing from case a) to case b) and from case b) to case c). Solid line reproduces the simple sum of individual lorentzian shapes. Dotted line is obtained by an exact diagonalization of the operator $L_0 + i\Gamma$ in Eq. (1). Dashed line is the shape obtained by the effective line method.

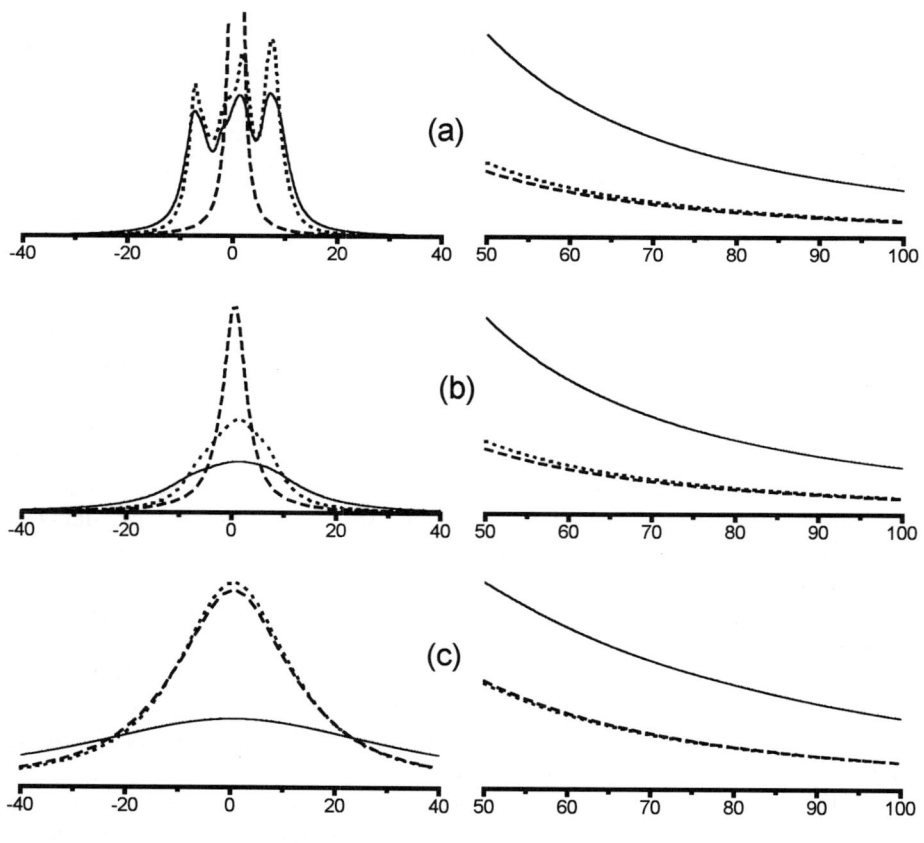

FIGURE 1. For explanation, see text.

Near to the band center one can see that the sum of lorentzians is wrong at high pressures, while effective line is wrong at low pressures and gives a good description when linewidth is larger than frequency spread.

In the band wings one can see that in any case the sum of lorentzians is in error, while effective line model gives good results. Hence, this simple method can account for line coupling effects in the wings of vibrational Q bands.

REFERENCES

1. G. Birnbaum and A. A. Mariott, *Phys. Rev.* **99**, 1868 (1955); *J. Chem. Phys.* **27**, 1422 (1958); G. Birnbaum, *J. Chem. Phys.* **27**, 360 (1957); *Phys. Rev.* **150**, 101 (1966).
2. G. Buffa, O. Tarrini, P. De Natale, M. Inguscio, F. S. Pavone, M. Prevedelli, K. M. Evenson, L. R. Zink, and G. W. Schwaab, *Phys. Rev.* A **45**, 6443, (1992).
3. G. Buffa and O. Tarrini, *Phys. Rev.* A **16**, 1612 (1977).
4. Krisnaji and G. P. Srivastava, *Phys. Rev.* **109**, 1560 (1958).
5. B. Bleaney and J. H. N. Loubser, *Proc. Roy. Soc.* A **63**, 483 (1950).
6. G. Birnbaum and A. A. Maryott, *Phys. Rev.* **92**, 270 (1953).

Collision-Induced Absorption Intensity Redistribution and the Atomic Pair Polarizabilities

M. O. Bulanin

*Institute of Physics, St.Petersburg University, Peterhof,
198904 St.Petersburg, Russia*

Abstract. A modified relation between the trace polarizability of a diatom and the $S(-2)$ dipole sum is proposed that accounts for the effect of atomic collisions on the dipole oscillator strength distribution. Contribution to the collision-induced trace due to redistribution in the ionization continuum of Ar is evaluated and is found to be significant.

Incremental dipole polarizability of a pair of colliding atoms plays a key role in a variety of the collision-induced phenomena including the density-dependent refractivity and light scattering spectra of monoatomic gases. Interactions underlying formation of the incremental trace and polarizability anisotropy are generally well understood at present, however, several aspects either remain unexplored or lack theoretical interpretation. For example, little is known on the dispersion of the induced polarizability tensor components, the reported temperature variation of the dielectric virial coefficients for the rare gases is not in accord with the theoretical predictions based on the available atomic pair polarizability models [1, 2]. As was suggested earlier [3], an insight into the nature of the processes contributing to the polarizability induced in atomic collisions could be gained from analysis of the collision-induced intensity redistribution in absorption spectra of dense gases.

The trace is related to the dipole oscillator strength distribution (DOSD), $D(E) = df/dE$, in a gas by

$$\alpha = a_0^3 E_h^2 S(-2) = a_0^3 E_h^2 \int_{E_0}^{E_{max}} \frac{D(E)}{E^2} dE, \qquad (1)$$

where E_0 is the resonance excitation threshold. Using for the upper integration limit E_{max} the high-energy limit of the first ionization continuum recovers 99–98% of the net trace for the rare gases [4]. This expression can be modified in the frame of the quasistatic theory of spectral line broadening in order to take

© 1997 American Institute of Physics

into account the effect of interatomic interactions. The basic assumption of the quasistatic theory is that the transitions in a pair of colliding species at any given separation R occur between the $V_{l(ower)}(R)$ and the $V_{u(pper)}(R)$ molecular potential energy curves (PECs) correlating with the terms of isolated species. The transition frequencies shift from that for the unperturbed resonance as $h\nu(R) = V_u(R) - V_l(R) = \Delta V(R)$, making the partial DOSD for transitions into different upper states to slide along the energy scale in accordance with the corresponding difference potentials $\Delta V(R)$. Contribution to the dipole sum $S(-2)$ from a given radiative channel i becomes then a function of R:

$$S_i(-2)(R) = \int_{E_{0i}+\Delta V_i(R)}^{E_{max}+\Delta V_i(R)} \frac{D_i(E - \Delta V_i(R))}{E^2} dE. \qquad (2)$$

This approach is tested by evaluating contribution to the incremental static trace for the argon diatom due to collisional redistribution in the ionization continuum, from the onset up to 50 eV, using the DOSD data tabulated by Chan et al. [5] and different model PECs for the Ar_2^+ dimer ion. The ionization continua in rare gases exhibit strong density dependence. Already at pressures as low as 40 Pa photoionization was detected at photon energies noticeably lower than the first atomic ionization potential [6]. As was shown in [3], downward energy shift of the ionization threshold with density results in a positive contribution to the collision-induced trace.

Of the six lowest molecular terms of Ar_2^+ resulting from the spin-orbit splitting, three are radiatively coupled to the ground state $0_g^+(^1\Sigma_g^+)$ of the Ar_2 dimer: $I(1/2)_u(^2\Sigma_{1/2u}^+)$ and $I(3/2)_u(^2\Pi_{3/2u})$ dissociate into $Ar^+(^2P_{3/2}) + Ar(^1S_0)$, and $II(1/2)_u(^2\Pi_{1/2u})$ which dissociates into $Ar^+(^2P_{1/2}) + Ar(^1S_0)$. These upper states are denoted here 1, 2, and 3, respectively. Contribution to the incremental trace due to transitions into the ionization continuum of Ar can thus be written as

$$\Delta\alpha(R) = a_0^3 E_h^2 \sum_{i=1}^{3} \sigma_i \left(\int_{I_i+\Delta V_i(R)}^{50+\Delta V_i(R)} \frac{D(E - \Delta V_i(R))}{E^2} dE - S_i(-2)(R \to \infty) \right), \qquad (3)$$

where σ_i are the statistical factors, $\sigma_2/\sigma_1 = 2$ and $(\sigma_1+\sigma_2)/\sigma_3 = 1.93$, assumed to be constant throughout the integration range [4], $I_1 = I_2 = 15.76$ eV, $I_3 = 15.94$ eV are the $^2P_{3/2}$ and $^2P_{1/2}$ ionization thresholds, $\Delta V_i(R) = V_{u,i}(R) - V_l(R)$. The ground state function $V_l(R)$ used in the calculation of the difference potentials was highly accurate Aziz-type potential [7]. Theoretical PECs for the strongly-bound state 1 are well established [8-10] and yield the dissociation energy close to the experimental value of $D_e = 1.226$ eV obtained from the photoelectron spectroscopy studies [11]. The equilibrium bond length, $R_e = 4.6\,a_0$, is to be taken from theory, because this parameter cannot be deduced

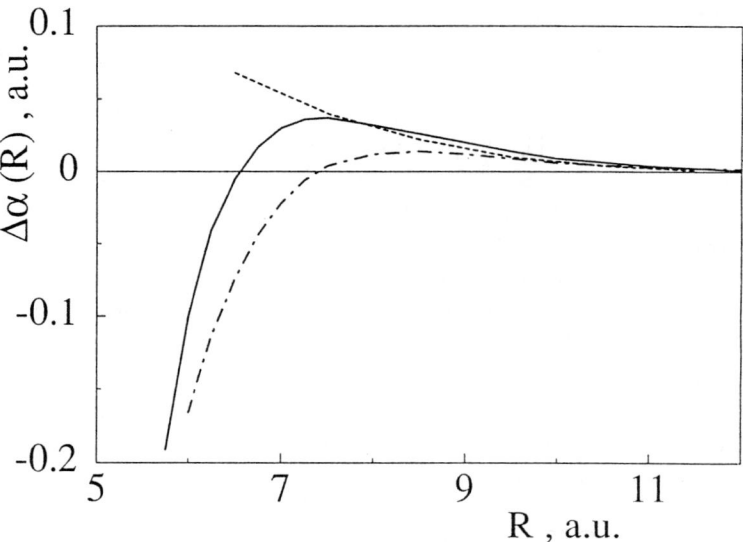

Incremental polarizability trace functions for argon computed with the PECs from ref. [9] (- - - -), with the PECs for the states 2 and 3 shifted to $R_e = 8\,a_0$ (———), and the trace function from ref. [12] (— · — · —·).

with confidence from rotationally-unresolved photoelectron spectra. Properties of the weakly-bound states 2 and 3 with shallow potential wells of 0.03–0.08 eV still seem to be rather uncertain. For example, the positions of the potential well minima reported for the state 3 range from $R_e = 6.4\,a_0$ [8] to $R_e = 8\,a_0$ [10]. The difference potentials were calculated from the data tabulated by Whitaker et al. [9] for all three upper states and also with the PECs for the states 2 and 3 shifted to match $R_e = 8\,a_0$ calculated by Mizukami et al. [10].

Incremental trace functions for the argon diatom obtained by numerical integration of eq. (3) are shown in the Figure. Because of the E^{-2} weighting factor, the results are insensitive to the choice of the upper integration limit. A mean value of DOSD was accepted in the autoionization region where the oscillatory structure of DOSD observed [4, 5] between the $^2P_{3/2}$ and $^2P_{1/2}$ thresholds is expected to flatten out at higher gas densities.

The trace computed with the difference potentials corresponding to the PECs from ref. [9] is positive at all relevant interatomic separations. When the potentials for the two upper states 2 and 3 are shifted to larger equilibrium bond lengths, the trace function acquires a correct profile, negative at small separations and qualitatively very similar to the Dacre's SCF trace [12] also shown for comparison. Meinander [2] developed a model pair trace with the short-range parametrized by the Dacre's SCF function. Using in the binary distribution function the same ground-state potential $V_l(R)$ [7] he found the second dielectric virial coefficient for Ar at 300 K to be $B_\epsilon = 0.0615 \, \text{cm}^6 \text{mol}^{-2}$. Two trace functions obtained in the present study yield $B_\epsilon = 2.64 \, \text{cm}^6 \text{mol}^{-2}$ for the first and $B_\epsilon = 1.56 \, \text{cm}^6 \text{mol}^{-2}$ for the second ("shifted") trace, the latter value being practically identical with the average of the latest experimental determinations at 300 K ($1.6 \pm 0.4 \, \text{cm}^6 \text{mol}^{-2}$ [2]).

Good agreement between calculated and experimental dielectric virial coefficients for the argon diatom should be fortuitous, both because the parameters of the upper PECs for the Ar_2^+ dimer ion are insufficiently well known and because the effect due to the discrete part of the spectrum has not been accounted for. The results of the present work do show however that the collisional intensity redistribution in the ionization continuum forms an important contribution to the incremental trace.

ACKNOWLEDGMENTS

Financial support provided by the Deutsche Forschungsgemeinschaft and by the Russian Foundation for Basic Research (Grant RFFI 96-02-1792a) is gratefully acknowledged.

REFERENCES

1. Bulanin, M. O., Hohm, U., Ladvishchenko, Yu. M., and Kerl, K., *Z. Naturforsch.* **49 a**, 890-894 (1994)
2. Meinander, N., *Chem. Phys. Lett.* **228**, 295-300 (1994).
3. Bulanin, M. O., *Opt. Spectrosc.* **75**, 574-578 (1993); *Chem. Phys. Lett.* **217**, 466-470 (1994).
4. Berkowitz, *Photoabsorption, Photoionization, and Photoelectron Spectroscopy*, New York: Academic Press, 1979.
5. Chan, W. F., Cooper, G., Guo, X., Burton, G. R., and Brion C. E., *Phys. Rev. A* **46**, 149-171 (1992).
6. Reininger, R., Saile, V., and Laporte, P., *Phys. Rev. Lett.* **54**, 1146-1149 (1985).
7. Aziz, R. A., Slaman, M. J., *J. Chem. Phys.* **92**, 1030-1035 (1990).
8. Michels, H. H., Hobbs, R.H., and Wright, L. A., *J. Chem. Phys.* **69**, 5151-5162 (1978).
9. Whitaker, B. J., Woodward, C. A., Knowles, P. J., and Stace A. J., *J. Chem. Phys.* **93**, 376-383 (1990).
10. Mizukami, Y., and Nakatsuji, H., *J. Chem. Phys.* **92**, 6084-6092 (1990).
11. Hall, R. I., Lu, Y., Morioka, Y., Matsui, T., Tanaka, T., Yoshii, H., Hayaishi, T., and Ito, K., *J. Phys. B: At. Mol. Opt. Phys.* **28**, 2435-2451 (1995).
12. Dacre, P. D., . *Mol. Phys.* **45**, 1-15 (1982).

Quantum analysis of experimental anisotropic and isotropic CILS spectra from Ar diatoms in the frequency band 0-400 cm^{-1}

Michael Chrysos, Omar Gaye, and Yves Le Duff

*Laboratoire des Propriétés Optiques des Matériaux et Applications, EP CNRS 0130
2, bd Lavoisier, Université d'Angers, 49045 Angers, Cedex, France*

Abstract. Absolute experimental intensities of the *isotropic* collision-induced light scattering binary spectrum of (Ar)$_2$ are for the first time measured over a very broad domain of Raman shifts (up to ω=400 cm^{-1}). Motivated by these measurements as well as by those for the *anisotropic* spectrum, a fully quantum analysis of spectral intensities is performed, from bound, metastable and free Ar diatoms. A novel methodology is developed for this purpose combining the advantages of both basis-set methods and propagative techniques. Different anisotropy and trace models are used and compared, while suggestions, when necessary, for new optimal models are made.

Keywords: Ar$_2$, collision-induced scattering, intermolecular interactions.

MEASUREMENTS

We report new experimental results on the absolute *isotropic* (I$_{ISO}$) collision-induced light scattering (CILS) spectrum from low pressure Ar gas at room temperature (1). This spectrum is for the first time measured over a so much extended Raman-shift domain, up to ω=400 cm^{-1} (Fig. 1). Set-up and experimental principles are identical to those previously used for the corresponding *anisotropic* (I$_\parallel$) spectrum (3). In contrast to the much smaller uncertainties characterizing the latter spectrum in the low-frequency region, I$_{ISO}$ exhibits error-bars attaining ±50-70% of the central intensities at $\omega \approx 60$ cm^{-1}. This comes from the fact that at such frequencies the depolarization

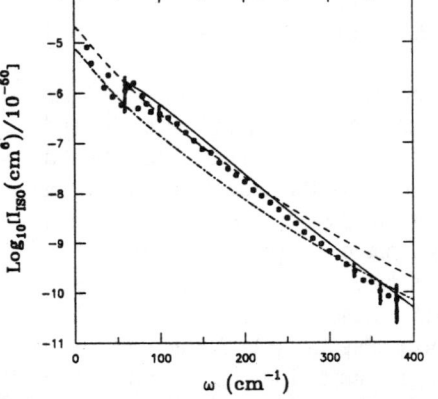

Figure 1. Comparison between experimental (points) and computed (lines) isotropic spectra. Solid line: Dacre's unscaled SCF (ref.2); broken line: A$_6$/r^6 (A$_6$=9665 a$_0^9$); dashed-dotted line: DID with the atomic polarizability α_0

ratio $I_\parallel/I_\perp$ gets high values (tending to its maximum value 6/7, as ω approaches zero). On the other hand, in the very far wings of the Raman spectra (ω>200 cm^{-1}), uncertainties of about ±50-70 % are again marked for both I_{ISO} and $I_\parallel$, as a consequence of the drastic attenuation of the scattered intensity (which may attain *seven* orders of magnitude within the interval 2<ω<400 cm^{-1}). The experimental measurements, taken from our group for both spectral components (1,3) in the range 2<ω<400 cm^{-1}, motivate a complete and systematic quantum analysis of the Ar-Ar system. Given that initial and final states are coupled by the tensorial invariants of the dimer polarizability, computed spectral shapes (in an as large as possible frequency band) can serve as sensitive probes to check the validity of the different interaction and coupling schemes, aiming at optimally modelling the process and quantitatively improving relevant properties.

COMPUTATIONS

Numerically, low and high ω give rise to difficulties of different nature: at low ω (< 20 cm^{-1}), stable and metastable dimers have an important contribution to the total cross section (Fig. 2); at very high ω (>200 cm^{-1}), although such contributions are no longer significant, accurate computations become the more and more expensive, since other effects appear (negligible in the low or intermediate frequency bands) which drastically influence the corresponding spectral wings (4). Among those effects, (i) the energy range of the colliding partners (> 3000 cm^{-1}), (ii) the number of rotational levels of the diatom (> 700) and (iii) the interparticle separations relevant to the process (>150 bohr) are quantities taking values over a broad domain and, therefore, playing decisive role to the convergence of the spectrum in the far wings (4).

An implicit basis-set method

Here, an exaustive quantum analysis of the Ar-Ar diatom is made and confronted to recent experimental data. To this purpose a novel methodology is developed for the systematic computation of frequency -resolved and -integrated Rayleigh and Raman spectra from stable, metastable and free Ar dimers over a large band of frequencies (up to 400 cm^{-1}), based on a suitable combination of a modern analytical

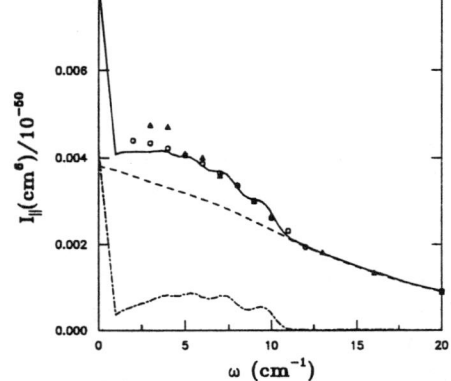

Figure 2. $I_\parallel$ in the low-ω Stokes side with a slit function of width 1 cm^{-1} at HM. Solid line: total $I_\parallel$ (ref.7); circles: experiment (ref.3); triangles: experiment (ref.8); broken line: $I_\parallel$, due to hf diatoms; dashed-dotted line: $I_\parallel$, due to b, m and lf diatoms.

discrete variable representation (DVR) (5) and a propagative (Fox-Goodwin) integrator (6). When transitions involving bound (b), metastable (m) or low-energy free (lf) states of the diatom are considered, the DVR approach provides an efficient means to systematically compute accurate eigenenergies and wavefunctions (7); propagative techniques, so far monopolizing CILS, can give rise to instabilities due to the unguided choice of the initializing energies in a multidiscrete level quantized system like the present one. Instead, a particularly simple and analytical DVR is used herein that can be programmed in a few lines (7). Within this representation the matrix-elements of the potential operator are diagonal while those of the kinetic operator are analytically known (5). It is noteworthy that, by means of this implicit basis-set method the low-energy continuum (lf) can be efficiently discretized as well.

A two-point propagative integrator

On the other hand, transitions involving high-energy free states (hf) of the dimer can be safely handled only within a propagative approach. In this work the two-point Fox-Goodwin algorithm (6) is used, building up the ratio of the wavefunction at adjacent grid points (Fig. 3). Apart from its simplicity and stability (especially when barriers are to be penetrated in a diatomic three-turning point effective potential) this integrator can be efficiently generalized to attack the much more complicated *multi-channel* problem where more than one physically relevant potential surfaces are coupled to each other, by either interatomic, intramolecular or photon-field interactions (9). In all computations, the Aziz-Slaman potential is used

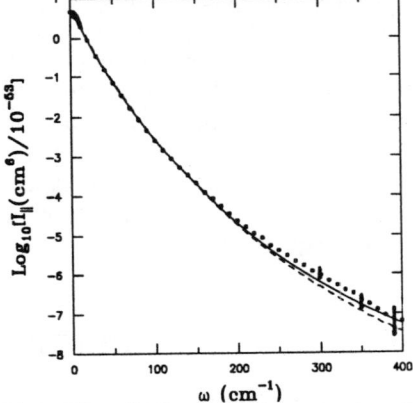

Figure 3. $I_\parallel$ as a function of ω. Solid curve: quantum calculation (ref.4) with the Dacres's SCF anisotropy (ref. 2); broken curve: classical calculation (ref.3); full circles: experiment (ref.3).

corresponding to one of the best Ar-Ar interactions ever constructed (10). Different anisotropy and trace models are checked and compared with spectra from both classical calculations and experimental measurements. Our conclusions are summarized as follows:

A. The anisotropic spectrum

The structure due to the bound dimers differs from that coming from the metastable diatoms, but both peak in the same frequency region (ω=4.5 cm^{-1}) and settle down to negligible values beyond 10 cm^{-1} (Fig. 2). Contributions of bound and metastable complexes amount about 8.5% and 3% of the integrated intensity respectively (11). Cross-term transitions induce relatively small effects on the spectral intensities. Their overall contribution does not exceed 2-3% of the zeroth-order moment. Bound, metastable and low-energy free dimers show a substantial contribution at zero frequency, thus evidencing the elastic Rayleigh peak responsible for 3.3% of the integrated intensity. In the range $2<|\omega|<10$ cm^{-1} an overall broad rovibrational structure is marked, well fitting the experimental data (7) (Fig. 2). Computed intensities using Dacre's SCF anisotropy (2) follow reasonably well the experimental spectrum even in the high frequency wings (4,7) (Fig. 3). Deviations from classical predictions are observed at the high frequencies (4) that should partly be attributed to the *ad hoc* desymmetrization of the classical spectrum (3).

B. The isotropic spectrum

The modified-asymptotic model A_6/r^6 (with $A_6=9665$ a_0^9) well reproduces the experimental spectrum at intermediate frequencies, becoming insufficient in the far wings. The asymptotic dipole-induced-dipole trace, displaying the same slope as that of the modified-asymptotic model in logarithmic scale, joins the experimental spectrum from below (Fig. 1). Surprisingly, Dacre's SCF trace (2) in logarithmic scale is seen to well reproduce the slope of the experimental spectrum all along the frequency domain (Fig. 1). This observation suggests a new model resulting from the Dacre's trace by simply a scaling constant. Its physical meaning should be searched in the notion of a negative contribution to the $1/r^6$ term of the dimer polarizability expansion (1).

REFERENCES

1. O. Gaye, M. Chrysos, V. Teboul and Y. Le Duff (to be published)
2. P. D. Dacre, *Mol. Phys.* **45**, 1 (1982)
3. F. Chapeau-Blondeau, V. Teboul, J. Berrué and Y. Le Duff, *Phys. Lett.* **173A**, 153 (1993)
4. M. Chrysos, O. Gaye and Y. Le Duff, *J. Phys.* **B29**, 583 (1996)
5. D. T. Colbert and W. H. Miller, *J. Chem. Phys.* **96**, 82 (1992)
6. D. W. Norcross and M. J. Seaton, *J. Phys.* **B6**, 614 (1973)
7. M. Chrysos, O. Gaye and Y. Le Duff, *J. Chem. Phys.* **105** (1996) (in press)
8. F. Barocchi, M. Zoppi, M. H. Proffitt and L. Frommhold, *Can. J. Phys.* **59**, 1418 (1981)
9. M. Chrysos and R. Lefebvre, *J. Phys.* **B26**, 2627 (1993)
10. R. A. Aziz and M. J. Slaman, *J. Chem. Phys.* **92**, 1030 (1990)
11. M. Chrysos and O. Gaye (to be published)

Measurements and Analysis of the He-Xe Absorption Spectrum in the Far Infrared

P. Dore*, L. Frommhold**, A. Nucara*, and P. Postorino*

*INFM - Dipartimento di Fisica, Universita' di Roma "La Sapienza",
Piazzale A.Moro 2, 00185 Roma, Italy
**Physics Department, University of Texas at Austin, Texas 78712-1081

Abstract. Accurate measurements of the translational band were recently performed in the far infrared region on low density He-Xe mixtures. Accurate values of intensity and range of the overlap induced dipole were obtained. A new analysis based on exact quantum line shape computations shows that a term representing the effect of dispersion must be included in the induced dipole, in order to obtain a translational profile which is consistent with the experimental one.

In low density mixtures of noble gases, the translational absorption band observed in the far infrared is due to dipole moments induced during collisions between dissimilar pairs of atoms [1]. These dipoles arise from small shifts of the centers of charge caused by exchange, overlap, and dispersion forces acting during the collision. By measuring shape and intensity of a collision induced absorption spectrum, in principle it is thus possible to derive information on both interatomic potential and induced dipole [1].

In the past, accurate absorption spectra were collected for He-Ar [2] and Kr-Ar [3] low density mixtures, but absorption wings were not measured. We recently measured the room temperature translational absorption band of the low density He-Xe mixture [4] and both low- and high-frequency wings were accurately measured. As usual, we described the shape of the translational band through a model profile. Owing to the accuracy of the measured spectral wings, for the first time it was possible to discriminate between the different model profiles reported in the literature [5,6,7] and to obtain reliable extrapolations of the measured spectrum down to zero frequency and towards high frequencies. A quite accurate determination of the spectral moments γ_0, γ_2, and γ_4 has thus been done. A standard spectral moment analysis was performed considering only the overlap contribution to the induced dipole, described by the usual exponential model:

$$\mu(R) = \mu_0 \exp[-(R-\sigma)/\rho] \qquad (1)$$

By employing a refined potential model for He-Xe interactions [8], we derived accurate values of intensity, $\mu_0 = 2.0 \times 10^{-20}$ e.s.u. and range, $\rho = 0.0372$ nm [4].

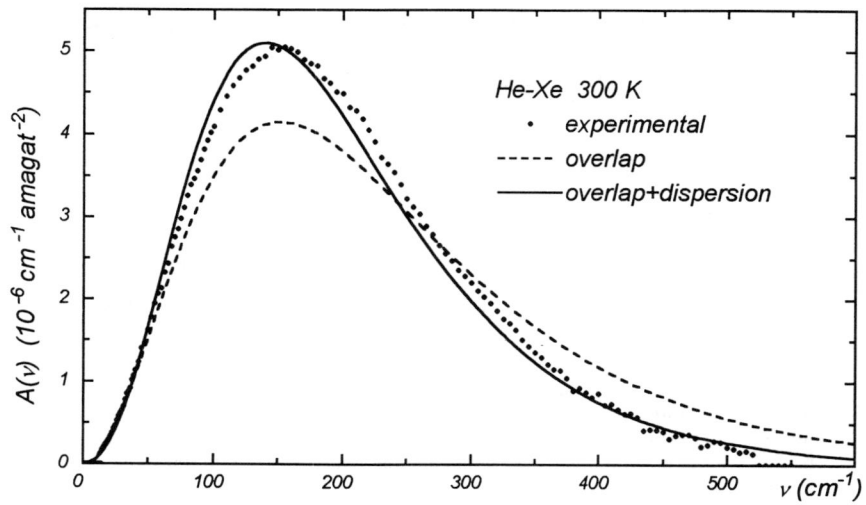

Fig.1 Comparison between experimental and computed absorption spectra

In general, the information which can be extracted from the spectral moments (i.e. from integrals which represent average quantities) is much less detailed than that provided by a direct analysis of the spectral profile. A new line shape analysis of the He-Xe translational band, based on exact quantum line shape computations, was thus performed. We found that the exponential dipole model (eq.1) does not reproduce the measured profile, as shown in Fig.1, in spite of the fact that the first spectral moments of the computed spectrum are in good agreement with the experimental ones. An absorption profile consistent with the experimental data is obtained [9] by introducing a dispersive term ($-D/R^7$) in eq.1 (see Fig.1). The main effect of this term is to remove the excess absorption at high frequencies falsely predicted by the exponential dipole model. The parameters we found for the induced dipole are $\mu_0 = 7.07 \cdot 10^{-20}$ e.s.u., $\rho = 0.0433$ nm, $D = 12000$ a.u.

REFERENCES

[1] L. Frommhold, *Collision-induced absorption in gases*, Cambridge University Press (1993).
[2] D.R. Bosomworth and H.P. Gush, Can.J.Phys. **43**, 751 (1965).
[3] U. Buontempo, S. Cunsolo, P. Dore, and P. Maselli, J.Chem.Phys. **66**, 1278 (1977).
[4] P. Dore, L. Finzi, A. Nucara, P. Postorino, and M. Rovere, Molec.Phys. **84**, 1065 (1995).
[5] H.B. Levine and G. Birnbaum, Phys.Rev. **154**, 86 (1967).
[6] V.F. Sears, Can.J.Phys. **46**, 1163 (1968).
[7] G. Birnbaum and E.R. Cohen, Can.J.Phys. **54**, 593 (1976).
[8] R.A. Aziz, U. Buck, H. Jonsson, J.C. Ruiz-Suarez, B. Schmidt, G. Scoles, J.M. Slaman, and J. Xu, J.Chem.Phys. **91**, 6477 (1989).
[9] L. Frommhold and P. Dore, to be published.

Collisional Interference in the Vibration-Rotation Spectrum of HD-He

B. McQuarrie and G.C. Tabisz

Department of Physics, University of Manitoba
Winnipeg, Manitoba, Canada, R3T 2N2

Abstract. The line shape parameters for the infrared vibration rotation spectrum of HD-He are calculated. They are found to be larger than those in the pure rotational case, strongly affected by collisional propagation among rotational levels in inelastic collisions and sensitive to the vibrational state dependence of the intermolecular potential.

INTRODUCTION

The infrared spectrum of HD is remarkable in that it displays effects due to interference between allowed and collision-induced transitions. The phenomenon was discovered by McKellar (1) and the first theoretical explanation was given by Herman, Tipping and Poll (2). In the simplest scenario, for interference to occur, both the allowed and induced dipole moments must be of the same symmetry, that is, arise from the same selection rules. The induced dipole moment in an interacting HD-X pair, where X is a perturber, can contain a component of the required symmetry, which does not depend on the orientation of the perturber or of the intermolecular axis. This component lies in the direction of the internuclear axis of HD, either parallel or antiparallel to the allowed moment, thereby leading to constructive or destructive interference. The induced moment dies away quickly, as it depends strongly on the magnitude of the intermolecular distance. Another collision can induce another dipole with a component in the same direction. The allowed dipole transition is perturbed by collisions and undergoes collision-broadening. The phase associated with the induced transition is locked to that of the allowed transition with a loss of phase correlation in time for both caused by collision-broadening type encounters (3). The resulting line shape is a narrow feature, with a width characteristic of collision-broadening, rather than a broad profile which typically accompanies pure collision-induced transitions.

THEORY

Gao, Tabisz, Trippenbach and Cooper have developed a general theoretical description of the phenomenon (4), which allows propagation among the HD rotational states during inelastic collisions. The including of an anisotropic intermolecular potential in the calculation allows several components of the induced dipole moment to participate in the interference effect (5). This treatment has been used to compute line shape parameters of the pure rotational transitions for HD-He and HD-Ar (5).

The expression obtained for the absorption coefficient $\alpha(\omega)$ at frequency ω is

$$\alpha(\omega) = n_R \left(\frac{4\pi^2 \omega}{3\hbar c} \right) (1 - e^{-\beta\hbar\omega}) P(J_g) |\mu_R|^2$$

$$\times \left(\frac{Bn_p/\pi}{(\Delta - Sn_p)^2 + (Bn_p)^2} (1 + an_p + bn_p^2) \right.$$

$$\left. - \frac{(\Delta - Sn_p)/\pi}{(\Delta - Sn_p)^2 + (Bn_p)^2} (cn_p + dn_p^2) \right). \quad (1)$$

Here n_R and n_p are the radiator and perturber number densities; β is $1/k_BT$; $P(J_g)$ is the Boltzmann distribution function for the initial state J_g; μ_R is the reduced matrix element for the allowed radiator dipole $\langle J_e \| \mu_R \| J_g \rangle$, where J_e refers to the final state. The parameters B and S are the line broadening and frequency shift coefficients in the impact approximation. The detuning $(\omega - \omega_{eg})$ is given by Δ. The total line shape is the sum of Lorentzian and dispersion components whose relative strengths are given, respectively, by the quantities a and b and c and d, which depend on the pair induced dipole moment. Explicit expressions are given for them in ref. 5.

The purpose of the present paper is to extend these calculations to the vibration-rotation spectrum. In this case collisional propagation can occur in principle among levels in the ground vibrational state, among levels in the excited vibrational states and between levels in the two vibrational states. The collision duration is too long and the temperature too low to make this last contribution significant. It is neglected in the following analysis.

CALCULATION

In order to perform the calculation three quantities are needed: allowed dipole transition elements, the induced dipole transition elements and the intermolecular potential. Allowed dipole elements have been calculated by Wolniewicz (6). Ab initio calculations of the induced dipole moment involved in the v = 0 to v = 1 transition for HD-He have been made by Borysow et al (7). Differences in the rotational energy level constants between vibrational states affect the degree of mixing within the rotational manifolds of the two vibrational states.

Vibrational dependence was introduced into the potential by following the method of Stefanov (8) which introduces different collisional diameters for head-on collisions in different vibrational states. The rotational constant B_v is inversely proportional to the internuclear distance r_v. If a head-on collision is assumed, the collisional diameters σ_1 and σ_0 for the two vibrational states are related by $\sigma_1 - \sigma_0 = (r_0/2)(\delta - 1)$ where δ is the ratio $(B_0/B_1)^{1/2}$. The potential used is that for H_2-He developed by Mulder et al (9). The isotropic part of this potential is shifted to produce a zero at σ_1 but remain unchanged at large R. The origin of this potential is shifted from the centre of mass of H_2 to the centre of mass of HD (5). This shift gives a vibrational state dependence to both the isotropic and anisotropic components of the resulting HD-He potential. When the collision partners enter the collision, the molecule HD is in the ground vibrational state and the intermolecular potential is denoted V (v = 0). After the induced dipole moment operator has acted, the molecule is in the excited state and the potential is V (v = 1). This situation affects the nature of the collisional propagation, calculated through matrix elements of the time evolution operator and the S-matrix.

The parameters a and c depend on $(\mu_R)^{-1}$ and b and d on $(\mu_R)^{-2}$. The value of μ_R is about -8.3×10^{-4} for the pure rotational transitions and $+5.5 \times 10^{-5}$ for the lines of the fundamental band (6). By this reckoning, a and c should differ by about a factor of -15 between the two types of transitions; similarly both b and d should differ by about 230 between these cases. In fact, a and c are always of opposite sign and have larger magnitude for the vibrational case; b and d usually have the same sign and again are larger for the vibrational lines. Their ratios for corresponding $R(J)$ transitions can however be considerably different from 15 and 230; collisional propagation in the excited vibrational state is thereby shown to be important.

Comparison with experiment is possible only for $R_1(0)$ and $R_1(1)$ at 77 K. Agreement for a is quite good and reasonable for b, the parameters affecting the total intensity in the line. The calculated asymmetry parameters c and d are, on the other hand, too small, especially for $R_1(1)$. The calculation does predict a large asymmetry, but not as large as experiment shows.

When an intermolecular potential independent of vibrational state is used in the calculation, the parameters do change, all of them decreasing: a by less than 1%, b by up to 5%, c by 2 to 25%, and d by 2 to 15%. The largest changes occur at 295 K. The broadening coefficient B changes uniformly by about 1-2%. Full details of the calculations are given in ref. 10.

DISCUSSION

Previous work (5) has shown that collisional propagation is important to the description of collisional interference, particularly to the line asymmetry. Investigations of the contributions of individual components of the induced dipole moment revealed that different components play major roles in the intensity interference and line asymmetry phenomena (11).

The intensity interference parameters (a and b) are determined by short range overlap components of the induced dipole, while the asymmetry parameters (c and d) are affected by both short range anisotropic and long-range multipole components. The present results indicate that the intensity effect is not very sensitive to the vibrational state dependence of the intermolecular potential but that the line asymmetry is. The collisional interference phenomenon seems to consist of two distinct effects which may be complementary sources of information on intermolecular interactions.

REFERENCES

1. A. R. W. McKellar, *Can. J. Phys.* **51**, 389 (1973).
2. R. Herman, R. H. Tipping and J. D. Poll, *Phys. Rev. A* **20**, 2006 (1979).
3. R. Herman, in *Spectral Line Shapes*, edited by R. Exton (Deepak, Hampton, 1987), Vol. 4, p. 351.
4. B. Gao, G. C. Tabisz, M. Trippenbach and J. Cooper, *Phys. Rev. A* **44**, 7379 (1992).
5. B. Gao, J. Cooper and G. C. Tabisz, *Phys. Rev. A* **46**, 5781 (1991).
6. L. Wolniewicz, *Can. J. Phys.* **54**, 672 (1976).
7. A. Borysow, L. Frommhold and W. Meyer, *J. Chem. Phys.* **88**, 4855 (1988).
8. B. Stefanov, *J. Phys. B* **25**, 4519 (1992).
9. F. Mulder, A.V.D. Avoird and P.E.S. Wormer, *Mol. Phys.* **37**, 159 (1979).
10. B. McQuarrie and G.C. Tabisz, submitted to *J. Molec. Liquids* (1996).
11. B. McQuarrie, G.C. Tabisz, B. Gao and J. Cooper, *Phys. Rev. A* **52**, 1976 (1995).

The Far-Infrared Continuum Spectrum of the Milky Way Explained by a Dust and Gas Model

Joachim Schaefer

Max-Planck-Institut für Astrophysik, 85740 Garching, Germany

Abstract. The far-infrared continuum spectra observed in the FIRAS-COBE mission are explained by a one-component dust source in the Galaxy and a cold excess due to collision induced radiation of the primordial mixture of normal molecular hydrogen and helium gas in the outer Galaxy.

The dust model is based on an empirical temperature-dependent mass absorption coefficient and a reasonable dust density distribution close to the Galactic plane. The temperature of the dust along the line of sight is determined by a fit curve.

The "dark matter" candidate cold molecular (normal-) hydrogen gas, in the primordial mixture with 14% helium, has been chosen for the gas model. It is conceivable as condensed gas in hierarchical fractal structured clouds with gravitationally bound elementary fragments in which enormous density gradients occur in gravitational turbulence providing the observed very low volume-filling high density inhomogeneities. The gas spectrum emitted in a small temperature interval at 11 Kelvin and at the estimated (equilibrium-) sublimation density of about $4.5 \cdot 10^{18}$ cm^{-3} has been found appropriate to accurately fit the remaining cold excess of the FIRAS-COBE spectra all over the sky. It was also found that the fitted column densities of the gas sources are closely correlated with the very specific HI (atomic H) column densities in the outer Galaxy (at $R_G > R_\odot$).

There is an alternative to the gas sources: the emission spectrum of a "cold" dust source of 7.0 K could equally well fit the remaining cold excess intensity.

Introduction

A rather small part of the collision induced radiation (CIR) of H_2 gas, the pure translational band of normal-H_2 at very low temperature, could possibly gain great interest in astrophysics if the dust and gas model presented in this paper to explain the continuum spectra observed in the FIRAS-COBE mission is proved valid. This translational band of H_2 gas has been calculated by using bound and scattering wave functions of the H_2–H_2 system obtained from a rigid-rotor non-spherical interaction potential described previously (Schaefer 1994) and a rigid-rotor dipole moment function obtained in ab initio calculations by Meyer et al.(1989). Quite satisfying agreement has been found in comparisons with laboratory measurements (Wishnow et al., 1996) at temperatures between 22 and 36 Kelvin and at frequencies from 20 to 320 cm^{-1}.

© 1997 American Institute of Physics

Validity of the calculated spectrum at 11 Kelvin, the temperature of interest here, and at frequencies down to 1 cm^{-1} can be assumed. Earlier measurements at 77 K have been reported by Birnbaum and Cohen (1976), Birnbaum (1978), and Dore et al. (1983).

There is no doubt about the CIR observability of H_2 in interstellar medium: high densities of the order of 0.1 amagat or more are required. This condition is fulfilled in the heavy planet atmospheres. Their infrared opacities have been predicted by Trafton (1966) to be mainly due to CIA rotational-translational spectra of H_2 pairs which has been observed by Hanel et al.(1979) in the VOYAGER-IRIS mission. Comparable densities occur at lower temperatures ($\approx$ 11 K) during sublimation of the gas. The question is: where and how could observable sublimation happen ?

Each fit of a FIRAS-COBE continuum spectrum is started by doing a reasonable fit of the dust radiation. There seems to be no problem with the dust model of the Galaxy ranging out to a Galactic radius of $\approx$15 kpc. The remaining intensity - as a cold excess - could be fitted successfully all over the sky with the translational emission spectrum of the primordial mixture of normal hydrogen and helium gas of 11 Kelvin. This is of course highly speculative, and a model of this source is not easily found. It is quite clear from the beginning that a homogeneous gas cloud cannot be the source. When strong correlation of the cold excess source with the well-known HI (atomic H) abundances in the outer Galaxy (at Galactic radii $R_G > R_\odot$) has been found, it was near at hand to interpret the gas source as being due to a small observable part of the dark-matter candidate cold molecular hydrogen. But an alternative dust source of 7 K obtained from the same dust model is shown to give also a spectrum so close to the spectrum of the proposed hydrogen gas model that the FIRAS-COBE observations can obviously not be decisive.

The Dust Model

More than 5000 spectra of the FIRAS-COBE mission have been released by NASA in 1995. They contain the radiation intensity of the microwave cosmic background and a rest called *Galactic Dust Continuum* after subtraction of the line spectra, the zodiacal light spectrum, and many types of systematic noise as well as "glitches" due to cosmic ray impacts on the bolometers. The main source of the rest is obviously dust in the Galaxy heated by star light. It is obvious as well to assume varying dust temperatures along the line of sight if the direction of observation is close to the Galactic plane. In order to account for this, an empirical temperature dependent mass absorption coefficient has been introduced (Schaefer, 1996) which has to be multiplied by the dust density along the line of sight. A reasonable density function with an exponential decrease at increasing Galactic radius and a Gaussian distribution perpendicular to the Galactic plane has been used for all fitted spectra. The temperature of the dust vs. Galactic radius is fitted for each spectrum to reach optimum agreement with the measured intensity at the high frequency side. The calculated flux is summed over a pixel distribution which is determined by the beam profile of the simulated instrument and, if the Galactic plane touches the field of view, according to the dust density distribution. All details of the dust model are described in a recently published paper (Schaefer 1996).

The Gas Model

After a satisfying dust spectrum has been fitted, the assumed validity of the Kramers-Kronig theorem (for spheroidal dust grains) makes sure that a cold excess intensity is always remaining which is fitted by the collision induced emission spectrum of the primordial mixture of (86%) normal hydrogen and (14%) helium gas of 11 K times a factor which is related to the gas column density – *and this holds for all* FIRAS-COBE *continuum spectra*. An example is shown in Fig. 1. Necessary conditions for the gas source model are:
1) It has to be gas, and the ortho-para ratio has to be normal (3:1) or close to normal. Solid hydrogen would become para-hydrogen after a short time.
2) The density of the observed gas should be as large as the equilibrium sublimation density of normal-H_2 at 11 K: 685 Pa or 0.168 amagat or $4.5 \cdot 10^{18}$ cm^{-3}. Significantly lower densities (e.g. sublimation densities) at lower temperatures are not observable. It would require an unreasonable amount of hydrogen mass. Thermodynamical equilibrium should not be assumed valid.

A hint for a source model was in sight when close correlation of the cold excess column densities with the well-known and very specific HI column densities (Burton and Te Lintel Hekkert, 1986) in the outer Galaxy was found, at least for scale lengths as given by the FIRAS 7° beam angle, at three carefully

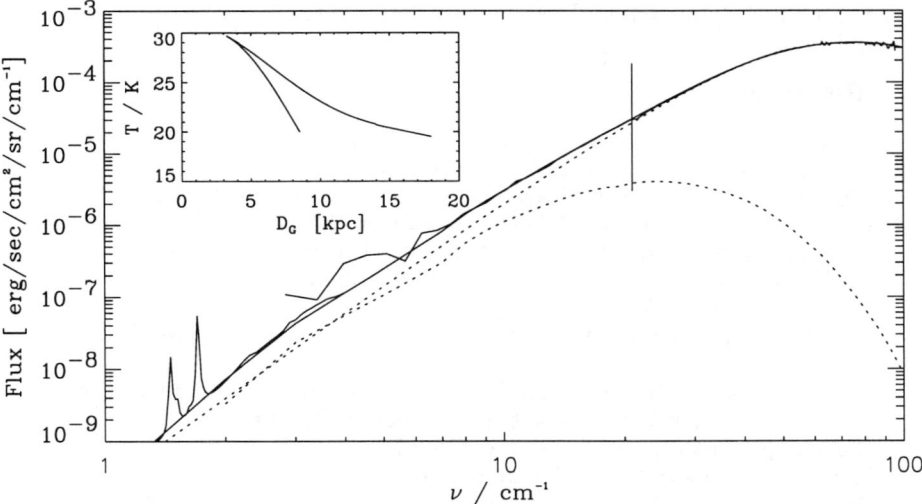

Fig. 1: The FIRAS-COBE continuum spectrum observed close to the Galactic plane at $(l, b) = (337.746°, -0295°)$, fitted by a dust spectrum with a temperature function vs. Galactic radius as shown in the insert and by the collision induced emission spectrum of the primordial mixture of normal-H_2 and He gas of 11 K, with theoretically predicted dimer features of H_2 in the lower left corner.

chosen intervals in Galactic longitude angle and over latitude angles inside ±40 - 50 degrees. Concomitance with HI could make the gas source part of dark matter (Pfenniger et al., 1994). In addition, much higher densities of molecular hydrogen than formerly assumed in ISM are subject of a dark matter model of Pfenniger and Combes (1994) who introduced condensed cold molecular hydrogen (of about 3 K and therefore not observable) in hierarchical fractal structures with gravitationally bound elementary fragments. In following these thoughts I may add two more conditions for the gas sources:

3) The H_2 and He gas could be correlated with HI in a way that mainly these three species together are condensed in hierarchies of fractal structured clouds with gravitationally bound elementary fragments and, in contrast to Pfenniger et al., further fractalization in gravitational turbulence with enormous density gradients in these fragments could provide the observable top densities occuring in very small, very low volume-filling inhomogenieties.

4) The relative large sublimation energy (≈ 70 cm^{-1} per molecule) is normally not dissipated in the small volume units of top densities. This kind of maintainance of the gas sources is provided by the familiar conditions of ISM in the outer Galaxy: not enough grains and too low metallicity.

There is an alternative cold dust source of 7 Kelvin which fits the cold excess intensity also very well, all over the sky, and therefore competes for "dark matter" with the proposed gas source. Since the same strong correlation with HI is valid, this kind of "extra" dust overlaps with the warmer "normal" dust and, because of the low temperature, has about 4 –7 times the opacity compared to the dust inside the Galaxy. The typical integrated emission intensity in the Galactic anticenter direction is $3.1 \cdot 10^{-5}$ erg/s/cm^2 for both sources. Could somebody offer a model for this ?

References

1. Birnbaum G. and Cohen E.R., (1976), Can.J.Phys. **54**, 593
2. Birnbaum G., (1978), J.Quant.Spectr.Rad.Transf. **19**, 51
3. Burton W.B. and te Lintel Hekkert P., Astron.&Astrophys. S. **65**, (1986), 427
4. *COBE Far Infrared Absolute Spectrophotometer (FIRAS) Explanatory Supplement*, ed. J. C. Mather, R. A. Shafer, R. E. Eplee, D. J. Fixsen, R. B. Isaacmen, and A. R. Trenholme, COBE Ref.Pub. No. 95-C (Greenbelt, MD: NASA/GSFC).
5. Dore P., Nencini L, and Birnbaum G., (1983), J.Quant.Spectr.Rad.Transf. **30**, 245
6. Hanel R., Conrath B., Flasar M., Kunde V., Lowman P., Maguire W., Pearl J., Pirraglia J., Samuelson R., Gautier D., Gierasch P., Kumar S., Ponnamperuma C., (1979), Science **204**, 972
7. Meyer W., Borysow A., and Frommhold L., (1989), Phys. Rev. A **40**, 6931
8. Pfenniger D., Combes F., and Martinet L., Astron.&Astrophys. **285** (1994), 79
9. Pfenniger D. and Combes F., Astron.&Astrophys. **285** (1994), 94
10. Schaefer J., (1994), Astron. Astrophys. **284**, 1015
11. Schaefer J., (1996), Europhys. Lett. **34**, 69
12. Trafton L.M., (1966), Astrophys.J. **146**, 558
13. Wishnow E.H., Ozier I., Gush H.P., and Schaefer J., to be published.

Frequency-Dependent Quenching in the Emission Spectra $3P\Lambda, 3D\Lambda \rightarrow 2P\Lambda$ of Lithium-Rare Gas Collision Molecules

W. Behmenburg[*], J. Bonsmann[*], A. Kaiser[*], A. Makonnen[*], W. Meyer[**]

[*] : Institut f. Experimentalphysik, University Düsseldorf, Universitätsstr. 1, D-40225 Düsseldorf; email : kaisera@mail.rz.uni-duesseldorf.de
[**] : Fachbereich Chemie, University Kaiserslautern, Erwin-Schrödinger Str., D-67663 Kaiserslautern ; email : meyer.chemie@uni-kl.de

Excitation- and emission spectra on the $2P\Lambda \leftrightarrow 3P\Lambda, 3D\Lambda$ transitions in LiX collision molecules (X = He, Ne, Ar) have been measured in the 15800 - 17600 cm^{-1} range around the atomic Li line at 16379 cm^{-1}, using two-step cw-laser excitation in cell experiments [1]. Absolute intensity scaling for both types of spectra is established by their normalisation to the separately measured total Li3D↔2P line intensities and rare gas number densities n_X.

Pronounced rainbow satellite bands are observed in the spectra; these are identified as related to potential barriers in the upper $3P\Sigma$ and $3D\Sigma$ states by using recent ab-initio potentials and transition moments [2,3].

In the LiNe and LiAr spectra, the emission intensity is found to be generally less than expected from the excitation intensity according to Kirchhoff's law at the experimental gas temperature (see figure). This quenching of the $3D\Sigma \rightarrow 2P\Lambda$ emission in the blue wings of the Li line is most pronounced and increases with detuning. In LiHe, no quenching in the emission spectrum is observed.

The observed quenching of molecular emission is probably related to collisional energy transfer from the Li3D level. In LiAr, the rate for 3D-3S transfer is $4 \cdot 10^8$ s^{-1} at $n_{Ar} = 4 \cdot 10^{18}$ cm^{-3}, which is large compared to the spontaneous decay rate $7 \cdot 10^7$ s^{-1}.

As quenching mechanism for the blue wing emission nonadiabatic transitions between 3DΣ and 3SΣ potentials at small internuclear distance is proposed. The rapid separation of these potentials as well as increasing 3DΣ barrier heights as one goes to lighter rare gases could then explain the different quenching behaviour in the rare gas series. The frequency dependence of the quenching can then be understood by assuming that only low-energy collisions, that cannot pass the 3DΣ barrier, contribute to the radiation.

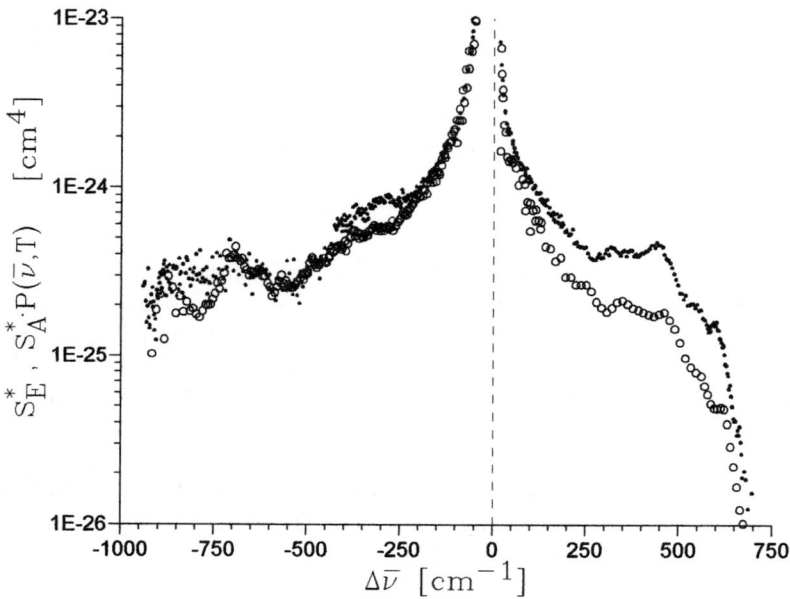

Exitation- and emission spectra on the 2PΛ↔3PΛ,3DΛ transitions in LiAr with reference to the Li-transition 2P-3D. Measurements at p_{Ar} = 340 mbar, T = 640 K.

- ∘ ∘ ∘ : normalised emission intensity S_E^*
- ••••• : normalised excitation intensity S_A^* multiplied by the Planck-function $P(\bar{\nu}, T)$ at gas temperature

References :

[1]: A.Makonnen, A. Kaiser, W.Behmenburg, Z.Phys.D **36**, 325 (1996)
[2]: W.Behmenburg, A.Makonnen, A.Kaiser, F.Rebentrost, V.Staemmler, M.Jungen, G.Peach, A.Devdariani, S.Tserkovnyi, A.Zagrebin, E.Czuchaj to appear in J. Phys. B
[3]: M.Jungen, private communication
[4]: G.Ennen, Ch.Ottinger, Chem.Phys.Lett. **88**, 487 (1982)

Molecular dynamics studies of light scattering from Ar$_{13}$ cluster: a two-body correlations

A. Dawid and Z. Gburski

Department of Physics, University of Silesia, Uniwersytecka 4, 40-007 Katowice, Poland,

Abstract. Molecular dynamics (MD) simulations have been used to calculate the two-particle correlation functions $C_2(t)$ which contribute to depolarized Rayleigh scattering from argon cluster Ar$_{13}$ at various liquid temperatures. The data obtained from this computer experiment can be used for the development and test of the appropriate theoretical model describing the two-body dynamics in a small cluster.

Depolarized light scattering is observed in monoatomic fluids because interatomic interactions induce time-dependent anisotropies, even though the atomic polarizability is isotropic. For atomic fluids, the dipole-induced-dipole (DID) interaction dominates. The DID interaction results from the fact that the incident laser beam induces a dipole on the ith atom and this dipole generates a "local field" at the jth atom. The pair anisotropy α_{ij} in the DID limit is (1)

$$\alpha_{ij}(t) \propto 3\, x_{ij}(t)\, z_{ij}(t)\, r^{-5}{}_{ij}(t) \tag{1}$$

where x_{ij} and z_{ij} are components of the separation vector $\mathbf{r}_{ij}$ between the ith and jth atoms. The depolarized light scattering spectrum is the Fourier transform of the polarizability anisotropy autocorrelation function $C(t)$ which can be decomposed into pair, triplet and quadruplet contributions (1), $C(t)=C_2(t)+C_3(t)+C_4(t)$. For the bulk monoatomic systems of N atoms the correlation functions $C(t)$ have been intensively studied (1), (2). In this contribution we would like to point out that by studying finite, small clusters one can better understand the origin of collective behaviour in bulk systems. We focus on the normalized two-body correlation function $C_2(t)$,

$$C_2(t) = \sum_{\substack{i,j=1 \\ (j>i)}}^{N} <\alpha_{ij}(t)\,\alpha_{ij}(0)> <\alpha^2{}_{ij}(0)>^{-1} \tag{2}$$

© 1997 American Institute of Physics

leaving the more elaborate three- and four-particle correlations for a further study. The DID interaction dies down relatively slowly (varies as r_{ij}^{-3}), this ensures that $C_2(t)$ is a sensitive probe of two-body dynamics in a cluster. The correlation functions $C_2(t)$ have been obtained by standard MD simulations of Lennard-Jones argon cluster Ar_{13} ($\sigma=3.405\ 10^{-8}$ cm, $\varepsilon/k_B=119.8$ K) at various liquid state temperatures T. A time step of 0.01 ps was used to numerically integrate the equation of motion (Verlet algorithm), the equilibrium configurations were collected after each 0.1 ps time step. The correlation functions were averaged over 200 time origins. The results are presented in the figure 1.

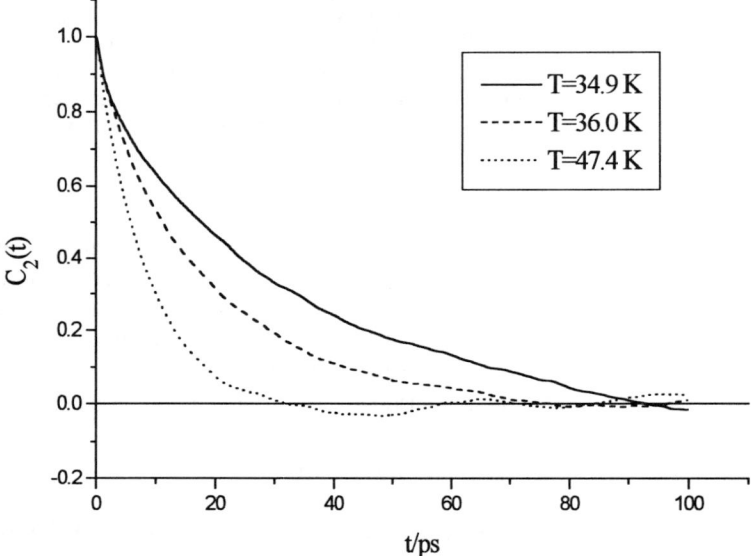

FIGURE 1. The simulated (MD method) two-body correlation functions $C_2(t)$ for the argon cluster Ar_{13} in the liquid state.

The simulations show that the correlation functions $C_2(t)$ in argon cluster decay slowly and quite regularly. The striking result of our simulations is the apparent sensitivity of $C_2(t)$ to the variation of temperature, a small increasing of the cluster temperature results in a significantly faster relaxation of $C_2(t)$. The results of the computer experiment provided for the cluster Ar_{13} may be used as a test of the future theoretical models of depolarized light scattering in a small atomic systems.

REFERENCES

1. Laad, A., Litovitz, T., and Montrose, C., J. Chem. Phys. **71**, 4242-4248 (1979).
2. Gburski, Z., Gruszka, M., and Dorfmuller, T., J. Mol. Liq. **54**, 63-68 (1992).

Contributions of multipolar polarizabilities to the anisotropic and isotropic light scattering induced by molecular interactions in gaseous CF_4

A. Elliasmine[◊], J.-L. Godet[◊], Y. Le Duff[◊], and T. Bancewicz[*]

[◊] *Laboratoire des Propriétés Optiques des Matériaux et Applications, EP CNRS 0130
2, bd Lavoisier, Université d'Angers, 49045 Angers, Cedex, France*

[*] *Nonlinear Optics Division, Institute of Physics, Adam Mickiewicz University,
Umultowska 85, 61-614 Poznan, Poland*

Abstract. Both *anisotropic* and *isotropic* scattering spectral components from pairs of gaseous CF_4 have been measured in the vicinity of the Rayleigh line in absolute intensities. Due to the large frequency shifts scanned (2-180cm^{-1}), thorough analysis of the influence of linear and nonlinear multipolar contributions has been performed. Comparison with experiment allows us to propose values for the irreducible spherical components of the dipole-quadrupole and dipole-octopole polarizability tensors. These values are compatible with recent quantum *ab initio* computations.

The analysis of the light scattered by an assembly of molecules can give information on collision-induced pair polarizabilities. In the case of gaseous CF_4, collision-induced scattering is not completely depolarized and has two components in the frequency range close to the Rayleigh line. The main one is anisotropic (1) while the weak one is isotropic. Before the present work, the latter component had not been observed (2). Likewise, the binary scattering intensities have been obtained in absolute units for the first time beyond 60 cm^{-1} (2-180 cm^{-1}). Therefore, the opportunity is presented to test the multipolar polarizability model and to compare its parameters with the corresponding *ab initio* computed values. Inspite of the fact that analysis of a pair of CF_4 molecules by means of quantum mechanics is a rather complicated task, the influence of the classical dipole-multipole mechanisms of tetrahedral molecules is well-known. In particular, it is expected that the dipole-quadrupole and dipole-octopole polarizability tensors **A** and **E** contribute significantly to the high frequency scattered intensities for this type of molecule (3). Nevertheless, due to the very large Raman shift domain examined, theoretical attention has been paid on the higher terms of the multipolar theory. In particular, the impact of the nonlinear polarizabilities combined with permanent multipoles on the anisotropic spectrum has been thoroughly studied (1). Figures 1 and 2 present the anisotropic and isotropic spectra, respectively, as well as a comparison with theoretical calculations. The dipole-induced-multipole components (DID, DIQ & DIO) have been obtained using the polarizability $\alpha = 2.93$ Å^3 and chosen values of the irreducible spherical tensor components A and E. The "cross" branch ATA+ATE+ETE includes also a nonlinear contribution using *ab initio* computed data of Maroulis (4). For each spectrum, maxima of A and E can be evaluated. The range values $A = 0.5 - 1.2$ Å^4 and $E = 1 - 3.5$ Å^5 are compatible with both spectra.

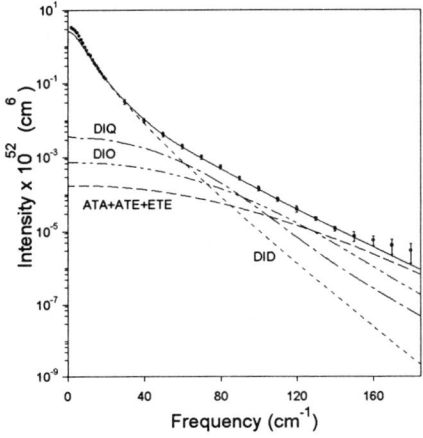 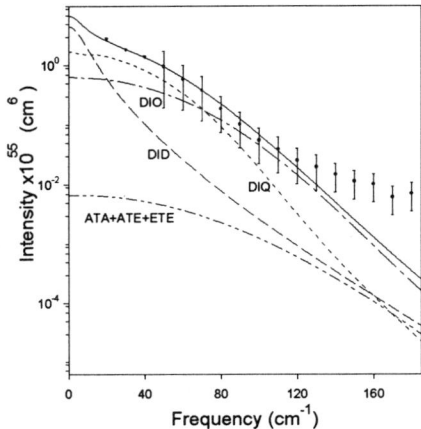

FIGURE 1. CIS binary anisotropic experimental spectrum of CF$_4$ (bars). The theoretical one (solid line) is computed for $A = 1.2$ Å^4 and $E = 3.5$ Å^5.

FIGURE 2. CIS binary isotropic experimental spectrum of CF$_4$ (bars). The theoretical one (solid line) is computed for $A = 0.7$ Å^4 and $E = 2.5$ Å^5.

Our value of A is close to that ($A = 0.97$ Å^4) computed by Maroulis (4). Use of the latter value implies that $E = 1.5 \pm 0.5$ Å^5, a value which is also compatible with the Maroulis' $E = 1.15$ Å^5. In this case, theoretical spectra lie lower than experimental ones in the high frequency range; this effect may be due to some short range contributions (such as molecular frame distortion, valence repulsion and/or exchange overlap effects) that have been neglected in our model. Nevertheless, collision-induced scattering provides a good agreement between both the A and E values deduced from experimental data and quantum *ab initio* computations.

REFERENCES

1. A. Elliasmine, J.-L. Godet, Y. Le Duff and T. Bancewicz, *Mol. Phys.* (in press)
2. A. Elliasmine, J.-L. Godet, Y. Le Duff and T. Bancewicz (in preparation)
3. A.D. Buckingham and G.C. Tabisz, *Opt. Lett.* 1, 220-222 (1977)
4. G. Maroulis, (to be published)

Triple Transitions $3Q_1$ in Compressed Hydrogen

Lothar Frommhold [*] and Massimo Moraldi [†]

[*] Physics Department, University of Texas at Austin, Texas 78712-1081 and
[†] Dipartimento di fisica, Universita' di Firenze and Istituto Nazionale di Fisica della Materia, Unita' di Firenze, Largo E.Fermi 2, I-50125 Firenze, Italy

Abstract. We have calculated the integrated intensity of the $3Q_1$ transition in hydrogen. The calculation is based on the one effective electron theory for the modelling of the irreducible three body dipole moment. The results of the calculation are in qualitative agreement with a recent measurement - the first measurement of that kind.

In a triple vibrational transition ($3Q_1$) three molecules absorp one single photon and as a result they make simultaneous transitions from the ground to the first vibrational state. Such a transition is possible if the three molecules are mutually interacting during the time in which the photon is absorbed. The microscopic property that interacts with the radiation field is an irreducible three-body electric dipole moment, that is a dipole that depends in an irreducible way on the relative configuration of the cluster of three molecules.

The $3Q_1$ transition in hydrogen is expected to appear around a frequency of 12,466 cm^{-1}, that is in the second overtone spectral region. Recently the absorption coefficient in that spectral region has been measured [1]. It is found to be composed of different contributions involving single and double rotovibrational transitions. Those features can be fitted by simple models which, however, fail to reproduce the spectrum around the frequency of 12,466 cm-1: a diffuse absorption line is missing just around that frequency at which the triple transition is expected. Moreover, the missing intensity, when divided by density to the third power, shows a linear behaviour with density with an intercept at zero density which determines the ternary absorption coefficient. The integrated intensity of such a contribution turns out to be about 10^{-14} cm^{-1} amagat^{-3} for temperatures ranging from 77 to 300 K [1].

In order to explain the previous measurements, a model for the irreducible three body dipole moment is needed. We have considered both the Quadrupole-Induced Dipole Induced Dipole (QDID) and the Overlap Induced Ternary Dipole (OITD) inducing mechanisms. The QDID results from the electric dipole created on a molecule by the field of the dipole on a second molecule created by the permanent quadrupole moment of a third molecule. Such a contribution can be easily evaluated by the use of the propagator of multipolar fields and the knowledge of molecular polarizabilities. On the other hand, OITD is a contribution due to exchange forces. Its evaluation requires a model for the electronic charge density of a system of three molecules. Moreover, such a model must also contain the dependence of the molecular vibrational degrees of freedom. That has been obtained by using the one effective electron model [2,3] after generalizing it in order to account for the vibrational dependence [4].

The results show that the QDID contribution to the integrated intensity accounts only for about 1% of the measured intensity. The OITD, on the other hand, reproduces the experimental results at different temperatures within a factor 3 or better[4,5]. This fact demonstrates that, at least qualitatively, the Jansen's model [2] is capable to describe ternary effects.

REFERENCES

1. Reddy S.P., Fan Xiang and Varghese G., *Phys. Rev. Lett.* **74**, 367 (1995)
2. Jansen L., *Phys. Rev.* **125**, 1798 (1962)
3. Guillot B., Mountain R.D.and Birnbaum G., *J. Chem. Phys.* **90**, 650 (1989)
4. Moraldi M.and Frommhold L., *J. Chem. Phys.*, **103**, 2377 (1995)
5. Moraldi M.and Frommhold L., *Phys. Rev. Lett.* **74**, 363 (1995)

Improved Analysis of the Spectral Moments of CIA in N_2 and CO_2 Gases

M. Gruszka*, and A. Borysow*,**

* Department of Physics, Michigan Technological University, Houghton, MI 49931, USA
** Niels Bohr Institute, Copenhagen University Observatory, Juliane Maries vej 30, DK-2100 Copenhagen, Denmark

Abstract: The zeroth (γ_1) and the second (α_1) classical spectral moment of the roto-translational collision - induced absorption (CIA) at the low density limit of gaseous N_2 and CO_2 were computed. Accurate formulas [1, 2] were used, without any approximations to the pair distribution function. Good agreement with existing experimental results was achieved by using advanced anisotropic intermolecular potentials and by adjusting the overlap correction to the induced dipole. In both cases, N_2 and CO_2, we found the perturbation series not to converge, when realistic potentials were used. The results presented here provide the correction to our earlier findings [3, 4], affected by a minor programming error.

Introduction: A common way of analyzing CIA spectra has been by comparing the experimental and theoretical values of the zeroth (γ_1) and the second (α_1) spectral moments. At the low density limit and in the classical approximation $\gamma_1 \sim \langle|\mu|^2\rangle$ and $\alpha_1 \sim \langle|\dot\mu|^2\rangle$. The dipole moment μ, induced between two colliding molecules can be written in terms of spherical components [5] as:

$$\mu_\nu(R,\hat{r}_1,\hat{r}_2,\widehat{R}) = \frac{(4\pi)^{3/2}}{\sqrt{3}} \sum_{\lambda_1\lambda_2\Lambda L} A_{\lambda_1\lambda_2\Lambda L}(R)\, Y^{1\nu}_{\lambda_1\lambda_2\Lambda L}(\hat{r}_1,\hat{r}_2,\widehat{R}),$$

where $A_{\lambda_1\lambda_2\Lambda L}(R)$ are real coefficients, the unit vector $\widehat{R}$ lies along the intermolecular axis, and the $\hat{r}_i$ denotes the orientation of each molecule.

CO_2: Assuming purely electrostatic induction mechanism γ_1 and α_1 were computed using 5q(CO_2), PQ(CO_2), B and CS potentials [2, 3]. None of the analyzed potentials reproduces the experimental values of the spectral moments. For the best model (5q(CO_2)) the discrepancy ranges form 25% to 40% for γ_1 and from 30% to 40% for α_1 over the range of temperatures from 230 to 350 K [2]. Having used best currently available CO_2-CO_2 potentials and eliminating other possible sources of this discrepancy like higher order multipole moment contribution or second order induction mechanism we propose an overlap correction to the five most significant terms of the induced dipole:

$$A_{2023} = -A_{0223} = \sqrt{3}\alpha\Theta/R^4 + \lambda_{23}\exp(-R/\rho_{23})$$
$$A_{4045} = -A_{0445} = \sqrt{5}\alpha\Phi/R^6 + \lambda_{45}\exp(-R/\rho_{45})$$
$$A_{2233} = -\sqrt{8/15}\gamma\Theta/R^4 + \lambda_{33}\exp(-R/\rho_{33}).$$

The values of the strength $\lambda_{\Lambda L}$ and range $\rho_{\Lambda L}$ parameters are based on our recent study of the CIA spectral lines of CO_2 [6], ($\lambda_{23} = -1.7, \rho_{23} = 1.44, \lambda_{45} =$

$-6.4, \rho_{45} = 0.95, \lambda_{33} = -0.81$ and $\rho_{33} = 1.39$; all data in a.u.). In Fig. 1 we present the γ_1 and α_1 moments computed using the $5q(CO_2)$ potential, with and without the overlap correction and compare them with existing experimental results [7]. Both γ_1 and α_1 are within the experimental accuracy.

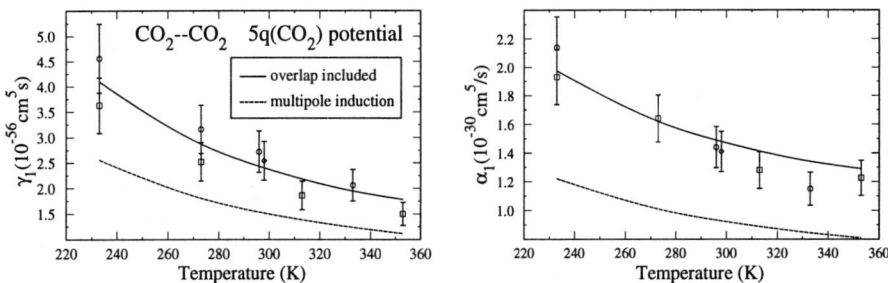

Figure 1: Computed and experimental values of γ_1 and α_1 for CO_2 pair.

N_2: The multipolar electrostatic induction was found to adequately represent the mechanism of dipole induction for N_2-N_2 complex. The relative difference between γ_1 and α_1 obtained with full anisotropic potential [8] and those computed using it's isotropic part only was less than 10%. However, the results obtained using the 1st order perturbation theory were from 25% to 30% too small. The results from the 2nd order perturbation theory were from 40% to 60% too large. This result presents a warning against indiscriminate use of perturbation theory with realistic intermolecular potentials, when applied to CIA.

Erratum: Due to a minor error in programming the $Y^{1\nu}_{\lambda_1\lambda_2\Lambda L}(\hat{r}_1,\hat{r}_2,\widehat{R})$ functions, discovered while developing a method of obtaining RT CIA spectra by using molecular dynamics [6], all previously reported values of γ_1 and α_1 were consistently too high [3, 4]. They accidently fell within the experimental uncertainty for CO_2 and were about two times too large for N_2 system.

In order to compensate for this spurious effect in N_2 we previously introduced an overlap correction to the induced dipole [4]. According to our latest findings [2], presented here, the overlap correction is crucial in correct representation of the spectral moments of CO_2 but is not necessary for N_2. The earlier results and discussion of the importance of the anisotropy of the intermolecular potential for CO_2 are still valid.

Acknowledgments: The support of the Planetary Atmospheres Division of NASA is gratefully acknowledged.

REFERENCES:
1. A. Borysow and M. Moraldi. *Phys. Rev. Lett.*, 68:3686, 1992
2. M. Gruszka and A. Borysow. *Mol. Phys., in press*, 1996
3. M. Gruszka and A. Borysow. *J. Chem. Phys.*, 101:3573, 1994
4. M. Gruszka and A. Borysow. *Spectral Line Shapes*, vol.8, p227
5. J.D. Poll and J.L. Hunt. *Can. J. Phys.*, 54:461, 1976
6. M. Gruszka and A. Borysow. *Mol. Phys., submitted*, 1996
7. W. Ho et al., *J. Chem. Phys.*, 55:1028, 1971, A. Afanasev et al., *Opt. Spectrosc*, 58:772, 1985, I.R. Dagg et al., *Can. J. Phys.*, 64:1475, 1986
8. A. van der Avoird et al., *J. Chem. Phys.*, 64:1475, 1986

Collision-Induced Absorption near the 227 nm Hg Line in the Hg+Ar Mixture

Teresa Grycuk*, Nicolay A. Kryukov[†] and Michael G. Lednev[‡]

*Institute of Experimental Physics, University of Warsaw,
00-681 Warszawa, Poland, [†]Institute of Physics, St. Petersburg State University,
and [‡]Baltic State Technical University, St. Petersburg, Russia.

Abstract Collision induced absorption band appearing on the shortwave-lenght side of the forbidden mercury line (λ 227 nm) in the presence of argon is observed for the first time. Absolute absorption coefficient determined experimentally is compared with the theoretical band profile calculated using the recently derived semiempirical potential energy curve for the excited D^31 state and the induced dipole transition moment.

Although the dipole forbidden transition ($6^3P_2 - 6^1S_0$) in mercury has been investigated by many authors, very few papers is devoted to studies of the collision induced spectral features associated with this transition. Up to now some data in this field have been derived only for the Hg+Xe mixture - from a flash lamp excited fluorescence (1, 2) and from emission spectra of discharge tubes (3). Quite recently the latter method has been extended to the Hg+Kr case (4).

In this work the first measurements of the band, attributed to the $D^31 \leftarrow {}^10^+$ molecular transition for Hg+Ar mixture were performed. Moreover, the results presented are the first absorption data obtained under the well defined physical conditions. The spectral profiles are determined on an absolute intensity scale, so they provide quantitative information on the interaction potentials and the induced dipole transition moment as a function of an interatomic separation.

We have measured the band profile with a high precise spectroscopic device. On the other hand, the shape of this band at the relevant temperatures were calculated using the recently determined semiempirical potential for the upper D^31 state (5) and the induced dipole transition moment (6). These functions have been obtained on the basis of Morse potentials derived in (7) from the analysis of the molecular beam vibronic spectra associated with the allowed transitions A^30 - X and B^31 - X of Hg-Ar.

Unfortunately, even accurate quantum calculations for such input data do not reproduce the experimental band profiles. On the other hand a considerable improvement of the situation is obtained (see figure 1) after some modification of the excited state potential and the transition moment in the direction which can be deduced from a knowledge of the B^31 potential for Hg-Ar obtained from the shape of the blue satellites of the 253.7 nm Hg line [8]. We believe that this is the way for determination of the more realistic potential for the D^31 excited state of Hg-Ar as well as the dipole transition moment involved.

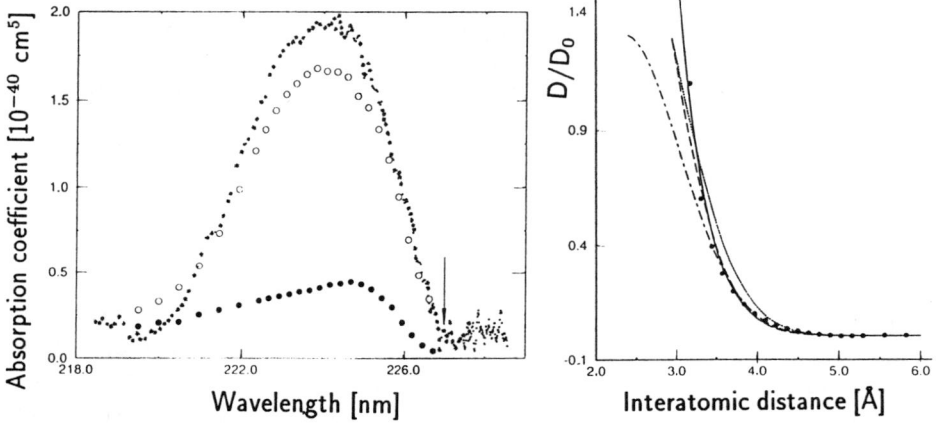

FIGURE 1. Collision induced absorption band near $6^3P_2 - 6^1S_0$ transition at 227 nm indicated by arrow (left plots) and the transition moment (right plots). Semiempirical data of Ref. (6) for D(R) are represented by points.
Upper spectrum: the experimental result; Bellow: (•) a theoretical profile calculated for the potentials and D(R) after (5) and (6).
(∘) - for some modification of D(R) represented by the dotted line in D(R).

REFERENCES

1. Callear, A.B., and Du K., *Chem. Phys. Lett.* **128**, 141-144 (1986).
2. Callear, A.B., and Du K., *Chem. Phys.* **113**, 73-86 (1987).
3. Kryukov, N.A., Penkin N.P., and Redko, T.P. *Opt. Spectrosc.* **66**, 1235-8 (1989).
4. Thaplyguine, M.A. *Thesis*, University of St. Petersburg, (1996).
5. Zagrebin, A.L., and Lednev M.G. *Opt. Spectrosc.* **77**, 544-551 (1994).
6. Zagrebin, A.L., and Lednev, M.G. *Opt. Spectrosc.* **78**, 758-769 (1995).
7. Yamanouchi, K. Isogai, S., Okunishi, M. and Tsuchiya, S. *J. Chem. Phys.* **88**, 205-212 (1988).
8. Grycuk, T., and Findeisen, M. to be submitted.

Bound Dimer Contribution in the Collision-Induced Pair Spectrum of Mercury Vapour

Niklas Meinander,# Lorenzo Ulivi,+ Gabriele Pratesi,*
Giovanni Cirnigliaro* and Fabrizio Barocchi*

Department of Physics P. O. Box 9, FIN-00014 University of Helsinki, Finland.
+ Istituto di Elettronica Quantistica, CNR, Via Panciatichi 56/30, 50127 Firenze, Italy.
* Dip. di Fisica dell'Università, and INFM, sez. di Firenze, L. E. Fermi 2, 50125 Firenze, Italy.

Abstract. We present recent measurements of CILS spectra in mercury vapour, at different temperatures, in the frequency region of the dimer band (0-15 cm^{-1}). The analysis of the experimental spectra, by comparison with classical calculations, including the dimers contribution, shows that such a precise measurement of the dimer band allows to discriminate between different potential models, without the necessity of modelling the pair polarizability.

Although the pair Collision Induced Light Scattering (CILS) spectrum has been used, with success, to study either the pair polarizability anisotropy, (as in the inert gases) or, conversely, the potential, (as in methane and SF$_6$ [1,2]), it has never been possible, in practice, to determine both these quantities. Mercury is an interesting (and challenging) case, since both the potential and the polarizability are poorly known, and the use of CILS for their determination, even if with less accuracy than in the case of inert gases, appears very promising. This is possible if the bound dimers band in the CILS spectrum is measured with great accuracy. Computations show, in fact, that the pair potential model influences mainly the shape of this band, while the polarizability model determines its intensity on an absolute scale. With this in mind we have measured the CILS spectrum of mercury, at T=753, 773, 793, and 853 K, in the range 0.2–30 cm^{-1}, using the high resolution SOPRA DPDM 2000 monochromator, and a high temperature cell which can reach a pressure of 30 bar. In all cases the dimer band is very well determined, and the CILS spectrum shows a maximum at about 4 cm^{-1}. Theoretical spectra have been computed, taking into account the bound dimer contribution, with four different potential models,[3,4,5,6] and assuming DID or "3-parameter" polarizability anisotropy $\beta(r)$, namely

$$\beta(r) = 6\alpha^2 r^{-3} + A r^{-6} + B\exp(-r/r_0). \tag{1}$$

The zero and second spectral moments, calculated assuming this polarizability model, (with A=1300 Å^9, B=37 Å^9 and r_0=1.2 Å) and the "Morse KAK" potential,[3] agree with the values derived from the experimental spectrum at three higher temperatures (873, 973 and 1073 K).[7]

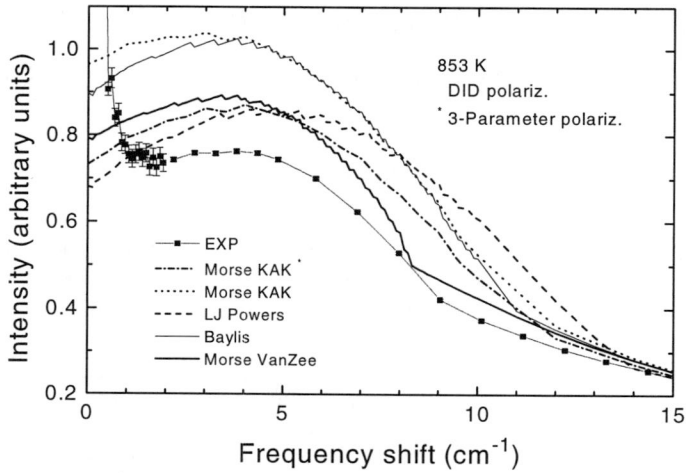

FIGURE 1. Experimentally measured CILS spectrum at 853 K (480 C), and calculated spectra for different potentials and polarizability models. The computed spectra have been normalised to match the experimental one in the region 20-30 cm-1.

To compare the shape of the experimental spectra with the computed ones, these latter have been normalised to best match the experimental data in the range 16–30 cm^{-1} (see fig. 1). It is observed that, even though a semiquantitative agreement can be obtained with some potentials models, (while others are immediately ruled out, like that of ref. 4), a precise match is far to be obtained yet. This is true also for the combination potential-polarizability which gives theoretical moments in good agreement with experimental ones. A change of curvature, around 9 cm^{-1}, is particularly evident in the experimental spectrum, while in the theoretical ones its frequency position is dependent on the shape of the pair potential. None of the computed spectra matches this frequency, thus indicating that a better potential may be found just using this information. Further study is under way in this direction in our laboratory.

REFERENCES

1. F. Barocchi, A. Guasti, M. Zoppi, S. El Sheikh, G.C. Tabisz, and N. Meinander, Phys. Rev. **A39**, 4537 (1989).
2. S. El Sheikh, N. Meinander, and G.C. Tabisz, Chem. Phys. Letters **118**, 151 (1985).
3. J. Koperski, J.B. Atkinson, and L. Krause, Chem. Phys. Lett. **219**, 161 (1994).
4. L. F. Epstein, and M. D. Powers, J. Phys. Chem. **57**, 336 (1953).
5. W. E. Baylis, J. Phys. **B10**, L583 (1977).
6. R. D. van Zee, S.C. Blankespoor, and T.S. Zwier, J. Chem. Phys. **88**, 4650 (1988).
7. F. Barocchi, F. Hensel, and M. Sampoli, Chem. Phys. Letters **232**, 445 (1995).

On the Nature of Collision-Induced Intensity and Bandshapes in the Region of N_2 and O_2 Fundamentals

Andrei A. Vigasin

*Institute of Atmospheric Physics, Russian Academy of Sciences,
Moscow 109017 Russia*

Abstract. Collision-induced intensity of absorption is divided into three parts which relate to contributions due to free collisional pairs, metastable compound pair states and firmly bound dimers. An importance of the coupled (metastable+bound) states formation is emphasized especially for the anisotropically interacting species. The model is suggested which enables one to reproduce overall collision-induced bandshape in account for the line-mixing effect. This model makes it possible to reveal the nature of weak regular oscillations found on the spectral envelopes of the nitrogen and oxygen collision-induced fundamental bands.

One of the issues of molecular scattering is a formation of quasibound and bound intermolecular states. Few attempts have been made up to now to reveal a role, played by these pair states in collision-induced absorption (CIA). This goal may be pursued by performing a subdivision of the CIA integrated intensity in the phase space into contributions which refer to strictly bound, quasibound and free pair states of the interacting monomers. General recipe for such consideration is suggested in [1] presupposing arbitrary anisotropic intermolecular potential. Decomposition of the integrated CIA intensity made it possible to conclude in [2] on the negligible role, played by the strictly bound pair states of diatomics in CIA absorption at normal temperatures. Metastable states are allowed, however, to occupy significant portion in the phase space even at elevated temperatures. For a selection of colliding moieties and temperature this may result in dimeric (bound+metastable) contribution to the overall CIA intensity being more pronounced, than that of free collisional pairs. It has been demonstrated in [3] e.g. that this occurs namely for CIA in the region of the carbon dioxide Fermi doublet at near room temperatures.

© 1997 American Institute of Physics

Recent accurate laboratory long-path measurements by McKellar [4] and Olson et al. [5] of the nitrogen fundamental band profile revealed small regular features, superimposed on diffuse collision-induced background and extending to higher and lower wave numbers relative to diatomic fundamental frequency. The local minima of absorption coincide with positions of individual vibro-rotational lines, while the maxima are situated in between of lines. Similar features were attributed earlier (see e.g. Henderson and Ewing [6]) to the combinations of the hindered rotation with ν_0 oxygen fundamental in van der Waals O_2...Ar molecule. Surprisingly regular character of these oscillations and their being almost independent on the sample temperature made us to doubt this attribution.

Present paper demonstrates that these small ripples may be considered in terms of the line-mixing effect, which manifests itself in collision-induced spectra. As a basis for our simplified model we have adopted the so-called strong collision model as it is formulated by Bulanin et al. [7]. The interbranch mixing has been completely neglected and the line-mixing was allowed only within O- and S-branches. The half-widths of vibro-rotational lines in Q-branch (γ_1) and in O- and S-branches (γ_2) as well as the relaxation time τ were found in the course of nonlinear least-square fitting of experimental spectra. The spectra taken at low temperatures (O_2 at 93 K and N_2 at 77 K) as well as N_2 spectra at six near-room temperatures over the range 228 to 296 K are treated successfully making use of the line-mixing model. The temperature dependencies of γ_1, γ_2, and τ are determined. This work shows that the line-mixing effect may be thought of to contribute significantly to the overall collision-induced bandshapes.

ACKNOWLEDGMENTS

The author gratefully acknowledges Drs A.R.W.McKellar and W.Lafferty for supplying their data in the digital form and valuable discussions. Author's thanks go also to Dr. J.-P.Bouanich for substantial help at the initial stage of this work.

REFERENCES

1. Vigasin, A. A., *Infrared Physics* **32**, 461-470 (1991).
2. Vigasin, A. A., *JQSRT*, in press, 1996.
3. Vigasin, A. A., Tarakanova, E. G., and Tchlenova, G. V., *JQSRT* **50**, 695-703 (1994).
4. McKellar, A. R. W., *J. chem. Phys.* **88**, 4190-4196 (1988).
5. Olson, W. B., Lafferty, W. J., Weber, A., and Solodov, A. M., "The collision-induced fundamental band of nitrogen at low temperatures," presented at the 49th Ohio State University International Symposium on Molecular Spectroscopy, Columbus, U.S.A., June 13-17, 1994.
6. Henderson, G., and Ewing, G. E., *J. chem. Phys.* **59**, 2280-2293 (1973).
7. Bulanin, M. O., Dokuchaev, A. B., Tonkov, M. V., and Filippov, N. N., *JQSRT* **31**, 521-543 (1984).

First Quantum Mechanical Computations of Collision-induced Absorption Spectra of H_2 Pairs in the Second Overtone Band

C. Zheng*, Y. Fu*, and A. Borysow*,**

* Physics Department, Michigan Technological University, U.S.A.
** Niels Bohr Institute, Copenhagen University Observatory,
Juliane Maries vej 30, DK-2100 Copenhagen, Denmark

abstract: The rotovibrational collision-induced absorption (CIA) spectra of H_2–H_2 pairs in the second overtone band ($\Delta v = 3$), which is very important in the studies of the stellar and the planetary atmospheres, are computed for the first time from the *ab initio* induced-dipole moments.

CIA spectra of H_2 pairs are of considerable interest for studies of the stellar, as well as the planetary atmospheres, because they constitute major components of the opacities of those atmospheres. CIA spectra of H_2 pairs in the second overtone band were used for the first positive identification of hydrogen in planetary atmospheres by the comparing the spectra of Uranus and Neptune with experimental data [1]. This band also plays significant role in determining the spherical albedo of Uranus and Neptune [2].

Quantum mechanical computations of the second overtone band CIA spectra of H_2 pairs are carried out for the first time in this work, based on the newly developed *ab initio* collision-induced H_2–H_2 dipole moments [3] and the ground state intermolecular interaction potential $<00|V_0(R)|00>$ of H_2–H_2 [4]. Two vibrational transitions are included, namely the single transition: $v_1 = 0 \rightarrow 0, v_2 = 0 \rightarrow 3$, and the double transitions $v_1 = 0 \rightarrow 1, v_2 = 0 \rightarrow 2$.

As for the dipole moments, because the second overtone band involves higher vibrational states of H_2 molecules, the relevant internuclear distances of H_2 become larger also. Thus, we have developed our induced dipole functions by combining the previous data [5] with new *ab initio* results [3] which account for one larger internuclear distance (2.15 a.u.).

As for the intermolecular potential, while the vibrational states dependencies of the intermolecular potential do affect the spectral lineshapes (the spectral intensities are not changed), the v-dependent potential $<03|V_0(R)|03>$

of H_2–H_2 is unfortunately not available at present.

A preliminary result of H_2–H_2 CIA spectra in the second overtone band has been obtained employing the so-called BC and K_0 model lineshapes [6], and the parameters of the lineshapes are based on the first three quantum mechanical spectral moments. The result at temperature 85 K is shown in the figure, also shown for comparison is the only available experimental data for the second overtone band at present [7].

In order to get a better agreement with the measurement, improvements in the accuracies of the dipole functions are needed. In further quantum mechanical spectra computations, the J-dependencies of the induced dipole functions will be taken into account, so will the v-dependencies of the intermolecular potential if possible. Greater accuracies are expected.

The ultimate goal of this work will be modeling the H_2–H_2 RV CIA in the second overtone band for a desired temperature range. The model will generate the CIA spectra for any given temperatures within the range, and thus facilitate the studies of radiative transfer in the planetary atmospheres.

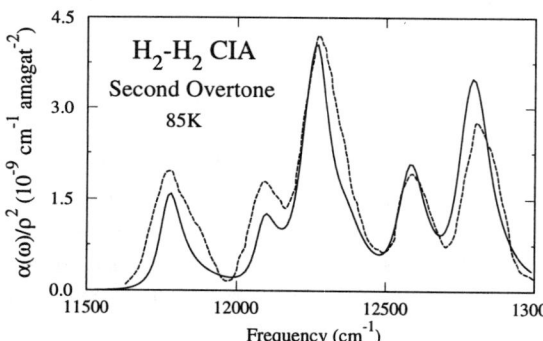

Rotovibrational CIA spectrum of H_2–H_2 in the second overtone band, obtained for normal H_2; solid line: this work; dashed line: experimental data.

The support of Planetary Atmospheres Division of NASA is gratefully acknowledged.

References

[1] G. Herzberg. *Astrophys. J.*, 115:337–340, 1952.

[2] D. Crisp. Private communication. 1996.

[3] C. Zheng and A. Borysow. Unpublished.

[4] J. Schaefer and W. E. Köhler. *Z. Phys. D; Atoms, Molecules and Clusters*, 13:217–229, 1989.

[5] W. Meyer, A. Borysow, and L. Frommhold. *Phys. Rev.*, A 40:6931 – 6949, 1989.

[6] J. Borysow, L. Trafton, L. Frommhold, and G. Birnbaum. *The Astrophy. J.*, 296:644–654, 1985.

[7] A. R. W. McKellar and H. L. Welsh. *Proc. Roy. Soc. Lond. A.*, 322:421 – 434, 1971.

Double Rotational Transitions of Molecular Hydrogen in the Condensed Phase

Marco Zoppi [*], Lorenzo Ulivi [*], Mario Santoro [†],
Massimo Moraldi [†] and Fabrizio Barocchi [†]

[*] Istituto di Elettronica Quantistica del CNR, via Panciatichi 56/30, I-50127 Firenze, Italy,

[†] Dipartimento di fisica, Universita' di Firenze and Istituto di Fisica della Materia, Unita' di Ricerca di Firenze, Largo E.Fermi 2, I-50125 Firenze, Italy

Abstract. We have measured the Raman spectra of the double rotational transition $S_0(0)+S_0(0)$ of liquid parahydrogen at different densities and temperatures. The integrated intensities of those spectra are then compared with numerical calculations based on quantum mechanical pair distribution function and involving certain components of the intermolecular potential and of the collision induced polarizability.

We have measured the Raman spectra of the double rotational transition $S_0(0)+S_0(0)$ in liquid parahydrogen in a range of pressure from 17.0 to 597.5 bar at a temperature T=28 K and in solid parahydrogen at the temperatures T=18.5 and 22.6 K. The scattered signal, dispersed by a double spectrometer with a typical resolution of 0.5 cm^{-1}, was detected by a high efficency photomultiplier with a low dark current level (0.5 counts/s). We have used a cryogenic cell designed for high pressures fluids with an internal volume of about 1 cm^3. Hydrogen was condensed directly in the scattering cell by cooling the sample under a moderate pressure.

© 1997 American Institute of Physics

We have also calculated [1] the integrated intensity of the double rotational spectra. Such an intensity is composed mainly of two terms: the contribution that arises from the mixing of the rotational states of the molecules due to the anisotropic part of the intermolecular potential and the contribution arsing from the collision induced polarizability. The first term is proportional to the thermal average of the component V_{224} [2] of the intermolecular potential squared whereas the second term is the thermal average of the square of the anisotropic component of the collision induced polarizability that transforms as V_{224} for rotations of the two molecules.

In order to perform the thermal averages we need the pair distribution functions. We have calculated such functions by performing Path Integral Monte Carlo simulations with a system of 108 particles interacting by means of the isotropic component of the potential described in ref.2. The sizeable quanticity of the system requires the Trotter number to be set equal to at least 32.

The comparison of numerical with experimental results depends greatly on the model used for the appropriate components of the collision induced polarizability. We have used three models. One is the first order dipole-induced dipole model and the other two are obtained by adding two terms to it: a term varying as the inverse sixth power of intermolecular distance and one as an exponential. In one of them (rare gas model), the ratio of the added terms to the dipole-induced dipole one and the width of the exponential were derived so to reproduce the pure translational collision induced Raman spectrum [3]. In the other model we have considered the width of the exponential as a free parameter and made it vary in order to give a best fit of the measured intensity of the double rotational spectra. It is found that the experimental results are best reproduced if the range of the exponential term in the rare gas model is increased by about 15% [1].

REFERENCES

1. Zoppi M., Ulivi L., Santoro M., Moraldi M. and Barocchi F., *Phys. Rev. A (Rap. Comm.)*, 53, 1935 (1996)

2. Norman M.J., Watts R.O. and Buck U., *J. Chem. Phys.*, **81**, 3500 (1984)

3. Meinander N., Tabisz G.C. and Zoppi M.,*J. Chem. Phys.*, **84**, 3005 (1986)

Double Rotovibrational Transitions in Solid Hydrogen

Marco Zoppi [*], Lorenzo Ulivi [*], Mario Santoro [†],
Massimo Moraldi [†] and Fabrizio Barocchi [†]

[*] Istituto di Elettronica Quantistica del CNR, via Panciatichi 56/30, I-50127 Firenze, Italy,

[†] Dipartimento di fisica, Universita' di Firenze and Istituto di Fisica della Materia, Unita' di Firenze, Largo E.Fermi 2, I-50125 Firenze, Italy

Abstract. We have measured the integrated intensity of the double rotovibrational transition $Q_1(0)+S_0(0)$ in solid hydrogen at different densities and temperatures. The experimental results are compared with numerical calculations based on a quantum mechanical pair distribution function and on a simple quadrupole-quadrupole interaction model for the anisotropic component of the intermolecular potential. The comparison shows the insufficiency of that model but indicates that double transition spectra are a valuable source of information for intermolecular interactions.

We have measured the Raman spectra of the double rotovibrational transition $Q_1(0)+S_0(0)$ together with the $S_1(0)$ in solid parahydrogen. The measurements have been performed at six different densities in the pressure range between 247 and 700 bar and at temperatures between 18.7 and 22.5 K. The scattered signal, dispersed by a double spectrometer with a typical resolution of 0.5 cm^{-1}, was detected by a high efficency photomultiplier with a low dark current level (0.5 counts/s). We have used a cryogenic cell designed for high pressures fluids with an internal volume of about 1 cm^3. Hydrogen was condensed directly in the scattering cell by cooling the sample under a moderate pressure.

We have also calculated the integrated intensity of the double transition spectrum. The calculations are based on a theory already described in the litterature [1] according to which the integrated intensity of the double transition spectrum is proportional to the average of the square of certain components of the intermolecular potential. In the case we are considering, the potential component responsible for the transition is the one experienced by one molecule in the ground and the other in the first excited vibrational state and such that it transforms as a quadrupole-quadrupole interaction under rotational transformations [1]. In ref. 1 the average was calculated by assuming a rigid lattice. Here, on the contrary, we have performed the calculations after introducing the pair distribution function. Such a quantity is evaluated numerically by means of a Path integral Monte Carlo (PIMC) simulation and assuming an existing model for the intermolecular potential [2].

Calculations of the intensity of the double transition spectrum have been carried out using the PIMC results for the pair distribution function and a pure quadrupole-quadrupole interaction as in ref.1. As an example, at a temperaure T=22.6 K and numerical density n=30.9 nm^{-3}, the calculated ratio f of the double transition to the single transition integrated intensity turns out to be about twice the experimental value (.66 compared to .32). Such a result is an indication of the insufficiency of pure quadrupole-quadrupole model, particularly of its dependence on the fith power of the inverse of the distance between the molecules. Actually, a comparison of a purely quadrupole-quadrupole interaction with a more realistic model [2] for two molecules in the ground vibrational state show that the latter, at short intermolecular distances, is roughly 80% of the former. If that was true also for the case of one molecule performing a vibrational transition, the theoretical value for f would reduce to .45 .

REFERENCES

1. Barocchi F., Guasti A.,.Zoppi M., Poll J.D. and Tipping R.H. *Phys. Rev. B* , **37**, 8377 (1988)
2. Norman M.J., Watts R.O. and Buck U., *J. Chem. Phys.*, **81**, 3500 (1984)

Interaction-Induced Dipoles and Polarizabilities In Diverse Phenomena

George Birnbaum* and Bertrand Guillot†

Physics Department, Catholic University of America, Washington, DC 20064, USA
*Guest Researcher, National Institute of Standards and Technology
Gaithersburg, MD 20899, USA
†Laboratoire de Physique Théorique des Liquides (CNRS, URA 765)
Université Pierre et Marie Curie, Boîte 121, 4 Place Jussieu
Paris, France

Abstract. This report briefly reviews the contributions made by the collision or interaction induced spectroscopies, i.e., far infrared and infrared absorption, and Rayleigh and Raman light scattering, to a variety of phenomena. The topics include: gaseous systems, planetary atmospheres; multipole moments; collision induced dipoles and polarizabilities; intermolecular potentials, dimers; three-body and liquid state interactions; double and triple-transitions; interference of allowed and induced transitions; solid state systems; halo-organic complexes; charge fluctuations in water; ionic melts; electrolyte solutions; viscoelastic relaxation in simple liquids; and interaction-induced emission.

It is nearly a half century since infrared absorption by molecules without permanent dipoles was discovered in 1947 (1). Such absorption, observed from microwave to optical wavelengths (2,3), was attributed to the transient dipoles produced by molecular (and atomic) electric fields during collisions. In 1968 the first depolarized light spectra from rare gases were observed (4). These spectra arise from transient anisotropic polarizability produced in colliding atoms by radiation induced dipoles, viz., the dipole-induced dipole (DID) effect (5,6). Later the spectrum due to the isotropic induced polarizability was observed. Just as infrared (IR) and far infrared (FIR) absorption spectra are normally forbidden in individual nonpolar molecules, depolarized light scattering is normally forbidden in individual atoms or spherical molecules.

The character of interaction induced[1] (I-I) spectra is quite different from that of

---

[1] The designations interaction-induced and collision-induced have frequently been used interchangeably. We will use the former since it appears to have a broader connotation.

allowed absorption and scattered light spectra. The former are weak and broad, usually without any resolved rotational line structure (the IR absorption spectra of H_2 is an exception). The breadth of I-I spectra can be understood by noting that the induced dipole moment or polarizability have a very short duration equal to that of an intermolecular collision. Thus the variation of the induced dipole and polarizability with intermolecular (interatomic) separation gives rise to a broad translational spectrum, a unique feature of I-I spectra. At relatively low densities (roughly less than tens of amagat) collisions predominantly involve only two particles, and in this regime the spectral shapes scale as the square of the density at all frequencies. At higher densities where three body collisions become significant, the resulting induced dipoles and polarizabilities cancel those induced in the pairs and profound changes in the spectral shapes can occur.

Although allowed infrared, Rayleigh, and Raman spectra are much more frequently encountered than these kinds of induced spectra, the latter are the more universal since they must occur in all molecular systems. Induced spectra are almost always much stronger than allowed spectra, however induced and allowed spectra can interfere to produce a contribution significantly greater than the pure induced part alone (7).

A great many experimental and theoretical studies on I-I spectra in gases, liquids and solids have appeared, and much progress has been made in understanding and describing such spectra theoretically. The early work emphasized compressed gases and the simplest (model) molecules and atoms in order to gain an understanding of the various induced quantities responsible for I-I spectra (2,8). However, as the field developed, increasing attention was paid to more complex molecules, mixtures, and liquid systems (9-12). In liquids where multibody interactions involving induced dipoles, polarizabilities, and potentials are of crucial importance, molecular dynamics (MD) computations have proven to be an indispensable tool in studying induced phenomena.

This report presents a brief survey on how the I-I spectroscopies, along with theory and MD computations, have been used to study a variety of phenomena in which induced dipoles and polarizabilities play important roles. No attempt is made to survey in any detail the very large literature that has developed, and wherever possible a review article is cited; otherwise one or a few typical references are given. More complete guides to the literature are given in Refs. (13,14) for interaction-induced absorption (I-IA), and Ref. (14a) for interaction-induced light scattering (I-ILS). The order of the topics that follow is somewhat arbitrary, although, in general, the arrangement is from simple to complex molecules and systems.

Gaseous Systems, Planetary Atmospheres (1,15). The investigation of the IR and FIR I-I spectra of molecular H_2, also N_2 and CH_4 and their mixtures has been central to the development of I-IA spectroscopy. Such studies which elucidated the nature of I-I spectra have also been of fundamental importance in investigating the atmospheres of the giant planets, since their bulk consists mostly of H_2, the rest being mainly He with smaller amounts of N_2 and CH_4. The I-I IR emission due to H_2-H_2, H_2-He, H_2-N_2 and H_2-CH_4 collisions has been extensively used as a diagnostic tool to determine some of the properties of planetary atmospheres, e.g., composition, temperature and pressure distributions. To accomplish this, absorption profiles need to be known accurately as functions of frequency, density, and temperature for the molecular pairs normally encountered in the planetary atmospheres. Moreover, these profiles need to be known for temperatures where laboratory measurements are not feasible. To meet this need, accurate quantum mechanical and model calculations of I-IA were developed (15) that were also extended to I-ILS (6).

Multipole Moments, Collision Induced Dipoles and Polarizabilities. It was appreciated from the outset that I-IA could provide an accurate way of determining molecular multipole moments, since the intensity of I-IA depends on the values of these moments. However, to deduce them from the measured integrated spectral intensities, it is necessary to know the intermolecular potential, since the magnitude of the induced dipole is a function of the intermolecular separation and must be integrated over this distance, multiplied by the pair distribution function. Fortunately, potential functions for simple nonpolar molecules such as H_2, N_2, and CO_2 were rather well known, and reliable values of the quadrupole moment of these molecules were obtained. For CH_4, both the octopole ad hexadecapole moments could be deduced from the FIR spectra (16,17). From an analysis of the I-I depolarized Rayleigh spectrum of spherical molecules such as CH_4, the dipole-quadrupole and dipole-octopole polarizability tensors, arising respectively from the first and second derivatives of the induced dipole fields, were deduced (18). Because classical multipole models do not include overlap, exchange, or electron correlation effects (dispersion), they frequently cannot give accurate total dipoles or polarizabilities for molecules interacting at short range. This need stimulated extensive *ab initio* calculations of these nonclassical induced dipoles and polarizabilites (19-21); such results also served as a test of the accuracy of the calculations themselves.

Dielectric and optical properties of compressed gases and liquids are also affected by changes in the dipole moment and molecular polarizability due to the distortion of electron clouds occurring when molecules collide (22). Thus, induced dipoles and polarizabilities appear in the higher order dielectric virial coefficients (23,24), and the Kerr effect and the electric-field induced birefringence (25).

Intermolecular Potentials, Dimers. Despite the fact that I-I spectra depend on both the potential function and the induced dipole and polarizability, useful information regarding the former can be obtained. For example, by noting that different experiments are sensitive to different regions of the intermolecular potential, an approximate anisotropic potential was developed for Xe-SF_6 mixtures by fitting an M3SV potential form to virial coefficient, diffusion coefficient, and I-ILS data which played a crucial role here (26). Several recent investigations employed I-ILS and

I-IA spectra to gain information regarding a poorly known potential function in Hg vapor (27-30).

By contrast, the spectroscopy of van der Waals complexes does provide very detailed information on intermolecular potentials, particularly in the region of the attractive well (31). In dimers formed of nonpolar constituents such as H_2-H_2, H_2-rare gas, and rare gas complexes, the dipole-moments producing the spectra are clearly induced (15,32). There is thus a connection between the spectroscopy of relatively stable complexes, which can be considered to be long-lived collisions, and I-I spectroscopy. However, the information regarding the potential is obtained from the rotational-vibrational frequencies of the complex, rather than spectral intensities.

Three-body and Liquid State Interactions (12, 33). I-I spectra at all frequencies and correlation functions (CF) at all times increase as the square of the density in the low-density bimolecular regime. But the increase with density is less rapid at higher densities because of the negative contribution of three-body interactions. Also, three-body interactions in the form of correlated bimolecular collisions produce a striking dip centered at zero frequency in the FIR spectrum of rare gas mixtures and in the Q-branch of the fundamental band of H_2-He mixtures.

At liquid densities induced intensities are very much less than would be the case if they were due to bimolecular collisions alone. Another way of regarding the cancellation of I-IA and I-ILS intensities in liquids is to observe that these intensities would be zero if the local liquid structure had spherical or cubic symmetry. It is thus the departure from such structure, produced by fluctuations in the positions of the liquid particles, that produce the I-I spectra. These are thus related to local structure and its relaxation, and such spectra are consequently a sensitive probe of this structure. Another striking multibody effect is the cancellation of the long time tails of the 2-, 3-, and 4-body CFs in I-ILS to produce a rapidly decaying total CF (34).

Two methods of analyzing I-I spectra have appeared that provide useful insights into the underlying phenomena. In one, fluctuations in the local structure are described by induced dipoles that are made up of a sum of contributions from each fourier component of the local density. The theory is thus cast in a form to make use of liquid structure factors obtained from neutron scattering experiments (12). Another approach is based on a direct application of the generalized Langevin equation to the induced CF. Here, approximate CFs based on the zeroth, second, and fourth classical spectral moments must be computed. To do this for the three and four body contributions, a lattice gas model was employed (11, 33).

Thus far, no distinction has been made between two types of three-body interactions; in one, where the interactions among the three particles are pair-wise additive, and in the other where such a decomposition is inapplicable and the three-body interactions are said to be irreducible. Such interactions in rare-gas fluids and in gaseous H_2-He (33) and H_2-H_2 (35) have been studied experimentally and theoretically by I-IA and I-ILS.

We previously emphasized that the sensitivity of I-I spectra to liquid structure arises through the cancellation effect whereby the contribution of the pair-wise additive induced dipoles cancel if they occur in a sufficiently symmetrical arrangement. It was recently suggested that a clue to the structural order of supercooled liquids, particularly those composed of molecules with tetrahedral symmetry, might be found in the temperature dependence of the broad spectral features observed in light scattering spectra, provided that this part of the spectrum arises from the DID interaction (36). However, extracting this quantity from the spectrum of liquids with much stronger allowed components is nontrivial. The significant aspect of this phenomenon is the observation that even where density changes are not great and the apparent structure relatively invariant, the DID CF seems to decrease markedly with small increases of density and decreases of temperature. By contrast, for atomic systems, the changes in the DID spectrum and CF that occur throughout the liquid regime for similar changes in density and temperature are rather small.

Double and Triple Transitions. Simultaneous transitions (ST) in which two molecules undergo rotational and rotovibrational transitions with the absorption or emission of a single photon have been extensively studied (1, 37). In the usual case of induced absorption and light scattering only one molecule of the interacting complex undergoes a transition. In general, in a ST, an active vibrating dipole or quadrupole in one molecule induces a dipole in a neighboring molecule with Raman active vibrating polarizability, as for two H_2 molecules, or $CO_2(v_3)$-H_2. The photon that is absorbed has a frequency which is the sum of the frequencies of the individual molecules, or in some cases their difference frequency. Double or (ST) transitions have been observed in many gaseous and liquid mixtures, as well as in solids. If the intermolecular potential is approximately isotopic and if the internal vibrational-rotational coupling of the interacting molecules is weak, the intensity of the ST band is simply due to additive two-molecule induced dipoles and the cancellation effect does not occur (38). This circumstance suggests that the study of double transitions in the liquid state could provide a method for studying the bimolecular induced dipole spectrum, isolated from the contributions of three and four particle induced dipoles.

Very recently the simultaneous transition in three H_2 molecules with the absorption of a single photon was observed (39). It was proposed that this interesting transition can be attributed to a ternary (irreducible) dipole arising from the very short-range overlap interaction (35).

Interference of Allowed and Induced Transitions. Although I-I spectra are most clearly identified when no allowed spectral components are present, such situations are relatively rare since most molecules do not have sufficiently high symmetry to eliminate the allowed components. The more common occurrence is the existence of weak I-I spectral components among a much stronger allowed spectrum. In general, the induced contributions consist of a pure I-I part and a stronger mixed contribution resulting from the interference of the induced and allowed parts. Although in favorable cases the allowed spectral components may be much sharper than I-I components and their time scale separation in the CFs may be good, in practice the weak induced components are very difficult if not essentially impossible to separate from allowed components. Thus, computer simulations have played an important role in the identification and interpretation of the I-I components of allowed absorption (40) and light scattering (41) spectra. Indeed, MD has been an indispensable tool for interpreting purely I-IA and I-ILS spectra as well.

A unique situation exists in the case of the HD molecule, which unlike H_2 possesses a dipole moment due to the non-adiabatic coupling of electronic and nuclear motions. This very small dipole moment makes the intensity of the allowed FIR and IR absorption comparable with that of the induced absorption. The possibility of simultaneous transitions to single-molecule discrete states due to the permanent dipole moment, and to the translational continuum of the interacting pair due to the induced dipole moment, creates interesting interference effects which reveal themselves through the asymmetric shape of the narrow allowed lines (42, 43). The theory of these HD spectra also includes a treatment of the finite collision duration in the allowed spectra and is interesting from this aspect; of course, induced spectra are entirely a result of interactions occuring during the finite collision time.

Solid State Systems. It is evident that I-I spectra must exist in solid state systems, although one should expect the spectral shapes to be quite different from those observed in compressed fluids where fluctuations in molecular position play an important role in shaping the CF and spectrum. In solids these motions are replaced by lattice vibrations and the phonon spectrum. The investigation of I-IA in the solid hydrogens, H_2, HD and D_2, played an important role in developing both an understanding I-IA phenomena in general, and in clarifying many subtle properties of these unique solids (44). A new type of experiment was initiated in which spectra were obtained in tritium and deuterium polycrystals from charges produced by high energy radiations (45, 46). New spectral lines were observed which suggested that charge fragments may become immobilized in the low temperature lattice, and by the effective radial field of these fragments, induce dipoles in the host molecules which thereby absorb radiation.

Other solid state studies include the infrared spectra of molecules adsorbed in zeolites. These show interesting induced effects among which are the appearance of normally forbidden bands due to the strong electric field created by the ions constituting the cavity walls (47). Also, an atom, ion or molecule can be trapped within a C_{60} cage, known as a buckminsterfullerene or buckyball cage. Vibrational and rotational motions of the inclusion lead to I-IA and I-ILS. These induced spectra are unusual in that they are predicted to be discrete (48), not broad and continuous as in usual induced spectra, since the guest particle is trapped inside a cage. Finally, we mention work on induced light scattering from electrically disordered solids as applied to mixed alkali halide crystals (49).

Halo-organic Complexes (50, 51). The color changes of iodine solutions with the nature of the solvent, coupled with the charge-transfer theory which gave a framework for the interpretation of such spectra, have stimulated many spectroscopic studies of these solutions, in particular, the FIR spectra of I_2 dissolved in benzene and pyridine. The liquid FIR spectrum of complex molecules such as benzene and pyridine (Py) are interesting in themselves; the former is a nonpolar symmetric top molecule giving rise to a collision-induced spectrum, the latter is anisotropic and polar producing strong interference between the allowed and induced spectra. By means of computer simulations, which gave results in accord with the FIR experiments, it was found that for I_2- C_6H_6 solutions the I_2 molecule changes partners so frequently within its first solvation shell that the concept of a complex is simply not applicable. On the other hand, I_2 and Py molecules associate to form long-lasting complexes, with a new relatively sharp vibrational band due to the I_2:Py complex. For both C_6H_6 and Py it was necessary to introduce overlapping dipoles between I_2 and solvent to obtain a satisfactory reproduction of the observed spectrum.

Charge Fluctuations in Water. Intermolecular motions in liquid water were investigated by MD simulations which showed that the experimental FIR absorption intensity and Rayleigh spectrum can be reproduced satisfactorily (52, 53). Indeed, the introduction of the DID mechanism and the various cancellation effects between the permanent and induced dipole moments give the correct FIR absorption intensity over a large frequency range. Moreover, the appearance of a shoulder in the 200 cm^{-1} region is mainly due to the induced dipoles modulated by the stretching of the OH..O hydrogen bonded units, whereas the high frequency band (~ 600 cm^{-1}) is directly related to the dynamics of the hydrogen bond network through the librational motions of the water molecules. Likewise MD simulations showed that the DID mechanism gives the most intense contribution to the Rayleigh spectrum at low and intermediate frequencies whereas the allowed spectrum which is more sensitive to libration motions gives rise essentially to the high frequency band.

Ionic Melts (54, 55). The light scattering spectrum of molten alkali and alkali earth halides was found to arise from the changes in ionic polarizability caused by interionic interactions, i.e., it is an I-I effect. The fluctuating polarizability measured by light scattering results from changes caused by short-range overlap forces, ionic field-induced distortion, and DID interactions. Since the I-I polarizabilities depend upon the relative positions of the ions, the line shape must directly reflect structural relaxation in the melt. The approach to this problem follows the I-ILS of atomic fluids where the light scattering spectrum is related to the spectrum of the density-density correlation function (the dynamical structure factor) which characterizes the dynamical properties of an atomic field. Although, as in the atomic case, the known functional form of the polarizability can be linked to the light scattering process via density fluctuations in the melt, the ionic case is more complicated than the atomic case, since both mass density and charge density CFs in the former can influence the line shape. Nevertheless, despite these complications, good accord of theory with experiment was obtained. The mass density and charge density CFs may be observed in dynamical neutron scattering and in computer simulations.

Electrolyte Solutions (56). This topic deals with the experimental and theoretical investigations of the far infrared spectra of solutions of alkali halide salts such as RbI and NaCI dissolved in protic (methanol) and dipolar (acetonitrile, acetone, and dimethyl sulfoxide) solvents. These spectra are characterized by several absorption peaks spread over the FIR spectral range from 25 to 650 cm^{-1}. The time CF at the origin is governed by essentially three process: (1) the ion-induced dipole mechanism; (2) ionic association, i.e., contact in pairs, tightly bound or solvent separated in pairs; and (3) relaxation of the highly polar solvent molecules within the ionic solvation shell. The polarization of the solvent molecules by the dipolar field of ion pairs and the mutual polarization between solvent molecules was neglected with respect to charge induced dipoles. A band shape based upon the Mori theory was used to describe the FIR absorption of the fast oscillations of an ion in its cage (primarily the first shell surrounding the ion) and the structural relaxation of the cage. The Mori treatment of the slower relaxation of ion pairs and the relaxation of solvent molecules within the ionic solvation shell is closely related to the investigation of induced absorption in nonpolar liquids (12). Three-body correlations can be estimated with the help of a lattice gas model. In this theoretical treatment, the role played by the time scale separation between the fast dynamics occurring within the first solvation shell of the ionic species and the time spent to interconvert inner and outer solvation shells was emphasized. The problem discussed here is evidently very complex and many assumptions and approximations were made. Much work should be done by computer simulation to test the validity of the above theory. We note that microscopic theories of solvation dynamics in polar solvents makes use of the ideas and theoretical approach that are similar to those used in I-IA.

Viscoelastic Relaxation in Simple Liquids. The dynamics of simple liquids composed of spherical particles was investigated within the theoretical framework employed for I-ILS, with only minor modifications. Thus, the stress relaxation function was written as a superposition of density correlation functions of all fourier components, and each of these was weighted by the liquid structure factor and a function related to the transform of the intermolecular interaction (57). In another approach, molecular dynamics simulations of the time CF functions that appear in the Green-Kubo expression for shear viscosity were obtained (58, 59). The potential part of the microscopic stress tensor was separated into contributions due to the repulsive and attractive parts of the (12-6) Lennard-Jones potential. The separate CFs for these parts were broken down into two-, three,- and four-body terms. Each of these parts of the L-J potential showed the same type of three-body cancellation that is found in studies of DID light scattering.

Interaction-Induced Emission. We mention a phenomenon that appears to have its roots in interaction-induced emission (I-IE). Consider a single bubble of air trapped in an acoustic standing wave in a water-filled container. If the proper drive frequency and amplitude are applied (the precise values are apparently not critical), the bubble may emit short bursts of light, an effect called sonoluminescence. It is supposed that a converging spherically symmetric shock wave is launched in the gas in the interior of the bubble as it collapses under the influence of the acoustic standing wave. This shock wave heats the trapped gas to high densities at very high temperatures. It was proposed that I-IE is responsible for the sonoluminescence (60). I-IE has been observed in H_2 gas in a standard shock tube experiment, and is thought to be an important source of electromagnetic radiation in the atmospheres of gaseous planets and cool stars (61).

CONCLUDING REMARKS

The interaction-induced spectroscopies (I-IS), absorption and light scattering, are unique among other spectroscopies, e.g., nuclear magnetic resonance, electron spin resonance, neutron scattering and indeed the usual allowed absorption and light scattering. I-IS require that a property (dipoles or polarizabilities) that does not exist in the isolated molecule (or atom) be induced by interactions with neighboring particles. Thus at the lowest densities, the effect of two particle interactions and dynamics are manifest directly in the spectra and CF rather than indirectly through the modification and broadening of internal molecular energy levels as in the allowed spectroscopies. With increasing density, three-body interactions become apparent in the induced properties, and even four-body interactions in the liquid state. The three-body interactions lead to cancellation effects which have profound effects on the I-I spectra and are manifest in a wide variety of phenomena. Because of the sensitivity of induced spectroscopies to molecular interactions, they are excellent probes of microscopic interactions and dynamics in condensed media.

ACKNOWLEDGMENTS

The authors thank Drs. Lothar Frommhold and George Tabisz for valuable comments regarding this manuscript.

REFERENCES

1. H. L. Welsh, Pressure-induced Absorption Spectra of Hydrogen," in *Physical Chemistry, Series one, V.3, Spectroscopy,* MTP International Review of Science, D. A. Ramsay, Ed., Butterworths, London, 1972, pp. 33-71.
2. *Intermolecular Spectroscopy and Dynamical properties of Dense Systems,* Proceedings of the International School of Physics "Enrico Fermi," Course LXXV, J. Van Kranendonk, Ed., North Holland Publishing Co., Amsterdam 1980.
3. *Collision-Induced Phenomena: Absorption, Light Scattering, and Static Properties,* Can. J. Phys. **59**, Special Issue, 1981.
4. J. P. McTague and G. Birnbaum, Phys. Rev. Letters $\underline{21}$, 661-664 (1968).
5. G. C. Tabisz, "Collision-Induced Rayleigh and Raman Scattering," in *Specialist Periodical Reports - Molecular Spectroscopy* VI, R. F. Barrow, D. A. Long, and J. Sheridan, Eds., Chemical Society, London, 1979, pp. 136-173.
6. L. Frommhold, "Collision-induced Scattering of Light and the Diatom Polarizabilities," in *Adv. Chem. Phys.,* Vol 6, I. Prigogine and S. A. Rice, Eds., John Wiley and Sons, Inc., New York, 1981, pp. 1-72.
7. J. Berrue, A. Chave, B. Dumon, and M. Thibeau, "Density Dependence of Light Scattering from Nitrogen: Permanent and Collision Effects," pp. 1510-1513, in Ref. 3.
8. *Phenomena Induced by Intermolecular Interactions,* NATO ASI Series B: Physics Vol. 127, G. Birnbaum, Ed., Plenum Press, New York, 1985.
9. *A Symposium on Interaction-induced Spectra in Dense Fluids and Disordered Solids,* J. Chem. Soc. Faraday Trans. II, Molecular and Chemical Physics, **83**, 1743-1940, 1987.
10. *Collision- and Interaction-Induced Spectroscopy,* NATO ASI Series C: *Mathematical and Physical Sciences,* Vol. 452, G. C. Tabisz and M. N. Neuman, Eds., Kluwer Academic Publishers, Dordrecht, 1995.
11. B. Guillot and G. Birnbaum, "Interaction-Induced Absorption in Simple to Complex Liquids," in *Reactive and Flexible Molecules in Liquids,* Th. Dorfmüller, Ed., Kluwer Academic Publishers, Dordrecht, 1989, pp. 1-36.
12. P. Madden, "Theory and Experimental Aspects of Collision Induced Processes,' in *Spectroscopy and Relaxation of Molecular Liquids, Studies in Physical and Theoretical Chemistry,* Vol. 74, D. Steele and J. Yarwood, Eds., Elsevier Publishers B. V., Amsterdam, 1991, pp. 124-173.
13. N. H. Rich and A. R. W. McKellar, Can. J. Phys. $\underline{54}$, 486 (1976).
14. J. L. Hunt and J. D. Poll, Molec. Phys. $\underline{59}$, 163-164 (1986).
14a. A. Borysow and L. Frommhold, "Collision-Induced Light Scattering: A Bibliography," in *Adv. Chem. Phys.,* Vol. 75, I. Prigogine and S. A. Rice, Eds., John Wiley and Sons, Inc. New York, 1989, pp. 439-505.
15. L. Frommhold, *Collision-induced Absorption in Gases,* Cambridge University Press, 1993.
16. G. Birnbaum, "Determination of Molecular Constants from Collision-Induced Far-Infrared Spectra and Related Methods," pp. 111-145, in ref. 2.
17. K. L. C. Hunt, "Classical Multipole Models; Comparison with *Ab Initio* and Experimental Results," pp. 1-28, in Ref. 8.
18. G. C. Tabisz, N. Meinander and A. R. Penner, "Interaction Induced Rotational Light Scattering in Molecular Gases," pp. 345-358, in Ref. 10.
19. K. L. C. Hunt, *"Ab Initio* and Approximate Calculations of Collision-Induced Polarizabilities," pp. 263-290, in Ref. 8.
20. K. L. C. Hunt and X. Li, "Collision-Induced Dipoles and Polarizabilities for S State Atoms or Diatomic Molecules," pp. 61-76, in Ref. 10.
21. W. Meyer and L. Frommhold, *"Ab initio* Interaction Induced Dipoles and Related Absorption Spectra," pp. 441-456, in Ref. 10.

22. A. D. Buckingham, "General Introduction," pp. 1743-1750, in Ref. 9.
23. T. K. Bose, "Dielectric Properties of Dense Fluids," pp. 77-86, in Ref. 10.
24. S. Kielich, *Mol. Phys.* **9**, 549-564 (1965).
25. B. M. Ladanyi and T. Keyes, "The Influence of Intermolecular Interactions on the Kerr Constant of Simple Liquids," pp. 1421-1429, in Ref. 3.
26. S. M. El-Sheikh, G. C. Tabisz, and R. T Pack, J. Chem. Phys. **92**, 4234-4238 (1990).
27. *Spectral Line Shapes*, Vol. 9, *Proceedings of the 13th International Conference*, Eds. M. Zoppi and L. Ulivi, American Institute of Physics.
28. F. Barocchi, M. Sampoli, and L. Ulivi, "The Interaction Induced Light Scattering Lineshape of Mercury Vapor at Low Density," in Ref. 27.
29. T. Grycuk, M. G. Lednev, and N. A. Kryukov, "Collision-Induced Absorption Near the 227 nm Line in the Hg+Ar Mixture," in Ref. 27.
30. N. Meinander, L. Ulivi, G. Pratesi, G. Cirnigliaro, and F. Barocchi, "Bound Dimer Contribution in the Collision-Induced Pair Spectrum of Mercury Vapor," in Ref. 27.
31. A. R. W. McKellar, "Infrared Studies of Van der Waals Complexes: The Low Temperature Limit of Collision-Induced Spectra," pp. 467-484, in Ref. 10.
32. J. Schaefer, "Dimer Features of H_2-H_2 and Isotopomers at Low Temperature," pp. 485-494, in Ref. 10.
33. G. Birnbaum and B. Guillot, "Cancellation Effects in Collision Induced Phenomena," pp. 1-30, in Ref. 10.
34. A. J. C. Ladd, T. A. Litovitz, and C. J. Montrose, J. Chem. Phys. **71**, 4242-4248 (1979).
35. L. Frommhold and M. Moraldi, "Triple Transitions in Hydrogen," to be published in Ref. 27.
36. D. Kivelson, X.-C. Zeng, H. Sakai, and G. Tarjus, "Interaction Induced Spectra as a Tool for the Study of Structure in Supercooled Liquids and Glasses," pp. 235-248, in Ref. 10.
37. C. Brodbeck and J.-P. Bouanich, "Simultaneous Transitions in Compressed Gas Mixtures," pp. 169-192, in Ref. 8.
38. J. Van Kranendonk, *Physica* **25**, 337-342 (1959).
39. S. P. Reddy, F. Xiang, and G. Varhese, *Phys. Rev. Letters* **74**, 367 (1995).
40. B. Ladanyi, M. S. Shaf, and Y. Q. Liang, "Interaction-Induced Contributions of Spectra of Polar Liquids," pp. 143-157 in Ref. 10.
41. M. Keller, A. Mueller, M. Reh, M. Röder, W. A. Steele, and H. Versmold, "Depolarized Light Scattering: The Influence of Induced Scattering on Allowed Raman and Rayleigh Bands," pp. 87-106, in Ref. 10.
42 J. D. Poll, "The Infrared Spectrum of HD," pp. 677-693, in Ref. 8.
43. L. Ulivi, Z. Lu, and G. C. Tabisz, "Interference of Allowed and Collision-Induced Transitions in HD: Experiment," pp. 407-416, in Ref. 10.
44. J. Van Kranendonk, *Solid Hydrogen*, Plenum, New York, 1993.
45. J. L. Hunt and J. D. Poll, "Charge Induced Effects in Solid Tritium and Deuterium," pp. 595-608, in Ref. 8.
46. R. L. Brooks, J. A. Forrest, and J. L. Hunt, "Atoms in Irradiated Solid Deuterium: Charge-Induced Spectra," pp. 297-306, in Ref. 10.
47. E. Cohen de Lara, "Electric Field Effects Observed on the Infrared Spectra of Molecules Adsorbed in Zeolites," pp. 287-296, in Ref. 10.
48. C. G. Joslin, C. G. Gray, J. D. Poll, S. Goldman, and A. D. Buckingham, "Interaction-Induced Spectra of Endohedral Complexes of Buckminsterfullerene," pp. 261-286, in Ref. 10.
49. P. Benassi, P. Gallo, G. Ruocco, G. Signorelli, and V. Mazzacurati, "Induced Light Scattering from Electrically Disordered Solids," pp. 307-321, in Ref. 10.
50. Y. Danten, B. Guillot and Y. Guissani, *J. Chem. Phys.* **96**, 3782-3810 (1992).
51. M. Besnard, Y. Danten, and T. Tassaing, "Far-Infrared Spectroscopic Investigation of Liquid Mixtures," pp. 201-213, in Ref. 10.
52. B. Guillot, *J. Chem. Phys.* **95**, 1543-1551 (1991).
53. M. A. Ricci, G. Ruocco, and M. Sampoli, *Mol. Phys.* **67**, 19-31 (1989).

54. P. A. Madden, K. O'Sullivan, J. A. Board, and P. W. Fowler, *J. Chem. Phys.* **94**, 918-927 (1991).
55. P. A. Madden and K. O'Sullivan, *J. Chem. Phys.* **95**, 1980-1990 (1991).
56. B. Guillot, P. Marteau, and J. Obriot, *J. Chem. Phys.* **93**, 6148-6164 (1990).
57. C. J. Montrose, T. A. Litovitz, G. Birnbaum, and R. Mennella, *J. Non-Crystalline Solids,* **131-133**, 177-181 (1991).
58. H. Stassen and W. A. Steele, *J. Chem. Phys.* **102**, 932-938 (1995).
59. H. Stassen and W. A. Steele, *J. Chem. Phys.* **102**, 8533-8540 (1995).
60. L. Frommhold and W. Meyer, "Collision-Induced Emission and Sonoluminescence," to be published in Ref. 27.
61. A. Borysow, "Collision-Induced Molecular Absorption in Stellar Atmospheres," pp. 529-539, in Ref. 10.

APPENDIX

MINUTES OF THE INTERNATIONAL COMMITTEE MEETING

The meeting of the International Committee for the 13th International Conference on Spectral Line Shapes was held in Firenze on June 20, 1996.

1) It was agreed that the next conference will be held at Pennsylvania State University and that Roger Herman will be the local Chairman. The committee invited Roger Herman to join in the International Committee.

2) The resignations of N. Konjevic and M. Zoppi from the International Committee were accepted. The committee accepted Dr. Milan Dimitrijevic of the Belgrade Astronomical Observatory and Dr. Massimo Moraldi of the University of Florence as new members of the International Committee.

3) It was confirmed that all appointments in the Committee should be reconsidered after four years. It was agreed that following the Firenze Conference, the International Committee will consist of the following members: C. Back, A. Devdariani, M. Dimitrijevic, N. Feautrier, R. Herman, J. Kielkopf, V. Lisitsa, D. May, M. Moraldi, H. Nguyen, E. Oks, G. Peach, G. Pichler,
J. Seidel, R. Stamm, G. Tabisz, R. Tipping.

4) It was agreed that it would be desirable to have one representative from Spain and one from Poland. Proposals will be examined at the next conference.

5) The committee approved the proposed budget of the 13th ICSLS. It was suggested that the organizer of each conference should be granted the registration fee and the hotel expenses for the following conference. It was agreed that this should become a rule for future meetings.

6) The proposal to hold the 15th ICSLS in Berlin, Germany, was accepted. The Committee expressed their gratitude to Dr. G. Roston for the kind offer to host the 15th ICSLS in Alexandria, Egypt.

Firenze, 20 June 1996
Marco Zoppi

LIST OF PARTICIPANTS

El Sheikh S.M.	Arab Emirates	
Gruber D.	Austria	icecube@fexphal01.tu-graz.ac.at
Li X.	Austria	xh@fexphal01.tu-graz.ac.at
Polly R.	Austria	bowy@fexphds03.tu-graz.ac.at
May A.D.	Canada	dmay@physics.utoronto.ca
Michaud G.	Canada	michaudg@cerca.umontreal.ca
Tabisz G.C.	Canada	tabisz@ccm.umanitoba.ca
Milosevic S.	Croatia	milosevic@olimp.irb.hr
Pichler G.	Croatia	pichler@olimp.irb.hr
Veza D.	Croatia	veza@olimp.irb.hr
Renner O.	Czech Rep.	renner@fzu.cz
Borysow A.	Denmark	aborysow@stella.nbi.dk
Bross M.	Deutschland	bross@ipf.uni-stuttgart.de
Günter S.	Deutschland	guenter@physik.uni-rostock.d400.de
Grosser J.	Deutschland	grosser2@ceres.amp.uni-hannover.de
Helbig V.	Deutschland	helbig@physik.uni-kiel.de
Kaiser A.	Deutschland	kaisera@mail.rz.uni-duesseldorf.de
Meyer W.	Deutschland	meyer@chemie.uni-kl.de
Neundorf D.	Deutschland	doerte.neundorf@etec.uni-karlsruhe.de
Pfendtner R.	Deutschland	reinard.pfendtner@etec.uni-karslsruhe.de
Schaefer J.	Deutschland	jas@mpa-garching.mpg.de
Schmidt M.	Deutschland	
Seidel J.	Deutschland	seidel@chbrb.berlin.ptb.de
Sorge S.	Deutschland	stefans@darss.mpg.uni-rostock.de
Steiger A.	Deutschland	
Wrubel T.	Deutschland	Thomas.Wrubel@rz.ruhr-uni-bochum.de
Galal A.A.	Egypt	astro2@frcu.eun.eg
Roston G.D.	Egypt	gdaniel@alex.eun.eg
Aparicio Calzada J.A.	Espana	apa@hp9000.uva.es
DeLaRosa Garcia M.I.	Espana	optica@cpd.uva.es
Perez M.C.	Espana	optica@cpd.uva.es
Meinander N.	Finland	meinander@phcu.helsinki.fi
Arranz J.P.	France	arranz@frcpn11.in2p3.fr
Bauer A.	France	bauer@lsh.univ-lille1.fr
Boulet C.	France	christian.boulet@lpma.u-psud.fr
Butaux J.	France	arranz@frcpn11.in2p3.fr
Chrysos M.	France	chrysos@babinet.univ-angers.fr
Feautrier N.	France	Nicole.Feautrier@obspm.fr
Ferri S.	France	

Gauthier P.	France	gauthier@moka.ccr.jussieu.fr
Gilles D.	France	gilles@limeil.cea.fr
Godbert L.	France	laure@piima1.univ-mrs.fr
Henry A.	France	anhenry@ccr.jussieu.fr
Henry L.R.	France	anhenry@ccr.jussieu.fr
Hurtmans D.	France	hurtmans@ccr.jussieu.fr
Le Duff Y.	France	
Leboucher-Dalimier E.	France	lebda@moka.ccr.jussieu.fr
Lesage A.	France	alain.lesage@obspm.fr
Mossé C.	France	caro@piima1.univ-mrs.fr
Mysyrowicz A.	France	mysy@enstay.ensta.fr
Nguyen H.N.	France	arranz@frcpn11.in2p3.fr
Peyrusse O.	France	peyrusse@limeil.cea.fr
Sahal-Brechot S.	France	sahal@obspm.fr
Stamm R.	France	rstamm@piima1.univ-mrs.fr
Stehlé C.	France	stehle@obspm.fr
Van Regemorter H.	France	hvr@obspm.fr
Alexiou S.	Israel	spiros@plasma-gate.weizmann.ac.il
Ben-Reuven A.	Israel	abr@taunivm.tau.ac.il
Halperin B.	Israel	wwilberg@weizmann.weizmann.ac.il
Allegrini M.	Italy	allegrin@ipifidpt.difi.unipi.it
Arimondo E.	Italy	arimondo@ipifidpt.difi.unipi.it
Bafile U.	Italy	bafile@ieq.fi.cnr.it
Balucani U.	Italy	balucani@ieq.fi.cnr.it
Bambini A.	Italy	bambini@ieq.fi.cnr.it
Barocchi F.	Italy	barocchi@firenze.infn.it
Bicchi P.	Italy	bicchi@unisi.it
Buffa G.	Italy	buffa@ipifidpt.difi.unipi.it
Buffa R.	Italy	buffa@fi.infn.it
Califano S.	Italy	califano@lens.inifi.it
Carli B.	Italy	carli@iroe.iroe.fi.cnr.it
Casini R.	Italy	casinir@arcetri.astro.it
Cavalieri S.	Italy	cavalieri@fi.infn.it
Cazzoli G.	Italy	
Cecchi G.	Italy	cecchi@iroe.iroe.fi.cnr.it
Celli R.M.	Italy	
Celli M.	Italy	celli@fi.infn.it
Ciucci A.	Italy	
Corsi C.	Italy	corsi@lens.unifi.it
DeRosa M.	Italy	
Dore L.	Italy	dore@ciam01.cineca.it
Dore P.	Italy	dorep@vxrm64.roma1.infn.it

Eramo R.	Italy	eramo@lens.unifi.it
Fini L.	Italy	fini@fi.infn.it
Fioretti A.	Italy	fioretti@ipifidpt.difi.unipi.it
Fort C.	Italy	fort@lens.unifi.it
Fuso F.	Italy	fuso@ipifidpt.difi.unipi.it
Gabbanini C.	Italy	carlo@risc.ifam.pi.cnr.it
Giulietti A.	Italy	tonino@risc.ifam.pi.cnr.it
Gizzi L.	Italy	
Gorelli F.	Italy	
Höpfner M.	Italy	Michael.Hoepfner@imk.fzk.de
Inguscio M.	Italy	
Landi E.	Italy	landi@arcetri.astro.it
Landi degli Innocenti E.	Italy	landie@arcetri.astro.it
Landini M.	Italy	landini@arcetri.astro.it
Lemaire V.	Italy	
Lucchesini A.	Italy	alex@risc.ifam.pi.cnr.it
Matera M.	Italy	matera@ieq.fi.cnr.it
Mazzinghi P.	Italy	mazzinghi@ieq.fi.cnr.it
Mazzoni M.	Italy	mazzoni@ieq.fi.cnr.it
Moraldi M.	Italy	moraldi@firenze.infn.it
Palla F.	Italy	palla@arcetri.astro.it
Pavone F.	Italy	pavone @lens.unifi.it
Pini R.	Italy	rpini@ieq.fi.cnr.t
Raspollini P.	Italy	piera@safire.iroe.fi.cnr.it
Ridolfi M.	Italy	ridolfi@iroe.iroe.fi.cnr.it
Santoro M.	Italy	santoro@ingfi1.ing.unifi.it
Siano S.	Italy	siano@ieq.fi.cnr.it
Tarrini O.	Italy	tarrini@ipifidpt.difi.unipi.it
Toci G.	Italy	toci@ieq.fi.cnr.it
Ulivi L.	Italy	ulivi@ieq.fi.cnr.it
Vallauri R.	Italy	vallauri@science.unitn.it
Vannini M.	Italy	vannini@ieq.fi.cnr.it
Xu J.H.	Italy	hua@sab.sns.it
Zoppi M.	Italy	zoppi@ieq.fi.cnr.it
Ciurylo R.	Poland	rciurylo@phys.uni.torun.pl
Domyslawska J.	Poland	jolka@phys.uni.torun.pl
Dzierzega K.	Poland	Krzycho@castor.if.uj.edu.pl
Gburski Z.	Poland	zgburski@rs6k.mphys.us.edu.pl
Golly A.	Poland	agoly@uni.opole.pl
Grycuk T.	Poland	tgrycuk@fuw.edu.pl
Konefal Z.	Poland	fizzk@halina.univ.gda.pl
Kusz J.	Poland	jkusz@uni.opole.pl

Olchawa W.	Poland	halenka@uni.opole.pl
Szonert J.	Poland	szonj@beta1.ifpan.edu.pl
Szudy J.	Poland	szudy@phys.uni.torun.pl
Wujec T.	Poland	wujec@sparc-1.uni.opole.pl
Zielinska-Kaniasty S.	Poland	sziel@mail.atr.bydgoszcz.pl
Bulanin M.O.	Russia	bulanin@niif.spb.su
Bureyeva L.A.	Russia	bureyeva@spectr.msk.su
Demura A.	Russia	demura@rec.msk.su
Devdariani A.Z.	Russia	ponik@devdar.spb.su
Lisitsa V.S.	Russia	lisitsa@qq.nfi.kiae.su
Magounov A.	Russia	faenov@glas.apc.org
Rubtsova N.N.	Russia	rubtsova@isph.nsk.su
Salakhov M. Kh.	Russia	msalakh@phys.ksu.ras.ru
Shapiro D.A.	Russia	shapiro@iae.nsk.su
Starostin A.N.	Russia	starostin@fly.triniti.troitsk.ru
Vigassin A.	Russia	root@iaph.msk.su
Schouten J.	Netherland	schouten@phys.uva.nl
Ben Nessib N.	Tunisia	
Leo P.	U.K.	p.leo@ucl.ac.uk
Peach G.	U.K.	g.peach@ucl.ac.uk
Ueda K.	U.K.	ueda@rism.tohoku.ac.jp
Back C.	U.S.A.	tinaback@llnl.gov
Baker H.	U.S.A.	hbaker@butler.edu
Bernheim R.	U.S.A.	r5b@psuvm.psu.edu
Birnbaum G.	U.S.A.	birnbaum@micf.nist.gov
De Lucia F.C.	U.S.A.	fcd@mps.ohio-state.edu
Frommhold L.	U.S.A.	frommhold@physics.utexas.edu
Gallagher A.	U.S.A.	alang@jila.colorado.edu
Griem H.R.	U.S.A.	welch@ipr.umd.edu
Herman R.M.	U.S.A.	rmh@phys.psu.edu
Kreye W.C.	U.S.A.	wkreye@desire.cc.wright.edu
Oks E.	U.S.A.	goks@physics.auburn.edu
Rothman L.	U.S.A.	LRothman@CfA.Harvard.edu
Sando K.	U.S.A.	kenneth-sando@uiowa.edu
Savchenko V.I.	U.S.A.	vsavchen@lyman.pppl.gov
Tipping R.	U.S.A.	rtipping@ua1vm.ua.edu
Varanasi P.	U.S.A.	pvaranasi@ccmail.sunysb.edu
Konjevic' N.	Yugoslavia	nikon@rudjer.ff.bg.ac.yu

AUTHOR INDEX

A

Alexiou, S., 19, 31, 61, 79
Allegrini, M., 203, 279
Allot, R., 49
Alman, V. B., 205
Amari, M., 21
Angelo, P., 27, 45, 61
Aparicio, J. A., 151, 153
Apolonsky, A. A., 157
Arranz, J. P., 21, 65
Asfaw, A., 35
Astapenko, V. A., 195

B

Babin, S. A., 157, 259
Back, C. A., 35, 45
Bacławski, A., 307
Baldacchini, G., 361
Balucani, U., 485
Bancewicz, T., 519
Barocchi, F., 453, 527, 533, 535
Barsotti, S., 203
Bassani, F., 255
Batani, D., 49
Bauer, A., 431
Behery, M., 423
Behmenburg, W., 515
Belli, S., 343, 365
Ben Lakhdar, Z., 117
Ben Nessib, N., 117
Ben-Reuven, A., 271
Benuzzi, A., 49
Berman, R., 331
Bernheim, R. A., 197
Beterov, I. M., 205
Beuc, R., 275
Bianchini, G., 255

Bicchi, P., 197, 201
Bielski, A., 313
Bieniak, B., 207
Birnbaum, G., 537
Blagoev, K., 209
Blagojević, B., 143
Bonsmann, J., 515
Borysow, A., 429, 489, 523, 531
Brill, J. U., 113
Brjazovsky, U. V., 205
Buffa, G., 343, 361, 365, 493
Buffa, R., 211
Bulanin, M. O., 497
Bulyshev, A. E., 23
Bureyeva, L., 161
Büscher, S., 71
Butaux, J., 21, 65

C

Calisti, A., 31, 35, 47, 61, 113
Cancio, P., 255
Carlier, J., 431
Casini, R., 125
Cataliotti, F. S., 255, 261
Cavalieri, S., 213, 309
Cazzoli, G., 141, 365
Cecchi, G., 433
Ceccotti, T., 45
Celli, M., 309
Celli, R. M., 269
Chernenko, A. A., 205
Chernykh, A. I., 157
Chrysos, M., 501
Cirnigliaro, G., 527
Ciucci, A., 367
Ciuryło, R., 311, 313
Colla, M., 279
Conti, A., 49
Corsi, C., 263

Costato, M., 49
Csambal, C., 265

D

D'Amato, F., 361
Dawid, A., 517
de la Rosa, I., 153
Degli Esposti, C., 141
De Lucia, F. C., 127
Demura, A. V., 23, 25, 67, 119
Depiesse, M., 155
Derevianko, A., 15
Derfoul, H., 45
De Rosa, M., 361, 367
Devdariani, A. Z., 27, 283
Dimitrijević, M. S., 143, 147, 149
Djaoui, A., 57
Djurović, S., 315
Domysławska, J., 317
Dore, L., 141, 365
Dore, P., 505
Dreischuh, A., 209
Drummond, J. R., 331
Duggan, P., 331

E

Elliasmine, A., 519
Eramo, R., 213
Evangelista, A., 277

F

Faenov, A., 49
Fakhr-Eslam, S. H., 331
Feautrier, N., 67
Ferri, S., 35, 47, 113
Fini, L., 213, 215
Fioretti, A., 279
Fisch, N. J., 33
Flora, F., 49
Forsman, J. W., 331

Förster, E., 45, 57
Fort, C., 261
Franco, M., 169
Frommhold, L., 441, 471, 505, 521
Fu, Y., 489, 531
Fuso, F., 203

G

Gabbanini, C., 277, 367
Gabrysch, M., 263
Galal, A., 423
Gamache, R. R., 431
Gauthier, P., 27, 45, 61
Gaye, O., 501
Gburski, Z., 363, 517
Gigosos, M. A., 151, 153
Gilles, D., 119
Giulietti, A., 53
Giulietti, D., 53
Gizzi, L. A., 53
Glenzer, S. H., 35, 71
Głódź, M., 207
Godbert, L., 31, 35
Godet, J. L., 519
Godon, M., 431
Gogić, S., 163
Goly, A., 307
Gonzáles, V. R., 151
Gorelli, F., 383
Gozzini, S., 277, 367
Grabowski, B., 29
Griem, H. R., 113
Grillon, G., 169
Grosser, J., 181
Gruber, D., 209, 219, 223
Gruszka, M., 429, 523
Grycuk, T., 229, 525
Guillot, B., 537
Günter, S., 99

H

Halenka, J., 29, 121
Hamid, R., 423
Hammel, B. A., 35
Hänsch, T. W., 255, 261
Helbig, V., 265, 267
Helmi, M. S., 227, 229
Henry, M. E., 251
Herman, R. M., 251
Heß, B. A., 219
Hoang Binh, D., 329
Hoffmann, O., 181
Höpfner, M., 435
Hung, Y. M., 297

I

Il'in, G. G., 123
Inguscio, M., 255, 261

K

Kablukov, S. I., 157, 259
Kachel, A., 363
Kaiser, A., 515
Keane, C. J., 35
Khorev, S. V., 157
Khvorostov, E. B., 357
Kielkopf, J. F., 321
Klein, L., 35, 47
Klose, S., 181
Kobilarov, R., 315
Koenig, M., 49
Kondratenko, M. I., 259
Konefał, Z., 217
Könies, A., 99
Konjević, N., 143, 315
Konovalova, O. A., 123
Korennoy, Ya. A., 319
Kosarev, I. N., 67
Krasik, Ya. E., 19
Kreye, W. C., 321
Krouský, E., 57
Kryukov, N. A., 233, 525
Książek, I., 307
Kukushkin, A. B., 195, 319
Kunze, H. J., 71
Kurochkin, V. L., 205
Kusz, J., 323

L

Landi Degl'Innocenti, E., 125
Landi, E., 419, 421
Landini, M., 411, 419, 421
Lange, R., 169
Leboucher-Dalimier, E., 27, 45, 61
Lednev, M. G., 283, 525
Le Duff, Y., 501, 519
Lee, R. W, 35, 47, 113
Lemaire, V., 365
Leo, P. J., 325
Lesage, A., 155
Letardi, T., 49
Li, X., 223
Liedahl, D., 35
Lipschultz, B., 113
Lisi, N., 49
Lisitsa, V. S, 23, 67, 161
Lucchesini, A., 277, 367
Lumma, D., 113

M

Magunov, A., 49
Makonnen, A., 515
Mar, S., 151, 153
Marinelli, C., 197, 201
Mariotti, E., 201
Maron, Y., 19
Masini, A., 49
Matera, M., 213
May, A. D., 331
Mazur, D., 323
Mazzinghi, P., 433, 437

Mazzoni, M., 215
McCracken, G., 113
McQuarrie, B., 507
Meftah, T., 31
Meinander, N., 527
Meiners, D., 155
Meucci, M., 201
Meyer, W., 471, 515
Michaud, G., 397
Michels, J. P. J., 369
Mijatović, Z., 315
Milani, M., 49
Milošević, S., 163, 275
Moi, L., 201
Molisch, A. F;, 203, 279
Monsignori Fossi, B., 421
Moraldi, M., 489, 521, 533, 535
Moreno, J. C., 35
Morozov, A., 209
Mossé, C., 35, 47
Motapon, O., 161
Mullamphy, D. F. T., 325
Müller, J. H., 279
Mysyrowicz, A., 169

N

Nash, J. K., 35
Nemtchinov, V., 425
Neundorf, D., 231
Nguyen, H., 21, 65
Nucara, A., 505

O

Oks, E., 3, 15, 25
Olchawa, W., 29, 121
Osterheld, A. L., 35

P

Pacini, G., 51
Palladino, L., 49

Panteleev, A. A., 33
Pavone, F. S., 255
Peach, G., 325
Pelagalli, F., 361
Pérez, C., 153
Pichler, G., 275
Pikuz, T., 49
Pini, R., 51
Pircher, P., 223
Podivilov, E. V., 157
Polly, R., 219
Popović, M. V., 143
Poquerusse, A., 45
Postorino, P., 505
Pozzi, A., 49
Prade, B., 169
Pratesi, G., 527
Prevedelli, M., 261, 263

R

Raimondi, V., 433
Raspollini, P., 435
Reale, A., 49
Reggadi, A., 65
Reiter-Domiaty, U., 209
Renner, O., 57
Richer, J., 397
Richou, J., 155
Ridolfi, M., 435
Rieper, T., 267
Ripoche, J. F., 169
Rosa-Clot, M., 263
Rose, T., 267
Roston, G. D., 227, 229
Rubtsova, N. N., 357

S

Sahal-Bréchot, S., 117, 147, 149
Salakhov, M. Kh., 123, 327
Salimbeni, R., 51
Salzmann, D., 57

Sampoli, M., 453
Sando, K. M., 297
Santoro, M., 533, 535
Sarandaev, E. V., 123, 327
Sarfaty, M., 19
Sauvan, P., 27, 45, 61
Savchenko, V. I., 33
Saveliev, P. A., 233
Schäfer, J., 511
Schouten, J. A., 369
Shapiro, D. A., 157, 259
Sheldon, G. D., 331
Shepard, T., 45
Siano, S., 51
Sinclair, P. M., 331
Skobelev, I., 49
Sondhauß, P., 57
Stamm, R., 31, 35, 113
Starostin, A. N., 23, 33
Stassen, H., 363
Stehlé, C., 67, 119
Suvorov, A. E., 23
Szonert, J., 207
Szudy, J., 311, 313, 385

T

Tabisz, G. C., 507
Talin, B., 31, 35, 47, 61, 113
Tarrini, O., 343, 361, 365, 493
Tchaplyguine, M. A., 233, 283
Terry, J., 113
Toci, G., 437
Trawiński, R. S., 313, 317
Turcu, E., 49

U

Ueda, K., 237
Ulivi, L., 383, 453, 527, 533, 535
Uschmann, I., 45

V

Vaia, R., 215
Vallauri, R., 485
Van Regemorter, H., 161, 329
Vannini, M., 437
Varanasi, P., 425
Vasilenko, L. S., 357
Venturi, V., 325
Veža, D., 267, 275
Vigasin, A. A., 529
Vollbrecht, M., 45

W

Weaver, J. L., 113
Weingarten, A., 19
Welch, B. L., 113
Whittingham, I. B., 325
Willi, O., 53
Windholz, L., 209, 219, 223
Wollschläger, F., 155
Woolsey, N. C., 35
Wrubel, Th., 71
Wujec, T., 121, 307

X

Xu, J. H., 269

Y

Yakunin, I. I., 23
Yousef, N., 423
Yurovsky, V., 271

Z

Zagrebin, A. L., 283
Zheng, C., 531
Zhidkov, A. G., 319
Zielinska, Kaniasty, S., 221
Zoppi, M., 383, 433, 533, 535

SUBJECT INDEX

A

ab initio calculations; 219
Abel inversion; 73, 315
absorption; *see also* far wing–, far infrared–
 –coefficient; 474*ff.*
 –profile; 229, 263
 –spectroscopy; 265, 437
action spectra; 297*ff.*
ADAS data base; 412
adiabatic approximation; 162
adiabatic wavefunction; 288
Al; 45
Al(ion); 53*ff.*, 57, 69
alkali atoms; 217
alkali–halide crystals; 542
amplitude of modulation; 375*ff.*
angular velocity coupling; 303
anisotropic interaction; 429, 523
anisotropic spectrum; 503
anomalous electric fields; 11*ff.*
APEX; –computer code; 93
Ar; 227, 232, 241, 315, 323*ff.*, 517
 L–S coupling in–; 327
Ar(ion); 36, 40, 151, 157
Ar$_2$; 497, 501
ARCETRI spectral code; 412
argon solid matrix; 383
associative ionisation; 210
astrophysics; 121
asymmetry; 99*ff.*
 collision-time–; 317
 –parameter; 100*ff.*, 317*ff.*
atmospheric spectra; 435
atomic beams; 182
atomic coherence; 261
atomic collisions; 203, 211, 297
atomic diffusion; 398

atomic drift effect; 222
atomic emission lines; 297
atomic kinetics; 36
atomic state interference; 67
atomic trap; 277
attractive contribution
 –to line shift; 369*ff.*
 –to line width; 369*ff.*
Autler–Townes splitting; 277
autocorrelation function; 19, 79*ff.*, 375*ff.*
 dipole–; 298
 field-field–; 114
autoionization; 408, 499
 –level; 205
 –spectra; 237*ff.*

B

B(ion); 85, 147
Ba; 247
Balmer γ; 121
Balmer series; 113
band wings; 496
Be(ion); 85, 147
benchmark calculations; 87
binary collisions; 162
birefringence; optical–; 255*ff.*
black–body radiation; 472*ff.*
blend; 406
bond length dependence;
 –of intermolecular forces; 374*ff.*
Born approximation; 102
Born–Oppenheimer approximation; 61
Br; 307
bremsstrahlung; 51
 –polarisation; 195*ff.*
 multiphoton stimulated–; 195

broadening; *see also* line–, Stark–, collisional–, impact–, Coulomb–, pressure–, ion–dynamical–, self–, thermal–
broadening; 398
   critical–; 378*ff.*
   electron impact–; 143
   ion impact–; 143*ff.*
   non–impact–; 321
   non–Lorentzian–; 386
   quantum–; 161
   quasistatic theory of line–; 498
   source–; 38

C

C(ion); 111
C_2H_4; 367
Ca; 247, 301
carbonyl sulphide; 363
Cd; 203, 227
^{114}Cd; 317
central limit theorem; 19
$(CF_2)_n$; 49
CF_4; 519
CH_3Br; 355
CH_3F; 354, 365
CH_3I; 355
charge exchange; 28
charge fluctuations; –in water; 543
CHF_2Cl; 355
CHIANTI data base; 412
chirp; 170*ff.*
   linear–; 170
   quadratic–; 170
classical oscillators; 319
classical path; 298*ff.*
close coupling; 84*ff.*
   –calculations; 148, 272*ff.*
   –method; 297*ff.*
   –model; 300
   –for Ne; 385
clusters; argon–; 517

CO; 128*ff.*, 331
CO_2; 331, 429, 523
   –combination band; 338
coherent atomic levels; 255
coherent rotation; –of molecules; 179
coherent transient phenomena; 357
coherent trapping; 269
cold atom traps; 271
cold collisions; 275
collision; finite duration of–; 491
collision dynamics; 272
collision molecules; 390
collision strengths; effective–; 415
collision time; 322
collisional broadening; 345, 365, 367, 398*ff.*, 426, 432, 507
   electron–; 71
collisional cooling; 127*ff.*
collisional coupling; 343, 494
collisional decay; 207
collisional effects; speed dependent–; 386
collisional energy transfer; 515
collisional interference; 507*ff.*
collisional line–coupling; 365
collisional lineshape analysis; 506
collisional process; 207
collisional products; alignment of–; 191
collisional relaxation; 357
collisional shift; 325, 367
collisional system; emission from–; 475*ff.*
collisional transfer; 207
collisional width; 325, 400
collision–induced absorption; 429, 523, 525, 529, 531
   –spectrum; 505
collision–induced emission; 447, 471*ff.*
collision–induced lineshape; 529
collision–induced phenomena; 497
collision–induced radiation; 511*ff.*

collision–induced scattering; 453*ff.*,
    501, 519, 527, 533, 535
    –polarized spectrum; 460
    computer simulation of–; 486
    density variation of–; 487
    depolarized–; 441*ff.*
    pair,triplet and quadruplet contribution; 486
    self vs distinct contribution; 486
    time correlation function of–; 485
collision–induced spectrum; 429, 490
collision–induced trace polarizability; 497
collisions; atomic–; 201
    cold atom–; 272
    energy pooling–; 202
    intermolecular–; 538
    state changing–; 297
complexes; halo–organic–; 543
computer code; *see under the code name*
Condon radius; 182
continuum emission; 421
Coriolis perturbation; 142
correlation function; angular velocity–; 363
    reorientational–; 363
    two body–; 517
correlation time; 374*ff.*
correspondence principle; 127
Coulomb broadening; 157*ff.*
Coulomb laser; 259
critical concentration fluctuations; 369*ff.*
cross phase modulation; 177*ff.*
cross–over resonance; 267
Cs; 261, 279
Cs_2; 275
CsHg; 219

D

D_2; 331

dark matter; 511
data base; 422
Debye line; 494
deconvolution; 155
degenerate structure of levels; 33
dense plasma; 21,27, 31, 45, 47, 61, 65
density diagnostic; –of astrophysical plasmas; 419
density fluctuations; 369*ff.*
density matrix; 33, 68, 221*ff.*, 345
density scaling; 114
dephasing; –of the dipole; 298
dephasing processes; collisional–; 179
depolarization; 298*ff.*
depolarized Q branch; 331*ff.*
depolarized spectrum; 485
depolarizing collisions; 357*ff.*
dessed ground state; 302
diamond anvil cell; 370
diatom polarizability tensor; 502
    anisotropy of–; 502
    trace of–; 502
Dicke narrowing; 338
dielectric virial coefficient; second–; 500
differential scattering; 181*ff.*
diffusion gas discharge; 293
diffusion model; 403
dimer polarizability; 502
dimers; 465*ff.*, 527, 529, 539
    alkali–; 275
    Ar–; 502
diode laser; 267, 361, 367, 437
    distributed feedback–; 263
dipolar matrix elements; 63
dipole forbidden transition; 525
dipole moment; 289
    collision induced–; 539
dipole oscillator strength distribution; 497*ff.*
dipole range; 475*ff.*
dipole strength; 475*ff.*

dipole–dipole interaction; 361
dipole–induced–dipole; 485, 490
dipole–induced–octopole; 490
discrete variable representation; 502*ff.*
dispersion interaction; 374*ff.*
Doppler broadening; 36, 113, 317, 398
Doppler shift corrrection; 58*ff.*
Doppler width; 155, 158, 264, 313, 338, 400, ,
Doppler–free spectra; 267
Doppler–free spectroscopy; 265
double transitions; 477*ff.*, 541
    rotational; 533
    rotovibrational; 535
dressed collisional pair; 184
dressed states; 211
duration of collision; 298
dynamic screening; 108
dynamics; Newtonian–; 300

E

Earth atmosphere; 431
effective Hamiltonian method; 286
effective line; 350
eigen channel representation; 240
electrolyte solutions; 544
electromagnetically induced transparency; 256
electron beams; 13
electron density; 35
electron drift effect; 222
electron quadrupole broadening; 72
electron temperature; 35, 55
electronic shift; 99
emission
    see also interaction induced–
emission spectrum; 35, 283
energy pooling; 203, 210
energy transfer; non–radiative–; 297
ENSTA; algoritm; 172*ff.*
equivalent width; 155

excimers; 219
    HgIn–; 197*ff.*
    intermetallic–; 163
    Xe–Kr–; 233
exciplex; 297*ff.*
excitation spectrum; 225
excitation transfer; 207
    radiationless–; 320
exp–six potential model; 227

F

F; 45
F(ion); 49, 62, 71, 75
Fabry Perot interferometer; 431
Fano theory; 206
far wings; 489
far–infrared absorption; 444
far–wing absorption; 309
FASCODE computer code; 431
fast modulation; 369*ff.*
Fe(ion); 420
Fermi–Omont model; 390
ferroelectric plasma source; 323
FFM; *see* frequency fluctuation model
fluctuation rate; 79*ff.*
fluorescence; –of Na D2 line; 269
fluorescence spectra; 204, 209
forbidden components; 403
formaldehyde; protonated; 141
formyl ion; 142
Fortrat diagram; 225
Frank–Condon treatment; unified–; 317
free–bound transitions; 275
frequency fluctuation model; 15, 32, 47, 75, 86, 113
frequency redistribution function; 23*ff.*
frequency separation technique; 87
friction force; 158*ff.*
fullerene; 542
fusion; inertial confinement–; 53

G

Galatry profile; speed dependent; 311
gas discharge plasma; 157
gasbag; 36
gas–liner pinch; 71
Gaussian distribution; 19
generalised Bohr radius; 231
generalised position; 231
giant planets; 425
glow discharge; 141, 313
gravitational settling; 403
Green's functions; 93

H

H; 29, 121
H_2; 305, 511, 521, 531, 533, 535
H_2O; 431
half–collision; –experiment; 302
harmonic generation; 205
HCN; 355
HCO(ion); 134*ff.*, 142
HD; 331
HD–He; 507*ff.*
He; 255, 301*ff.*, 303, 385
He(ion); 31, 72, 110, 120
heat–pipe oven; 201, 203, 215, 224
He–Xe; 505
Hg; 197, 209, 227, 232, 283, 453*ff.*, 525, 527
Hg+Ar; –mixture; 525
Hg+Ar; –potential; 525
high pressure; 369*ff.*, 383
HNO_3; 437
hohlraum; 36
Holtsmark distribution; 16
hot plasma; 31, 47
HSTRK; computer code; 93
Hubble space telescope; 388
hydrodynamic simulation; 59
hydrogen lines; 3*ff.*, 99
    effective Landé factor; 126
    Lyman α; 104
    Lyman β; 104*ff.*
    Paschen α; 108*ff.*
    Paschen β; 109
    polarization profiles of–; 125
hydrogen–like lines; 3*ff.*
hyperfine components; 353
hyperfine optical pumping; 269
hyperfine structure; 267, 277
hyperpolarizability; dipole-dipole-quadrupole–; 490

I

impact approximation; 88*ff.*, 117, 161, 309, 317, 331, 346
impact broadening; 8*ff.*, 15*ff.*, 85, electron–; 79*ff.*
impact limit; 311
impact line shape; 9
impact parameter; 352
impact shift; 149
impact theory; 81*ff.*
impact width; 147, 149
In; 197, 201
induced dipole; 525, 507*ff.*
    overlap–; 505
induced transparency; 211, 256*ff.*
    electromagnetic–; 261
inelastic collision; 508
inert gas; 227
infrared Q band; 493
interaction potential (energy); 231, 527
    –curves; 285
interaction–induced emission; 545
interatomic potential; 221, 229
intercombination rules; 227
interference; 183
    –of allowed and induced transitions; 541
    –of atomic states; 23

interference terms; 8*ff.*
    elastic contribution to–; 8*ff.*
    inelastic contribution to–; 8*ff.*
intermolecolar interaction; 501
intermolecular forces; 369*ff.*
intermolecular potential; 227, 539*ff.*
    anisotropic–; 491*ff.*
    vibrational dependence of–; 509
International Ultraviolet Explorer; 388
interstellar chemistry; 141
interstellar medium; 512
interstellar molecular clouds; 127
ion; *see also* low temperature–
ion; higly charged–; 195
    multiply charged–; 148
ion-dipole interaction; 6
ion dynamical broadening; 15
    simulation models of–; 15
    analytical models of–; 15
ion dynamics; 11, 23, 32, 35, 43, 47, 67, 79, 99*ff.*, 113*ff.*, 119
ion octupole interaction; 7
ion production; 129
ion quadrupole interaction; 6*ff.*
ion quadrupole shift; 103
ion sphere model; 57
ionic melts; 543
ionic shift; 99
ionic transition; 259
ionization; multiphoton–; 205
ionization balance; 414, 421
ionization continuum; 498
IRON project; 412
isoelectronic sequence; –of Li, B and Be; 143
isolated lines; 81
ISOLINE; computer code; 92
isotropic spectrum; 504

J

J–K coupling; 246

Jovian atmosphere; 426

K

K; 251
Kerr–coefficients; 169
    –effect; 171
    –nonlinearity; 175
Kr; 227, 233, 241, 293
K–shell; –lines; 40
    –spectra; 39*ff.*

L

Landau–Zener approximation; 271
Langevin theory; 136
Langmuir waves; 13
laser ablation; 51, 163
laser excitation; –of HgIn; 197*ff.*
    –of In; 197*ff.*
    –spectroscopy; 390
laser fusion; 8, 35
    –targets; 24
laser produced plasma; 45
laser spectroscopy; 255
    Doppler free–; 256
laser without inversion; 261
laser–assisted collisions; 211
laser–induced collisional autoionization; 211
laser–induced continuum structure; 213
LASNEX; computer code; 39, 45, 64
Li; 163, 515
    isoelectronic sequence of–; 147*ff.*
Li(ion); 85
LICET; 212
Lidar; 433
lifetime; 200
    radiative–; 279
LiHg; 223*ff.*
line asymmetry; 29, 509*ff.*

line broadening; 331*ff.*
   –and shift; 361
line emission; 413, 421
line interference; 491
line mixing; 334*ff.*, 386, 530
line profile; isolated–; 75
line shift; 99*ff.*, 331*ff.*, 369*ff.*
line width; 369*ff.*, 398
lineshape; *see also* Rydberg–, Raman–
lineshape; 445*ff.*
   dynamical approach to–; 27
   Gross–; 431
   Lorentzian; 311
   Lorentzian–; 71
   van Vleck–Weisskopf–; 431
   Voigt–; 435
   Zhevvakin–Naumov–; 431
lineshape asymmetry; 119
lineshape calculation; 473*ff.*
Liouville scattering matrix; 344
liquid state; 540
liquid systems; 538
Lorentz profile; 399
Lorentzian width; 264, 313
low pressure pulsed arc; 143
low temperature ions; 127
lower stratosphere; 437
L–shell specra; 39*ff.*
LS–type wavefunctions; 288
Lu–Fano plot; 239
Lyman α; 387
Lyman β; 390
Lyman series; –of Al; 57*ff.*
Ly–radiation; 27

M

magnetic degeneracy; 309
magnetic field; diagnostics of–; 125
magnetic field probe; 25
magneto–optical trap; 277, 279
magnetron discharge; 265

MARA; computer code; 93
MD; *see* simulation, MD–
MEDUSA; computer code; 58
memory loss time; 79
metal vapours; 453*ff.*
meta–particle; 231
Mg; 305
Mg(ion); 149, 330
microwave; –spectral region; 129
mid infrared; 435, 437
milky way; 511
millimeter; –spectral region; 129
millimeter–wave spectroscopy; 141
mixture; 224, 378*ff.*
model microfield method; 15, 67*ff.*
modified semi–empirical
   –approach; 150
   –approximation; 406
molecular beam; 163
molecular gas; 357
molecular ions; 127*ff.*, 141
molecular rotations; 363
molecular solids; 383
molecular spectra; 198
MOLSCAT; computer code; 128
momentum space representation; 329
multibody interactions; 538
multichannel quantum defect;
   –theory; 237*ff.*
multipolar polarizability; 519
multipole moments; 539

N

N(ion); 153
N_2; 304, 369*ff.*, 489, 523, 529
N_2–He; 369*ff.*, 375*ff.*
Na; 269, 272, 299, 304
Ne; 197, 227, 253, 313, 325, 385
Ne(ion); 65, 71, 80
near–resonant collision; 309
negative glow; 130
NH_2(ion); 135*ff.*

NH$_3$; 361, 425, 495
Ni(ion); 69
non resonant absorption; 442*ff.*
non–adiabatic–coupling; 184*ff.*
 –transitions; 28
non–linear interference effect; 67*ff.*
non–linear propagation; 169
non–linear susceptibility; 256
non–resonant absorption; 493

O

O(ion); 85
O$_2$; 263, 367, 383, 529
opacity; 404
 –of light source; 477
optical collision; 181*ff.*
optical depth; measurements of–; 74
optical shielding; 271
optical theorem; 351
optical thickness; 158*ff.*
orientational correlation; 491
overlapping lines; 349
overtone; second–; 531
 rotovibrational–; 367
oxygen; 407

P

pair polarizability; 486
 DID model for–; 458
pair potential; 453*ff.*
para-hydrogen; liquid–; 533
 solid–; 535
Paschen α; 31
Paschen series; 113
phase modulation; non linear–; 172
 self–; 169*ff.*
phase–shifted representation; 240
photoassociation; 275
 –experiment; 192
photoionization; 213
photon echo; 357

PIM/PAM/POUM; computer code; 92
planetary atmospheres; 425, 531, 538
planetary spectra; 426
plasma; *see also* dense–, solar–, hot–
 –opacity effect; 55
 astrophysical–; 161
 astrophysical–; 411, 419, 421
 confined–; 36
 laboratory–; 161
 laser–; 35*ff.*, 49, 51, 53, 59, 163
 low density–; 411
 low temperature–; 323
 moderately coupled–; 61
 optically dense–; 23
 optically thin–; 421
 undercritical–; 51
 weakly coupled; 79
plasma broadening; 67, 315
plasma diagnostics; 37*ff.*, 52, 123,
 151, 153, 265, 416
plasma microfield; 29
plasma opening switch; 20
plasma polarization shift; 57*ff.*
plasma quasimolecular features; 61*ff.*
polarizability; *see also* interaction
 induced–, diatom–tensor, dimer–
 anisotropic–; 177
 atomic pair–; 497
 collision induced–; 539
 induced–; 489
 interaction–induced–; 537
 pair–; 453*ff.*
 revival of rotational–; 178
polarizability anisotropy; 527
polarizability tensor;
 dipole–octopole–; 539
polarizability tensor;
 dipole–quadrupole; 539
polarization; 11*ff.*
polarization difference profile; 25
polarization experiment; 188
polarization of radiation; 33
polymer coated cell; 269

population; –distribution; 221
    –inversion; 75
    –kinetics; 67
potential; *see also* interaction potential, interaction energy
potential curve; 223*ff.*, 230, 275
potential energy; 200, 219, 321, 498
potential energy surface; 298*ff.*
    crossing of–; 298
pressure broadening; 128, 229, 267, 317, 325, 361, 438, 448
pressure shift; 361
profile; absorption–; 539
    Fano–; 241
pseudo–hamiltonian; –impact theory; 251
pulse characterisation; 169
pulse–probe experiment; 192

Q

quadratic Stark effect; 103
quadrupolar interaction; 119
quadrupole corrections; 4*ff.*
quantum coherence effect; 255*ff.*
quantum defect theory; 329
quantum interference; 211
quantum Monte Carlo simulation; 534
quantum oscillations; 252
quantum vacuum radiation; 482
quantum–mechanical treatment; 321
quasi–molecular; –model; 66
    –radiation; 283
    –spectra; 289
    –theory; 229
quasi–static approximation; 15*ff.*, 19, 65, 79, 117, 283
quasi–static model; 306
quenching; 298, 304
    –of molecular emission; 515
    non reactive–; 304
    reactive–; 305

R

Rabi frequency; 211, 256
radiation probabilities; 285
radiation transfer; 33
radiation trapping; 203, 209, 215, 279
radiation widths; 288
radiative acceleration; 404
radiative association; 297
radiative decay; 207, 298
radiative transfer; 23, 35
rainbow satellites; 515
Raman lineshape; 363
Raman scattering; *see also* stimulated–
383, 502
Raman spectrum; 533, 535
    –diagnostics; 433
    rototranslational–; 489
rate equations; 39
RATION; computer code; 54*ff.*
Rb; 207, 267, 277
reactance matrix; 240
recombination emission; 51
red shift; 177, 370*ff.*
redistribution; 47
    –of resonance radiation; 33
refractive fringe diagnostic; 51
refractive index; 171
    non–linear–; 175*ff.*
relativistic effects; 219
relaxation matrix; 345
    sum rule of–; 347
    symmetry of–; 347
relaxation theory; 87
relaxational rate; 359
remote sensing; 433
repulsive forces; 371*ff.*
resonance coupling; 373
resonance emission; broadening of–; 304
resonant collision; 309
resonant interaction; 221

rotational analysis; 223
rotational constant; 225*ff.*
rotational coupling; 303
rotational motion; 177
rotational spectrum; allowed–; 489
rotovibrational spectrum; 531
run–away electrons; 13
Rydberg levels; 238*ff.*
Rydberg lineshape; 161
Rydberg series; 238*ff.*
Rydberg states; 184, 329
Rydberg transitions; 251

S

satellite like features; 63
satellites; 12
satellites; rainbow–; 387
saturation; 158*ff.*
screening potential; 21
sea–water temperature; 433
selective radiative acceleration; 397
self broadening; 325
self broadening; –in Rydberg transitions; 251
self resonance; 232
self reversed profile; 123
selfconsistent field routine; 62
self–reversed spectral line; 123
self–similar range; 19
semi–classical approximation; 348
semi–classical calculation; 81
; –of cross section; 183
semi–classical perturbation formalism; 147, 149
semi–empirical approach; 284
SF_6; 358
shilft; *see also* line–, Stark–, pressure–, ionic–, collisional–
   center of gravity–; 3*ff.*
   center of mass–; 103
   dip–; 107*ff.*
   ion–; 3*ff.*

non–impact–; 321
plasma–; 143
shock–wave; 51
shock–wave; –heating; 472
shock–wave; spherically convergent–; 473
Si(ion); 155, 409, 420
simulation; MD–; 61, 363, 369*ff.*, 429, 487, 517
   Monte Carlo–; 280
simultaneous transitions; 477*ff.*, 541
slow modulation; 380
SOHO mission; 412
solar corona; 411
solar flares; 11
solar photosphere; 423
solar plasma; 125
solar spectroscopy; 423
solid state systems; 542
sonoluminescence; 472*ff.*, 543
spectral chirp; 170
spectral density; 474
spectral distribution; 289
spectral moments; 523
spectroscopy; non–linear–; 309
spectroscopy; second harmonic modulation–; 437
spectrum; atmospheric emission–; 435
   computed absorption–; 506
   depolarized ligth–; 537*ff.*
   emission–; 515
   excitation–; 515
   interaction–induced–; 537*ff.*
   Raman–; 383
   roto–vibrational–; 367
   syntetic–; 414
   syntetic–of sun; 423
   translational–; 538
speed dependent width; 338
spin–orbit interaction; 219
spin–orbit splitting; 303
splitting; *see also* Stark–, spin–orbit–

Sr; 215, 303
stabilized argon arc; 315
STARCODE; computer code; 92
Stark broadening; 19, 23, 29, 31, 35,
    47, 79, 99, 113, 117, 121, 147*ff.*,
    149, 151, 153, 155, 315, 327
    linear–; 164
Stark components; 353
Stark effect; *see also* quadratic Stark effect
Stark effect; 155, 307, 323, 365
    linear–; 125
    linear–; 406*ff.*
    quadratic–; 71*ff.*
Stark shift; 29, 117, 121, 146, 147*ff.*,
    151, 315, 324
Stark splitting; 25, 47
Stark width; 117, 144*ff.*, 155, 324
stars; chemical compositions of–; 397
stellar atmosphere; 531, 155
stellar plasmas; 149
stimulated emission; multiphoton–; 195
stimulated fluorescence; 217
stimulated Raman scattering; 217
strong collisions; 82, 99
    –model; 529
strong coupling; 147
    –quantum calculations; 149
submillimeter; –spectral region; 129
sun; 397
superfluid helium; 391
supramolecular spectra; 471
susceptibility tensor; third order–; 171
symmetric–top molecules; 495

T

targets
    close geometry–; 36
    open geometry–; 36
temperature dependence; 315
temperature variation; 361

ternary dipoles; 447
thermal broadening; 19
thin plasma emission; 419, 421
Thomson scattering; diagnostic by–; 71
three–body; –interactions; 540
    –electric moment; 521
three–level system; atomic–; 255
time behavior; 210
time of interest; 79*ff.*
time scales; 85
time–of–flight spectrometry; 214
Tl(isotope); 401
tokamak; 113
transition frequency; 475
transition probabilities; 151
translational band; 511
    absorption; 505
triple transitions; 521, 541
turbulence; fully developped–; 19
turbulent broadening; 19
two–center approach; 61
two–dimensional effects; 45
two–ion center model; 21
two–ion center model; 65
two–level atom; 221

U

ultrashort pulses; 169
unitarity criterion; 83
unlike interaction; 370
UV emission; –of astrophysical plasmas; 411

V

van der Waals; –complexes; 540
van der Waals; –forces; 231
velocity changing collisions; 221
velocity dependence; 357*ff.*
velocity redistribution; 222
velocity selective laser excitation; 221

Venus atmosphere; 429
vibrational coupling; 383
vibration–rotation coupling; 375*ff.*
vibration–rotation spectra; 508
viscoelastic relaxation; –in simple liquid; 544
Voigt profile; 311

zeolites; 542
Zn; 163, 227
Z–pinch; 11

W

wall–stabilized arc; 307
water continuum; 431
wavelength modulation; 367
Weisskopf sphere; 16
white dwarfs; 387*ff.*, 402
white dwarfs; –spectra; 121
width; *see* Doppler–, Stark–, Lorentzian–, collisional–, equivalent–, speed dependent–

X

Xe; 227, 233, 241, 293,
Xe(ion); 36
x–ray emission; –of astrophysical plasmas; 411
–spectroscopy; 45
x–ray line profile; 11*ff.*
x–ray measurements; 57
x–ray spectroscopy; 35, 53
real time–; 39
time resolved–; 53*ff.*

Y

Yb; 205

Z

Zeeman effect; 113, 125, 265
Zeeman laser; 259
Zeeman splitting; 25

AIP Conference Proceedings

	Title	L.C. Number	ISBN
No. 355	Eleventh Topical Conference on Radio Frequency Power in Plasmas (Palm Springs, CA 1995)	95-80867	1-56396-536-4
No. 356	The Future of Accelerator Physics (Austin, TX 1994)	96-83292	1-56396-541-0
No. 357	10th Topical Workshop on Proton-Antiproton Collider Physics (Batavia, IL 1995)	95-83078	1-56396-543-7
No. 358	The Second NREL Conference on Thermophotovoltaic Generation of Electricity	95-83335	1-56396-509-7
No. 359	Workshops and Particles and Fields and Phenomenology of Fundamental Interactions (Puebla, Mexico 1995)	96-85996	1-56396-548-8
No. 360	The Physics of Electronic and Atomic Collisions XIX International Conference (Whistler, Canada, 1995)	95-83671	1-56396-440-6
No. 361	Space Technology and Applications International Forum (Albuquerque, NM 1996)	95-83440	1-56396-568-2
No. 362	Two-Center Effects in Ion-Atom Collisions (Lincoln, NE 1994)	96-83379	1-56396-342-6
No. 363	Phenomena in Ionized Gases XXII ICPIG (Hoboken, NJ, 1995)	96-83294	1-56396-550-X
No. 364	Fast Elementary Processes in Chemical and Biological Systems (Villeneuve d'Ascq, France, 1995)	96-83624	1-56396-564-X
No. 365	Latin-American School of Physics XXX ELAF Group Theory and Its Applications (México City, México, 1995)	96-83489	1-56396-567-4
No. 366	High Velocity Neutron Stars and Gamma-Ray Bursts (La Jolla, CA 1995)	96-84067	1-56396-593-3
No. 367	Micro Bunches Workshop (Upton, NY, 1995)	96-83482	1-56396-555-0
No. 368	Acoustic Particle Velocity Sensors: Design, Performance and Applications (Mystic, CT, 1995)	96-83548	1-56396-549-6
No. 369	Laser Interaction and Related Plasma Phenomena (Osaka, Japan 1995)	96-85009	1-56396-445-7
No. 370	Shock Compression of Condensed Matter-1995 (Seattle, WA 1995)	96-84595	1-56396-566-6

	Title	L.C. Number	ISBN
No. 371	Sixth Quantum 1/f Noise and Other Low Frequency Fluctuations in Electronic Devices Symposium (St. Louis, MO, 1994)	96-84200	1-56396-410-4
No. 372	Beam Dynamics and Technology Issues for + - Colliders 9th Advanced ICFA Beam Dynamics Workshop (Montauk, NY, 1995)	96-84189	1-56396-554-2
No. 373	Stress-Induced Phenomena in Metallization (Palo Alto, CA 1995)	96-84949	1-56396-439-2
No. 374	High Energy Solar Physics (Greenbelt, MD 1995)	96-84513	1-56396-542-9
No. 375	Chaotic, Fractal, and Nonlinear Signal Processing (Mystic, CT 1995)	96-85356	1-56396-443-0
No. 376	Chaos and the Changing Nature of Science and Medicine: An Introduction (Mobile, AL 1995)	96-85220	1-56396-442-2
No. 377	Space Charge Dominated Beams and Applications of High Brightness Beams (Bloomington, IN 1995)	96-85165	1-56396-625-7
No. 379	Physical Origin of Homochirality in Life (Santa Monica, CA 1995)	96-86631	1-56396-507-0
No. 378	Surfaces, Vacuum, and Their Applications (Cancun, Mexico 1994)	96-85594	1-56396-418-X
No. 380	Production and Neutralization of Negative Ions and Beams / Production and Application of Light Negative Ions (Upton, NY 1995)	96-86435	1-56396-565-8
No. 381	Atomic Processes in Plasmas (San Francisco, CA 1996)	96-86304	1-56396-552-6
No. 382	Solar Wind Eight (Dana Point, CA 1995)	96-86447	1-56396-551-8
No. 383	Workshop on the Earth's Trapped Particle Environment (Taos, NM 1994)	96-86619	1-56396-540-2
No. 384	Gamma-Ray Bursts (Huntsville, AL 1995)	96-79458	1-56396-685-9
No. 385	Robotic Exploration Close to the Sun: Scientific Basis (Marlboro, MA 1996)	96-79560	1-56396-618-2
No. 386	Spectral Line Shapes, Volume 9 13th ICSLS (Firenze, Italy 1996)		1-56396-656-5
No. 387	Space Technology and Applications International Forum (Albuquerque, NM 1997)	96-80254	1-56396-679-4 (Case set) 1-56396-691-3